Elasticity

Elasticity
Theory, Applications, and Numerics

Fourth Edition

Martin H. Sadd

Professor Emeritus, University of Rhode Island
Department of Mechanical Engineering and Applied Mechanics
Kingston, Rhode Island

ELSEVIER

ACADEMIC PRESS
An imprint of Elsevier

Academic Press is an imprint of Elsevier
125 London Wall, London EC2Y 5AS, United Kingdom
525 B Street, Suite 1650, San Diego, CA 92101, United States
50 Hampshire Street, 5th Floor, Cambridge, MA 02139, United States
The Boulevard, Langford Lane, Kidlington, Oxford OX5 1GB, United Kingdom

MATLAB® is a trademark of The MathWorks, Inc. and is used with permission. The MathWorks does not
warrant the accuracy of the text or exercises in this book.

This book's use or discussion of MATLAB® software or related products does not constitute endorsement or
sponsorship by The MathWorks of a particular pedagogical approach or particular use of the MATLAB®
software.

Notices
Knowledge and best practice in this field are constantly changing. As new research and experience broaden our
understanding, changes in research methods, professional practices, or medical treatment may become
necessary.

Practitioners and researchers must always rely on their own experience and knowledge in evaluating and using
any information, methods, compounds, or experiments described herein. In using such information or
methods they should be mindful of their own safety and the safety of others, including parties for whom they
have a professional responsibility.

To the fullest extent of the law, neither the Publisher nor the authors, contributors, or editors, assume any
liability for any injury and/or damage to persons or property as a matter of products liability, negligence or
otherwise, or from any use or operation of any methods, products, instructions, or ideas contained in the
material herein.

Library of Congress Cataloging-in-Publication Data
A catalog record for this book is available from the Library of Congress

British Library Cataloguing-in-Publication Data
A catalogue record for this book is available from the British Library

ISBN: 978-0-12-815987-3

For information on all Academic Press publications visit our website at
https://www.elsevier.com/books-and-journals

Publisher: Katey Birtcher
Acquisition Editor: Steve Merken
Editorial Project Manager: Susan Ikeda
Production Project Manager: Sruthi Satheesh
Cover Designer: Greg Harris

Typeset by TNQ Technologies

Contents

Preface xi
Acknowledgments.. xv
About the Author xvii

PART 1 Foundations and elementary applications

CHAPTER 1 Mathematical preliminaries 3
1.1 Scalar, vector, matrix, and tensor definitions 3
1.2 Index notation 4
1.3 Kronecker delta and alternating symbol 7
1.4 Coordinate transformations.. 8
1.5 Cartesian tensors 10
1.6 Principal values and directions for symmetric second-order tensors 12
1.7 Vector, matrix, and tensor algebra... 16
1.8 Calculus of Cartesian tensors 17
    1.8.1 Divergence or Gauss theorem 20
    1.8.2 Stokes theorem 20
    1.8.3 Green's theorem in the plane 20
    1.8.4 Zero-value or localization theorem 21
1.9 Orthogonal curvilinear coordinates 21
References.. 26
Exercises 26

CHAPTER 2 Deformation: displacements and strains 31
2.1 General deformations.. 31
2.2 Geometric construction of small deformation theory 34
2.3 Strain transformation 40
2.4 Principal strains.. 41
2.5 Spherical and deviatoric strains... 42
2.6 Strain compatibility 42
2.7 Curvilinear cylindrical and spherical coordinates............................ 48
References.. 49
Exercises 50

CHAPTER 3 Stress and equilibrium 57
3.1 Body and surface forces 57
3.2 Traction vector and stress tensor 58
3.3 Stress transformation 62
3.4 Principal stresses.. 63
3.5 Spherical, deviatoric, octahedral, and von mises stresses................ 66

3.6 Stress distributions and contour lines...68
3.7 Equilibrium equations...71
3.8 Relations in curvilinear cylindrical and spherical coordinates..............................73
References...76
Exercises 76

CHAPTER 4 **Material behavior—linear elastic solids**..............................**83**
4.1 Material characterization 83
4.2 Linear elastic materials—Hooke's law...85
4.3 Physical meaning of elastic moduli..88
  4.3.1 Simple tension 89
  4.3.2 Pure shear 89
  4.3.3 Hydrostatic compression (or tension) 89
4.4 Thermoelastic constitutive relations...92
References...93
Exercises 93

CHAPTER 5 **Formulation and solution strategies** **97**
5.1 Review of field equations 97
5.2 Boundary conditions and fundamental problem classifications 98
5.3 Stress formulation 103
5.4 Displacement formulation 104
5.5 Principle of superposition 106
5.6 Saint-Venant's principle..107
5.7 General solution strategies..109
  5.7.1 Direct method 109
  5.7.2 Inverse method 110
  5.7.3 Semi-inverse method 111
  5.7.4 Analytical solution procedures 111
  5.7.5 Approximate solution procedures 112
  5.7.6 Numerical solution procedures 113
5.8 Singular elasticity solutions...114
References..116
Exercises 117

CHAPTER 6 **Strain energy and related principles** **123**
6.1 Strain energy 123
6.2 Uniqueness of the elasticity boundary-value problem...........................128

6.3 Bounds on the elastic constants 129
    6.3.1 Uniaxial tension 129
    6.3.2 Simple shear 129
    6.3.3 Hydrostatic compression 129
6.4 Related integral theorems 130
    6.4.1 Clapeyron's theorem 130
    6.4.2 Betti/Rayleigh reciprocal theorem 130
    6.4.3 Integral formulation of elasticity—Somigliana's identity 131
6.5 Principle of virtual work 132
6.6 Principles of minimum potential and complementary energy 134
6.7 Rayleigh–Ritz method 138
References 140
Exercises 141

CHAPTER 7 Two-dimensional formulation **145**
7.1 Plane strain 145
7.2 Plane stress 148
7.3 Generalized plane stress 151
7.4 Antiplane strain 153
7.5 Airy stress function 154
7.6 Polar coordinate formulation 156
References 158
Exercises 158

CHAPTER 8 Two-dimensional problem solution **163**
8.1 Cartesian coordinate solutions using polynomials 163
8.2 Cartesian coordinate solutions using Fourier methods 174
    8.2.1 Applications involving Fourier series 177
8.3 General solutions in polar coordinates 182
    8.3.1 General Michell solution 182
    8.3.2 Axisymmetric solution 183
8.4 Example polar coordinate solutions 184
    8.4.1 Pressurized hole in an infinite medium 186
    8.4.2 Stress-free hole in an infinite medium under equal biaxial loading
         at infinity 187
    8.4.3 Biaxial and shear loading cases 191
    8.4.4 Quarter-plane example 193
    8.4.5 Half-space examples 194
    8.4.6 Half-space under uniform normal stress over $x \leq 0$ 194
    8.4.7 Half-space under concentrated surface force system
         (Flamant problem) 195
    8.4.8 Half-space under a surface concentrated moment 200

8.4.9 Half-space under uniform normal loading over $-a \geq x \geq a$ 201
8.4.10 Notch and crack problems 205
8.4.11 Pure bending example 207
8.4.12 Curved cantilever under end loading 208
8.5 Simple plane contact problems..217
References...223
Exercises 223

CHAPTER 9 Extension, torsion, and flexure of elastic cylinders.......................**241**
9.1 General formulation..242
9.2 Extension formulation...242
9.3 Torsion formulation...243
9.3.1 Stress–stress function formulation 244
9.3.2 Displacement formulation 248
9.3.3 Multiply connected cross-sections 249
9.3.4 Membrane analogy 251
9.4 Torsion solutions derived from boundary equation 253
9.5 Torsion solutions using Fourier methods 259
9.6 Torsion of cylinders with hollow sections 264
9.7 Torsion of circular shafts of variable diameter 267
9.8 Flexure formulation...270
9.9 Flexure problems without twist..274
References...278
Exercises 278

PART 2 Advanced applications

CHAPTER 10 Complex variable methods **291**
10.1 Review of complex variable theory 291
10.2 Complex formulation of the plane elasticity problem...........................298
10.3 Resultant boundary conditions 301
10.4 General structure of the complex potentials 303
10.4.1 Finite simply connected domains..303
10.4.2 Finite multiply connected domains..304
10.4.3 Infinite domains..305
10.5 Circular domain examples...305
10.6 Plane and half-planc problems..310
10.7 Applications using the method of conformal mapping 315
10.8 Applications to fracture mechanics 320
10.9 Westergaard method for crack analysis 323
References...324
Exercises 325

CHAPTER 11 Anisotropic elasticity **331**

11.1 Basic concepts 331

11.2 Material symmetry...333

    11.2.1 Plane of symmetry (monoclinic material) 334

    11.2.2 Three perpendicular planes of symmetry (orthotropic material) 335

    11.2.3 Axis of symmetry (transversely isotropic material) 336

    11.2.4 Cubic symmetry 337

    11.2.5 Complete symmetry (isotropic material) 338

11.3 Restrictions on elastic moduli 339

11.4 Torsion of a solid possessing a plane of material symmetry 341

    11.4.1 Stress formulation..342

    11.4.2 Displacement formulation 344

    11.4.3 General solution to the governing equation.........................345

11.5 Plane deformation problems..347

    11.5.1 Uniform pressure loading case..357

11.6 Applications to fracture mechanics 360

11.7 Curvilinear anisotropic problems 364

    11.7.1 Two-dimensional polar-orthotropic problem 365

    11.7.2 Three-dimensional spherical-orthotropic problem 368

References..372

Exercises 372

CHAPTER 12 Thermoelasticity **379**

12.1 Heat conduction and the energy equation......................................379

12.2 General uncoupled formulation..381

12.3 Two-dimensional formulation 381

    12.3.1 Plane strain 381

    12.3.2 Plane stress 383

12.4 Displacement potential solution 384

12.5 Stress function formulation 385

12.6 Polar coordinate formulation...388

12.7 Radially symmetric problems...389

12.8 Complex variable methods for plane problems 394

References..401

Exercises 402

CHAPTER 13 Displacement potentials and stress functions: applications to three-dimensional problems **407**

13.1 Helmholtz displacement vector representation 407

13.2 Lamé's strain potential 408

13.3 Galerkin vector representation 409

13.4 Papkovich−Neuber representation 414

13.5 Spherical coordinate formulations 418

13.6 Stress functions...421

    13.6.1 Maxwell stress function representation 424

    13.6.2 Morera stress function representation 425

References...425

Exercises 426

CHAPTER 14   Nonhomogeneous elasticity **433**

    14.1 Basic concepts 433

    14.2 Plane problem of a hollow cylindrical domain under uniform pressure...........437

    14.3 Rotating disk problem 444

    14.4 Point force on the free surface of a half-space..................................449

    14.5 Antiplane strain problems 458

    14.6 Torsion problem 462

    References...469

    Exercises 471

CHAPTER 15   Micromechanics applications **477**

    15.1 Dislocation modeling..478

    15.2 Singular stress states...482

    15.3 Elasticity theory with distributed cracks..491

    15.4 Micropolar/couple-stress elasticity..494

        15.4.1 Two-dimensional couple-stress theory................................496

    15.5 Elasticity theory with voids...503

    15.6 Doublet mechanics 508

    15.7 Higher gradient elasticity theories 513

    References...518

    Exercises 520

CHAPTER 16   Numerical finite and boundary element methods **523**

    16.1 Basics of the finite element method...524

    16.2 Approximating functions for two-dimensional linear triangular elements 525

    16.3 Virtual work formulation for plane elasticity 527

    16.4 FEM problem application..532

    16.5 FEM code applications 535

    16.6 Boundary element formulation..540

    References...546

    Exercises 546

Appendix A: Basic field equations in Cartesian, cylindrical, and spherical coordinates 549

Appendix B: Transformation of field variables between Cartesian, cylindrical,
        and spherical components 555

Appendix C: MATLAB® Primer... 559

Appendix D: Review of mechanics of materials 577

Index 593

Preface

As with the previous works, this fourth edition continues the author's efforts to present linear elasticity with complete and concise theoretical development, numerous and contemporary applications, and enriching numerics to aid in problem solution and understanding. Over the years the author has given much thought on what should be taught to students in this field and what educational outcomes would be expected. Theoretical topics that are related to the foundations of elasticity should be presented in sufficient detail that will allow students to read and generally understand contemporary research papers. Related to this idea, students should acquire necessary vector and tensor notational skills and understand fundamental development of the basic field equations. Students should also have a solid understanding of the formulation and solution of various elasticity boundary-value problems that include a variety of domain and loading geometries. Finally, students should be able to apply modern engineering software (MATLAB, Maple or Mathematica) to aid in the solution, evaluation and graphical display of various elasticity problem solutions. These points are all emphasized in this text.

In addition to making numerous small corrections and clarifications, several new items have been added. A new section in Chapter 5 on singular elasticity solutions has been introduced to generally acquaint students with this type of behavior. Cubic anisotropy has now been presented in Chapter 11 as another particular form of elastic anisotropy. Inequality elastic moduli restrictions for various anisotropic material models have been better organized in a new table in Chapter 11. The general Naghdi-Hsu solution has now been introduced in Chapter 13. An additional micromechanical model of gradient elasticity has been added in Chapter 15. A couple of new MATLAB codes in Appendix C have been added and all codes are now referenced in the text where they are used. With the addition of 31 new exercises, the fourth edition now has 441 total exercises. These problems should provide instructors with many new and previous options for homework, exams, or material for in-class presentations or discussions. The online Solutions Manual has been updated and corrected and includes solutions to all exercises in the new edition. All text editions follow the original lineage as an outgrowth of lecture notes that I have used in teaching a two-course sequence in the theory of elasticity. Part I of the text is designed primarily for the first course, normally taken by beginning graduate students from a variety of engineering disciplines. The purpose of the first course is to introduce students to theory and formulation, and to present solutions to some basic problems. In this fashion students see how and why the more fundamental elasticity model of deformation should replace elementary strength of materials analysis. The first course also provides foundation for more advanced study in related areas of solid mechanics. Although the more advanced material included in Part II has normally been used for a second course, I often borrow selected topics for use in the first course. The elasticity presentation in this book reflects the words used in the title - theory, applications, and numerics. Because theory provides the fundamental cornerstone of this field, it is important to first provide a sound theoretical development of elasticity with sufficient rigor to give students a good foundation for the development of solutions to a broad class of problems. The theoretical development is carried out in an organized and concise manner in order to not lose the attention of the less mathematically inclined students or the focus of applications. With a primary goal of solving problems of engineering interest, the text offers numerous applications in contemporary areas, including anisotropic composite and functionally graded materials, fracture mechanics, micromechanics modeling, thermoelastic problems, and computational finite and boundary element methods. Numerous solved example problems and exercises are included in all chapters.

The new edition continues the special use of integrated numerics. By taking the approach that applications of theory need to be observed through calculation and graphical display, numerics is accomplished through the use of MATLAB, one of the most popular engineering software packages. This software is used throughout the text for applications such as stress and strain transformation, evaluation and plotting of stress and displacement distributions, finite element calculations, and comparisons between strength of materials and analytical and numerical elasticity solutions. With numerical and graphical evaluations, application problems become more interesting and useful for student learning. Other software such as Maple or Mathematica could also be used.

Contents summary

Part I of the book emphasizes formulation details and elementary applications. Chapter 1 provides a mathematical background for the formulation of elasticity through a review of scalar, vector, and tensor field theory. Cartesian tensor notation is introduced and is used throughout the book's formulation sections. Chapter 2 covers the analysis of strain and displacement within the context of small deformation theory. The concept of strain compatibility is also presented in this chapter. Forces, stresses, the equilibrium concept and various stress contour lines are developed in Chapter 3. Linear elastic material behavior leading to the generalized Hooke's law is discussed in Chapter 4, which also briefly presents nonhomogeneous, anisotropic, and thermoelastic constitutive forms. Later chapters more fully investigate these types of applications. Chapter 5 collects the previously derived equations and formulates the basic boundary value problems of elasticity theory. Displacement and stress formulations are constructed and general solution strategies are identified. This is an important chapter for students to put the theory together. Chapter 6 presents strain energy and related principles, including the reciprocal theorem, virtual work, and minimum potential and complementary energy. Two-dimensional formulations of plane strain, plane stress, and antiplane strain are given in Chapter 7. An extensive set of solutions for specific two dimensional problems is then presented in Chapter 8, and many applications employing MATLAB are used to demonstrate the results. Analytical solutions are continued in Chapter 9 for the Saint-Venant extension, torsion, and flexure problems. The material in Part I provides a logical and orderly basis for a sound one-semester beginning course in elasticity. Selected portions of the text's second part could also be incorporated into such a course. Part II delves into more advanced topics normally covered in a second course. The powerful method of complex variables for the plane problem is presented in Chapter 10, and several applications to fracture mechanics are given. Chapter 11 extends the previous isotropic theory into the behavior of anisotropic solids with emphasis on composite materials. This is an important application and, again, examples related to fracture mechanics are provided. Curvilinear anisotropy including both cylindrical and spherical orthotropy is included in this chapter to explore some basic problem solutions with this type of material structure. An introduction to thermoelasticity is developed in Chapter 12, and several specific application problems are discussed, including stress concentration and crack problems. Potential methods, including both displacement potentials and stress functions, are presented in Chapter 13. These methods are used to develop several three-dimensional elasticity solutions.

Chapter 14 covers nonhomogeneous elasticity, and this material is unique among current standard elasticity texts. After briefly covering theoretical formulations, several two-dimensional solutions are generated along with comparison field plots with the corresponding homogeneous cases. Chapter 15 presents a collection of elasticity applications to problems involving micromechanics modeling. Included are applications for dislocation modeling, singular stress states, solids with distributed cracks, micropolar, distributed voids, doublet mechanics and higher gradient theories. Chapter 16 provides a brief introduction to the powerful numerical methods of finite and boundary element techniques. Although only two-dimensional theory is developed, the numerical results in the example problems provide interesting comparisons with previously generated analytical solutions from earlier chapters. This fourth edition of Elasticity concludes with four appendices that contain a concise summary listing of basic field equations; transformation relations between Cartesian, cylindrical, and spherical coordinate systems; a MATLAB primer; and a self-contained review of mechanics of materials.

The subject

Elasticity is an elegant and fascinating subject that deals with determination of the stress, strain, and displacement distribution in an elastic solid under the influence of external forces. Following the usual assumptions of linear, small-deformation theory, the formulation establishes a mathematical model that allows solutions to problems that have applications in many engineering and scientific fields such as:

- Civil engineering applications include important contributions to stress and deflection analysis of structures, such as rods, beams, plates, and shells. Additional applications lie in geomechanics involving the stresses in materials such as soil, rock, concrete, and asphalt.
- Mechanical engineering uses elasticity in numerous problems in analysis and design of machine elements. Such applications include general stress analysis, contact stresses, thermal stress analysis, fracture mechanics, and fatigue.
- Materials engineering uses elasticity to determine the stress fields in crystalline solids, around dislocations, and in materials with microstructure.
- Applications in aeronautical and aerospace engineering typically include stress, fracture, and fatigue analysis in aerostructures.
- Biomechanical engineering uses elasticity to study the mechanics of bone and various types of soft tissue.

The subject also provides the basis for more advanced work in inelastic material behavior, including plasticity and viscoelasticity, and the study of computational stress analysis employing finite and boundary element methods. Since elasticity establishes a mathematical model of the deformation problem, it requires mathematical knowledge to understand formulation and solution procedures. Governing partial differential field equations are developed using basic principles of continuum mechanics commonly formulated in vector and tensor language. Techniques used to solve these field equations can encompass Fourier methods, variational calculus, integral transforms, complex variables, potential theory, finite differences, finite elements, and so forth. To prepare students for this subject, the text provides reviews of many mathematical topics, and additional references are given for further study. It is important for students to be adequately prepared for the theoretical developments, or else they will not be able to understand necessary formulation details. Of course, with emphasis on applications, the text concentrates on theory that is most useful for problem solution.

The concept of the elastic force—deformation relation was first proposed by Robert Hooke in 1678. However, the major formulation of the mathematical theory of elasticity was not developed until the nineteenth century. In 1821 Navier presented his investigations on the general equations of equilibrium; this was quickly followed by Cauchy, who studied the basic elasticity equations and developed the concept of stress at a point. A long list of prominent scientists and mathematicians continued development of the theory, including the Bernoullis, Lord Kelvin, Poisson, Lame', Green, Saint-Venant, Betti, Airy, Kirchhoff, Rayleigh, Love, Timoshenko, Kolosoff, Muskhelishvilli, and others.

During the two decades after World War II, elasticity research produced a large number of analytical solutions to specific problems of engineering interest. The 1970s and 1980s included considerable work on numerical methods using finite and boundary element theory. Also during this period, elasticity applications were directed at anisotropic materials for applications to composites. More recently, elasticity has been used in modeling materials with internal microstructures or heterogeneity and in inhomogeneous, graded materials. The rebirth of modern continuum mechanics in the 1960s led to a review of the foundations of elasticity and established a rational place for the theory within the general framework. Historical details can be found in the texts by Todhunter and Pearson, *History of the Theory of Elasticity*; Love, A *Treatise on the Mathematical Theory of Elasticity*; and Timoshenko, *A History of Strength of Materials.*

Exercises and web support

Of special note in regard to this text is the use of exercises and the publisher's website, www.textbooks. elsevier.com. Numerous exercises are provided at the end of each chapter for homework assignments to engage students with the subject matter. The exercises also provide an ideal tool for the instructor to present additional application examples during class lectures. Many places in the text make reference to specific exercises that work out details to a particular topic. Exercises marked with an asterisk (*) indicate problems that require numerical and plotting methods using the suggested MATLAB software. Solutions to all exercises are provided to registered instructors online at the publisher's website, thereby providing instructors with considerable help in using this material. In addition, downloadable MATLAB software is available to aid both students and instructors in developing codes for their own particular use to allow easy integration of the numerics. As with the previous edition, an on-line collection of PowerPoint slides is available for Chapters 1-9. This material includes graphical figures and summaries of basic equations that have proven to be useful during class presentations.

Feedback

The author is strongly interested in continual improvement of engineering education and welcomes feedback from users of the book. Please feel free to send comments concerning suggested improvements or corrections via surface mail or email (saddm@uri.edu). It is likely that such feedback will be shared with the text's user community via the publisher's website.

Acknowledgments

Many individuals deserve acknowledgment for aiding in the development of this textbook. First, I would like to recognize the many graduate students who sat in my elasticity classes. They have been a continual source of challenge and inspiration, and certainly influenced my efforts to find more effective ways to present this material. A special recognition goes to one particular student, Qingli Dai, who developed most of the original exercise solutions and did considerable proofreading. . I would also like to acknowledge the support of my institution, the University of Rhode Island, for providing me time, resources and the intellectual climate to complete this and previous editions. Several photoelastic pictures have been graciously provided by the URI Dynamic Photomechanics Laboratory (Professor Arun Shukla, director). Development and production support from several Elsevier staff was greatly appreciated.

As with the previous editions, this book is dedicated to the late Professor Marvin Stippes of the University of Illinois; he was the first to show me the elegance and beauty of the subject. His neatness, clarity, and apparently infinite understanding of elasticity will never be forgotten by his students.

Martin H. Sadd

About the Author

Martin H. Sadd is Professor Emeritus of Mechanical Engineering at the University of Rhode Island. He received his Ph.D. in mechanics from the Illinois Institute of Technology in 1971 and began his academic career at Mississippi State University. In 1979 he joined the faculty at Rhode Island and served as department chair from 1991 to 2000. He is a member of Phi Kappa Phi, Pi Tau Sigma, Tau Beta Pi, Sigma Xi, and is a Fellow of ASME. Professor Sadd's teaching background has been in the area of solid mechanics with emphasis in elasticity, continuum mechanics, wave propagation, and computational methods. His research has included analytical and computational modeling of materials under static and dynamic loading conditions. He has authored over 75 publications and is the author of *Continuum Mechanics Modeling of Material Behavior* (Elsevier, 2019).

Foundations and elementary applications

Mathematical preliminaries

Similar to other field theories such as fluid mechanics, heat conduction, and electromagnetics, the study and application of elasticity theory requires knowledge of several areas of applied mathematics. The theory is formulated in terms of a variety of variables including scalar, vector, and tensor fields, and this calls for the use of tensor notation along with tensor algebra and calculus. Through the use of particular principles from continuum mechanics, the theory is developed as a system of partial differential field equations that are to be solved in a region of space coinciding with the body under study. Solution techniques used on these field equations commonly employ Fourier methods, variational techniques, integral transforms, complex variables, potential theory, finite differences, and finite and boundary elements. Therefore, to develop proper formulation methods and solution techniques for elasticity problems, it is necessary to have an appropriate mathematical background. The purpose of this initial chapter is to provide a background primarily for the formulation part of our study. Additional review of other mathematical topics related to problem solution technique is provided in later chapters where they are to be applied.

1.1 Scalar, vector, matrix, and tensor definitions

Elasticity theory is formulated in terms of many different types of variables that are either specified or sought at spatial points in the body under study. Some of these variables are *scalar quantities*, representing a single magnitude at each point in space. Common examples include the material density ρ and temperature T. Other variables of interest are *vector quantities* that are expressible in terms of components in a two- or three-dimensional coordinate system. Examples of vector variables are the displacement and rotation of material points in the elastic continuum. Formulations within the theory also require the need for *matrix variables*, which commonly require more than three components to quantify. Examples of such variables include stress and strain. As shown in subsequent chapters, a three-dimensional formulation requires nine components (only six are independent) to quantify the stress or strain at a point. For this case, the variable is normally expressed in a matrix format with three rows and three columns. To summarize this discussion, in a three-dimensional Cartesian coordinate system, scalar, vector, and matrix variables can thus be written as follows

$$\text{mass density scalar} = \rho$$

$$\text{displacement vector} = \boldsymbol{u} = u\boldsymbol{e}_1 + v\boldsymbol{e}_2 + w\boldsymbol{e}_3$$

$$\text{stress matrix} = [\boldsymbol{\sigma}] = \begin{bmatrix} \sigma_x & \tau_{xy} & \tau_{xz} \\ \tau_{yx} & \sigma_y & \tau_{yz} \\ \tau_{zx} & \tau_{zy} & \sigma_z \end{bmatrix}$$

Elasticity. https://doi.org/10.1016/B978-0-12-815987-3.00001-3

where e_1, e_2, e_3 are the usual unit basis vectors in the coordinate directions. Thus, scalars, vectors, and matrices are specified by one, three, and nine components respectively.

The formulation of elasticity problems not only involves these types of variables, but also incorporates additional quantities that require even more components to characterize. Because of this, most field theories such as elasticity make use of a *tensor formalism* using index notation. This enables efficient representation of all variables and governing equations using a single standardized scheme. The tensor concept is defined more precisely in a later section, but for now we can simply say that scalars, vectors, matrices, and other higher-order variables can all be represented by tensors of various orders. We now proceed to a discussion on the notational rules of order for the tensor formalism. Additional information on tensors and index notation can be found in many texts such as Goodbody (1982), Simmons (1994), Itskov (2015) and Sadd (2019).

1.2 Index notation

Index notation is a shorthand scheme whereby a whole set of numbers (elements or components) is represented by a single symbol with subscripts. For example, the three numbers a_1, a_2, a_3 are denoted by the symbol a_i, where index i has range 1, 2, 3. In a similar fashion, a_{ij} represents the nine numbers a_{11}, a_{12}, a_{13}, a_{21}, a_{22}, a_{23}, a_{31}, a_{32}, a_{33}. Although these representations can be written in any manner, it is common to use a scheme related to vector and matrix formats such that

$$a_i = \begin{bmatrix} a_1 \\ a_2 \\ a_3 \end{bmatrix}, \quad a_{ij} = \begin{bmatrix} a_{11} & a_{12} & a_{13} \\ a_{21} & a_{22} & a_{23} \\ a_{31} & a_{32} & a_{33} \end{bmatrix} \tag{1.2.1}$$

In the matrix format, a_{1j} represents the first row, while a_{i1} indicates the first column. Other columns and rows are indicated in similar fashion, and thus the first index represents the row, while the second index denotes the column.

In general a symbol $a_{ij...k}$ with N distinct indices represents 3^N distinct numbers. It should be apparent that a_i and a_j represent the same three numbers, and likewise a_{ij} and a_{mn} signify the same matrix. Addition, subtraction, multiplication, and equality of index symbols are defined in the normal fashion. For example, addition and subtraction are given by

$$a_i \pm b_i = \begin{bmatrix} a_1 \pm b_1 \\ a_2 \pm b_2 \\ a_3 \pm b_3 \end{bmatrix}, \quad a_{ij} \pm b_{ij} = \begin{bmatrix} a_{11} \pm b_{11} & a_{12} \pm b_{12} & a_{13} \pm b_{13} \\ a_{21} \pm b_{21} & a_{22} \pm b_{22} & a_{23} \pm b_{23} \\ a_{31} \pm b_{31} & a_{32} \pm b_{32} & a_{33} \pm b_{33} \end{bmatrix} \tag{1.2.2}$$

and scalar multiplication is specified as

$$\lambda a_i = \begin{bmatrix} \lambda a_1 \\ \lambda a_2 \\ \lambda a_3 \end{bmatrix}, \quad \lambda a_{ij} = \begin{bmatrix} \lambda a_{11} & \lambda a_{12} & \lambda a_{13} \\ \lambda a_{21} & \lambda a_{22} & \lambda a_{23} \\ \lambda a_{31} & \lambda a_{32} & \lambda a_{33} \end{bmatrix} \tag{1.2.3}$$

The multiplication of two symbols with different indices is called *outer multiplication*, and a simple example is given by

$$a_i b_j = \begin{bmatrix} a_1 b_1 & a_1 b_2 & a_1 b_3 \\ a_2 b_1 & a_2 b_2 & a_2 b_3 \\ a_3 b_1 & a_3 b_2 & a_3 b_3 \end{bmatrix} \tag{1.2.4}$$

The previous operations obey usual commutative, associative, and distributive laws, for example

$$a_i + b_i = b_i + a_i$$
$$a_{ij}b_k = b_k a_{ij}$$
$$a_i + (b_i + c_i) = (a_i + b_i) + c_i \qquad (1.2.5)$$
$$a_i(b_{jk}c_l) = (a_i b_{jk})c_l$$
$$a_{ij}(b_k + c_k) = a_{ij}b_k + a_{ij}c_k$$

Note that the simple relations $a_i = b_i$ and $a_{ij} = b_{ij}$ imply that $a_1 = b_1, a_2 = b_2, \ldots$ and $a_{11} = b_{11}, a_{12} = b_{12},$ ... However, relations of the form $a_i = b_j$ or $a_{ij} = b_{kl}$ have ambiguous meaning because the distinct indices on each term are not the same, and these types of expressions are to be avoided in this notational scheme. In general, the distinct subscripts on all individual terms in an equation should match.

It is convenient to adopt the convention that if a subscript appears twice in the same term, then *summation* over that subscript from one to three is implied, for example

$$a_{ii} = \sum_{i=1}^{3} a_{ii} = a_{11} + a_{22} + a_{33}$$
$$a_{ij}b_j = \sum_{j=1}^{3} a_{ij}b_j = a_{i1}b_1 + a_{i2}b_2 + a_{i3}b_3 \qquad (1.2.6)$$

It should be apparent that $a_{ii} = a_{jj} = a_{kk} = \ldots$, and therefore the *repeated* subscripts or indices are sometimes called *dummy* subscripts. Unspecified indices that are not repeated are called *free* or *distinct* subscripts. The summation convention may be suspended by underlining one of the repeated indices or by writing *no sum*. The use of three or more repeated indices in the same term (e.g., a_{iii} or $a_{iij}b_{ij}$) has ambiguous meaning and is to be avoided. On a given symbol, the process of setting two free indices equal is called *contraction*. For example, a_{ii} is obtained from a_{ij} by contraction on i and j. The operation of outer multiplication of two indexed symbols followed by contraction with respect to one index from each symbol generates an *inner multiplication*; for example, $a_{ij}b_{jk}$ is an inner product obtained from the outer product $a_{ij}b_{mk}$ by contraction on indices j and m.

A symbol $a_{ij\ldots m\ldots n\ldots k}$ is said to be *symmetric* with respect to index pair mn if

$$a_{ij\ldots m\ldots n\ldots k} = a_{ij\ldots n\ldots m\ldots k} \qquad (1.2.7)$$

while it is *antisymmetric* or *skewsymmetric* if

$$a_{ij\ldots m\ldots n\ldots k} = -a_{ij\ldots n\ldots m\ldots k} \qquad (1.2.8)$$

Note that if $a_{ij\ldots m\ldots n\ldots k}$ is symmetric in mn while $b_{pq\ldots m\ldots n\ldots r}$ is antisymmetric in mn, then the product is zero

$$a_{ij\ldots m\ldots n\ldots k}b_{pq\ldots m\ldots n\ldots r} = 0 \qquad (1.2.9)$$

A useful identity may be written as

$$a_{ij} = \frac{1}{2}(a_{ij} + a_{ji}) + \frac{1}{2}(a_{ij} - a_{ji}) = a_{(ij)} + a_{[ij]} \qquad (1.2.10)$$

The first term $a_{(ij)} = 1/2(a_{ij} + a_{ji})$ is symmetric, while the second term $a_{[ij]} = 1/2(a_{ij} - a_{ji})$ is antisymmetric, and thus an arbitrary symbol a_{ij} can be expressed as the sum of symmetric and antisymmetric pieces. Note that if a_{ij} is symmetric, it has only six independent components. On the other hand, if a_{ij} is antisymmetric, its diagonal terms a_{ii} (no sum on i) must be zero, and it has only three independent components. Since $a_{[ij]}$ has only three independent components, it can be related to a quantity with a single index, for example a_i (see Exercise 1.15).

Example 1.1 Index notation examples

The matrix a_{ij} and vector b_i are specified by

$$a_{ij} = \begin{bmatrix} 1 & 2 & 0 \\ 0 & 4 & 3 \\ 2 & 1 & 2 \end{bmatrix}, \quad b_i = \begin{bmatrix} 2 \\ 4 \\ 0 \end{bmatrix}$$

Determine the following quantities: a_{ii}, $a_{ij}a_{ij}$, $a_{ij}a_{jk}$, $a_{ij}b_j$, $a_{ij}b_ib_j$, b_ib_i, b_ib_j, $a_{(ij)}$, $a_{[ij]}$, and indicate whether they are a scalar, vector, or matrix.

Following the standard definitions given in Section 1.2

$$a_{ii} = a_{11} + a_{22} + a_{33} = 7 \text{ (scalar)}$$

$$a_{ij}a_{ij} = a_{11}a_{11} + a_{12}a_{12} + a_{13}a_{13} + a_{21}a_{21} + a_{22}a_{22} + a_{23}a_{23} + a_{31}a_{31} + a_{32}a_{32} + a_{33}a_{33}$$
$$= 1 + 4 + 0 + 0 + 16 + 9 + 4 + 1 + 4 = 39 \text{ (scalar)}$$

$$a_{ij}a_{jk} = a_{i1}a_{1k} + a_{i2}a_{2k} + a_{i3}a_{3k} = \begin{bmatrix} 1 & 10 & 6 \\ 6 & 19 & 18 \\ 6 & 10 & 7 \end{bmatrix} \text{ (matrix)}$$

$$a_{ij}b_j = a_{i1}b_1 + a_{i2}b_2 + a_{i3}b_3 = \begin{bmatrix} 10 \\ 16 \\ 8 \end{bmatrix} \text{ (vector)}$$

$$a_{ij}b_ib_j = a_{11}b_1b_1 + a_{12}b_1b_2 + a_{13}b_1b_3 + a_{21}b_2b_1 + \ldots = 84 \text{ (scalar)}$$

$$b_ib_i = b_1b_1 + b_2b_2 + b_3b_3 = 4 + 16 + 0 = 20 \text{ (scalar)}$$

$$b_ib_j = \begin{bmatrix} 4 & 8 & 0 \\ 8 & 16 & 0 \\ 0 & 0 & 0 \end{bmatrix} \text{ (matrix)}$$

$$a_{(ij)} = \frac{1}{2}(a_{ij} + a_{ji}) = \frac{1}{2}\begin{bmatrix} 1 & 2 & 0 \\ 0 & 4 & 3 \\ 2 & 1 & 2 \end{bmatrix} + \frac{1}{2}\begin{bmatrix} 1 & 0 & 2 \\ 2 & 4 & 1 \\ 0 & 3 & 2 \end{bmatrix} = \begin{bmatrix} 1 & 1 & 1 \\ 1 & 4 & 2 \\ 1 & 2 & 2 \end{bmatrix} \text{ (matrix)}$$

$$a_{[ij]} = \frac{1}{2}(a_{ij} - a_{ji}) = \frac{1}{2}\begin{bmatrix} 1 & 2 & 0 \\ 0 & 4 & 3 \\ 2 & 1 & 2 \end{bmatrix} - \frac{1}{2}\begin{bmatrix} 1 & 0 & 2 \\ 2 & 4 & 1 \\ 0 & 3 & 2 \end{bmatrix} = \begin{bmatrix} 0 & 1 & -1 \\ -1 & 0 & 1 \\ 1 & -1 & 0 \end{bmatrix} \text{ (matrix)}$$

1.3 Kronecker delta and alternating symbol

A useful special symbol commonly used in index notational schemes is the *Kronecker delta* defined by

$$\delta_{ij} = \begin{cases} 1, & if\ i = j\ (no\ sum) \\ 0, & if\ i \neq j \end{cases} = \begin{bmatrix} 1 & 0 & 0 \\ 0 & 1 & 0 \\ 0 & 0 & 1 \end{bmatrix} \tag{1.3.1}$$

Within usual matrix theory, it is observed that this symbol is simply the unit matrix. Note that the Kronecker delta is a symmetric symbol. Particularly useful properties of the Kronecker delta include the following

$$\delta_{ij} = \delta_{ji}$$
$$\delta_{ii} = 3,\ \delta_{i\underline{i}} = 1$$
$$\delta_{ij}a_j = a_i,\ \delta_{ij}a_i = a_j \tag{1.3.2}$$
$$\delta_{ij}a_{jk} = a_{ik},\ \delta_{jk}a_{ik} = a_{ij}$$
$$\delta_{ij}a_{ij} = a_{ii},\ \delta_{ij}\delta_{ij} = 3$$

Another useful special symbol is the *alternating* or *permutation symbol* defined by

$$\varepsilon_{ijk} = \begin{cases} +1, & if\ ijk\ is\ an\ even\ permutation\ of\ 1,\ 2,\ 3 \\ -1, & if\ ijk\ is\ an\ odd\ permutation\ of\ 1,\ 2,3 \\ 0, & otherwise \end{cases} \tag{1.3.3}$$

Consequently, $\varepsilon_{123} = \varepsilon_{231} = \varepsilon_{312} = 1, \varepsilon_{321} = \varepsilon_{132} = \varepsilon_{213} = -1, \varepsilon_{112} = \varepsilon_{131} = \varepsilon_{222} = \ldots = 0$. Therefore, of the 27 possible terms for the alternating symbol, three are equal to $+1$, three are equal to -1, and all others are 0. The alternating symbol is antisymmetric with respect to any pair of its indices.

This particular symbol is useful in evaluating determinants and vector cross products, and the determinant of an array a_{ij} can be written in two equivalent forms

$$\det[a_{ij}] = |a_{ij}| = \begin{vmatrix} a_{11} & a_{12} & a_{13} \\ a_{21} & a_{22} & a_{23} \\ a_{31} & a_{32} & a_{33} \end{vmatrix} = \varepsilon_{ijk}a_{1i}a_{2j}a_{3k} = \varepsilon_{ijk}a_{i1}a_{j2}a_{k3} \tag{1.3.4}$$

where the first index expression represents the row expansion, while the second form is the column expansion. Using the property

$$\varepsilon_{ijk}\varepsilon_{pqr} = \begin{vmatrix} \delta_{ip} & \delta_{iq} & \delta_{ir} \\ \delta_{jp} & \delta_{jq} & \delta_{jr} \\ \delta_{kp} & \delta_{kq} & \delta_{kr} \end{vmatrix} \tag{1.3.5}$$

another form of the determinant of a matrix can be written as

$$\det[a_{ij}] = \frac{1}{6}\varepsilon_{ijk}\varepsilon_{pqr}a_{ip}a_{jq}a_{kr} \tag{1.3.6}$$

1.4 Coordinate transformations

It is convenient and in fact necessary to express elasticity variables and field equations in several different coordinate systems (see Appendix A). This situation requires the development of particular transformation rules for scalar, vector, matrix, and higher-order variables. This concept is fundamentally connected with the basic definitions of tensor variables and their related tensor transformation laws. We restrict our discussion to transformations only between Cartesian coordinate systems, and thus consider the two systems shown in Fig. 1.1. The two Cartesian frames (x_1, x_2, x_3) and (x_1', x_2', x_3') differ only by orientation, and the unit basis vectors for each frame are $\{e_i\} = \{e_1, e_2, e_3\}$ and $\{e_i'\} = \{e_1', e_2', e_3'\}$.

Let Q_{ij} denote the cosine of the angle between the x_i'-axis and the x_j-axis

$$Q_{ij} = \cos(x_i', x_j) \tag{1.4.1}$$

Using this definition, the basis vectors in the primed coordinate frame can be easily expressed in terms of those in the unprimed frame by the relations

$$e_1' = Q_{11}e_1 + Q_{12}e_2 + Q_{13}e_3$$
$$e_2' = Q_{21}e_1 + Q_{22}e_2 + Q_{23}e_3 \tag{1.4.2}$$
$$e_3' = Q_{31}e_1 + Q_{32}e_2 + Q_{33}e_3$$

or in index notation

$$e_i' = Q_{ij}e_j \tag{1.4.3}$$

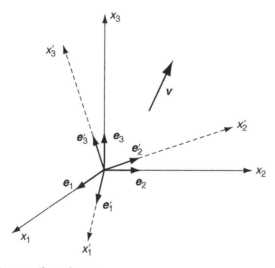

FIG. 1.1 Change of Cartesian coordinate frames.

Likewise, the opposite transformation can be written using the same format as

$$e_i = Q_{ji}e'_j \tag{1.4.4}$$

Now an arbitrary vector v can be written in either of the two coordinate systems as

$$\begin{aligned} v &= v_1 e_1 + v_2 e_2 + v_3 e_3 = v_i e_i \\ &= v'_1 e'_1 + v'_2 e'_2 + v'_3 e'_3 = v'_i e'_i \end{aligned} \tag{1.4.5}$$

Substituting form (1.4.4) into (1.4.5)₁ gives

$$v = v_i Q_{ji} e'_j$$

but from (1.4.5)₂, $v = v'_j e'_j$, and so we find that

$$v'_i = Q_{ij} v_j \tag{1.4.6}$$

In similar fashion, using (1.4.3) in (1.4.5)₂ gives

$$v_i = Q_{ji} v'_j \tag{1.4.7}$$

Relations (1.4.6) and (1.4.7) constitute the transformation laws for the Cartesian components of a vector under a change of rectangular Cartesian coordinate frame. It should be understood that under such transformations, the vector is unaltered (retaining original length and orientation), and only its components are changed. Consequently, if we know the components of a vector in one frame, relation (1.4.6) and/or relation (1.4.7) can be used to calculate components in any other frame.

The fact that transformations are being made only between orthogonal coordinate systems places some particular restrictions on the transformation or direction cosine matrix Q_{ij}. These can be determined by using (1.4.6) and (1.4.7) together to get

$$v_i = Q_{ji} v'_j = Q_{ji} Q_{jk} v_k \tag{1.4.8}$$

From the properties of the Kronecker delta, this expression can be written as

$$\delta_{ik} v_k = Q_{ji} Q_{jk} v_k \quad \text{or} \quad (Q_{ji} Q_{jk} - \delta_{ik}) v_k = 0$$

and since this relation is true for all vectors v_k, the expression in parentheses must be zero, giving the result

$$Q_{ji} Q_{jk} = \delta_{ik} \tag{1.4.9}$$

In similar fashion, relations (1.4.6) and (1.4.7) can be used to eliminate v_i (instead of v'_i) to get

$$Q_{ij} Q_{kj} = \delta_{ik} \tag{1.4.10}$$

Relations (1.4.9) and (1.4.10) comprise the *orthogonality conditions* that Q_{ij} must satisfy. Taking the determinant of either relation gives another related result

$$\det[Q_{ij}] = \pm 1 \tag{1.4.11}$$

Matrices that satisfy these relations are called orthogonal, and the transformations given by (1.4.6) and (1.4.7) are therefore referred to as orthogonal transformations.

1.5 Cartesian tensors

Scalars, vectors, matrices, and higher-order quantities can be represented by a general index notational scheme. Using this approach, all quantities may then be referred to as tensors of different orders. The previously presented transformation properties of a vector can be used to establish the general transformation properties of these tensors. Restricting the transformations to those only between Cartesian coordinate systems, the general set of transformation relations for various orders can be written as

$$
\begin{aligned}
a' &= a, \text{ zero order (scalar)} \\
a'_i &= Q_{ip}a_p, \text{ first order (vector)} \\
a'_{ij} &= Q_{ip}Q_{jq}a_{pq}, \text{ second order (matrix)} \\
a'_{ijk} &= Q_{ip}Q_{jp}Q_{kr}a_{pqr}, \text{ third order} \\
a'_{ijkl} &= Q_{ip}Q_{jq}Q_{kr}Q_{ls}a_{pqrs}, \text{ fourth order} \\
&\ \vdots \\
a'_{ijk\ldots m} &= Q_{ip}Q_{jq}Q_{kr}\cdots Q_{mt}a_{pqr\ldots t}, \text{ general order}
\end{aligned}
\tag{1.5.1}
$$

Note that, according to these definitions, a scalar is a zero-order tensor, a vector is a tensor of order one, and a matrix is a tensor of order two. Relations (1.5.1) then specify the transformation rules for the components of Cartesian tensors of any order under the rotation Q_{ij}. This transformation theory proves to be very valuable in determining the displacement, stress, and strain in different coordinate directions. Some tensors are of a special form in which their components remain the same under all transformations, and these are referred to as *isotropic tensors*. It can be easily verified (see Exercise 1.8) that the Kronecker delta δ_{ij} has such a property and is therefore a second-order isotropic tensor. The alternating symbol ε_{ijk} is found to be the third-order isotropic form. The fourth-order case (Exercise 1.9) can be expressed in terms of products of Kronecker deltas, and this has important applications in formulating isotropic elastic constitutive relations in Section 4.2.

The distinction between the components and the tensor should be understood. Recall that a vector v can be expressed as

$$
\begin{aligned}
v &= v_1 e_1 + v_2 e_2 + v_3 e_3 = v_i e_i \\
&= v'_1 e'_1 + v'_2 e'_2 + v'_3 e'_3 = v'_i e'_i
\end{aligned}
\tag{1.5.2}
$$

In a similar fashion, a second-order tensor A can be written

$$
\begin{aligned}
A &= A_{11}e_1 e_1 + A_{12}e_1 e_2 + A_{13}e_1 e_3 \\
&\quad + A_{21}e_2 e_1 + A_{22}e_2 e_2 + A_{23}e_2 e_3 \\
&\quad + A_{31}e_3 e_1 + A_{32}e_3 e_2 + A_{33}e_3 e_3 \\
&= A_{ij}e_i e_j = A'_{ij}e'_i e'_j
\end{aligned}
\tag{1.5.3}
$$

and similar schemes can be used to represent tensors of higher order. The representation used in Eq. (1.5.3) is commonly called *dyadic notation*, and some authors write the dyadic products $e_i e_j$ using a *tensor product* notation $e_i \otimes e_j$. Additional information on dyadic notation can be found in Weatherburn (1948) and Chou and Pagano (1967).

Relations (1.5.2) and (1.5.3) indicate that any tensor can be expressed in terms of components in any coordinate system, and it is only the components that change under coordinate transformation. For example, the state of stress at a point in an elastic solid depends on the problem geometry and applied loadings. As is shown later, these stress components are those of a second-order tensor and therefore obey transformation law (1.5.1)$_3$. Although the components of the stress tensor change with choice of coordinates, the stress tensor (representing the state of stress) does not.

An important property of a tensor is that if we know its components in one coordinate system, we can find them in any other coordinate frame by using the appropriate transformation law. Because the components of Cartesian tensors are representable by indexed symbols, the operations of equality, addition, subtraction, multiplication, and so forth are defined in a manner consistent with the indicial notation procedures previously discussed. The terminology *tensor* without the adjective *Cartesian* usually refers to a more general scheme in which the coordinates are not necessarily rectangular Cartesian and the transformations between coordinates are not always orthogonal. Such general tensor theory is not discussed or used in this text.

Example 1.2 Transformation examples

The components of a first- and second-order tensor in a particular coordinate frame are given by

$$a_i = \begin{bmatrix} 1 \\ 4 \\ 2 \end{bmatrix}, \; a_{ij} = \begin{bmatrix} 1 & 0 & 3 \\ 0 & 2 & 2 \\ 3 & 2 & 4 \end{bmatrix}$$

Determine the components of each tensor in a new coordinate system found through a rotation of 60° ($\pi/3$ radians) about the x_3-axis. Choose a counterclockwise rotation when viewing down the negative x_3-axis (see Fig. 1.2).

The original and primed coordinate systems shown in Fig. 1.2 establish the angles between the various axes. The solution starts by determining the rotation matrix for this case

$$Q_{ij} = \begin{bmatrix} \cos 60° & \cos 30° & \cos 90° \\ \cos 150° & \cos 60° & \cos 90° \\ \cos 90° & \cos 90° & \cos 0° \end{bmatrix} = \begin{bmatrix} 1/2 & \sqrt{3}/2 & 0 \\ -\sqrt{3}/2 & 1/2 & 0 \\ 0 & 0 & 1 \end{bmatrix}$$

The transformation for the vector quantity follows from Eq. (1.5.1)$_2$

$$a_i' = Q_{ij}a_j = \begin{bmatrix} 1/2 & \sqrt{3}/2 & 0 \\ -\sqrt{3}/2 & 1/2 & 0 \\ 0 & 0 & 1 \end{bmatrix}\begin{bmatrix} 1 \\ 4 \\ 2 \end{bmatrix} = \begin{bmatrix} 1/2 + 2\sqrt{3} \\ 2 - \sqrt{3}/2 \\ 2 \end{bmatrix}$$

and the second-order tensor (matrix) transforms according to (1.5.1)$_3$

$$a_{ij}' = Q_{ip}Q_{jq}a_{pq} = \begin{bmatrix} 1/2 & \sqrt{3}/2 & 0 \\ -\sqrt{3}/2 & 1/2 & 0 \\ 0 & 0 & 1 \end{bmatrix}\begin{bmatrix} 1 & 0 & 3 \\ 0 & 2 & 2 \\ 3 & 2 & 4 \end{bmatrix}\begin{bmatrix} 1/2 & \sqrt{3}/2 & 0 \\ -\sqrt{3}/2 & 1/2 & 0 \\ 0 & 0 & 1 \end{bmatrix}^T$$

$$= \begin{bmatrix} 7/4 & \sqrt{3}/4 & 3/2 + \sqrt{3} \\ \sqrt{3}/4 & 5/4 & 1 - 3\sqrt{3}/2 \\ 3/2 + \sqrt{3} & 1 - 3\sqrt{3}/2 & 4 \end{bmatrix}$$

where $[\ ]^T$ indicates transpose (defined in Section 1.7). Although simple transformations can be worked out by hand, for more general cases it is more convenient to use a computational scheme to evaluate the necessary matrix multiplications required in the transformation laws (1.5.1). MATLAB® software is ideally suited to carry out such calculations, and an example program to evaluate the transformation of second-order tensors is given in Example C.1 in Appendix C.

1.6 Principal values and directions for symmetric second-order tensors

Considering the tensor transformation concept previously discussed, it should be apparent that there might exist particular coordinate systems in which the components of a tensor take on maximum or minimum values. This concept is easily visualized when we consider the components of a vector as shown in Fig. 1.1. If we choose a particular coordinate system that has been rotated so that the x_3-axis lies along the direction of the vector, then the vector will have components $v = \{0, 0, |v|\}$. For this case, two of the components have been reduced to zero, while the remaining component becomes the largest possible (the total magnitude).

This situation is most useful for symmetric second-order tensors that eventually represent the stress and/or strain at a point in an elastic solid. The direction determined by the unit vector n is said to be a *principal direction* or *eigenvector* of the symmetric second-order tensor a_{ij} if there exists a parameter λ such that

$$a_{ij}n_j = \lambda n_i$$
$$a_{11}n_1 + a_{12}n_2 + a_{13}n_3 = \lambda n_1$$
$$a_{21}n_1 + a_{22}n_2 + a_{23}n_3 = \lambda n_2 \tag{1.6.1}$$
$$a_{31}n_1 + a_{32}n_2 + a_{33}n_3 = \lambda n_3$$

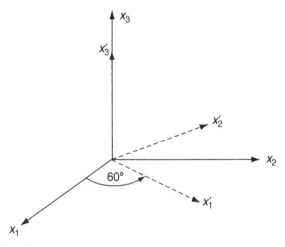

FIG. 1.2 Coordinate transformation.

where λ is called the *principal value* or *eigenvalue* of the tensor. Relation (1.6.1) can be rewritten as

$$(a_{ij} - \lambda\delta_{ij})n_j = 0$$

and this expression is simply a homogeneous system of three linear algebraic equations in the unknowns n_1, n_2, n_3. The system possesses a nontrivial solution if and only if the determinant of its coefficient matrix vanishes

$$\det[a_{ij} - \lambda\delta_{ij}] = 0$$

Expanding the determinant produces a cubic equation in terms of λ

$$\det[a_{ij} - \lambda\delta_{ij}] = -\lambda^3 + I_a\lambda^2 - II_a\lambda + III_a = 0 \tag{1.6.2}$$

where

$$I_a = a_{ii} = a_{11} + a_{22} + a_{33}$$

$$II_a = \frac{1}{2}(a_{ii}a_{jj} - a_{ij}a_{ij}) = \begin{vmatrix} a_{11} & a_{12} \\ a_{21} & a_{22} \end{vmatrix} + \begin{vmatrix} a_{22} & a_{23} \\ a_{32} & a_{33} \end{vmatrix} + \begin{vmatrix} a_{11} & a_{13} \\ a_{31} & a_{33} \end{vmatrix} \tag{1.6.3}$$

$$III_a = \det[a_{ij}]$$

The scalars I_a, II_a, and III_a are called the *fundamental invariants* of the tensor a_{ij}, and relation (1.6.2) is known as the *characteristic equation*. As indicated by their name, the three invariants do not change value under coordinate transformation. The roots of the characteristic equation determine the allowable values for λ, and each of these may be back-substituted into relation (1.6.1) to solve for the associated principal direction $\boldsymbol{n}$. Usually these principal directions are normalized using $\boldsymbol{n} \cdot \boldsymbol{n} = 1$.

Under the condition that the components a_{ij} are real, it can be shown that all three roots λ_1, λ_2, λ_3 of the cubic equation (1.6.2) must be real. Furthermore, if these roots are distinct, the principal directions associated with each principal value are orthogonal. Thus, we can conclude that every symmetric second-order tensor has at least three mutually perpendicular principal directions and at most three distinct principal values that are the roots of the characteristic equation. By denoting the principal directions $\boldsymbol{n}^{(1)}$, $\boldsymbol{n}^{(2)}$, $\boldsymbol{n}^{(3)}$ corresponding to the principal values λ_1, λ_2, λ_3, three possibilities arise:

1. All three principal values are distinct; the three corresponding principal directions are unique (except for sense).
2. Two principal values are equal ($\lambda_1 \neq \lambda_2 = \lambda_3$); the principal direction $\boldsymbol{n}^{(1)}$ is unique (except for sense), and every direction perpendicular to $\boldsymbol{n}^{(1)}$ is a principal direction associated with λ_2, λ_3.
3. All three principal values are equal; every direction is principal, and the tensor is isotropic, as per discussion in the previous section.

Therefore, according to what we have presented, it is always possible to identify a right-handed Cartesian coordinate system such that each axis lies along the principal directions of any given symmetric second-order tensor. Such axes are called the *principal axes* of the tensor. For this case, the basis vectors are actually the unit principal directions $\{n^{(1)}, n^{(2)}, n^{(3)}\}$, and it can be shown that with respect to principal axes the tensor reduces to the diagonal form

$$a'_{ij} = \begin{bmatrix} \lambda_1 & 0 & 0 \\ 0 & \lambda_2 & 0 \\ 0 & 0 & \lambda_3 \end{bmatrix} \tag{1.6.4}$$

Note that the fundamental invariants defined by relations (1.6.3) can be expressed in terms of the principal values as

$$I_a = \lambda_1 + \lambda_2 + \lambda_3$$
$$II_a = \lambda_1\lambda_2 + \lambda_2\lambda_3 + \lambda_3\lambda_1 \tag{1.6.5}$$
$$III_a = \lambda_1\lambda_2\lambda_3$$

The eigenvalues have important extremal properties. If we arbitrarily rank the principal values such that $\lambda_1 > \lambda_2 > \lambda_3$, then λ_1 will be the largest of all possible diagonal elements, while λ_3 will be the smallest diagonal element possible. This theory is applied in elasticity as we seek the largest stress or strain components in an elastic solid.

Example 1.3 Principal value problem

Determine the invariants and principal values and directions of the following symmetric second-order tensor

$$a_{ij} = \begin{bmatrix} 2 & 0 & 0 \\ 0 & 3 & 4 \\ 0 & 4 & -3 \end{bmatrix}$$

The invariants follow from relations (1.6.3)

$$I_a = a_{ii} = 2 + 3 - 3 = 2$$

$$II_a = \begin{vmatrix} 2 & 0 \\ 0 & 3 \end{vmatrix} + \begin{vmatrix} 3 & 4 \\ 4 & -3 \end{vmatrix} + \begin{vmatrix} 2 & 0 \\ 0 & -3 \end{vmatrix} = 6 - 25 - 6 = -25$$

$$III_a = \begin{vmatrix} 2 & 0 & 0 \\ 0 & 3 & 4 \\ 0 & 4 & -3 \end{vmatrix} = 2(-9 - 16) = -50$$

The characteristic equation then becomes

$$\det[a_{ij} - \lambda\delta_{ij}] = -\lambda^3 + 2\lambda^2 + 25\lambda - 50 = 0$$
$$\Rightarrow (\lambda - 2)(\lambda^2 - 25) = 0$$
$$\therefore \lambda_1 = 5, \quad \lambda_2 = 2, \quad \lambda_3 = -5$$

Thus, for this case all principal values are distinct.

For the $\lambda_1 = 5$ root, Eq. (1.6.1) gives the system

$$-3n_1^{(1)} = 0$$
$$-2n_2^{(1)} + 4n_3^{(1)} = 0$$
$$4n_2^{(1)} - 8n_3^{(1)} = 0$$

which gives a normalized solution $n^{(1)} = \pm(2e_2 + e_3)/\sqrt{5}$. In similar fashion, the other two principal directions are found to be $n^{(2)} = \pm e_1$, $n^{(3)} = \pm(e_2 - 2e_3)/\sqrt{5}$. It is easily verified that these directions are mutually orthogonal. Fig. 1.3 illustrates their directions with respect to the given coordinate system, and this establishes the right-handed principal coordinate axes $\left(x_1', x_2', x_3'\right)$ For this case, the transformation matrix Q_{ij} defined by (1.4.1) becomes

$$Q_{ij} = \begin{bmatrix} 0 & 2/\sqrt{5} & 1/\sqrt{5} \\ 1 & 0 & 0 \\ 0 & 1/\sqrt{5} & -2/\sqrt{5} \end{bmatrix}$$

Notice the eigenvectors actually form the rows of the Q matrix.
Using this in the transformation law $(1.5.1)_3$, the components of the given second-order tensor become

$$a_{ij}' = \begin{bmatrix} 5 & 0 & 0 \\ 0 & 2 & 0 \\ 0 & 0 & -5 \end{bmatrix}$$

This result then validates the general theory given by relation (1.6.4), indicating that the tensor should take on diagonal form with the principal values as the elements.

    Only simple second-order tensors lead to a characteristic equation that is factorable, thus allowing solution by hand calculation. Most other cases normally develop a general cubic equation and a more complicated system to solve for the principal directions. Again, particular routines within the MATLAB® package offer convenient tools to solve these more general problems. Example C.2 in Appendix C provides a simple code to determine the principal values and directions for symmetric second-order tensors.

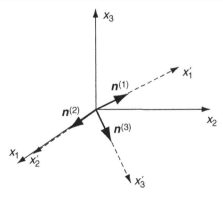

FIG. 1.3 Principal axes for Example 1.3.

1.7 Vector, matrix, and tensor algebra

Elasticity theory requires the use of many standard algebraic operations among vector, matrix, and tensor variables. These operations include dot and cross products of vectors and numerous matrix/tensor products. All of these operations can be expressed efficiently using compact tensor index notation. First, consider some particular vector products. Given two vectors a and b, with Cartesian components a_i and b_i, the *scalar* or *dot product* is defined by

$$a \cdot b = a_1 b_1 + a_2 b_2 + a_3 b_3 = a_i b_i \qquad (1.7.1)$$

Because all indices in this expression are repeated, the quantity must be a scalar, that is, a tensor of order zero. The magnitude of a vector can then be expressed as

$$|a| = (a \cdot a)^{1/2} = (a_i a_i)^{1/2} \qquad (1.7.2)$$

The *vector* or *cross product* between two vectors a and b can be written as

$$a \times b = \begin{vmatrix} e_1 & e_2 & e_3 \\ a_1 & a_2 & a_3 \\ b_1 & b_2 & b_3 \end{vmatrix} = \varepsilon_{ijk} a_j b_k e_i \qquad (1.7.3)$$

where e_i are the unit basis vectors for the coordinate system. Note that the cross product gives a vector resultant whose components are $\varepsilon_{ijk} a_j b_k$. Another common vector product is the *scalar triple product* defined by

$$a \cdot b \times c = \begin{vmatrix} a_1 & a_2 & a_3 \\ b_1 & b_2 & b_3 \\ c_1 & c_2 & c_3 \end{vmatrix} = \varepsilon_{ijk} a_i b_j c_k \qquad (1.7.4)$$

Next consider some common matrix products. Using the usual direct notation for matrices and vectors, common products between a matrix $A = [A]$ with a vector a can be written as

$$\begin{aligned} Aa &= [A]\{a\} = A_{ij} a_j = a_j A_{ij} \\ a^T A &= \{a\}^T [A] = a_i A_{ij} = A_{ij} a_i \end{aligned} \qquad (1.7.5)$$

where a^T denotes the *transpose*, and for a vector quantity this simply changes the column matrix (3×1) into a row matrix (1×3). Note that each of these products results in a vector resultant. These types of expressions generally involve various inner products within the index notational scheme and, as noted, once the summation index is properly specified, the order of listing the product terms does not change the result. We will encounter several different combinations of products between two matrices A and B

$$\begin{aligned} AB &= [A][B] = A_{ij} B_{jk} \\ AB^T &= A_{ij} B_{kj} \\ A^T B &= A_{ji} B_{jk} \\ tr(AB) &= A_{ij} B_{ji} \\ tr(AB^T) &= tr(A^T B) = A_{ij} B_{ij} \end{aligned} \qquad (1.7.6)$$

where A^T indicates the *transpose* and $tr(A)$ is the *trace* of the matrix defined by

$$A_{ij}^T = A_{ji}$$
$$tr(A) = A_{ii} = A_{11} + A_{22} + A_{33}$$

(1.7.7)

Similar to vector products, once the summation index is properly specified, the results in (1.7.6) do not depend on the order of listing the product terms. Note that this does not imply that $AB = BA$, which is certainly not true.

1.8 Calculus of Cartesian tensors

Most variables within elasticity theory are field variables, that is, functions depending on the spatial coordinates used to formulate the problem under study. For time-dependent problems, these variables could also have temporal variation. Thus, our scalar, vector, matrix, and general tensor variables are functions of the spatial coordinates (x_1, x_2, x_3). Because many elasticity equations involve differential and integral operations, it is necessary to have an understanding of the calculus of Cartesian tensor fields. Further information on vector differential and integral calculus can be found in Hildebrand (1976) and Kreyszig (2011).

The field concept for tensor components can be expressed as

$$a = a(x_1, x_2, x_3) = a(x_i) = a(\boldsymbol{x})$$
$$a_i = a_i(x_1, x_2, x_3) = a_i(x_i) = a_i(\boldsymbol{x})$$
$$a_{ij} = a_{ij}(x_1, x_2, x_3) = a_{ij}(x_i) = a_{ij}(\boldsymbol{x})$$
$$\vdots$$

It is convenient to introduce the *comma notation* for partial differentiation

$$a_{,i} = \frac{\partial}{\partial x_i} a; \quad a_{i,j} = \frac{\partial}{\partial x_j} a_i; \quad a_{ij,k} = \frac{\partial}{\partial x_k} a_{ij}; \quad \cdots$$

It can be shown that if the differentiation index is distinct, the order of the tensor is increased by one. For example, the derivative operation on a vector $a_{i,j}$ produces a second-order tensor or matrix given by

$$a_{i,j} = \begin{bmatrix} \dfrac{\partial a_1}{\partial x_1} & \dfrac{\partial a_1}{\partial x_2} & \dfrac{\partial a_1}{\partial x_3} \\[2mm] \dfrac{\partial a_2}{\partial x_1} & \dfrac{\partial a_2}{\partial x_2} & \dfrac{\partial a_2}{\partial x_3} \\[2mm] \dfrac{\partial a_3}{\partial x_1} & \dfrac{\partial a_3}{\partial x_2} & \dfrac{\partial a_3}{\partial x_3} \end{bmatrix}$$

Using Cartesian coordinates (x, y, z), consider the *directional derivative* of a scalar field function f with respect to a direction s

$$\frac{df}{ds} = \frac{\partial f}{\partial x} \frac{dx}{ds} + \frac{\partial f}{\partial y} \frac{dy}{ds} + \frac{\partial f}{\partial z} \frac{dz}{ds}$$

Note that the unit vector in the direction of s can be written as

$$n = \frac{dx}{ds}e_1 + \frac{dy}{ds}e_2 + \frac{dz}{ds}e_3$$

Therefore, the directional derivative can be expressed as the following scalar product

$$\frac{df}{ds} = n \cdot \nabla f \tag{1.8.1}$$

where ∇f is called the *gradient* of the scalar function f and is defined by

$$\nabla f = grad\, f = e_1 \frac{\partial f}{\partial x} + e_2 \frac{\partial f}{\partial y} + e_3 \frac{\partial f}{\partial z} \tag{1.8.2}$$

and the symbolic vector operator ∇ is called the *del operator*

$$\nabla = e_1 \frac{\partial}{\partial x} + e_2 \frac{\partial}{\partial y} + e_3 \frac{\partial}{\partial z} \tag{1.8.3}$$

These and other useful operations can be expressed in Cartesian tensor notation. Given the scalar field ϕ and vector field u, the following common differential operations can be written in index notation

$$
\begin{aligned}
&\textit{Gradient of a Scalar} \quad \nabla\phi = \phi_{,i}e_i \\
&\textit{Gradient of a Vector} \quad \nabla u = u_{i,j}e_i e_j \\
&\textit{Laplacian of a Scalar} \quad \nabla^2\phi = \nabla \cdot \nabla\phi = \phi_{,ii} \\
&\textit{Divergence of a Vector} \quad \nabla \cdot u = u_{i,i} \\
&\textit{Curl of a Vector} \quad \nabla \times u = \varepsilon_{ijk}u_{k,j}e_i \\
&\textit{Laplacian of a Vector} \quad \nabla^2 u = u_{i,kk}e_i
\end{aligned}
\tag{1.8.4}
$$

If ϕ and ψ are scalar fields and u and v are vector fields, several useful identities exist

$$
\begin{aligned}
&\nabla(\phi\psi) = (\nabla\phi)\psi + \phi(\nabla\psi) \\
&\nabla^2(\phi\psi) = (\nabla^2\phi)\psi + \phi(\nabla^2\psi) + 2\nabla\phi \cdot \nabla\psi \\
&\nabla \cdot (\phi u) = \nabla\phi \cdot u + \phi(\nabla \cdot u) \\
&\nabla \times (\phi u) = \nabla\phi \times u + \phi(\nabla \times u) \\
&\nabla \cdot (u \times v) = v \cdot (\nabla \times u) - u \cdot (\nabla \times v) \\
&\nabla \times \nabla\phi = 0 \\
&\nabla \cdot \nabla\phi = \nabla^2\phi \\
&\nabla \cdot \nabla \times u = 0 \\
&\nabla \times (\nabla \times u) = \nabla(\nabla \cdot u) - \nabla^2 u \\
&u \times (\nabla \times u) = \frac{1}{2}\nabla(u \cdot u) - u \cdot \nabla u
\end{aligned}
\tag{1.8.5}
$$

Each of these identities can be easily justified by using index notation from definition relations (1.8.4).

Example 1.4 Scalar and vector field examples

Scalar and vector field functions are given by $\phi = x^2 - y^2$ and $\boldsymbol{u} = 2x\boldsymbol{e}_1 + 3yz\boldsymbol{e}_2 + xy\boldsymbol{e}_3$. Calculate the following expressions, $\nabla\phi$, $\nabla^2\phi$, $\nabla\cdot\boldsymbol{u}$, $\nabla\boldsymbol{u}$, $\nabla \times \boldsymbol{u}$.

Using the basic relations (1.8.4)

$$\nabla\phi = 2x\boldsymbol{e}_1 - 2y\boldsymbol{e}_2$$

$$\nabla^2\phi = 2 - 2 = 0$$

$$\nabla\cdot\boldsymbol{u} = 2 + 3z + 0 = 2 + 3z$$

$$\nabla\boldsymbol{u} = u_{i,j} = \begin{bmatrix} 2 & 0 & 0 \\ 0 & 3z & 3y \\ y & x & 0 \end{bmatrix}$$

$$\nabla \times \boldsymbol{u} = \begin{vmatrix} \boldsymbol{e}_1 & \boldsymbol{e}_2 & \boldsymbol{e}_3 \\ \partial/\partial x & \partial/\partial y & \partial/\partial z \\ 2x & 3yz & xy \end{vmatrix} = (x - 3y)\boldsymbol{e}_1 - y\boldsymbol{e}_2$$

Using numerical methods, some of these variables can be conveniently computed and plotted in order to visualize the nature of the field distribution. For example, contours of $\phi =$ constant can easily be plotted using MATLAB® software, and vector distributions of $\nabla\phi$ can be shown as plots of vectors properly scaled in magnitude and orientation. Fig. 1.4 shows these two types of plots, and it is observed that the vector field $\nabla\phi$ is orthogonal to the ϕ-contours, a result that is true in general for all scalar fields. Numerically generated plots such as these will prove to be useful in understanding displacement, strain, and/or stress distributions found in the solution to a variety of problems in elasticity.

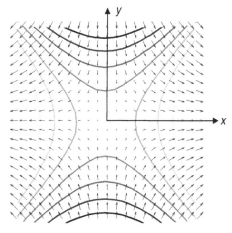

FIG. 1.4 Contours of $\phi =$ constant and vector distributions of $\nabla\phi$.

Next consider some results from vector/tensor integral calculus. We simply list some theorems that have later use in the development of elasticity theory.

1.8.1 Divergence or Gauss theorem

Let S be a piecewise continuous surface bounding the region of space V. If a vector field u is continuous and has continuous first derivatives in V, then

$$\iint_S u \cdot n \, dS = \iiint_V \nabla \cdot u \, dV \tag{1.8.6}$$

where n is the outer unit normal vector to surface S. This result is also true for tensors of any order

$$\iint_S a_{ij\ldots k} n_k \, dS = \iiint_V a_{ij\ldots k, k} \, dV \tag{1.8.7}$$

1.8.2 Stokes theorem

Let S be an open two-sided surface bounded by a piecewise continuous simple closed curve C. If u is continuous and has continuous first derivatives on S, then

$$\oint_C u \cdot dr = \iint_S (\nabla \times u) \cdot n \, dS \tag{1.8.8}$$

where the positive sense for the line integral is for the region S to lie to the left as one traverses curve C and n is the unit normal vector to S. Again, this result is also valid for tensors of arbitrary order, and so

$$\oint_C a_{ij\ldots k} dx_k = \iint_S \varepsilon_{rsk} a_{ij\ldots k, s} n_r \, dS \tag{1.8.9}$$

It can be shown that both divergence and Stokes theorems can be generalized so that the dot product in (1.8.6) and/or (1.8.8) can be replaced with a cross product.

1.8.3 Green's theorem in the plane

Applying Stokes theorem to a planar domain S with the vector field selected as $u = f e_1 + g e_2$ gives the result

$$\iint_S \left(\frac{\partial g}{\partial x} - \frac{\partial f}{\partial y} \right) dx \, dy = \int_C (f dx + g dy) \tag{1.8.10}$$

Further, special choices with either $f = 0$ or $g = 0$ imply

$$\iint_S \frac{\partial g}{\partial x} dx \, dy = \int_C g n_x ds, \quad \iint_S \frac{\partial f}{\partial y} dx \, dy = \int_C f n_y ds \tag{1.8.11}$$

1.8.4 Zero-value or localization theorem

Let $f_{ij...k}$ be a continuous tensor field of any order defined in an arbitrary region V. If the integral of $f_{ij...k}$ over V vanishes, then $f_{ij...k}$ must vanish in V

$$\iiint_V f_{ij...k}\,dV = 0 \Rightarrow f_{ij...k} = 0 \in V \qquad (1.8.12)$$

1.9 Orthogonal curvilinear coordinates

Many applications in elasticity theory involve domains that have curved boundary surfaces, commonly including circular, cylindrical, and spherical surfaces. To formulate and develop solutions for such problems, it is necessary to use curvilinear coordinate systems. This requires redevelopment of some previous results in orthogonal curvilinear coordinates. Before pursuing these general steps, we review the two most common curvilinear systems, cylindrical and spherical coordinates. The cylindrical coordinate system shown in Fig. 1.5 uses (r, θ, z) coordinates to describe spatial geometry. Relations between the Cartesian and cylindrical systems are given by

$$x_1 = r\cos\theta, \quad x_2 = r\sin\theta, \quad x_3 = z$$

$$r = \sqrt{x_1^2 + x_2^2}, \quad \theta = \tan^{-1}\left(\frac{x_2}{x_1}\right), \quad z = x_3 \qquad (1.9.1)$$

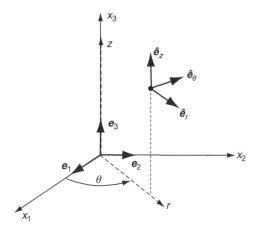

FIG. 1.5 Cylindrical coordinate system.

The spherical coordinate system is shown in Fig. 1.6 and uses (R, ϕ, θ) coordinates to describe geometry. The relations between Cartesian and spherical coordinates are

$$x_1 = R\cos\theta\sin\phi, \quad x_2 = R\sin\theta\sin\phi, \quad x_3 = R\cos\phi$$

$$R = \sqrt{x_1^2 + x_2^2 + x_3^2}, \quad \phi = \cos^{-1}\frac{x_3}{\sqrt{x_1^2 + x_2^2 + x_3^2}}, \quad \theta = \tan^{-1}\left(\frac{x_2}{x_1}\right) \qquad (1.9.2)$$

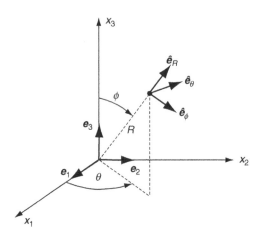

FIG. 1.6 Spherical coordinate system.

The unit basis vectors for each of these curvilinear systems are illustrated in Figs. 1.5 and 1.6. These represent unit tangent vectors along each of the three orthogonal coordinate curves.

Although primary use of curvilinear systems employs cylindrical and spherical coordinates, we briefly present a discussion valid for arbitrary coordinate systems. Consider the general case in which three orthogonal curvilinear coordinates are denoted by ξ^1, ξ^2, ξ^3, while the Cartesian coordinates are defined by x^1, x^2, x^3 (see Fig. 1.7). We assume invertible coordinate transformations exist between these systems specified by

$$\xi^m = \xi^m\left(x^1, x^2, x^3\right), \quad x^m = x^m\left(\xi^1, \xi^2, \xi^3\right) \tag{1.9.3}$$

In the curvilinear system, an arbitrary differential length in space can be expressed by

$$(ds)^2 = \left(h_1 d\xi^1\right)^2 + \left(h_2 d\xi^2\right)^2 + \left(h_3 d\xi^3\right)^2 \tag{1.9.4}$$

where h_1, h_2, h_3 are called *scale factors* that are in general non-negative functions of position. Let e_k be the fixed Cartesian basis vectors and $\hat{e}_k$ the curvilinear basis (see Fig. 1.7). By using similar concepts from the transformations discussed in Section 1.4, the curvilinear basis can be expressed in terms of the Cartesian basis as

$$\begin{aligned}
\hat{e}_1 &= \frac{dx^k}{ds_1} e_k = \frac{1}{h_1} \frac{\partial x^k}{\partial \xi^1} e_k \\[2mm]
\hat{e}_2 &= \frac{dx^k}{ds_2} e_k = \frac{1}{h_2} \frac{\partial x^k}{\partial \xi^2} e_k \\[2mm]
\hat{e}_3 &= \frac{dx^k}{ds_3} e_k = \frac{1}{h_3} \frac{\partial x^k}{\partial \xi^3} e_k
\end{aligned} \tag{1.9.5}$$

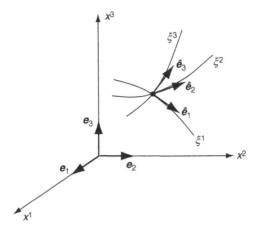

FIG. 1.7 Curvilinear coordinates.

where we have used (1.9.4). By using the fact that $\widehat{\boldsymbol{e}}_i \cdot \widehat{\boldsymbol{e}}_j = \delta_{ij}$, relation (1.9.5) gives

$$(h_1)^2 = \frac{\partial x^k}{\partial \xi^1}\frac{\partial x^k}{\partial \xi^1}$$

$$(h_2)^2 = \frac{\partial x^k}{\partial \xi^2}\frac{\partial x^k}{\partial \xi^2} \qquad (1.9.6)$$

$$(h_3)^2 = \frac{\partial x^k}{\partial \xi^3}\frac{\partial x^k}{\partial \xi^3}$$

It follows from (1.9.5) that the quantity

$$Q^k_r = \frac{1}{h_r}\frac{\partial x^k}{\partial \xi^r} \text{ (no sum on } r) \qquad (1.9.7)$$

represents the transformation tensor giving the curvilinear basis in terms of the Cartesian basis. This concept is similar to the transformation tensor Q_{ij} defined by (1.4.1) that is used between Cartesian systems.

The *physical components* of a vector or tensor are simply the components in a local set of Cartesian axes tangent to the curvilinear coordinate curves at any point in space. Thus, by using transformation relation (1.9.7), the physical components of a tensor $\boldsymbol{a}$ in a general curvilinear system are given by

$$a_{<ij...k>} = Q^p_i Q^q_j \cdots Q^s_k a_{pq...s} \qquad (1.9.8)$$

where $a_{pq...s}$ are the components in a fixed Cartesian frame. Note that the tensor can be expressed in either system as

$$\boldsymbol{a} = a_{ij...k}\boldsymbol{e}_i\boldsymbol{e}_j\cdots\boldsymbol{e}_k$$
$$= a_{<ij...k>}\widehat{\boldsymbol{e}}_i\widehat{\boldsymbol{e}}_j\cdots\widehat{\boldsymbol{e}}_k \qquad (1.9.9)$$

Because many applications involve differentiation of tensors, we must consider the differentiation of the curvilinear basis vectors. The Cartesian basis system e_k is fixed in orientation and therefore $\partial e_k/\partial x^i = \partial e_k/\partial \xi^j = 0$. However, derivatives of the curvilinear basis do not in general vanish, and differentiation of relations (1.9.5) gives the following results

$$\frac{\partial \widehat{e}_m}{\partial \xi^m} = -\frac{1}{h_n}\frac{\partial h_m}{\partial \xi^n}\widehat{e}_n - \frac{1}{h_r}\frac{\partial h_m}{\partial \xi^r}\widehat{e}_r; \quad m \neq n \neq r$$

$$\frac{\partial \widehat{e}_m}{\partial \xi^n} = \frac{1}{h_m}\frac{\partial h_n}{\partial \xi^m}\widehat{e}_n; \quad m \neq n, \quad no\ sum\ on\ repeated\ indices$$

(1.9.10)

Using these results, the derivative of any tensor can be evaluated. Consider the first derivative of a vector u

$$\frac{\partial}{\partial \xi^n}u = \frac{\partial}{\partial \xi^n}(u_{<m>}\widehat{e}_m) = \frac{\partial u_{<m>}}{\partial \xi^n}\widehat{e}_m + u_{<m>}\frac{\partial \widehat{e}_m}{\partial \xi^n}$$

(1.9.11)

The last term can be evaluated using (1.9.10), and thus the derivative of u can be expressed in terms of curvilinear components. Similar patterns follow for derivatives of higher-order tensors.

All vector differential operators of gradient, divergence, curl, and so forth can be expressed in any general curvilinear system by using these techniques. For example, the vector differential operator previously defined in Cartesian coordinates in (1.8.3) is given by

$$\nabla = \widehat{e}_1\frac{1}{h_1}\frac{\partial}{\partial \xi^1} + \widehat{e}_2\frac{1}{h_2}\frac{\partial}{\partial \xi^2} + \widehat{e}_3\frac{1}{h_3}\frac{\partial}{\partial \xi^3} = \sum_i \widehat{e}_i\frac{1}{h_i}\frac{\partial}{\partial \xi^i}$$

(1.9.12)

and this leads to the construction of the other common forms

Gradient of a Scalar $\nabla f = \widehat{e}_1\frac{1}{h_1}\frac{\partial f}{\partial \xi^1} + \widehat{e}_2\frac{1}{h_2}\frac{\partial f}{\partial \xi^2} + \widehat{e}_3\frac{1}{h_3}\frac{\partial f}{\partial \xi^3} = \sum_i \widehat{e}_i\frac{1}{h_i}\frac{\partial f}{\partial \xi^i}$ (1.9.13)

Divergence of a Vector $\nabla \cdot u = \frac{1}{h_1h_2h_3}\sum_i \frac{\partial}{\partial \xi^i}\left(\frac{h_1h_2h_3}{h_i}u_{<i>}\right)$ (1.9.14)

Laplacian of a Scalar $\nabla^2 \phi = \frac{1}{h_1h_2h_3}\sum_i \frac{\partial}{\partial \xi^i}\left(\frac{h_1h_2h_3}{(h_i)^2}\frac{\partial \phi}{\partial \xi^i}\right)$ (1.9.15)

Curl of a Vector $\nabla \times u = \sum_i \sum_j \sum_k \frac{\varepsilon_{ijk}}{h_jh_k}\frac{\partial}{\partial \xi^j}(u_{<k>}h_k)\widehat{e}_i$ (1.9.16)

Gradient of a Vector $\nabla u = \sum_i \sum_j \frac{\widehat{e}_i}{h_i}\left(\frac{\partial u_{<j>}}{\partial \xi^i}\widehat{e}_j + u_{<j>}\frac{\partial \widehat{e}_j}{\partial \xi^i}\right)$ (1.9.17)

Laplacian of a Vector $\nabla^2 u = \left(\sum_i \frac{\widehat{e}_i}{h_i}\frac{\partial}{\partial \xi^i}\right)\cdot\left(\sum_j \sum_k \frac{\widehat{e}_k}{h_k}\left[\frac{\partial u_{<j>}}{\partial \xi^k}\widehat{e}_j + u_{<j>}\frac{\partial \widehat{e}_j}{\partial \xi^k}\right]\right)$ (1.9.18)

It should be noted that these forms are significantly different from those previously given in relations (1.8.4) for Cartesian coordinates. Curvilinear systems add additional terms not found in rectangular coordinates. Other operations on higher-order tensors can be developed in a similar fashion (see Malvern, 1969, app. II). Specific transformation relations and field equations in cylindrical and spherical coordinate systems are given in Appendices A and B. Further discussion of these results is taken up in later chapters.

Example 1.5 Polar coordinates

Consider the two-dimensional case of a polar coordinate system as shown in Fig. 1.8. The differential length relation (1.9.4) for this case can be written as

$$(ds)^2 = (dr)^2 + (rd\theta)^2$$

and thus $h_1 = 1$ and $h_2 = r$. By using relations (1.9.5) or simply by using the geometry shown in Fig. 1.8

$$\widehat{e}_r = \cos\theta e_1 + \sin\theta e_2$$
$$\widehat{e}_\theta = -\sin\theta e_1 + \cos\theta e_2 \tag{1.9.19}$$

and so

$$\frac{\partial \widehat{e}_r}{\partial \theta} = \widehat{e}_\theta, \quad \frac{\partial \widehat{e}_\theta}{\partial \theta} = -\widehat{e}_r, \quad \frac{\partial \widehat{e}_r}{\partial r} = \frac{\partial \widehat{e}_\theta}{\partial r} = 0 \tag{1.9.20}$$

The basic vector differential operations then follow to be

$$\boldsymbol{\nabla} = \widehat{e}_r\frac{\partial}{\partial r} + \widehat{e}_\theta\frac{1}{r}\frac{\partial}{\partial \theta}$$

$$\boldsymbol{\nabla}\phi = \widehat{e}_r\frac{\partial\phi}{\partial r} + \widehat{e}_\theta\frac{1}{r}\frac{\partial\phi}{\partial \theta}$$

$$\boldsymbol{\nabla}\cdot\boldsymbol{u} = \frac{1}{r}\frac{\partial}{\partial r}(ru_r) + \frac{1}{r}\frac{\partial u_\theta}{\partial \theta}$$

$$\nabla^2\phi = \frac{1}{r}\frac{\partial}{\partial r}\left(r\frac{\partial\phi}{\partial r}\right) + \frac{1}{r^2}\frac{\partial^2\phi}{\partial \theta^2} \tag{1.9.21}$$

$$\boldsymbol{\nabla}\times\boldsymbol{u} = \left(\frac{1}{r}\frac{\partial}{\partial r}(ru_\theta) - \frac{1}{r}\frac{\partial u_r}{\partial \theta}\right)\widehat{e}_z$$

$$\boldsymbol{\nabla}\boldsymbol{u} = \frac{\partial u_r}{\partial r}\widehat{e}_r\widehat{e}_r + \frac{\partial u_\theta}{\partial r}\widehat{e}_r\widehat{e}_\theta + \frac{1}{r}\left(\frac{\partial u_r}{\partial \theta} - u_\theta\right)\widehat{e}_\theta\widehat{e}_r + \frac{1}{r}\left(\frac{\partial u_\theta}{\partial \theta} - u_r\right)\widehat{e}_\theta\widehat{e}_\theta$$

$$\nabla^2\boldsymbol{u} = \left(\nabla^2 u_r - \frac{2}{r^2}\frac{\partial u_\theta}{\partial \theta} - \frac{u_r}{r^2}\right)\widehat{e}_r + \left(\nabla^2 u_\theta + \frac{2}{r^2}\frac{\partial u_r}{\partial \theta} - \frac{u_\theta}{r^2}\right)\widehat{e}_\theta$$

where $\boldsymbol{u} = u_r\widehat{e}_r + u_\theta\widehat{e}_\theta$, $\widehat{e}_z = \widehat{e}_r \times \widehat{e}_\theta$. Notice that the Laplacian of a vector does not simply pass through and operate on each of the individual components as in the Cartesian case. Additional terms are generated because of the curvature of the particular coordinate system. Similar relations can be developed for cylindrical and spherical coordinate systems (see Exercises 1.17 and 1.18).

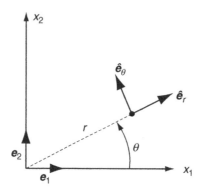

FIG. 1.8 Polar coordinate system.

The material reviewed in this chapter is used in many places for formulation development of elasticity theory. Throughout the text, notation uses scalar, vector, and tensor formats depending on the appropriateness to the topic under discussion. Most of the general formulation procedures in Chapters 2–5 use tensor index notation, while later chapters commonly use vector and scalar notation. Additional review of mathematical procedures for problem solution is supplied in chapter locations where they are applied.

References

Chou PC, Pagano NJ: *Elasticity—tensor, dyadic and engineering approaches*, Princeton, NJ, 1967, D. Van Nostrand.

Goodbody AM: *Cartesian tensors: with applications to mechanics, fluid mechanics and elasticity*, New York, 1982, Ellis Horwood.

Hildebrand FB: *Advanced calculus for applications*, ed 2, Englewood Cliffs, NJ, 1976, Prentice Hall.

Itskov M: *Tensor algebra and tensor analysis for engineers*, ed 4, New York, 2015, Springer.

Kreyszig E: *Advanced engineering mathematics*, ed 10, New York, 2011, John Wiley.

Malvern LE: *Introduction to the mechanics of a continuous medium*, Englewood Cliffs, NJ, 1969, Prentice Hall.

Sadd MH: Continuum mechanics modeling of material behavior, 2019, Elsevier.

Simmons JG: *A brief on tensor analysis*, New York, 1994, Springer.

Weatherburn CE: *Advanced vector analysis*, LaSalle, IL, 1948, Open Court.

Exercises

1.1 For the given matrix/vector pairs, compute the following quantities: a_{ii}, $a_{ij}a_{ij}$, $a_{ij}a_{jk}$, $a_{ij}b_j$, $a_{ij}b_ib_j$, b_ib_j, b_ib_i. For each case, point out whether the result is a scalar, vector or matrix. Note that $a_{ij}b_j$ is actually the matrix product $[a]\{b\}$, while $a_{ij}a_{jk}$ is the product $[a][a]$.

(a) $a_{ij} = \begin{bmatrix} 1 & 1 & 0 \\ 0 & 4 & 2 \\ 0 & 1 & 1 \end{bmatrix}$, $b_i = \begin{bmatrix} 0 \\ 1 \\ 2 \end{bmatrix}$ (b) $a_{ij} = \begin{bmatrix} 1 & 2 & 0 \\ 0 & 1 & 1 \\ 0 & 4 & 2 \end{bmatrix}$, $b_i = \begin{bmatrix} 2 \\ 1 \\ 0 \end{bmatrix}$

(c) $a_{ij} = \begin{bmatrix} 1 & 0 & 1 \\ 1 & 0 & 2 \\ 0 & 1 & 3 \end{bmatrix}$, $b_i = \begin{bmatrix} 1 \\ 0 \\ 1 \end{bmatrix}$ (d) $a_{ij} = \begin{bmatrix} 1 & 0 & 0 \\ 0 & 2 & 1 \\ 0 & 2 & 0 \end{bmatrix}$, $b_i = \begin{bmatrix} 2 \\ 0 \\ 1 \end{bmatrix}$

1.2 Use the decomposition result (1.2.10) to express a_{ij} from Exercise 1.1 in terms of the sum of symmetric and antisymmetric matrices. Verify that $a_{(ij)}$ and $a_{[ij]}$ satisfy the conditions given in the last paragraph of Section 1.2.

1.3 If a_{ij} is symmetric and b_{ij} is antisymmetric, prove in general that the product $a_{ij}b_{ij}$ is zero. Verify this result for the specific case by using the symmetric and antisymmetric terms from Exercise 1.2.

1.4 Explicitly verify the following properties of the Kronecker delta

$$\delta_{ij}a_j = a_i$$
$$\delta_{ij}a_{jk} = a_{ik}$$

1.5 Formally expand the expression (1.3.4) for the determinant and justify that either index notation form yields a result that matches the traditional form for $\det[a_{ij}]$.

1.6 Determine the components of the vector b_i and matrix a_{ij} given in Exercise 1.1 in a new coordinate system found through a rotation of $45°$ ($\pi/4$ radians) about the x_1-axis. The rotation direction follows the positive sense presented in Example 1.2.

1.7 Consider the two-dimensional coordinate transformation shown in Fig. 1.8. Through the counterclockwise rotation θ, a new polar coordinate system is created. Show that the transformation matrix for this case is given by

$$Q_{ij} = \begin{bmatrix} \cos\theta & \sin\theta \\ -\sin\theta & \cos\theta \end{bmatrix}$$

If $b_i = \begin{bmatrix} b_1 \\ b_2 \end{bmatrix}$, $a_{ij} = \begin{bmatrix} a_{11} & a_{21} \\ a_{12} & a_{22} \end{bmatrix}$ are the components of a first- and second-order tensor in the x_1, x_2 system, calculate their components in the rotated polar coordinate system.

1.8 Show that the second-order tensor $a\delta_{ij}$, where a is an arbitrary constant, retains its form under any transformation Q_{ij}. This form is then an isotropic second-order tensor.

1.9 The most general form of a fourth-order isotropic tensor can be expressed by

$$\alpha\delta_{ij}\delta_{kl} + \beta\delta_{ik}\delta_{jl} + \gamma\delta_{il}\delta_{jk}$$

where α, β, and γ are arbitrary constants. Verify that this form remains the same under the general transformation given by $(1.5.1)_5$.

1.10 For the fourth-order isotropic tensor given in Exercise 1.9, show that if $\beta = \gamma$, then the tensor will have the following symmetry $C_{ijkl} = C_{klij}$.

1.11 Show that the fundamental invariants can be expressed in terms of the principal values as given by relations (1.6.5).

1.12 Determine the invariants, and principal values and directions of the following matrices. Use the determined principal directions to establish a principal coordinate system, and following the procedures in Example 1.3, formally transform (rotate) the given matrix into the principal system to arrive at the appropriate diagonal form.

(a) $\begin{bmatrix} -1 & 1 & 0 \\ 1 & -1 & 0 \\ 0 & 0 & 1 \end{bmatrix}$ (Answer : $\lambda_i = -2, 0, 1$) (b) $\begin{bmatrix} -2 & 1 & 0 \\ 1 & -2 & 0 \\ 0 & 0 & 0 \end{bmatrix}$ (Answer : $\lambda_i = -3, -1, 0$)

(c) $\begin{bmatrix} -1 & 1 & 0 \\ 1 & -1 & 0 \\ 0 & 0 & 0 \end{bmatrix}$ (Answer : $\lambda_i = -2, 0, 0$) (d) $\begin{bmatrix} 6 & -3 & 0 \\ -3 & 6 & 0 \\ 0 & 0 & 6 \end{bmatrix}$ (Answer : $\lambda_i = 3, 6, 9$)

1.13* A second-order symmetric tensor field is given by

$$a_{ij} = \begin{bmatrix} 2x_1 & x_1 & 0 \\ x_1 & -6x_1^2 & 0 \\ 0 & 0 & 5x_1 \end{bmatrix}$$

Using MATLAB® (or similar software), investigate the nature of the variation of the principal values and directions over the interval $1 \le x_1 \le 2$. Formally plot the variation of the absolute value of each principal value over the range $1 \le x_1 \le 2$.

1.14 Calculate the quantities $\nabla \cdot \boldsymbol{u}$, $\nabla \times \boldsymbol{u}$, $\nabla^2 \boldsymbol{u}$, $\nabla \boldsymbol{u}$, $tr(\nabla \boldsymbol{u})$ for the following Cartesian vector fields:

(a) $\boldsymbol{u} = 2x_1 \boldsymbol{e}_1 + x_1 x_2 \boldsymbol{e}_2 + 2x_1 x_3 \boldsymbol{e}_3$

(b) $\boldsymbol{u} = x_1^2 \boldsymbol{e}_1 + 2x_1 x_2 x_3 \boldsymbol{e}_2 + 2x_3^2 \boldsymbol{e}_3$

(c) $\boldsymbol{u} = x_2 \boldsymbol{e}_1 + 2x_2 x_3 \boldsymbol{e}_2 + 4x_1^2 x_3 \boldsymbol{e}_3$

1.15 The *dual vector* a_i of an antisymmetric second-order tensor a_{ij} is defined by $a_i = -1/2 \varepsilon_{ijk} a_{jk}$. Show that this expression can be inverted to get $a_{jk} = -\varepsilon_{ijk} a_i$.

1.16 Using index notation, explicitly verify the vector identities:

(a) $(1.8.5)_{1,2,3}$

(b) $(1.8.5)_{4,5,6,7}$

(c) $(1.8.5)_{8,9,10}$

1.17 Extend the results found in Example 1.5, and determine the forms of ∇f, $\nabla \cdot u$, $\nabla^2 f$, and $\nabla \times u$ for a three-dimensional cylindrical coordinate system (see Fig. 1.5).

1.18 For the spherical coordinate system (R, ϕ, θ) in Fig. 1.6, show that

$$h_1 = 1, \quad h_2 = R, \quad h_3 = R \sin \phi$$

and the standard vector operations are given by

$$\nabla f = \widehat{e}_R \frac{\partial f}{\partial R} + \widehat{e}_\phi \frac{1}{R} \frac{\partial f}{\partial \phi} + \widehat{e}_\theta \frac{1}{R \sin \phi} \frac{\partial f}{\partial \theta}$$

$$\nabla \cdot u = \frac{1}{R^2} \frac{\partial}{\partial R} \left(R^2 u_R \right) + \frac{1}{R \sin \phi} \frac{\partial}{\partial \phi} \left(\sin \phi u_\phi \right) + \frac{1}{R \sin \phi} \frac{\partial u_\theta}{\partial \theta}$$

$$\nabla^2 f = \frac{1}{R^2} \frac{\partial}{\partial R} \left(R^2 \frac{\partial f}{\partial R} \right) + \frac{1}{R^2 \sin \phi} \frac{\partial}{\partial \phi} \left(\sin \phi \frac{\partial f}{\partial \phi} \right) + \frac{1}{R^2 \sin^2 \phi} \frac{\partial^2 f}{\partial \theta^2}$$

$$\nabla \times u = \widehat{e}_R \left[\frac{1}{R \sin \phi} \left(\frac{\partial}{\partial \phi} (\sin \phi u_\theta) - \frac{\partial u_\phi}{\partial \theta} \right) \right] + \widehat{e}_\phi \left[\frac{1}{R \sin \phi} \frac{\partial u_R}{\partial \theta} - \frac{1}{R} \frac{\partial}{\partial R} (R u_\theta) \right]$$

$$+ \widehat{e}_\theta \left[\frac{1}{R} \left(\frac{\partial}{\partial R} (R u_\phi) - \frac{\partial u_R}{\partial \phi} \right) \right]$$

Deformation: displacements and strains

2

We begin development of the basic field equations of elasticity theory by first investigating the kinematics of material deformation. As a result of applied loadings, elastic solids will change shape or deform, and these deformations can be quantified by knowing the displacements of material points in the body. The continuum hypothesis establishes a displacement field at all points within the elastic solid. Using appropriate geometry, particular measures of deformation can be constructed leading to the development of the strain tensor. As expected, the strain components are related to the displacement field. The purpose of this chapter is to introduce the basic definitions of displacement and strain, establish relations between these two field quantities, and finally investigate requirements to ensure single-valued, continuous displacement fields. As appropriate for linear elasticity, these kinematical results are developed under the conditions of small deformation theory. Developments in this chapter lead to two fundamental sets of field equations: the strain–displacement relations and the compatibility equations. Further field equation development, including internal force and stress distribution, equilibrium, and elastic constitutive behavior, occurs in subsequent chapters.

2.1 General deformations

Under the application of external loading, elastic solids deform. A simple two-dimensional cantilever beam example is shown in Fig. 2.1. The undeformed configuration is taken with the rectangular beam in the vertical position, and the end loading displaces material points to the deformed shape as shown. As is typical in most problems, the deformation varies from point to point and is thus said to be *nonhomogeneous*. A superimposed square mesh is shown in the two configurations, and this indicates how elements within the material deform locally. It is apparent that elements within the mesh undergo extensional and shearing deformation. An elastic solid is said to be deformed or strained when the *relative displacements* between points in the body are changed. This is in contrast to *rigid-body motion* where the distance between points remains the same.

In order to quantify deformation, consider the general example shown in Fig. 2.2. In the undeformed configuration, we identify two neighboring material points P_o and P connected with the *relative position vector* r as shown. Through a general deformation, these points are mapped to locations P'_o and P' in the deformed configuration. For *finite* or *large deformation theory*, the undeformed and deformed configurations can be significantly different, and a distinction between these two configurations must be maintained leading to Lagrangian and Eulerian descriptions; see, for example, Sadd (2019). However, since we are developing linear elasticity, which uses only small deformation theory, the distinction between undeformed and deformed configurations can be dropped.

Elasticity. https://doi.org/10.1016/B978-0-12-815987-3.00002-5

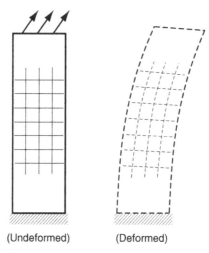

FIG. 2.1 Two-dimensional deformation example.

Using Cartesian coordinates, define the displacement vectors of points P_0 and P to be u^o and u respectively. Since P and P_o are neighboring points, we can use a Taylor series expansion around point P_0 to express the components of u as

$$u = u^o + \frac{\partial u}{\partial x}r_x + \frac{\partial u}{\partial y}r_y + \frac{\partial u}{\partial z}r_z$$

$$v = v^o + \frac{\partial v}{\partial x}r_x + \frac{\partial v}{\partial y}r_y + \frac{\partial v}{\partial z}r_z \qquad (2.1.1)$$

$$w = w^o + \frac{\partial w}{\partial x}r_x + \frac{\partial w}{\partial y}r_y + \frac{\partial w}{\partial z}r_z$$

where u, v, w are the Cartesian components of the displacement vector.

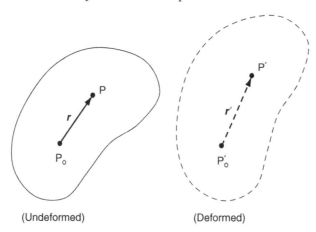

FIG. 2.2 General deformation between two neighboring points.

Note that the higher-order terms of the expansion have been dropped since the components of **r** are small. The change in the relative position vector **r** can be written as

$$\Delta r = r' - r = u - u^o \tag{2.1.2}$$

and using (2.1.1) gives

$$\Delta r_x = \frac{\partial u}{\partial x} r_x + \frac{\partial u}{\partial y} r_y + \frac{\partial u}{\partial z} r_z$$

$$\Delta r_y = \frac{\partial v}{\partial x} r_x + \frac{\partial v}{\partial y} r_y + \frac{\partial v}{\partial z} r_z \tag{2.1.3}$$

$$\Delta r_z = \frac{\partial w}{\partial x} r_x + \frac{\partial w}{\partial y} r_y + \frac{\partial w}{\partial z} r_z$$

or in index notation

$$\Delta r_i = u_{i,j} r_j \tag{2.1.4}$$

The tensor $u_{i,j}$ is called the *displacement gradient tensor*, and may be written out as

$$u_{i,j} = \begin{bmatrix} \dfrac{\partial u}{\partial x} & \dfrac{\partial v}{\partial y} & \dfrac{\partial u}{\partial z} \\[2mm] \dfrac{\partial v}{\partial x} & \dfrac{\partial v}{\partial y} & \dfrac{\partial v}{\partial z} \\[2mm] \dfrac{\partial w}{\partial x} & \dfrac{\partial v}{\partial y} & \dfrac{\partial w}{\partial z} \end{bmatrix} \tag{2.1.5}$$

From relation (1.2.10), this tensor can be decomposed into symmetric and antisymmetric parts as

$$u_{i,j} = e_{ij} + \omega_{ij} \tag{2.1.6}$$

where

$$e_{ij} = \frac{1}{2}\left(u_{i,j} + u_{j,i}\right)$$
$$\omega_{ij} = \frac{1}{2}\left(u_{i,j} - u_{j,i}\right) \tag{2.1.7}$$

The tensor e_{ij} is called the *strain tensor*, while ω_{ij} is referred to as the *rotation tensor*. Relations (2.1.4) and (2.1.6) thus imply that for small deformation theory, the change in the relative position vector between neighboring points can be expressed in terms of a *sum* of strain and rotation components. Combining relations (2.1.2), (2.1.4), and (2.1.6), and choosing $r_i = dx_i$, we can also write the general result in the form

$$u_i = u_i^o + e_{ij} dx_j + \omega_{ij} dx_j \tag{2.1.8}$$

Because we are considering a general displacement field, these results include both strain deformation and rigid-body motion. Recall from Exercise 1.15 that a dual vector ω_i can be associated with the rotation tensor such that $\omega_i = -1/2\varepsilon_{ijk}\omega_{jk}$. Using this definition, it is found that

$$\omega_1 = \omega_{32} = \frac{1}{2}\left(\frac{\partial u_3}{\partial x_2} - \frac{\partial u_2}{\partial x_3}\right)$$

$$\omega_2 = \omega_{13} = \frac{1}{2}\left(\frac{\partial u_1}{\partial x_3} - \frac{\partial u_3}{\partial x_1}\right) \qquad (2.1.9)$$

$$\omega_3 = \omega_{21} = \frac{1}{2}\left(\frac{\partial u_2}{\partial x_1} - \frac{\partial u_1}{\partial x_2}\right)$$

which can be expressed collectively in vector format as $\boldsymbol{\omega} = (1/2)(\nabla \times \boldsymbol{u})$. As is shown in the next section, these components represent rigid-body rotation of material elements about the coordinate axes. These general results indicate that the strain deformation is related to the strain tensor e_{ij}, which in turn is related to the displacement gradients. We next pursue a more geometric approach and determine specific connections between the strain tensor components and geometric deformation of material elements.

2.2 Geometric construction of small deformation theory

Although the previous section developed general relations for small deformation theory, we now wish to establish a more geometrical interpretation of these results. Typically, elasticity variables and equations are field quantities defined at each point in the material continuum. However, particular field equations are often developed by first investigating the behavior of infinitesimal elements (with co-ordinate boundaries), and then a limiting process is invoked that allows the element to shrink to a point. Thus, consider the common deformational behavior of a rectangular element as shown in Fig. 2.3. The usual types of motion include rigid-body rotation and extensional and shearing deformations as

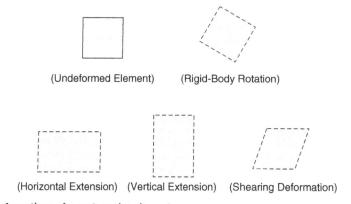

(Undeformed Element)   (Rigid-Body Rotation)

(Horizontal Extension)   (Vertical Extension)   (Shearing Deformation)

FIG. 2.3 Typical deformations of a rectangular element.

illustrated. Rigid-body motion does not contribute to the strain field, and thus also does not affect the stresses. We therefore focus our study primarily on the extensional and shearing deformation.

Fig. 2.4 illustrates the two-dimensional deformation of a rectangular element with original dimensions dx by dy. After deformation, the element takes a parallelogram form as shown in the dotted outline. The displacements of various corner reference points are indicated in the figure. Reference point A is taken at location (x,y), and the displacement components of this point are thus $u(x,y)$ and $v(x,y)$. The corresponding displacements of point B are $u(x+dx,y)$ and $v(x+dx,y)$, and the displacements of the other corner points are defined in an analogous manner. According to small deformation theory, $u(x+dx,y) \approx u(x,y)+(\partial u/\partial x)dx$, with similar expansions for all other terms.

The *normal* or *extensional strain component* in a direction n is defined as the change in length per unit length of fibers oriented in the n-direction. Normal strain is positive if fibers increase in length and negative if the fiber is shortened. In Fig. 2.4, the normal strain in the x-direction can thus be defined by

$$\varepsilon_x = \frac{A'B' - AB}{AB}$$

From the geometry in Fig. 2.4

$$A'B' = \sqrt{\left(dx + \frac{\partial u}{\partial x}dx\right)^2 + \left(\frac{\partial v}{\partial x}dx\right)^2} = \sqrt{1 + 2\frac{\partial u}{\partial x} + \left(\frac{\partial u}{\partial x}\right)^2 + \left(\frac{\partial v}{\partial x}\right)^2}\, dx \approx \left(1 + \frac{\partial u}{\partial x}\right)dx$$

where, consistent with small deformation theory, we have dropped the higher-order terms. Using these results and the fact that $AB = dx$, the normal strain in the x-direction reduces to

$$\varepsilon_x = \frac{\partial u}{\partial x} \tag{2.2.1}$$

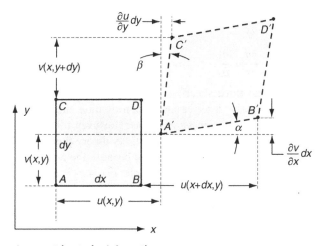

FIG. 2.4 Two-dimensional geometric strain deformation.

In similar fashion, the normal strain in the y-direction becomes

$$\varepsilon_y = \frac{\partial v}{\partial y} \tag{2.2.2}$$

A second type of strain is shearing deformation, which involves angle changes (see Fig. 2.3). *Shear strain* is defined as the change in angle between two originally orthogonal directions in the continuum material. This definition is actually referred to as the *engineering shear strain*. Theory of elasticity applications generally use a tensor formalism that requires a shear strain definition corresponding to one-half the angle change between orthogonal axes; see previous relation $(2.1.7)_1$. Measured in radians, shear strain is positive if the right angle between the positive directions of the two axes decreases. Thus, the sign of the shear strain depends on the coordinate system. In Fig. 2.4, the engineering shear strain with respect to the x- and y-directions can be defined as

$$\gamma_{xy} = \frac{\pi}{2} - \angle C'A'B' = \alpha + \beta$$

For small deformations, $\alpha \approx \tan \alpha$ and $\beta \approx \tan \beta$, and the shear strain can then be expressed as

$$\gamma_{xy} = \frac{\frac{\partial v}{\partial x}dx}{dx + \frac{\partial u}{\partial x}dx} + \frac{\frac{\partial u}{\partial y}dy}{dy + \frac{\partial v}{\partial y}dy} = \frac{\partial u}{\partial y} + \frac{\partial v}{\partial x} \tag{2.2.3}$$

where we have again neglected higher-order terms in the displacement gradients. Note that each derivative term is positive if lines AB and AC rotate inward as shown in the figure. By simple interchange of x and y and u and v, it is apparent that $\gamma_{xy} = \gamma_{yx}$.

By considering similar behaviors in the y-z and x-z planes, these results can be easily extended to the general three-dimensional case, giving the results

$$\varepsilon_x = \frac{\partial u}{\partial x}, \quad \varepsilon_y = \frac{\partial v}{\partial y}, \quad \varepsilon_z = \frac{\partial w}{\partial z}$$

$$\gamma_{xy} = \frac{\partial u}{\partial y} + \frac{\partial v}{\partial x}, \quad \gamma_{yz} = \frac{\partial v}{\partial z} + \frac{\partial w}{\partial y}, \quad \gamma_{zx} = \frac{\partial w}{\partial x} + \frac{\partial u}{\partial z} \tag{2.2.4}$$

Thus, we define three normal and three shearing strain components, leading to a total of six independent components that completely describe small deformation theory. This set of equations is normally referred to as the *strain–displacement relations*. However, these results are written in terms of the engineering strain components, and tensorial elasticity theory prefers to use the strain tensor e_{ij} defined by $(2.1.7)_1$. This represents only a minor change because the normal strains are identical and shearing strains differ by a factor of one-half; for example, $e_{11} = e_x = \varepsilon_x$ and $e_{12} = e_{xy} = 1/2\gamma_{xy}$, and so forth.

Therefore, using the strain tensor e_{ij}, the strain–displacement relations can be expressed in component form as

$$e_x = \frac{\partial u}{\partial x}, \quad e_y = \frac{\partial v}{\partial y}, \quad e_z = \frac{\partial w}{\partial z}$$

$$e_{xy} = \frac{1}{2}\left(\frac{\partial u}{\partial y} + \frac{\partial v}{\partial x}\right), \quad e_{yz} = \frac{1}{2}\left(\frac{\partial v}{\partial z} + \frac{\partial w}{\partial y}\right), \quad e_{zx} = \frac{1}{2}\left(\frac{\partial w}{\partial x} + \frac{\partial u}{\partial z}\right) \tag{2.2.5}$$

Using the more compact tensor notation, these relations are written as

$$e_{ij} = \frac{1}{2}\left(u_{i,j} + u_{j,i}\right)$$ (2.2.6)

while in direct vector/matrix notation the form reads

$$e = \frac{1}{2}\left[\nabla u + (\nabla u)^T\right]$$ (2.2.7)

where e is the strain matrix, ∇u is the displacement gradient matrix, and $(\nabla u)^T$ is its transpose.

The strain is a symmetric second-order tensor ($e_{ij} = e_{ji}$) and is commonly written in matrix format

$$e = [e] = \begin{bmatrix} e_x & e_{xy} & e_{xz} \\ e_{xy} & e_y & e_{yz} \\ e_{xz} & e_{yz} & e_z \end{bmatrix}$$ (2.2.8)

Before we conclude this geometric presentation, consider the rigid-body rotation of our two-dimensional element in the x-y plane, as shown in Fig. 2.5. If the element is rotated through a small rigid-body angular displacement about the z-axis, using the bottom element edge, the rotation angle is determined as $\partial v/\partial x$, while using the left edge, the angle is given by $-\partial u/\partial y$. These two expressions are of course the same; that is, $\partial v/\partial x = -\partial u/\partial y$ and note that this would imply $e_{xy} = 0$. The rotation can then be expressed as $\omega_z = [(\partial v/\partial x) - (\partial u/\partial y)]/2$, which matches with the expression given earlier in (2.1.9)$_3$. The other components of rotation follow in an analogous manner.

Relations for the constant rotation ω_z can be integrated to give the result

$$\begin{aligned} u^* &= u_o - \omega_z y \\ v^* &= v_o + \omega_z x \end{aligned}$$ (2.2.9)

where u_o and v_o are arbitrary constant translations in the x- and y-directions. This result then specifies the general form of the displacement field for two-dimensional rigid-body motion. We can easily

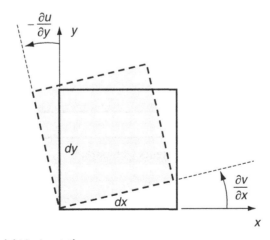

FIG. 2.5 Two-dimensional rigid-body rotation.

verify that the displacement field given by (2.2.9) yields zero strain. For the three-dimensional case, the most general form of rigid-body displacement can be expressed as

$$u^* = u_o - \omega_z y + \omega_y z$$
$$v^* = v_o - \omega_x z + \omega_z x \qquad (2.2.10)$$
$$w^* = w_o - \omega_y x + \omega_x y$$

As shown later, integrating the strain−displacement relations to determine the displacement field produces arbitrary constants and functions of integration, which are related to rigid-body motion terms of the form given by (2.2.9) or (2.2.10). Thus, it is important to recognize such terms because we normally want to drop them from the analysis since they do not contribute to the strain or stress fields.

Example 2.1 Strain and rotation examples

Determine the displacement gradient, strain, and rotation tensors for the following displacement field: $u = Ax^2y$, $v = Byz$, $w = Cxz^3$, where A, B, and C are arbitrary constants. Also calculate the dual rotation vector $\boldsymbol{\omega} = (1/2)(\nabla \times \boldsymbol{u})$.

$$u_{i,j} = \begin{bmatrix} 2Axy & Ax^2 & 0 \\ 0 & Bz & By \\ Cz^3 & 0 & 3Cxz^2 \end{bmatrix}$$

$$e_{ij} = \frac{1}{2}\left(u_{i,j} + u_{j,i}\right) = \begin{bmatrix} 2Axy & Ax^2/2 & Cz^3/2 \\ Ax^2/2 & Bz & By/2 \\ Cz^3/2 & By/2 & 3Cxz^2 \end{bmatrix}$$

$$\omega_{ij} = \frac{1}{2}\left(u_{i,j} - u_{j,i}\right) = \begin{bmatrix} 0 & Ax^2/2 & -Cz^3/2 \\ -Ax^2/2 & 0 & By/2 \\ Cz^3/2 & -By/2 & 0 \end{bmatrix}$$

$$\omega = \frac{1}{2}(\nabla \times \boldsymbol{u}) = \frac{1}{2}\begin{vmatrix} e_1 & e_2 & e_3 \\ \partial/\partial x & \partial/\partial y & \partial/\partial z \\ Ax^2y & Byz & Cxz^3 \end{vmatrix} = \frac{1}{2}\left(-Bye_1 - Cz^3e_2 - Ax^2e_3\right)$$

Example 2.2 Determination of displacements from strains

Given the following strain field, $e_{ij} = \begin{bmatrix} Ax & 0 & 0 \\ 0 & By^2 & 0 \\ 0 & 0 & 0 \end{bmatrix}$, with A and B constants. Determine the corresponding displacements assuming they only depend on x and y, and that $w = 0$ (two-dimensional problem).

To determine displacements we have to integrate the strain-displacement relations

$$e_x = \frac{\partial u}{\partial x} = Ax \Rightarrow u = \frac{A}{2}x^2 + f(y); \text{ note that integration of the partial derivative creates an arbitrary}$$

function $f(y)$

$$e_y = \frac{\partial v}{\partial y} = By^2 \Rightarrow v = \frac{B}{3}y^3 + g(x); \text{ note integration creates arbitrary function } g(x)$$

$$e_{xy} = \frac{1}{2}\left(\frac{\partial u}{\partial y} + \frac{\partial v}{\partial x}\right) = 0 \Rightarrow \frac{df(y)}{dy} + \frac{dg(x)}{dx} = 0 \Rightarrow \frac{df(y)}{dy} = -\frac{dg(x)}{dx} = \text{constant}$$

The last relation now involves only ordinary derivatives and must equal a constant since x and y are independent variables. Choosing the constant to be a and integrating gives $f(y) = ay + b$, $g(x) = -ax + c$ where a, b, c are all arbitrary constants.

Thus the displacements are given by $u = \frac{A}{2}x^2 + ay + b$, $u_2 = \frac{B}{3}y^3 - ax + c$

Note that portions from $f(y)$ and $g(x)$ are actually rigid body motion terms.

Example 2.3 Line length and angle changes from given strain tensor

Consider a two-dimensional deformation problem where a rectangular region with original dimensions $a \times b$ as shown in Fig. 2.6 is subjected to the following strain field,

$$e_{ij} = \begin{bmatrix} 2 & 1 & 0 \\ 1 & 4 & 0 \\ 0 & 0 & 0 \end{bmatrix} \times 10^{-3}. \text{ Determine the region's new dimensions and the angle change } \theta.$$

For this homogeneous deformation problem, changes in length can be computed by relation preceding (2.2.1)

$$e = \frac{l_{final} - l_{inital}}{l_{inital}} \Rightarrow l_{final} = (1+e)l_{inital}$$

$$\therefore a_f = (1 + e_x)a = 1.002a$$

$$b_f = (1 + e_y)b = 1.004b$$

Likewise the angle change can be determined from

$$\theta = \gamma_{xy} = 2e_{xy} = 0.002 \text{ radians} = 0.115 \text{ degrees}$$

As expected, small strains lead to small changes in lengths and angles.

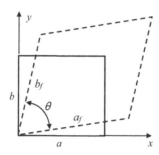

FIG. 2.6 Two-dimensional deformation example.

2.3 Strain transformation

Because the strains are components of a second-order tensor, the transformation theory discussed in Section 1.5 can be applied. Transformation relation $(1.5.1)_3$ is applicable for second-order tensors, and applying this to the strain gives

$$e'_{ij} = Q_{ip}Q_{jq}e_{pq} \tag{2.3.1}$$

where the rotation matrix $Q_{ij} = \cos(x'_i, x_j)$. Thus, given the strain in one coordinate system, we can determine the new components in any other rotated system. For the general three-dimensional case, define the rotation matrix as

$$Q_{ij} = \begin{bmatrix} l_1 & m_1 & n_1 \\ l_2 & m_2 & n_2 \\ l_3 & m_3 & n_3 \end{bmatrix} \tag{2.3.2}$$

Using this notational scheme, the specific transformation relations from Eq. (2.3.1) become

$$e'_x = e_x l_1^2 + e_y m_1^2 + e_z n_1^2 + 2(e_{xy}l_1 m_1 + e_{yz}m_1 n_1 + e_{zx}n_1 l_1)$$

$$e'_y = e_x l_2^2 + e_y m_2^2 + e_z n_2^2 + 2(e_{xy}l_2 m_2 + e_{yz}m_2 n_2 + e_{zx}n_2 l_2)$$

$$e'_z = e_x l_3^2 + e_y m_3^2 + e_z n_3^2 + 2(e_{xy}l_3 m_3 + e_{yz}m_3 n_3 + e_{zx}n_3 l_3)$$

$$e'_{xy} = e_x l_1 l_2 + e_y m_1 m_2 + e_z n_1 n_2 + e_{xy}(l_1 m_2 + m_1 l_2) + e_{yz}(m_1 n_2 + n_1 m_2) + e_{zx}(n_1 l_2 + l_1 n_2)$$

$$e'_{yz} = e_x l_2 l_3 + e_y m_2 m_3 + e_z n_2 n_3 + e_{xy}(l_2 m_3 + m_2 l_3) + e_{yz}(m_2 n_3 + n_2 m_3) + e_{zx}(n_2 l_3 + l_2 n_3)$$

$$e'_{zx} = e_x l_3 l_1 + e_y m_3 m_1 + e_z n_3 n_1 + e_{xy}(l_3 m_1 + m_3 l_1) + e_{yz}(m_3 n_1 + n_3 m_1) + e_{zx}(n_3 l_1 + l_3 n_1)$$

$$\tag{2.3.3}$$

For the two-dimensional case shown in Fig. 2.7, the transformation matrix can be expressed as

$$Q_{ij} = \begin{bmatrix} \cos\theta & \sin\theta & 0 \\ -\sin\theta & \cos\theta & 0 \\ 0 & 0 & 1 \end{bmatrix} \tag{2.3.4}$$

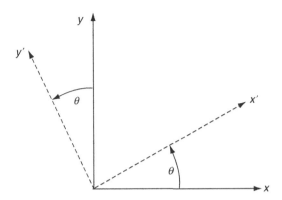

FIG. 2.7 Two-dimensional rotational transformation.

Under this transformation, the in-plane strain components transform according to

$$e'_x = e_x \cos^2 \theta + e_y \sin^2 \theta + 2e_{xy} \sin \theta \cos \theta$$
$$e'_y = e_x \sin^2 \theta + e_y \cos^2 \theta - 2e_{xy} \sin \theta \cos \theta \qquad (2.3.5)$$
$$e'_{xy} = -e_x \sin \theta \cos \theta + e_y \sin \theta \cos \theta + e_{xy}\left(\cos^2 \theta - \sin^2 \theta\right)$$

which is commonly rewritten in terms of the double angle

$$e'_x = \frac{e_x + e_y}{2} + \frac{e_x - e_y}{2} \cos 2\theta + e_{xy} \sin 2\theta$$
$$e'_y = \frac{e_x + e_y}{2} - \frac{e_x - e_y}{2} \cos 2\theta - e_{xy} \sin 2\theta \qquad (2.3.6)$$
$$e'_{xy} = \frac{e_y - e_x}{2} \sin 2\theta + e_{xy} \cos 2\theta$$

Transformation relations (2.3.6) can be directly applied to establish transformations between Cartesian and polar coordinate systems (see Exercise 2.6). Additional applications of these results can be found when dealing with experimental strain gage measurement systems. For example, standard experimental methods using a rosette strain gage allow the determination of extensional strains in three different directions on the surface of a structure. Using this type of data, relation (2.3.6)$_1$ can be repeatedly used to establish three independent equations that can be solved for the state of strain (e_x, e_y, e_{xy}) at the surface point under study (see Exercise 2.7). Note that double angle forms can sometimes provide better computational accuracy by avoiding squaring of trig functions.

Both two- and three-dimensional transformation equations can be easily incorporated in MATLAB® to provide numerical solutions to problems of interest. Such examples are given in Exercises 2.8 and 2.9.

2.4 Principal strains

From the previous discussion in Section 1.6, it follows that because the strain is a symmetric second-order tensor, we can identify and determine its principal axes and values. According to this theory, for any given strain tensor we can establish the principal value problem and solve the characteristic equation to explicitly determine the principal values and directions. The general characteristic equation for the strain tensor can be written as

$$\det[e_{ij} - e\delta_{ij}] = -e^3 + \vartheta_1 e^2 - \vartheta_2 e + \vartheta_3 = 0 \tag{2.4.1}$$

where e is the *principal strain* and the fundamental invariants of the strain tensor can be expressed in terms of the three principal strains e_1, e_2, e_3 as

$$\begin{aligned} \vartheta_1 &= e_1 + e_2 + e_3 \\ \vartheta_2 &= e_1 e_2 + e_2 e_3 + e_3 e_1 \\ \vartheta_3 &= e_1 e_2 e_3 \end{aligned} \tag{2.4.2}$$

The first invariant $\vartheta_1 = \vartheta$ is normally called the *cubical dilatation*, because it is related to the change in volume of material elements (see Exercise 2.11).

The strain matrix in the principal coordinate system takes the special diagonal form

$$e_{ij} = \begin{bmatrix} e_1 & 0 & 0 \\ 0 & e_2 & 0 \\ 0 & 0 & e_3 \end{bmatrix} \tag{2.4.3}$$

Notice that for this principal coordinate system, the deformation does not produce any shearing and thus is only extensional. Therefore, a rectangular element oriented along principal axes of strain will retain its orthogonal shape and undergo only extensional deformation of its sides.

2.5 Spherical and deviatoric strains

In particular applications it is convenient to decompose the strain tensor into two parts called *spherical* and *deviatoric strain tensors*. The spherical strain is defined by

$$\tilde{e}_{ij} = \frac{1}{3} e_{kk} \delta_{ij} = \frac{1}{3} \vartheta \delta_{ij} \tag{2.5.1}$$

while the deviatoric strain is specified as

$$\hat{e}_{ij} = e_{ij} - \frac{1}{3} e_{kk} \delta_{ij} \tag{2.5.2}$$

Note that the total strain is then simply the sum

$$e_{ij} = \tilde{e}_{ij} + \hat{e}_{ij} \tag{2.5.3}$$

The spherical strain represents only volumetric deformation and is an isotropic tensor, being the same in all coordinate systems (as per the discussion in Section 1.5). The deviatoric strain tensor then accounts for changes in shape of material elements. It can be shown that the principal directions of the deviatoric strain are the same as those of the strain tensor.

2.6 Strain compatibility

We now investigate in more detail the nature of the strain—displacement relations (2.2.5), and this will lead to the development of some additional relations necessary to ensure continuous, single-valued displacement field solutions. Relations (2.2.5), or the index notation form (2.2.6), represent six equations for the six strain components in terms of three displacements. If we specify continuous, single-valued displacements u, v, w, then through differentiation the resulting strain field will be equally well behaved. However, the converse is not necessarily true; given the six strain components, integration of the strain—displacement relations (2.2.5) does not necessarily produce continuous, single-valued displacements. This should not be totally surprising since we are trying to solve six equations for only three unknown displacement components. In order to ensure continuous, single-valued displacements, the strains must satisfy additional relations called *integrability* or *compatibility equations*.

Before we proceed with the mathematics to develop these equations, it is instructive to consider a geometric interpretation of this concept. A two-dimensional example is shown in Fig. 2.8 whereby an elastic solid is first divided into a series of elements in case (a). For simple visualization, consider only four such elements. In the undeformed configuration shown in case (b), these elements of course fit together perfectly. Next, let us arbitrarily specify the strain of each of the four elements and attempt to reconstruct the solid. For case (c), the elements have been carefully strained, taking into consideration neighboring elements so that the system fits together yielding continuous, single-valued displacements. However, for case (d), the elements have been individually deformed without any concern for neighboring deformations. It is observed for this case that the system will not fit together without voids and gaps, and this situation produces a discontinuous displacement field. So, we again conclude that

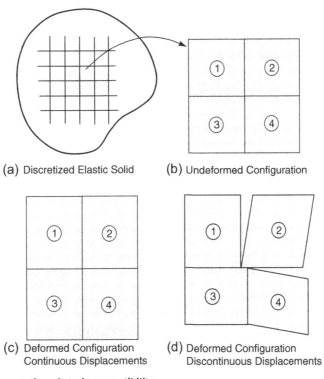

(a) Discretized Elastic Solid   (b) Undeformed Configuration

(c) Deformed Configuration
Continuous Displacements

(d) Deformed Configuration
Discontinuous Displacements

FIG. 2.8 Physical interpretation of strain compatibility.

the strain components must be somehow related to yield continuous, single-valued displacements. We now pursue these particular relations.

The process to develop these equations is based on eliminating the displacements from the strain—displacement relations. Working in index notation, we start by differentiating (2.2.6) twice with respect to x_k and x_l

$$e_{ij,kl} = \frac{1}{2}\left(u_{i,jkl} + u_{j,ikl}\right)$$

Through simple interchange of subscripts, we can generate the following additional relations

$$e_{kl,ij} = \frac{1}{2}\left(u_{k,lij} + u_{l,kij}\right)$$

$$e_{jl,ik} = \frac{1}{2}\left(u_{j,lik} + u_{l,jik}\right)$$

$$e_{ik,jl} = \frac{1}{2}\left(u_{i,kjl} + u_{k,ijl}\right)$$

Working under the assumption of continuous displacements, we can interchange the order of differentiation on $\boldsymbol{u}$, and the displacements can be eliminated from the preceding set to get

$$e_{ij,kl} + e_{kl,ij} - e_{ik,jl} - e_{jl,ik} = 0 \qquad (2.6.1)$$

These are called the *Saint–Venant compatibility equations*. Although the system would lead to 81 individual equations, most are either simple identities or repetitions, and only six are meaningful. These six relations may be determined by letting $k = l$, and in scalar notation they become

$$\frac{\partial^2 e_x}{\partial y^2} + \frac{\partial^2 e_y}{\partial x^2} = 2\frac{\partial^2 e_{xy}}{\partial x \partial y}$$

$$\frac{\partial^2 e_y}{\partial z^2} + \frac{\partial^2 e_z}{\partial y^2} = 2\frac{\partial^2 e_{yz}}{\partial y \partial z}$$

$$\frac{\partial^2 e_z}{\partial x^2} + \frac{\partial^2 e_x}{\partial z^2} = 2\frac{\partial^2 e_{zx}}{\partial z \partial x}$$

$$\frac{\partial^2 e_x}{\partial y \partial z} = \frac{\partial}{\partial x}\left(-\frac{\partial e_{yz}}{\partial x} + \frac{\partial e_{zx}}{\partial y} + \frac{\partial e_{xy}}{\partial z}\right) \qquad (2.6.2)$$

$$\frac{\partial^2 e_y}{\partial z \partial x} = \frac{\partial}{\partial y}\left(-\frac{\partial e_{zx}}{\partial y} + \frac{\partial e_{xy}}{\partial z} + \frac{\partial e_{yz}}{\partial x}\right)$$

$$\frac{\partial^2 e_z}{\partial x \partial y} = \frac{\partial}{\partial z}\left(-\frac{\partial e_{xy}}{\partial z} + \frac{\partial e_{yz}}{\partial x} + \frac{\partial e_{zx}}{\partial y}\right)$$

It can be shown that these six equations are not all independent. Exercise 2.15 illustrates that certain differential relations exist between these compatibility equations, and Exercise 2.16 shows that the six equations can be reduced to three independent fourth-order relations. Although there has been some discussion in the literature about trying to reduce the number of equations in (2.6.2), almost all applications simply use the entire set of six equations.

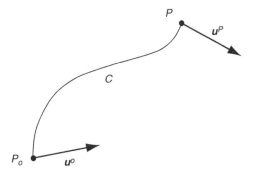

FIG. 2.9 Continuity of displacements.

In the development of the compatibility relations, we assumed that the displacements were continuous, and thus the resulting equations (2.6.2) are actually only a necessary condition. In order to show that they are also sufficient, consider two arbitrary points P and P_o in an elastic solid, as shown in Fig. 2.9. Without loss in generality, the origin may be placed at point P_o.

The displacements of points P and P_o are denoted by u_i^P and u_i^o, and the displacement of point P can be expressed as

$$u_i^P = u_i^o + \int_C du_i = u_i^o + \int_C \frac{\partial u_i}{\partial x_j} dx_j \tag{2.6.3}$$

where C is any continuous curve connecting points P_o and P. Using relation (2.1.6) for the displacement gradient, (2.6.3) becomes

$$u_i^P = u_i^o + \int_C (e_{ij} + \omega_{ij}) dx_j \tag{2.6.4}$$

Integrating the last term by parts gives

$$\int_C \omega_{ij} dx_j = \omega_{ij}^P x_j^P - \int_C x_j \omega_{ij,k} dx_k \tag{2.6.5}$$

where ω_{ij}^P is the rotation tensor at point P. Using relation $(2.1.7)_2$

$$\omega_{ij,k} = \frac{1}{2}\left(u_{i,jk} - u_{j,ik}\right) = \frac{1}{2}\left(u_{i,jk} - u_{j,ik}\right) + \frac{1}{2}\left(u_{k,ji} - u_{k,ji}\right)$$

$$= \frac{1}{2}\frac{\partial}{\partial x_j}\left(u_{i,k} + u_{k,i}\right) - \frac{1}{2}\frac{\partial}{\partial x_i}\left(u_{j,k} + u_{k,j}\right) = e_{ik,j} - e_{jk,i} \tag{2.6.6}$$

Substituting results (2.6.5) and (2.6.6) into (2.6.4) yields

$$u_i^P = u_i^o + \omega_{ij}^P x_j^P + \int_C U_{ik} dx_k \tag{2.6.7}$$

where $U_{ik} = e_{ik} - x_j(e_{ik,j} - e_{jk,i})$.

Now if the displacements are to be continuous, single-valued functions, the line integral appearing in Eq. (2.6.7) must be the same for any curve C; that is, the integral must be independent of the path of integration. This implies that the integrand must be an exact differential, so that the value of the integral depends only on the end points. Invoking Stokes theorem, we can show that if the region is *simply connected* (definition of the term *simply connected* is postponed for the moment), a necessary and sufficient condition for the integral to be path independent is for $U_{ik,l} = U_{il,k}$. Using this result yields

$$e_{ik,l} - \delta_{jl}\left(e_{ik,j} - e_{jk,i}\right) - x_j\left(e_{ik,jl} - e_{jk,il}\right) = e_{il,k} - \delta_{jk}\left(e_{il,j} - e_{jl,i}\right) - x_j\left(e_{il,jk} - e_{ji,ik}\right)$$

which reduces to

$$x_j\left(e_{ik,jl} - e_{jk,il} - e_{il,jk} + e_{jl,ik}\right) = 0$$

Because this equation must be true for all values of x_j, the terms in parentheses must vanish, and after some index renaming this gives the identical result previously stated by the compatibility relations (2.6.1)

$$e_{ij,kl} + e_{kl,ij} - e_{ik,jl} - e_{jl,ik} = 0$$

Thus, relations (2.6.1) or (2.6.2) are the necessary and sufficient conditions for continuous, single-valued displacements in simply connected regions.

Now let us get back to the term *simply connected*. This concept is related to the topology or geometry of the region under study. There are several places in elasticity theory where the connectivity of the region fundamentally affects the formulation and solution method. The term *simply connected* refers to regions of space for which all simple closed curves drawn in the region can be continuously shrunk to a point without going outside the region. Domains not having this property are called *multiply connected*. Several examples of such regions are illustrated in Fig. 2.10. A general simply connected two-dimensional region is shown in case (a), and clearly this case allows any contour within the region to be shrunk to a point without going out of the domain. However, if we create a hole in the region as shown in case (b), a closed contour surrounding the hole cannot be shrunk to a point without going into the hole and thus outside of the region. For two-dimensional regions, the presence of one or more holes makes the region multiply connected. Note that by introducing a cut between the outer and inner boundaries in case (b), a new region is created that is now simply connected. Thus, multiply connected regions can be made simply connected by introducing one or more cuts between appropriate boundaries. Case (c) illustrates a simply connected three-dimensional example of a solid circular cylinder. If a spherical cavity is placed inside this cylinder as shown in case (d), the region is still simply connected because any closed contour can still be shrunk to a point by sliding around the interior cavity. However, if the cylinder has a through hole as shown in case (e), then an interior contour encircling the axial through hole cannot be reduced to a point without going into the hole and outside the body. Thus, case (e) is an example of the multiply connected three-dimensional region.

It was found that the compatibility equations are necessary and sufficient conditions for continuous, single-valued displacements only for simply connected regions. Thus for multiply connected domains, relations (2.6.1) or (2.6.2) provide only necessary but not sufficient conditions. For this case, further relations can be developed and imposed on the problem, and these are found through the introduction of cuts within the region to make it simply connected as per our earlier discussion. For the two-dimensional case shown in Fig. 2.10B, this process will lead to a relation commonly called a

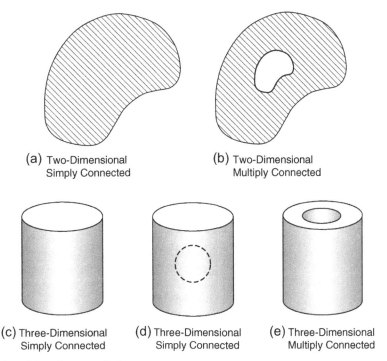

(a) Two-Dimensional
Simply Connected

(b) Two-Dimensional
Multiply Connected

(c) Three-Dimensional
Simply Connected

(d) Three-Dimensional
Simply Connected

(e) Three-Dimensional
Multiply Connected

FIG. 2.10 Examples of domain connectivity.

Cesàro integral taken around any closed irreducible curve enclosing the internal cavity. Thus, for multiply connected domains, strain compatibility is guaranteed if compatibility relations (2.6.2) are satisfied and all Cesàro integrals vanish. Details on this topic are given by Fung (1965), Fung and Tong (2001), or Asaro and Lubarda (2006).

Although the compatibility relations guarantee (under appropriate conditions) continuous displacements, they do not ensure uniqueness of the displacement field. At the end of Section 2.2 we mentioned that, through integration of the strain—displacement relations, the displacements can be determined only up to an arbitrary rigid-body motion (see Example 2.2). In some elasticity problems (e.g., thermal stress, crack problems, and dislocation modeling), it is necessary to use multivalued displacement fields to properly model the problem. Chapters 10, 12, and 15 contain a few examples of such problems, and a specific case is given in Exercise 2.19.

The compatibility equations clearly come from a mathematical construct. If the displacements are part of the overall formulation, compatibility will automatically be satisfied and thus relations (2.6.2) will not be needed. If however, we formulate the elasticity problem in terms of just the stresses or strains, then it will be necessary to explicitly incorporate compatibility relations into the problem formulation. These issues will be more clearly illustrated in subsequent chapters.

2.7 Curvilinear cylindrical and spherical coordinates

The solution to many problems in elasticity requires the use of curvilinear cylindrical and spherical coordinates. It is therefore necessary to have the field equations expressed in terms of such coordinate systems. We now pursue the development of the strain–displacement relations in cylindrical and spherical coordinates. Starting with form (2.2.7)

$$e = \frac{1}{2} = \left[\nabla u + (\nabla u)^T \right]$$

the desired curvilinear relations can be determined using the appropriate forms for the displacement gradient term ∇u.

The cylindrical coordinate system previously defined in Fig. 1.5 establishes new components for the displacement vector and strain tensor

$$u = u_r e_r + u_\theta e_\theta + u_z e_z$$

$$e = \begin{bmatrix} e_r & e_{r\theta} & e_{rz} \\ e_{r\theta} & e_\theta & e_{\theta z} \\ e_{rz} & e_{\theta z} & e_z \end{bmatrix} \tag{2.7.1}$$

Notice that the symmetry of the strain tensor is preserved in this orthogonal curvilinear system. Using results (1.9.17) and (1.9.10), the derivative operation in cylindrical coordinates can be expressed by

$$\nabla u = \frac{\partial u_r}{\partial r} e_r e_r + \frac{\partial u_\theta}{\partial r} e_r e_\theta + \frac{\partial u_z}{\partial r} e_r e_z$$

$$+ \frac{1}{r}\left(\frac{\partial u_r}{\partial \theta} - u_\theta \right) e_\theta e_r + \frac{1}{r}\left(u_r + \frac{\partial u_\theta}{\partial \theta} \right) e_\theta e_\theta + \frac{1}{r}\frac{\partial u_z}{\partial \theta} e_\theta e_z \tag{2.7.2}$$

$$+ \frac{\partial u_r}{\partial z} e_z e_r + \frac{\partial u_\theta}{\partial z} e_z e_\theta + \frac{\partial u_z}{\partial z} e_z e_z$$

Placing this result into the strain–displacement form (2.2.7) gives the desired relations in cylindrical coordinates. The individual scalar equations are given by

$$e_r = \frac{\partial u_r}{\partial r}, \quad e_\theta = \frac{1}{r}\left(u_r + \frac{\partial u_\theta}{\partial \theta} \right), \quad e_z = \frac{\partial u_z}{\partial z}$$

$$e_{r\theta} = \frac{1}{2}\left(\frac{1}{r}\frac{\partial u_r}{\partial \theta} + \frac{\partial u_\theta}{\partial r} - \frac{u_\theta}{r} \right)$$

$$e_{\theta z} = \frac{1}{2}\left(\frac{\partial u_\theta}{\partial z} + \frac{1}{r}\frac{\partial u_z}{\partial \theta} \right) \tag{2.7.3}$$

$$e_{zr} = \frac{1}{2}\left(\frac{\partial u_r}{\partial z} + \frac{\partial u_z}{\partial r} \right)$$

For spherical coordinates defined by Fig. 1.6, the displacement vector and strain tensor can be written as

$$u = u_R e_R + u_\phi e_\phi + u_\theta e_\theta$$

$$e = \begin{bmatrix} e_R & e_{R\phi} & e_{R\theta} \\ e_{R\phi} & e_\phi & e_{\phi\theta} \\ e_{R\theta} & e_{\phi\theta} & e_\theta \end{bmatrix}$$

(2.7.4)

Following identical procedures as used for the cylindrical equation development, the strain–displacement relations for spherical coordinates become

$$e_R = \frac{\partial u_R}{\partial R}, e_\phi = \frac{1}{R}\left(u_R + \frac{\partial u_\phi}{\partial \phi}\right)$$

$$e_\theta = \frac{1}{R \sin \phi}\left(\frac{\partial u_\theta}{\partial \theta} + \sin \phi u_R + \cos \phi u_\phi\right)$$

$$e_{R\phi} = \frac{1}{2}\left(\frac{1}{R}\frac{\partial u_R}{\partial \phi} + \frac{\partial u_\phi}{\partial R} - \frac{u_\phi}{R}\right)$$

(2.7.5)

$$e_{\phi\theta} = \frac{1}{2R}\left(\frac{1}{\sin \phi}\frac{\partial u_\phi}{\partial \theta} + \frac{\partial u_\theta}{\partial \phi} - \cot \phi u_\theta\right)$$

$$e_{\theta R} = \frac{1}{2}\left(\frac{1}{R \sin \phi}\frac{\partial u_R}{\partial \theta} + \frac{\partial u_\theta}{\partial R} - \frac{u_\theta}{R}\right)$$

We can observe that these relations in curvilinear systems contain additional terms that do not include derivatives of individual displacement components. For example, in spherical coordinates a simple uniform radial displacement u_R gives rise to transverse extensional strains $e_\phi = e_\theta = u_R/R$. This deformation can be simulated by blowing up a spherical balloon and observing the separation of points on the balloon's surface. Such terms were not found in the Cartesian forms given by (2.2.5), and their appearance is thus related to the curvature of the spatial coordinate system. A more physical interpretation can be found by redeveloping these equations using the geometric procedures of Section 2.2 on an appropriate differential element. A two-dimensional polar coordinate example of this technique is given in Exercise 2.20. Clearly, the curvilinear forms (2.7.3) and (2.7.5) appear more complicated than the corresponding Cartesian relations. However, for particular problems, the curvilinear relations, when combined with other field equations, allow analytical solutions to be developed that could not be found using a Cartesian formulation. Many examples of this are demonstrated in later chapters. Appendix A lists the complete set of elasticity field equations in cylindrical and spherical coordinates.

References

Asaro R, Lubarda V: *Mechanics of solids and materials*, New York, 2006, Cambridge University Press.
Fung YC: *Foundations of solid mechanics*, Englewood Cliffs, NJ, 1965, Prentice Hall.
Fung YC, Tong P: *Classical and computational solid mechanics*, Singapore, 2001, World Scientific.
Sadd MH: *Continuum mechanics modeling of material behavior*, Elsevier, 2019.

Exercises

2.1 Determine the strain and rotation tensors e_{ij} and ω_{ij} for the following displacement fields:

$$\text{(a) } u = Axy, \quad v = Byz^2, \quad w = C(x^2 + z^2)$$

$$\text{(b) } u = Ay^2, \quad v = Bx^2y, \quad w = Cyz$$

$$\text{(c) } u = Axz^3, \quad v = Bzy^2, \quad w = C(x^2 + y^2)$$

where A, B, and C are arbitrary constants.

2.2 A two-dimensional displacement field is given by $u = k(x^2 + y^2)$, $v = k(2x - y)$, $w = 0$, where k is a constant. Determine and plot the deformed shape of a differential rectangular element originally located with its left bottom corner at the origin as shown. Finally, calculate the rotation component ω_z.

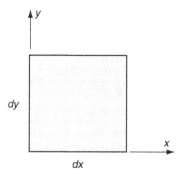

2.3 Assume that we are dealing with a two-dimensional deformation problem in the x,y-plane where all field variables depend only on x and y. For the following strain fields (A and B are constants), integrate the strain-displacement relations to determine the displacements, and identify any rigid-body terms.

$$\text{(a) } e_x = A, e_y = -B, e_{xy} = 0$$

$$\text{(b) } e_x = Ax, e_y = -By, e_{xy} = 0$$

$$\text{(c) } e_x = Ay, e_y = -Bx, e_{xy} = 0$$

2.4 A three-dimensional elasticity problem of a uniform bar stretched under its own weight gives the following strain field

$$e_{ij} = \begin{bmatrix} Az & 0 & 0 \\ 0 & Az & 0 \\ 0 & 0 & Bz \end{bmatrix}$$

where A and B are constants. Integrate the strain–displacement relations to determine the displacement components and identify all rigid-body motion terms.

2.5 Explicitly verify that the general rigid-body motion displacement field given by (2.2.10) yields zero strains. Next, assuming that all strains vanish, formally integrate relations (2.2.5) to develop the general form (2.2.10).

2.6 For polar coordinates defined by Fig. 1.8, show that the transformation relations can be used to determine the normal and shear strain components e_r, e_θ, and $e_{r\theta}$ in terms of the corresponding Cartesian components

$$e_r = \frac{e_x + e_y}{2} + \frac{e_x - e_y}{2}\cos 2\theta + e_{xy}\sin 2\theta$$

$$e_\theta = \frac{e_x + e_y}{2} - \frac{e_x - e_y}{2}\cos 2\theta - e_{xy}\sin 2\theta$$

$$e_{r\theta} = \frac{e_y - e_x}{2}\sin 2\theta + e_{xy}\cos 2\theta$$

2.7 A rosette strain gage is an electromechanical device that can measure relative surface elongations in three directions. Bonding such a device to the surface of a structure allows determination of elongational strains in particular directions. A schematic of one such gage is shown in the following figure, and the output of the device will provide data on the strains along the gage arms a, b, and c. During one application, it is found that $e_a = 0.001$, $e_b = 0.002$, and $e_c = 0.004$. Using the two-dimensional strain transformation relations, calculate the surface strain components e_x, e_y, and e_{xy}.

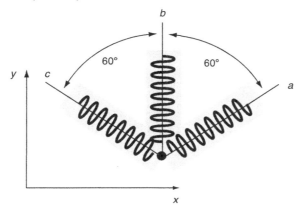

2.8* A two-dimensional strain field is found to be given by $e_x = 0.002$, $e_y = -0.004$, and $e_{xy} = 0.001$. Incorporating the transformation relations (2.3.6) into a MATLAB® code, calculate and plot the new strain components in a rotated coordinate system as a function of the rotation angle θ. Determine the particular angles at which the new components take on maximum values.

2.9* A three-dimensional strain field is specified by

$$e_{ij} = \begin{bmatrix} 1 & -2 & 0 \\ -2 & -4 & 0 \\ 0 & 0 & 5 \end{bmatrix} \times 10^{-3}$$

Determine information on the strains in the shaded plane in the following figure that makes equal angles with the x- and z-axes as shown. Use MATLAB_ to calculate and plot the normal and inplane shear strain along line AB (in the plane) as a function of angle θ in the interval $0 \le \theta \le \pi/2$.

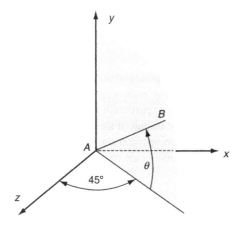

2.10* Using MATLAB®, determine the principal values and directions of the following state of strain

$$e_{ij} = \begin{bmatrix} 2 & -2 & 0 \\ -2 & -4 & 1 \\ 0 & 1 & 6 \end{bmatrix} \times 10^{-3}$$

2.11 A rectangular parallelepiped with original volume V_o is oriented such that its edges are parallel to the principal directions of strain as shown in the following figure. For small strains, show that the dilatation is given by

$$\vartheta = e_{kk} = \frac{\text{change in volume}}{\text{original volume}} = \frac{\Delta V}{V_o}$$

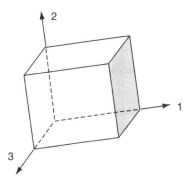

2.12 Determine the spherical and deviatoric strain tensors for the strain field given in Exercise 2.10. Justify that the first invariant or dilatation of the deviatoric strain tensor is zero. In light of the results from Exercise 2.11, what does the vanishing of the dilatation imply?

2.13 Using scalar methods, differentiate the individual strain–displacement relations for e_x, e_y, and e_{xy}, and independently develop the first compatibility equation of set (2.6.2).

2.14 Using relation (1.3.5), show that the compatibility relations (2.6.1) with $l = k$ can be expressed by $\eta_{ij} = \varepsilon_{ikl}\varepsilon_{jmp}e_{lp,km} = 0$, which can also be written in vector notation as $\nabla \times e \times \nabla = 0$.

2.15 In light of Exercise 2.14, the compatibility equations (2.6.2) can be expressed as $\eta_{ij} = \varepsilon_{ikl} \varepsilon_{jmp}e_{lp,km} = 0$, where η_{ij} is sometimes referred to as the *incompatibility tensor* (Asaro and Lubarda, 2006). It is observed that η_{ij} is symmetric, but its components are not independent from one another. Since the divergence of a curl vanishes, show that they are related through the differential *Bianchi relations* $\eta_{ij,j} = 0$, which can be expanded to

$$\eta_{11,1} + \eta_{12,2} + \eta_{13,3} = 0$$
$$\eta_{21,1} + \eta_{22,2} + \eta_{23,3} = 0$$
$$\eta_{31,1} + \eta_{32,2} + \eta_{33,3} = 0$$

Thus we see again that the six compatibility relations are not all independent.

2.16 Show that the six compatibility equations (2.6.2) may also be represented by the three independent fourth-order equations

$$\frac{\partial^4 e_x}{\partial y^2 \partial z^2} = \frac{\partial^3}{\partial x \partial y \partial z}\left(-\frac{\partial e_{yz}}{\partial x} + \frac{\partial e_{zx}}{\partial y} + \frac{\partial e_{xy}}{\partial z} \right)$$

$$\frac{\partial^4 e_y}{\partial z^2 \partial x^2} = \frac{\partial^3}{\partial x \partial y \partial z}\left(-\frac{\partial e_{zx}}{\partial y} + \frac{\partial e_{xy}}{\partial z} + \frac{\partial e_{yz}}{\partial x} \right)$$

$$\frac{\partial^4 e_z}{\partial x^2 \partial y^2} = \frac{\partial^3}{\partial x \partial y \partial z}\left(-\frac{\partial e_{xy}}{\partial z} + \frac{\partial e_{yz}}{\partial x} + \frac{\partial e_{zx}}{\partial y} \right)$$

2.17 Determine if the following strains satisfy the compatibility equations (2.6.2)

(a) $e_x = Ay, e_y = e_z = 0, e_{xy} = \left(Ax + Bz^2\right)/2, \; e_{yz} = Bxz + Cy, \; e_{zx} = Cx$

(b) $e_x = 2Ax, e_y = Bx, \; e_z = Cxy, e_{xy} = By/2, \; e_{yz} = Cxz/2, \; e_{zx} = Cyz/2$

(c) $e_x = 2Ax, e_y = By^2, \; e_z = Cxz, e_{xy} = By, \; e_{yz} = Cx^3, \; e_{zx} = Cyz^2$

where A, B and C are constants.

2.18 Show that the following strain field

$$e_x = Ay^3, \quad e_y = Ax^3, \quad e_{xy} = Bxy(x+y), \quad e_z = e_{xz} = e_{yz} = 0$$

gives continuous, single-valued displacements in a simply connected region only if the constants are related by $A = 2B/3$.

2.19 In order to model dislocations in elastic solids, multivalued displacement fields are necessary. As shown later in Chapter 15, for the particular case of a *screw dislocation* the displacements are given by

$$u = v = 0, w = \frac{b}{2\pi} \tan^{-1}\frac{y}{x}$$

where b is a constant called the *Burgers vector.* Show that the strains resulting from these displacements are given by

$$e_x = e_y = e_z = e_{xy} = 0, \quad e_{xz} = -\frac{b}{4\pi}\frac{y}{x^2+y^2}, \quad e_{yz} = \frac{b}{4\pi}\frac{x}{x^2+y^2}$$

2.20 Consider the plane deformation of the differential element $ABCD$ defined by polar coordinates r, θ as shown in the following figure. Using the geometric methods outlined in Section 2.2, investigate the changes in line lengths and angles associated with the deformation to a configuration $A'B'C'D'$, and develop the strain–displacement relations

$$e_r = \frac{\partial u_r}{\partial r}, \quad e_\theta = \frac{1}{r}\left(u_r + \frac{\partial u_\theta}{\partial \theta}\right), \quad e_{r\theta} = \frac{1}{2}\left(\frac{1}{r}\frac{\partial u_r}{\partial \theta} + \frac{\partial u_\theta}{\partial r} - \frac{u_\theta}{r}\right)$$

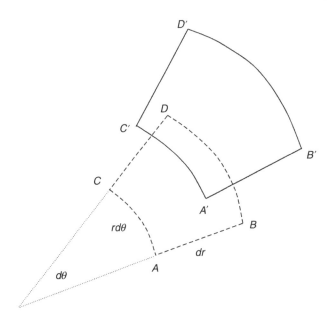

2.21 Using the results from Exercise 2.20, determine the two-dimensional strains e_r, e_θ, $e_{r\theta}$ for the following displacement fields:

$$(a) \; u_r = \frac{A}{r}, \quad u_\theta = B \cos \theta$$

$$(b) \; u_r = Ar^2, \quad u_\theta = Br \sin \theta$$

$$(c) \; u_r = A \sin \theta + B \cos \theta, \quad u_\theta = A \cos \theta - B \sin \theta + Cr$$

where A, B, and C are arbitrary constants.

2.22 Consider the cylindrical coordinate description for problems with axial symmetry such that all fields depend only on the radial coordinate r and the axial coordinate z. Determine the reduced strain displacement relations for this case. Further reduce these equations for the case with $u_\theta = 0$.

2.23 Consider the spherical coordinate description for problems with radial symmetry such that all fields depend only on the radial coordinate R. Determine the reduced strain displacement relations for this case. Further reduce these equations for the case with $u_\phi = u_\theta = 0$.

2.24* In Section 2.2, before employing any small deformation assumptions, the two-dimensional extensional strain component e_x was expressed as

$$e_x = \frac{A'B' - AB}{AB} = \sqrt{1 + 2\frac{\partial u}{\partial x} + \left(\frac{\partial u}{\partial x}\right)^2 + \left(\frac{\partial v}{\partial x}\right)^2} - 1$$

We now wish to investigate the nature of small deformation theory by exploring this non-linear relation for the special case with linear displacement field distributions $u = Ax$, $v = Bx$ (A and B constants). Clearly this case gives constant displacement gradients $\partial u/\partial x = A$, $\partial v/\partial x = B$. Numerically evaluate this strain component and make plots of e_x/A versus A for $0.001 \leq A \leq 0.01$ with particular values of $B = 0.01, 0.02, 0.03$. Realizing that under small deformation assumptions, $e_x/A = 1$, comment on the behavior shown in these plots.

Stress and equilibrium

3

The previous chapter investigated the kinematics of deformation without regard to the force or stress distribution within the elastic solid. We now wish to examine these issues and explore the transmission of forces through deformable materials. Our study leads to the definition and use of the traction vector and stress tensor. Each provides a quantitative method to describe both boundary and internal force distributions within a continuum solid. Because it is commonly accepted that maximum stresses are a major contributing factor to material failure, primary application of elasticity theory is used to determine the distribution of stress within a given structure. Related to these force distribution issues is the concept of equilibrium. Within a deformable solid, the force distribution at each point must be balanced. For the static case, the summation of forces on an infinitesimal element is required to be zero, while for a dynamic problem the resultant force must equal the mass times the element's acceleration. In this chapter, we establish the definitions and properties of the traction vector and stress tensor and develop the equilibrium equations, which become another set of field equations necessary in the overall formulation of elasticity theory. It should be noted that the developments in this chapter do not require that the material be elastic, and thus in principle these results apply to a broader class of material behavior.

3.1 Body and surface forces

When a structure is subjected to applied external loadings, internal forces are induced inside the body. Following the philosophy of continuum mechanics, these internal forces are distributed continuously within the solid. In order to study such forces, it is convenient to categorize them into two major groups, commonly referred to as body forces and surface forces.

Body forces are proportional to the body's mass and are reacted with an agent outside of the body. Examples of these include gravitational-weight forces, magnetic forces, and inertial forces. Fig. 3.1A shows an example body force of an object's self-weight. By using continuum mechanics principles, a *body force density* (force per unit volume) $F(x)$ can be defined such that the total resultant body force of an entire solid can be written as a volume integral over the body

$$F_R = \iiint_V F(x)dV \tag{3.1.1}$$

Surface forces always act on the surface and generally result from physical contact with another body. Fig. 3.1B illustrates surface forces existing in a beam section that has been created by sectioning the body into two pieces. For this particular case, the surface S is a virtual one in the sense that it was

Elasticity. https://doi.org/10.1016/B978-0-12-815987-3.00003-7

(a)

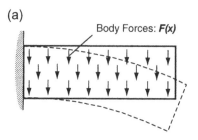

Cantilever Beam Under Self-Weight Loading

(b)

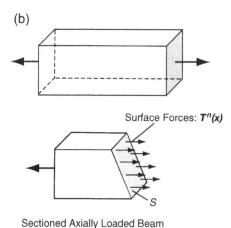

Sectioned Axially Loaded Beam

FIG. 3.1 Examples of body and surface forces.

artificially created to investigate the nature of the internal forces at this location in the body. Again, the resultant surface force over the entire surface S can be expressed as the integral of a *surface force density function* $T^n(x)$

$$F_S = \iint_S T^n(x)dS \qquad (3.1.2)$$

The surface force density is normally referred to as the traction vector and is discussed in more detail in the next section. In the development of classical elasticity, distributions of body or surface couples are normally not included. Theories that consider such force distributions have been constructed in an effort to extend the classical formulation for applications in micromechanical modeling. Such approaches are normally called *micropolar* or *couple-stress theory* (see Eringen, 1968) and are briefly presented in Chapter 15.

3.2 Traction vector and stress tensor

In order to quantify the nature of the internal distribution of forces within a continuum solid, consider a general body subject to arbitrary (concentrated and distributed) external loadings, as shown in Fig. 3.2. To investigate the internal forces, a section is made through the body as shown. On this section

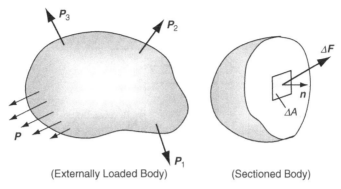

(Externally Loaded Body)   (Sectioned Body)

FIG. 3.2 Sectioned solid under external loading.

consider a small area ΔA with unit outward normal vector $\boldsymbol{n}$. The resultant surface force acting on ΔA is defined by $\Delta \boldsymbol{F}$. Consistent with our earlier discussion, no resultant surface couple is included. The *stress* or *traction vector* is defined by

$$T^n(\boldsymbol{x}, \boldsymbol{n}) = \lim_{\Delta A \to 0} \frac{\Delta \boldsymbol{F}}{\Delta A} \qquad (3.2.1)$$

Notice that the traction vector depends on both the spatial location and the unit normal vector to the surface under study. Thus, even though we may be investigating the same point, the traction vector still varies as a function of the orientation of the surface normal. Because the traction is defined as force per unit area, the total surface force is determined through integration as per relation (3.1.2). Note, also, the simple action–reaction principle (Newton's third law)

$$T^n(\boldsymbol{x}, \boldsymbol{n}) = -T^n(\boldsymbol{x}, -\boldsymbol{n})$$

Consider now the special case in which ΔA coincides with each of the three coordinate planes with the unit normal vectors pointing along the positive coordinate axes. This concept is shown in Fig. 3.3, where the three coordinate surfaces for ΔA partition off a cube of material. For this case, the traction vector on each face can be written as

$$\begin{aligned}
T^n(\boldsymbol{x}, \boldsymbol{n} = \boldsymbol{e}_1) &= \sigma_x \boldsymbol{e}_1 + \tau_{xy} \boldsymbol{e}_2 + \tau_{xz} \boldsymbol{e}_3 \\
T^n(\boldsymbol{x}, \boldsymbol{n} = \boldsymbol{e}_2) &= \tau_{yx} \boldsymbol{e}_1 + \sigma_y \boldsymbol{e}_2 + \tau_{yz} \boldsymbol{e}_3 \\
T^n(\boldsymbol{x}, \boldsymbol{n} = \boldsymbol{e}_3) &= \tau_{zx} \boldsymbol{e}_1 + \tau_{zy} \boldsymbol{e}_2 + \sigma_z \boldsymbol{e}_3
\end{aligned} \qquad (3.2.2)$$

where $\boldsymbol{e}_1, \boldsymbol{e}_2, \boldsymbol{e}_3$ are the unit vectors along each coordinate direction, and the nine quantities $\{\sigma_x, \sigma_y, \sigma_z, \tau_{xy}, \tau_{yx}, \tau_{yz}, \tau_{zy}, \tau_{zx}, \tau_{xz}\}$ are the components of the traction vector on each of three coordinate planes as illustrated. These nine components are called the *stress components*, with $\sigma_x, \sigma_y, \sigma_z$ referred to as *normal stresses* and $\tau_{xy}, \tau_{yx}, \tau_{yz}, \tau_{zy}, \tau_{zx}, \tau_{xz}$ called the *shearing stresses*. The components of stress σ_{ij} are commonly written in matrix format

$$\boldsymbol{\sigma} = [\boldsymbol{\sigma}] = \begin{bmatrix} \sigma_x & \tau_{xy} & \tau_{xz} \\ \tau_{yx} & \sigma_y & \tau_{yz} \\ \tau_{zx} & \tau_{zy} & \sigma_z \end{bmatrix} \qquad (3.2.3)$$

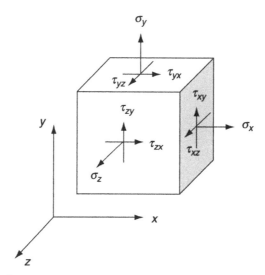

FIG. 3.3 Components of the stress.

and it can be formally shown that the stress is a second-order tensor that obeys the appropriate transformation law $(1.5.3)_3$.

The positive directions of each stress component are illustrated in Fig. 3.3. Regardless of the coordinate system, positive normal stress always acts in tension out of the face, and only one subscript is necessary because it always acts normal to the surface. The shear stress, however, requires two subscripts, the first representing the plane of action and the second designating the direction of the stress. Similar to shear strain, the sign of the shear stress depends on coordinate system orientation. For example, on a plane with a normal in the positive x direction, positive τ_{xy} acts in the positive y direction. Similar definitions follow for the other shear stress components. In subsequent chapters, proper formulation of elasticity problems requires knowledge of these basic definitions, directions, and sign conventions for particular stress components.

Consider next the traction vector on an oblique plane with arbitrary orientation, as shown in Fig. 3.4. The unit normal to the surface can be expressed by

$$\boldsymbol{n} = n_x \boldsymbol{e}_1 + n_y \boldsymbol{e}_2 + n_z \boldsymbol{e}_3 \tag{3.2.4}$$

where n_x, n_y, n_z are the direction cosines of the unit vector $\boldsymbol{n}$ relative to the given coordinate system. We now consider the equilibrium of the pyramidal element interior to the oblique and coordinate planes. Invoking the force balance between tractions on the oblique and coordinate faces gives

$$\boldsymbol{T}^n = n_x \boldsymbol{T}^n(\boldsymbol{n} = \boldsymbol{e}_1) + n_y \boldsymbol{T}^n(\boldsymbol{n} = \boldsymbol{e}_2) + n_z \boldsymbol{T}^n(\boldsymbol{n} = \boldsymbol{e}_3)$$

and by using relations (3.2.2), this can be written as

$$
\begin{aligned}
\boldsymbol{T}^n = {} & \left(\sigma_x n_x + \tau_{yx} n_y + \tau_{zx} n_z\right)\boldsymbol{e}_1 \\
& + \left(\tau_{xy} n_x + \sigma_y n_y + \tau_{zy} n_z\right)\boldsymbol{e}_2 \\
& + \left(\tau_{xz} n_x + \tau_{yz} n_y + \sigma_z n_z\right)\boldsymbol{e}_3
\end{aligned}
\tag{3.2.5}
$$

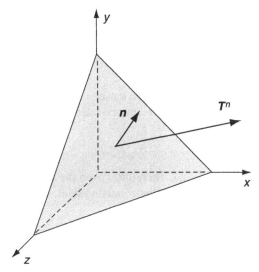

FIG. 3.4 Traction on an oblique plane.

or in index notation

$$T_i^n = \sigma_{ji} n_j \tag{3.2.6}$$

Relation (3.2.5) or (3.2.6) provides a simple and direct method to calculate the forces on oblique planes and surfaces. This technique proves to be very useful to specify general boundary conditions during the formulation and solution of elasticity problems. Based on these previous definitions, the distinction between the traction vector and stress tensor should be carefully understood. Although each quantity has the same units of force per unit area, they are fundamentally different since the traction is a vector while the stress is a second-order tensor (matrix). Components of traction can be defined on any surface, but particular stress components only exist on coordinate surfaces, as shown in Fig. 3.3 for the Cartesian case. Clearly, Eq. (3.2.6) establishes the relation between the two variables, thereby indicating that each traction component can be expressed as a linear combination of particular stress components. Further discussion on this topic will be given in Section 5.2 when boundary condition development is presented.

Following the principles of small deformation theory, the previous definitions for the stress tensor and traction vector do not make a distinction between the deformed and undeformed configurations of the body. As mentioned in the previous chapter, for small deformations such a distinction only leads to small modifications that are considered *higher-order* effects and are normally neglected. However, for large deformation theory, sizeable differences exist between these configurations, and the undeformed configuration (commonly called the reference configuration) is often used in problem formulation. This gives rise to the definition of an additional stress called the *Piola–Kirchhoff stress tensor* that represents the force per unit area in the reference configuration (see Sadd, 2019). In the more general scheme, the stress σ_{ij} is referred to as the *Cauchy* stress tensor. Throughout the text only small deformation theory is considered, and thus the distinction between these two definitions of stress disappears, thereby eliminating any need for this additional terminology.

3.3 Stress transformation

Analogous to our previous discussion with the strain tensor, the stress components must also follow the standard transformation rules for second-order tensors established in Section 1.5. Applying transformation relation $(1.5.1)_3$ for the stress gives

$$\sigma'_{ij} = Q_{ip}Q_{jq}\sigma_{pq} \tag{3.3.1}$$

where the rotation matrix $Q_{ij} = \cos(x'_i, x_j)$. Therefore, given the stress in one coordinate system, we can determine the new components in any other rotated system. For the general three-dimensional case, the rotation matrix may be chosen in the form

$$Q_{ij} = \begin{bmatrix} l_1 & m_1 & n_1 \\ l_2 & m_2 & n_2 \\ l_3 & m_3 & n_3 \end{bmatrix} \tag{3.3.2}$$

Using this notational scheme, the specific transformation relations for the stress then become

$$\begin{aligned}
\sigma'_x &= \sigma_x l_1^2 + \sigma_y m_1^2 + \sigma_z n_1^2 + 2(\tau_{xy}l_1 m_1 + \tau_{yz}m_1 n_1 + \tau_{zx}n_1 l_1) \\
\sigma'_y &= \sigma_x l_2^2 + \sigma_y m_2^2 + \sigma_z n_2^2 + 2(\tau_{xy}l_2 m_2 + \tau_{yz}m_2 n_2 + \tau_{zx}n_2 l_2) \\
\sigma'_z &= \sigma_x l_3^2 + \sigma_y m_3^2 + \sigma_z n_3^2 + 2(\tau_{xy}l_3 m_3 + \tau_{yz}m_3 n_3 + \tau_{zx}n_3 l_3) \\
\tau'_{xy} &= \sigma_x l_1 l_2 + \sigma_y m_1 m_2 + \sigma_z n_1 n_2 + \tau_{xy}(l_1 m_2 + m_1 l_2) + \tau_{yz}(m_1 n_2 + n_1 m_2) + \tau_{zx}(n_1 l_2 + l_1 n_2) \\
\tau'_{yz} &= \sigma_x l_2 l_3 + \sigma_y m_2 m_3 + \sigma_z n_2 n_3 + \tau_{xy}(l_2 m_3 + m_2 l_3) + \tau_{yz}(m_2 n_3 + n_2 m_3) + \tau_{zx}(n_2 l_3 + l_2 n_3) \\
\tau'_{zx} &= \sigma_x l_3 l_1 + \sigma_y m_3 m_1 + \sigma_z n_3 n_1 + \tau_{xy}(l_3 m_1 + m_3 l_1) + \tau_{yz}(m_3 n_1 + n_3 m_1) + \tau_{zx}(n_3 l_1 + l_3 n_1)
\end{aligned} \tag{3.3.3}$$

For the two-dimensional case originally shown in Fig. 2.6, the transformation matrix was given by relation (2.3.4). Under this transformation, the in-plane stress components transform according to

$$\begin{aligned}
\sigma'_x &= \sigma_x \cos^2 \theta + \sigma_y \sin^2\theta + 2\tau_{xy} \sin \theta \cos \theta \\
\sigma'_y &= \sigma_x \sin^2\theta + \sigma_y \cos^2\theta - 2\tau_{xy} \sin \theta \cos \theta \\
\tau'_{xy} &= -\sigma_x \sin \theta \cos \theta + \sigma_y \sin \theta \cos \theta + \tau_{xy}\left(\cos^2 \theta - \sin^2 \theta\right)
\end{aligned} \tag{3.3.4}$$

which is commonly rewritten in terms of the double angle

$$\begin{aligned}
\sigma'_x &= \frac{\sigma_x + \sigma_y}{2} + \frac{\sigma_x - \sigma_y}{2}\cos 2\theta + \tau_{xy} \sin 2\theta \\
\sigma'_y &= \frac{\sigma_x + \sigma_y}{2} - \frac{\sigma_x - \sigma_y}{2}\cos 2\theta - \tau_{xy} \sin 2\theta \\
\tau'_{xy} &= \frac{\sigma_y - \sigma_x}{2}\sin 2\theta + \tau_{xy} \cos 2\theta
\end{aligned} \tag{3.3.5}$$

Similar to our discussion on strain in the previous chapter, relations (3.3.5) can be directly applied to establish stress transformations between Cartesian and polar coordinate systems (see Exercise 3.3). Both two- and three-dimensional stress transformation equations can be easily incorporated in MATLAB$^®$ to provide numerical solutions to problems of interest (see Exercise 3.2).

3.4 Principal stresses

We can again use the previous developments from Section 1.6 to discuss the issues of principal stresses and directions. It is shown later in the chapter that the stress is a symmetric tensor. Using this fact, appropriate theory has been developed to identify and determine principal axes and values for the stress. For any given stress tensor we can establish the principal value problem and solve the characteristic equation to explicitly determine the principal values and directions. The general characteristic equation for the stress tensor becomes

$$\det[\sigma_{ij} - \sigma\delta_{ij}] = -\sigma^3 + I_1\sigma^2 - I_2\sigma + I_3 = 0 \tag{3.4.1}$$

where σ are the *principal stresses* and the fundamental invariants of the stress tensor can be expressed in terms of the three principal stresses σ_1, σ_2, σ_3 as

$$\begin{aligned} I_1 &= \sigma_1 + \sigma_2 + \sigma_3 \\ I_2 &= \sigma_1\sigma_2 + \sigma_2\sigma_3 + \sigma_3\sigma_1 \\ I_3 &= \sigma_1\sigma_2\sigma_3 \end{aligned} \tag{3.4.2}$$

In the principal coordinate system, the stress matrix takes the special diagonal form

$$\sigma_{ij} = \begin{bmatrix} \sigma_1 & 0 & 0 \\ 0 & \sigma_2 & 0 \\ 0 & 0 & \sigma_3 \end{bmatrix} \tag{3.4.3}$$

A comparison of the general and principal stress states is shown in Fig. 3.5. Notice that for the principal coordinate system, all shearing stresses vanish and thus the state includes only normal

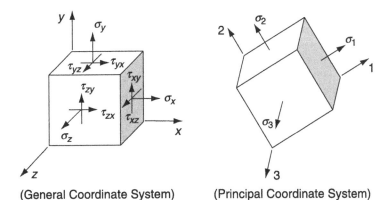

(General Coordinate System)   (Principal Coordinate System)

FIG. 3.5 Comparison of general and principal stress states.

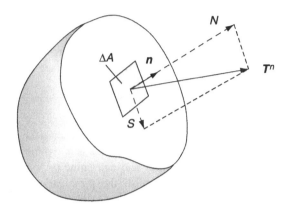

FIG. 3.6 Traction vector decomposition.

stresses. These issues should be compared to the equivalent comments made for the strain tensor at the end of Section 2.4.

We now wish to go back to investigate another issue related to stress and traction transformation that makes use of principal stresses. Consider the general traction vector T^n that acts on an arbitrary surface as shown in Fig. 3.6. The issue of interest is to determine the traction vector's normal and shear components N and S. The normal component is simply the traction's projection in the direction of the unit normal vector n, while the shear component is found by Pythagorean theorem

$$N = T^n \cdot n$$
$$S = \left(|T^n|^2 - N^2 \right)^{1/2} \tag{3.4.4}$$

Using the relationship for the traction vector (3.2.5) into (3.4.4)$_1$ gives

$$N = T^n \cdot n = T_i^n n_i = \sigma_{ji} n_j n_i$$
$$= \sigma_1 n_1^2 + \sigma_2 n_2^2 + \sigma_3 n_3^2 \tag{3.4.5}$$

where, in order to simplify the expressions, we have used the principal axes for the stress tensor. In a similar manner

$$|T^n|^2 = T^n \cdot T^n = T_i^n T_i^n = \sigma_{ji} n_j \sigma_{ki} n_k$$
$$= \sigma_1^2 n_1^2 + \sigma_2^2 n_2^2 + \sigma_3^2 n_3^2 \tag{3.4.6}$$

Using these results back in relation (3.4.4) yields

$$N = \sigma_1 n_1^2 + \sigma_2 n_2^2 + \sigma_3 n_3^2$$
$$S^2 + N^2 = \sigma_1^2 n_1^2 + \sigma_2^2 n_2^2 + \sigma_3^2 n_3^2 \tag{3.4.7}$$

In addition, we add the condition that the vector n has unit magnitude

$$1 = n_1^2 + n_2^2 + n_3^2 \tag{3.4.8}$$

Relations (3.4.7) and (3.4.8) can be viewed as three linear algebraic equations for the unknowns n_1^2, n_2^2, n_3^2. Solving this system gives the following result:

$$n_1^2 = \frac{S^2 + (N - \sigma_2)(N - \sigma_3)}{(\sigma_1 - \sigma_2)(\sigma_1 - \sigma_3)}$$

$$n_2^2 = \frac{S^2 + (N - \sigma_3)(N - \sigma_1)}{(\sigma_2 - \sigma_3)(\sigma_2 - \sigma_1)} \tag{3.4.9}$$

$$n_3^2 = \frac{S^2 + (N - \sigma_1)(N - \sigma_2)}{(\sigma_3 - \sigma_1)(\sigma_3 - \sigma_2)}$$

Without loss in generality, we can rank the principal stresses as $\sigma_1 > \sigma_2 > \sigma_3$. Noting that the expression given by (3.4.9) must be greater than or equal to zero, we can conclude:

$$S^2 + (N - \sigma_2)(N - \sigma_3) \geq 0$$
$$S^2 + (N - \sigma_3)(N - \sigma_1) \leq 0 \tag{3.4.10}$$
$$S^2 + (N - \sigma_1)(N - \sigma_2) \geq 0$$

For the equality case, Eqs. (3.4.10) represent three circles in an S–N coordinate system, and Fig. 3.7 illustrates the location of each circle. These results were originally generated by Otto Mohr over a century ago, and the circles are commonly called *Mohr's circles of stress*. The three inequalities given in (3.4.10) imply that all admissible values of N and S lie in the shaded regions bounded by the three

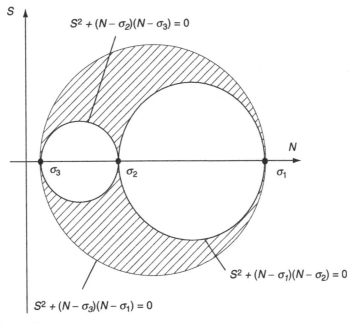

FIG. 3.7 Mohr's circles of stress.

circles. Note that, for the ranked principal stresses, the largest shear component is easily determined as $S_{max} = 1/2|\sigma_1 - \sigma_3|$. Although these circles can be effectively used for two-dimensional stress transformation, the general tensorial-based equations (3.3.1) or (3.3.3) are normally used for general transformation computations.

Example 3.1 Stress transformation

For the following state of stress, determine the principal stresses and directions and find the traction vector on a plane with unit normal $n = (0, 1, 1)/\sqrt{2}$

$$\sigma_{ij} = \begin{bmatrix} 3 & 1 & 1 \\ 1 & 0 & 2 \\ 1 & 2 & 0 \end{bmatrix}$$

The principal stress problem is first solved by calculating the three invariants, giving the result $I_1 = 3, I_2 = -6, I_3 = -8$. This yields the following characteristic equation

$$-\sigma^3 + 3\sigma^2 + 6\sigma - 8 = 0$$

The roots of this equation are found to be $\sigma = 4, 1, -2$. Back-substituting the first root into the fundamental system [see (1.6.1)] gives

$$-n_1^{(1)} + n_2^{(1)} + n_3^{(1)} = 0$$
$$n_1^{(1)} - 4n_2^{(1)} + 2n_3^{(1)} = 0$$
$$n_1^{(1)} + 2n_2^{(1)} - 4n_3^{(1)} = 0$$

Solving this system, the normalized principal direction is found to be $n^{(1)} = (2, 1, 1)/\sqrt{6}$. In similar fashion the other two principal directions are $n^{(2)} = (-1, 1, 1)/\sqrt{3}, n^{(3)} = (0, -1, 1)/\sqrt{2}$.
The traction vector on the specified plane is calculated by using the relation

$$T_i^n = \begin{bmatrix} 3 & 1 & 1 \\ 1 & 0 & 2 \\ 1 & 2 & 0 \end{bmatrix} \begin{bmatrix} 0 \\ 1/\sqrt{2} \\ 1/\sqrt{2} \end{bmatrix} = \begin{bmatrix} 2/\sqrt{2} \\ 2/\sqrt{2} \\ 2/\sqrt{2} \end{bmatrix}$$

3.5 Spherical, deviatoric, octahedral, and von Mises stresses

As mentioned in our previous discussion on strain, it is often convenient to decompose the stress into two parts called the *spherical* and *deviatoric stress tensors*. Analogous to relations (2.5.1) and (2.5.2), the spherical stress is defined by

$$\tilde{\sigma}_{ij} = \frac{1}{3}\sigma_{kk}\delta_{ij} \tag{3.5.1}$$

while the deviatoric stress becomes

$$\hat{\sigma}_{ij} = \sigma_{ij} - \frac{1}{3}\sigma_{kk}\delta_{ij} \tag{3.5.2}$$

Note that the total stress is then simply the sum

$$\sigma_{ij} = \widetilde{\sigma}_{ij} + \widehat{\sigma}_{ij} \tag{3.5.3}$$

The spherical stress is an isotropic tensor, being the same in all coordinate systems (as per the discussion in Section 1.5). It can be shown that the principal directions of the deviatoric stress are the same as those of the stress tensor itself (see Exercise 3.15).

We next briefly explore a couple of particular stress components or combinations that have been defined in the literature and are commonly used in formulating failure theories related to inelastic deformation. It has been found that ductile materials normally exhibit inelastic yielding failures that can be characterized by these particular stresses.

Consider first the normal and shear stresses (tractions) that act on a special plane whose normal makes equal angles with the three principal axes. This plane is commonly referred to as the *octahedral plane*. Determination of these normal and shear stresses is straightforward if we use the principal axes of stress. Since the unit normal vector to the octahedral plane makes equal angles with the principal axes, its components are given by $n_i = \pm(1, 1, 1)/\sqrt{3}$. Referring to Fig. 3.6 and using the results of the previous section, relations (3.4.7) give the desired normal and shear stresses as

$$N = \sigma_{oct} = \frac{1}{3}(\sigma_1 + \sigma_2 + \sigma_3) = \frac{1}{3}\sigma_{kk} = \frac{1}{3}I_1$$

$$S = \tau_{oct} = \frac{1}{3}\left[(\sigma_1 - \sigma_2)^2 + (\sigma_2 - \sigma_3)^2 + (\sigma_3 - \sigma_1)^2\right]^{1/2} \tag{3.5.4}$$

$$= \frac{1}{3}(2I_1^2 - 6I_2)^{1/2}$$

It can be shown that the octahedral shear stress τ_{oct} is directly related to the *distortional strain energy* [defined by equation (6.1.17)], which is often used in failure theories for ductile materials.

Another specially defined stress also related to the distortional strain energy failure criteria is known as the *effective* or *von Mises stress* and is given by the expression

$$\sigma_e = \sigma_{vonMises} = \sqrt{\frac{3}{2}\widehat{\sigma}_{ij}\widehat{\sigma}_{ij}} = \frac{1}{\sqrt{2}}\left[(\sigma_x - \sigma_y)^2 + (\sigma_y - \sigma_z)^2 + (\sigma_z - \sigma_x)^2 + 6\left(\tau_{xy}^2 + \tau_{yz}^2 + \tau_{zx}^2\right)\right]^{1/2}$$

$$= \frac{1}{\sqrt{2}}\left[(\sigma_1 - \sigma_2)^2 + (\sigma_2 - \sigma_3)^2 + (\sigma_3 - \sigma_1)^2\right]^{1/2} \tag{3.5.5}$$

Note that although the von Mises stress is not really a particular stress or traction component in the usual sense, it is obviously directly related to the octahedral shear stress by the relation $\sigma_e = (3/\sqrt{2})\tau_{oct}$. If at some point in the structure the von Mises stress equals the yield stress, then the material is considered to be at the failure condition. Because of this fact, many finite element computer codes commonly plot von Mises stress distributions based on the numerically generated stress field. It should be noted that the von Mises and octahedral shear stresses involve only the *differences* in the principal stresses and not the individual values. Thus, increasing each principal stress by the same amount will not change the value of σ_e or τ_{oct}. This result also implies that these values are independent of the hydrostatic stress. We will not further pursue failure criteria, and the interested reader is referred to Ugural and Fenster (2003) for details on this topic.

It should be pointed out that the spherical, octahedral, and von Mises stresses are all expressible in terms of the stress invariants and thus are independent of the coordinate system used to calculate them.

3.6 Stress distributions and contour lines

Over the years the stress analysis community has developed a large variety of schemes to help visualize and understand the nature of the stress distribution in elastic solids. Of course these efforts are not limited solely to stress because such information is also needed for strain and displacement distribution. Much of this effort is aimed at determining the magnitude and location of maximum stresses within the structure. Simple schemes involve just plotting the distribution of particular stress components along chosen directions within the body under study. Other methods focus on constructing contour plots of principal stress, maximum shear stress, von Mises stress, and other stress variables or combinations. Some techniques have been constructed to compare with optical experimental methods that provide photographic data of particular stress variables (Shukla and Dally, 2010). We now will briefly explore some of these schemes as they relate to two-dimensional plane stress distributions defined by the field: $\sigma_x = \sigma_x(x,y)$, $\sigma_y = \sigma_y(x,y)$, $\tau_{xy} = \tau_{xy}(x,y)$, $\sigma_z = \tau_{xz} = \tau_{yz} = 0$. Note for this case, the principal stresses and maximum shear stress are given in Exercise 3.5.

By passing polarized light through transparent model samples under load, the experimental method of *photoelasticity* can provide full field photographic stress data of particular stress combinations. The method can generate *isochromatic* fringe patterns that represent lines of constant difference in the principal stresses, i.e. $\sigma_1 - \sigma_2 = $ constant, which would also be lines of maximum shearing stress. Examples of isochromatic fringe patterns are shown in Figs. 8.28 and 8.36, and in Exercises 8.40 and 8.46. Photoelasticity can also generate another series of fringe lines called *isoclinics*, along which the principal stresses have a constant orientation. Still another set of contour lines often used in optical experimental stress analysis are *isopachic* contours, which are lines of $\sigma_x + \sigma_y = \sigma_1 + \sigma_2 = $ constant. These contours are related to the out-of-plane strain and displacement; see, for example, relation $(7.2.2)_3$.

Another useful set of lines are *isostatics*, sometimes referred to as *stress trajectories*. Such lines are oriented along the direction of a particular principal stress. For the two-dimensional plane stress case, the principal stresses σ_1 and σ_2 give rise to two families of stress trajectories that form an orthogonal network composed of lines free of shear stress. These trajectories have proven to be useful aids for understanding load paths, i.e. how external loadings move through a structure to the reaction points (Kelly and Tosh, 2000). Stress trajectories are also related to structural optimization and Michell structures composed of frameworks of continuous members in tension and compression.

Considering a particular stress trajectory, the orientation angle θ_p with respect to the x-axis can be found using the relation (see Exercise 3.21)

$$\tan 2\theta_p = \frac{2\tau_{xy}}{\sigma_x - \sigma_y} \tag{3.6.1}$$

Now for a given trajectory specified by $y(x)$, $\tan\theta_p = dy/dx$ and combining these results with a standard trigonometric identity gives

$$\tan 2\theta_p = \frac{2\tan\theta_p}{1-\tan^2\theta_p} = \frac{2\dfrac{dy}{dx}}{1-\left(\dfrac{dy}{dx}\right)^2} = \frac{2\tau_{xy}}{\sigma_x - \sigma_y}$$

This relation is easily solved for the trajectory slope

$$\frac{dy}{dx} = -\frac{\sigma_x - \sigma_y}{2\tau_{xy}} \pm \sqrt{1 + \left(\frac{\sigma_x - \sigma_y}{2\tau_{xy}}\right)^2} \qquad (3.6.2)$$

So given in-plane stress components, the differential equation (3.6.2) can be integrated to generate the stress trajectories, $y(x)$. Although some special cases can be done analytically (Molleda et al., 2005), most stress distributions will generate complicated forms that require numerical integration (Breault, 2012). A particular example will now be explored and several of the previously discussed stress contours and lines will be generated and plotted.

Example 3.2 Stress distributions in disk under diametrical compression

Let us now explore a specific two-dimensional problem of a circular disk loaded by equal but opposite concentrated forces along a given diameter as shown in Fig. 3.8A. This problem is solved in Example 8.10, and with respect to the given axes the in-plane stresses are found to be

$$\sigma_x = -\frac{2P}{\pi}\left[\frac{(R-y)x^2}{r_1^4} + \frac{(R+y)x^2}{r_2^4} - \frac{1}{2R}\right]$$

$$\sigma_y = -\frac{2P}{\pi}\left[\frac{(R-y)^3}{r_1^4} + \frac{(R+y)^3}{r_2^4} - \frac{1}{2R}\right] \qquad (3.6.3)$$

$$\tau_{xy} = \frac{2P}{\pi}\left[\frac{(R-y)^2 x}{r_1^4} - \frac{(R+y)^2 x}{r_2^4}\right]$$

where $r_{1,2} = \sqrt{x^2 + (R\mp y)^2}$. Numerical results are presented for the case with unit radius and unit loading, $R = P = 1$. For this case, Fig. 3.8 illustrates several contour distributions and the stress trajectories that have been previously discussed. These contour plots were made using MATLAB, and an example code is shown in EXAMPLE C-3 in Appendix C. It should be apparent that each of these contour distributions is in general different from one another, and each will convey particular information about the nature of the stress field under study. Later chapters in the text will make considerable use of various distribution plots such as these.

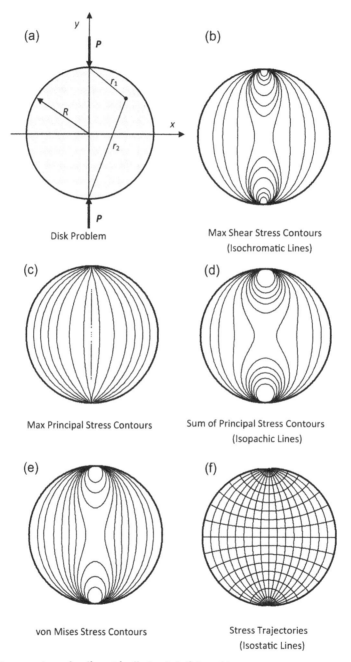

(a)

Disk Problem

(b)

Max Shear Stress Contours
(Isochromatic Lines)

(c)

Max Principal Stress Contours

(d)

Sum of Principal Stress Contours
(Isopachic Lines)

(e)

von Mises Stress Contours

(f)

Stress Trajectories
(Isostatic Lines)

FIG. 3.8 Example stress contours for diametrically loaded disk problem.

3.7 Equilibrium equations

The stress field in an elastic solid is continuously distributed within the body and uniquely determined from the applied loadings. Because we are dealing primarily with bodies in equilibrium, the applied loadings satisfy the equations of static equilibrium; the summation of forces and moments is zero. If the entire body is in equilibrium, then all parts must also be in equilibrium. Thus, we can partition any solid into an appropriate subdomain and apply the equilibrium principle to that region. Following this approach, equilibrium equations can be developed that express the vanishing of the resultant force and moment at a continuum point in the material. These equations can be developed by using either an arbitrary finite subdomain or a special differential region with boundaries coinciding with coordinate surfaces. We shall formally use the first method in the text, and the second scheme is included in Exercises 3.22 and 3.23.

Consider a closed subdomain with volume V and surface S within a body in equilibrium. The region has a general distribution of surface tractions T^n and body forces F as shown in Fig. 3.9. For static equilibrium, conservation of linear momentum implies that the forces acting on this region are balanced and thus the resultant force must vanish. This concept can be easily written in index notation as follows

$$\iint_S T_i^n \, dS + \iiint_V F_i \, dV = 0 \tag{3.7.1}$$

Using relation (3.2.6) for the traction vector, we can express the equilibrium statement in terms of stress

$$\iint_S \sigma_{ji} n_j \, dS + \iiint_V F_i \, dV = 0 \tag{3.7.2}$$

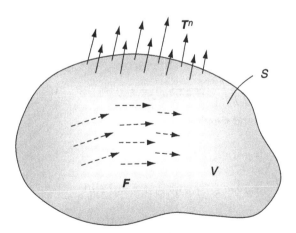

FIG. 3.9 Body and surface forces acting on an arbitrary portion of a continuum.

Applying the divergence theorem (1.8.7) to the surface integral allows the conversion to a volume integral, and relation (3.7.2) can then be expressed as

$$\iiint_V (\sigma_{ji,j} + F_i)dV = 0 \tag{3.7.3}$$

Because the region V is arbitrary (any part of the medium can be chosen) and the integrand in (3.7.3) is continuous, then by the zero-value theorem (1.8.12), the integrand must vanish

$$\sigma_{ji,j} + F_i = 0 \tag{3.7.4}$$

This result represents three scalar relations called the *equilibrium equations*. Written in scalar notation, they are

$$\frac{\partial \sigma_x}{\partial x} + \frac{\partial \tau_{yx}}{\partial y} + \frac{\partial \tau_{zx}}{\partial z} + F_x = 0$$

$$\frac{\partial \tau_{xy}}{\partial x} + \frac{\partial \sigma_y}{\partial y} + \frac{\partial \tau_{zy}}{\partial z} + F_y = 0 \tag{3.7.5}$$

$$\frac{\partial \tau_{xz}}{\partial x} + \frac{\partial \tau_{yz}}{\partial y} + \frac{\partial \sigma_z}{\partial z} + F_z = 0$$

Thus, all elasticity stress fields must satisfy these relations in order to be in static equilibrium.

Next consider the angular momentum principle that states that the moment of all forces acting on any portion of the body must vanish. Note that the point about which the moment is calculated can be chosen arbitrarily. Applying this principle to the region shown in Fig. 3.9 results in a statement of the vanishing of the moments resulting from surface and body forces

$$\iint_S \varepsilon_{ijk} x_j T_k^n dS + \iiint_V \varepsilon_{ijk} x_j F_k dV = 0 \tag{3.7.6}$$

Again using relation (3.2.6) for the traction, (3.7.6) can be written as

$$\iint_S \varepsilon_{ijk} x_j \sigma_{lk} n_l dS + \iiint_V \varepsilon_{ijk} x_j F_k dV = 0$$

and application of the divergence theorem gives

$$\iiint_V \left[(\varepsilon_{ijk} x_j \sigma_{lk})_{,l} + \varepsilon_{ijk} x_j F_k \right] dV = 0$$

This integral can be expanded and simplified as

$$\iiint_V \left[\varepsilon_{ijk} x_{j,l} \sigma_{lk} + \varepsilon_{ijk} x_j \sigma_{lk,l} + \varepsilon_{ijk} x_j F_k \right] dV =$$

$$\iiint_V \left[\varepsilon_{ijk} \delta_{jl} \sigma_{lk} + \varepsilon_{ijk} x_j \sigma_{lk,l} + \varepsilon_{ijk} x_j F_k \right] dV =$$

$$\iiint_V \left[\varepsilon_{ijk} \sigma_{jk} - \varepsilon_{ijk} x_j F_k + \varepsilon_{ijk} x_j F_k \right] dV = \iiint_V \varepsilon_{ijk} \sigma_{jk} dV$$

where we have used the equilibrium equations (3.7.4) to simplify the final result. Thus, (3.7.6) now gives

$$\iiint_V \varepsilon_{ijk}\sigma_{jk}dV = 0$$

As per our earlier arguments, because the region V is arbitrary, the integrand must vanish, giving $\varepsilon_{ijk}\,\sigma_{jk} = 0$. However, because the alternating symbol is antisymmetric in indices jk, the other product term σ_{jk} must be symmetric, implying

$$\sigma_{ij} = \sigma_{ji} \Rightarrow \begin{matrix} \tau_{xy} = \tau_{yx} \\ \tau_{yz} = \tau_{zy} \\ \tau_{zx} = \tau_{xz} \end{matrix} \tag{3.7.7}$$

We thus find that, similar to the strain, the stress tensor is also symmetric and therefore has only six independent components in three dimensions. Under these conditions, the equilibrium equations can then be written as

$$\sigma_{ij,j} + F_i = 0 \tag{3.7.8}$$

3.8 Relations in curvilinear cylindrical and spherical coordinates

As mentioned in the previous chapter, in order to solve many elasticity problems, formulation must be done in curvilinear coordinates typically using cylindrical or spherical systems. Thus, by following similar methods as used with the strain–displacement relations, we now wish to develop expressions for the equilibrium equations in curvilinear cylindrical and spherical coordinates. By using a direct vector/matrix notation, the equilibrium equations can be expressed as

$$\boldsymbol{\nabla}\cdot\boldsymbol{\sigma} + \boldsymbol{F} = 0 \tag{3.8.1}$$

where $\boldsymbol{\sigma} = \sigma_{ij}\boldsymbol{e}_i\boldsymbol{e}_j$ is the stress matrix or dyadic, $\boldsymbol{e}_i$ are the unit basis vectors in the curvilinear system, and $\boldsymbol{F}$ is the body force vector. The desired curvilinear expressions can be obtained from (3.8.1) by using the appropriate form for $\boldsymbol{\nabla}\cdot\boldsymbol{\sigma}$ from our previous work in Section 1.9.

Cylindrical coordinates were originally presented in Fig. 1.5. For such a system, the stress components are defined on the differential element shown in Fig. 3.10, and thus the stress matrix is given by

$$\boldsymbol{\sigma} = \begin{bmatrix} \sigma_r & \tau_{r\theta} & \tau_{rz} \\ \tau_{r\theta} & \sigma_\theta & \tau_{\theta z} \\ \tau_{rz} & \tau_{\theta z} & \sigma_z \end{bmatrix} \tag{3.8.2}$$

Now the stress can be expressed in terms of the traction components as

$$\boldsymbol{\sigma} = \boldsymbol{e}_r\boldsymbol{T}_r + \boldsymbol{e}_\theta\boldsymbol{T}_\theta + \boldsymbol{e}_z\boldsymbol{T}_z \tag{3.8.3}$$

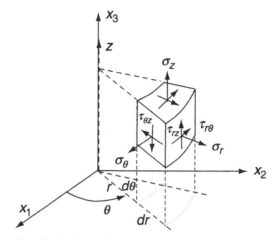

FIG. 3.10 Stress components in cylindrical coordinates.

where

$$T_r = \sigma_r e_r + \tau_{r\theta} e_\theta + \tau_{rz} e_z$$
$$T_\theta = \tau_{r\theta} e_r + \sigma_\theta e_\theta + \tau_{\theta z} e_z \tag{3.8.4}$$
$$T_z = \tau_{rz} e_r + \tau_{\theta z} e_\theta + \sigma_z e_z$$

Using relations (1.9.10) and (1.9.14), the divergence operation in the equilibrium equations can be written as

$$
\begin{aligned}
\nabla \cdot \sigma &= \frac{\partial T_r}{\partial r} + \frac{1}{r} T_r + \frac{1}{r}\frac{\partial T_\theta}{\partial \theta} + \frac{\partial T_z}{\partial z} \\[2mm]
&= \frac{\partial \sigma_r}{\partial r} e_r + \frac{\partial \tau_{r\theta}}{\partial r} e_\theta + \frac{\partial \tau_{rz}}{\partial r} e_z + \frac{1}{r}(\sigma_r e_r + \tau_{r\theta} e_\theta + \tau_{rz} e_z) \\[2mm]
&\quad + \frac{1}{r}\left(\frac{\partial \tau_{r\theta}}{\partial \theta} e_r + \tau_{r\theta} e_\theta + \frac{\partial \sigma_\theta}{\partial \theta} e_\theta - \sigma_\theta e_r + \frac{\partial \tau_{\theta z}}{\partial \theta} e_z\right) \\[2mm]
&\quad + \frac{\partial \tau_{rz}}{\partial z} e_r + \frac{\partial \tau_{\theta z}}{\partial z} e_\theta + \frac{\partial \sigma_z}{\partial z} e_z
\end{aligned}
\tag{3.8.5}
$$

Combining this result into (3.8.1) gives the vector equilibrium equation in cylindrical coordinates. The three scalar equations expressing equilibrium in each coordinate direction then become

$$\frac{\partial \sigma_r}{\partial r} + \frac{1}{r}\frac{\partial \tau_{r\theta}}{\partial \theta} + \frac{\partial \tau_{rz}}{\partial z} + \frac{1}{r}(\sigma_r - \sigma_\theta) + F_r = 0$$

$$\frac{\partial \tau_{r\theta}}{\partial r} + \frac{1}{r}\frac{\partial \sigma_\theta}{\partial \theta} + \frac{\partial \tau_{\theta z}}{\partial z} + \frac{2}{r}\tau_{r\theta} + F_\theta = 0 \tag{3.8.6}$$

$$\frac{\partial \tau_{rz}}{\partial r} + \frac{1}{r}\frac{\partial \tau_{\theta z}}{\partial \theta} + \frac{\partial \sigma_z}{\partial z} + \frac{1}{r}\tau_{rz} + F_z = 0$$

We now wish to repeat these developments for the spherical coordinate system, as previously shown in Fig. 1.6. The stress components in spherical coordinates are defined on the differential element illustrated in Fig. 3.11, and the stress matrix for this case is

$$\boldsymbol{\sigma} = \begin{bmatrix} \sigma_R & \tau_{R\phi} & \tau_{R\theta} \\ \tau_{R\phi} & \sigma_\phi & \tau_{\phi\theta} \\ \tau_{R\theta} & \tau_{\phi\theta} & \sigma_\theta \end{bmatrix} \tag{3.8.7}$$

Following similar procedures as used for the cylindrical equation development, the three scalar equilibrium equations for spherical coordinates become

$$\frac{\partial \sigma_R}{\partial R} + \frac{1}{R}\frac{\partial \tau_{R\phi}}{\partial \phi} + \frac{1}{R\sin\phi}\frac{\partial \tau_{R\theta}}{\partial \theta} + \frac{1}{R}\left(2\sigma_R - \sigma_\phi - \sigma_\theta + \tau_{R\phi}\cot\phi\right) + F_R = 0$$

$$\frac{\partial \tau_{r\phi}}{\partial R} + \frac{1}{R}\frac{\partial \sigma_\phi}{\partial \phi} + \frac{1}{R\sin\phi}\frac{\partial \tau_{\phi\theta}}{\partial \theta} + \frac{1}{R}\left[(\sigma_\phi - \sigma_\theta)\cot\phi + 3\tau_{R\phi}\right] + F_\phi = 0 \tag{3.8.8}$$

$$\frac{\partial \tau_{r\theta}}{\partial R} + \frac{1}{R}\frac{\partial \tau_{\phi\theta}}{\partial \phi} + \frac{1}{R\sin\phi}\frac{\partial \sigma_\theta}{\partial \theta} + \frac{1}{R}\left(2\tau_{\phi\theta}\cot\phi + 3\tau_{R\theta}\right) + F_\theta = 0$$

It is interesting to note that the equilibrium equations in curvilinear coordinates contain additional terms not involving derivatives of the stress components. The appearance of these terms can be explained mathematically due to the curvature of the space. However, a more physical interpretation can be found by redeveloping these equations through a simple force balance analysis on the appropriate differential element. This analysis is proposed for the less demanding two-dimensional polar coordinate case in Exercise 3.24. In general, relations (3.8.6) and (3.8.8) look much more

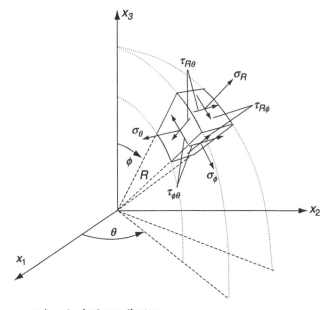

FIG. 3.11 Stress components in spherical coordinates.

complicated when compared to the Cartesian form (3.7.5). However, under particular conditions the curvilinear forms will lead to an analytical solution that could not be reached using Cartesian coordinates. For easy reference, Appendix A lists the complete set of elasticity field equations in cylindrical and spherical coordinates.

References

Breault S: *Improving load distributions in cellular materials using stress trajectory topology* (MS thesis). 2012, University of Rhode Island.
Eringen AC: Theory of micropolar elasticity. In Liebowitz H, editor: *Fracture*, vol. 2. New York, 1968, Academic Press, pp 662−729.
Kelly D, Tosh M: Interpreting load paths and stress trajectories in elasticity, *Eng Comput* 17:117−135, 2000.
Molleda F, Mora J, Molleda FJ, Carrillo E, Mellor BG: Stress trajectories for Mode I fracture, *Mater Char* 54: 9−12, 2005.
Sadd MH: *Continuum mechanics modeling of material behavior*, 2019, Elsevier.
Shukla A, Dally JW: *Experimental solid mechanics*, Knoxville, TN, 2010, College House Enterprises.
Ugural AC, Fenster SK: *Advanced strength and applied elasticity*, Englewood Cliffs, NJ, 2003, Prentice Hall.

Exercises

3.1 The illustrated rectangular plate is under uniform biaxial loading which yields the following state of stress

$$\sigma_{ij} = \begin{bmatrix} X & 0 & 0 \\ 0 & Y & 0 \\ 0 & 0 & 0 \end{bmatrix}$$

Determine the traction vector and the normal and shearing stresses on the oblique plane S.

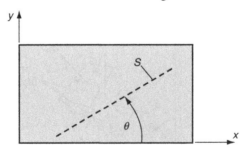

3.2* Using suitable units, the stress at a particular point in a solid is found to be

$$\text{(a)}\quad \sigma_{ij} = \begin{bmatrix} 2 & 1 & -4 \\ 1 & 4 & 0 \\ -4 & 0 & 1 \end{bmatrix} \qquad \text{(b)}\quad \sigma_{ij} = \begin{bmatrix} 4 & 1 & 0 \\ 1 & -6 & 2 \\ 0 & 2 & 1 \end{bmatrix}$$

Determine the traction vector on a surface with unit normal $(\cos\theta, \sin\theta, 0)$, where θ is a general angle in the range $0 \le \theta \le \pi$. Plot the variation of the magnitude of the traction vector $|T^n|$ as a function of θ.

3.3 Show that the general two-dimensional stress transformation relations can be used to generate relations for the normal and shear stresses in a polar coordinate system in terms of Cartesian components

$$\sigma_r = \frac{\sigma_x + \sigma_y}{2} + \frac{\sigma_x - \sigma_y}{2}\cos 2\theta + \tau_{xy}\sin 2\theta$$

$$\sigma_\theta = \frac{\sigma_x + \sigma_y}{2} - \frac{\sigma_x - \sigma_y}{2}\cos 2\theta - \tau_{xy}\sin 2\theta$$

$$\tau_{r\theta} = \frac{\sigma_y - \sigma_x}{2}\sin 2\theta + \tau_{xy}\cos 2\theta$$

3.4 Verify that the two-dimensional transformation relations giving Cartesian stresses in terms of polar components are given by

$$\sigma_x = \sigma_r \cos^2\theta + \sigma_\theta \sin^2\theta - 2\tau_{r\theta}\sin\theta\cos\theta$$
$$\sigma_y = \sigma_r \sin^2\theta + \sigma_\theta \cos^2\theta - 2\tau_{r\theta}\sin\theta\cos\theta$$
$$\tau_{xy} = \sigma_r \sin\theta\cos\theta - \sigma_\theta \sin\theta\cos\theta + \tau_{r\theta}(\cos^2\theta - \sin^2\theta)$$

3.5 A two-dimensional state of *plane stress* in the x, y-plane is defined by $\sigma_z = \tau_{yz} = \tau_{zx} = 0$. Using general principal value theory, show that for this case the in-plane principal stresses and maximum shear stress are given by

$$\sigma_{1,2} = \frac{\sigma_x + \sigma_y}{2} \pm \sqrt{\left(\frac{\sigma_x - \sigma_y}{2}\right)^2 + \tau_{xy}^2}$$

$$\tau_{max} = \sqrt{\left(\frac{\sigma_x - \sigma_y}{2}\right)^2 + \tau_{xy}^2}$$

3.6 Explicitly verify relations (3.5.4) for the octahedral stress components. Also show that they can be expressed in terms of the general stress components by

$$\sigma_{oct} = \frac{1}{3}(\sigma_x + \sigma_y + \sigma_z)$$

$$\tau_{oct} = \frac{1}{3}\left[(\sigma_x - \sigma_y)^2 + (\sigma_y - \sigma_z)^2 + (\sigma_z - \sigma_x)^2 + 6\tau_{xy}^2 + 6\tau_{yz}^2 + 6\tau_{zx}^2\right]^{1/2}$$

3.7 For the plane stress case in Exercise 3.5, demonstrate the invariant nature of the principal stresses and maximum shear stresses by showing that

$$\sigma_{1,2} = \frac{1}{2}I_1 \pm \sqrt{\frac{1}{4}I_1^2 - I_2} \text{ and } \tau_{max} = \sqrt{\frac{1}{4}I_1^2 - I_2}$$

Thus, conclude that

$$\sigma_{1,2} = \frac{\sigma_x + \sigma_y}{2} \pm \sqrt{\left(\frac{\sigma_x - \sigma_y}{2}\right)^2 + \tau_{xy}^2} = \frac{\sigma_r + \sigma_\theta}{2} \pm \sqrt{\left(\frac{\sigma_r - \sigma_\theta}{2}\right)^2 + \tau_{r\theta}^2}$$

$$\tau_{max} = \sqrt{\left(\frac{\sigma_x - \sigma_y}{2}\right)^2 + \tau_{xy}^2} = \sqrt{\left(\frac{\sigma_r - \sigma_\theta}{2}\right)^2 + \tau_{r\theta}^2}$$

3.8 Exercise 8.2 provides the plane stress (see Exercise 3.5) solution for a cantilever beam of unit thickness, with depth $2c$, and carrying an end load of P with stresses given by

$$\sigma_x = \frac{3P}{2c^3} xy, \quad \sigma_y = 0, \quad \tau_{xy} = \frac{3P}{4c}\left[1 - \frac{y^2}{c^2}\right]$$

Show that the principal stresses are given by

$$\sigma_{1,2} = \frac{3P}{4c^3}\left[xy \pm \sqrt{(c^2 - y^2)^2 + x^2 y^2}\right]$$

and the principal directions are

$$n^{(1,2)} = \left[\left(xy \pm \sqrt{(c^2 - y^2)^2 + x^2 y^2}\right)e_1 + (c^2 - y^2)e_2\right] \times \text{constant}$$

Note that the principal directions do not depend on the loading P.

3.9* Plot contours of the maximum principal stress σ_1 in Exercise 3.8 in the region $0 \le x \le L, -c \le y \le c$, with $L = 1$, $c = 0.1$, and $P = 1$.

3.10 We wish to generalize the findings in Exercise 3.8, and thus consider a stress field of the general form $\sigma_{ij} = Pf_{ij}(x_k)$, where P is a loading parameter and the tensor function f_{ij} specifies only the field distribution. Show that the principal stresses will be a linear form in P, that is, $\sigma_{1,2,3} = Pg_{1,2,3}(x_k)$. Next demonstrate that the principal directions will not depend on P.

3.11* The plane stress solution for a semi-infinite elastic solid under a concentrated point loading is developed in Chapter 8. With respect to the axes shown in the following figure, the Cartesian stress components are found to be

$$\sigma_x = -\frac{2Px^2 y}{\pi(x^2 + y^2)^2}$$

$$\sigma_y = -\frac{2Py^3}{\pi(x^2 + y^2)^2}$$

$$\tau_{xy} = -\frac{2Pxy^2}{\pi(x^2 + y^2)^2}$$

Using results from Exercise 3.5, calculate the maximum shear stress at any point in the body and plot contours of τ_{max}. You can compare your results with the corresponding photoelastic contours shown in Fig. 8.28. Example MATLAB Code C-3 will be useful to develop the contour plotting code.

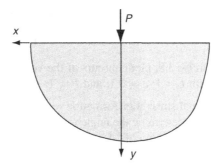

3.12 Show that shear stress S acting on a plane defined by the unit normal vector $\mathbf{n}$ (see Fig. 3.6) can be written as

$$S = \left[n_1^2 n_2^2 (\sigma_1 - \sigma_2)^2 + n_2^2 n_3^2 (\sigma_2 - \sigma_3)^2 + n_3^2 n_1^2 (\sigma_3 - \sigma_1)^2 \right]^{1/2}$$

3.13 It was discussed in Section 3.4 that for the case of ranked principal stresses $(\sigma_1 > \sigma_2 > \sigma_3)$, the maximum shear stress was given by $S_{\max} = (\sigma_1 - \sigma_3)/2$, which was the radius of the largest Mohr circle shown in Fig. 3.7. For this case, show that the normal stress acting on the plane of maximum shear is given by $N = (\sigma_1 + \sigma_3)/2$. Finally, using relations (3.4.9) show that the components of the unit normal vector to this plane are $n_i = \pm (1, 0, 1)/\sqrt{2}$. This result implies that the maximum shear stress acts on a plane that bisects the angle between the directions of the largest and the smallest principal stress.

3.14 Explicitly show that the stress state given in Example 3.1 will reduce to the proper diagonal form under transformation to principal axes.

3.15 Show that the principal directions of the deviatoric stress tensor coincide with the principal directions of the stress tensor. Also show that the principal values of the deviatoric stress σ_d can be expressed in terms of the principal values σ of the total stress by the relation $\sigma_d = \sigma - \frac{1}{3}\sigma_{kk}$.

3.16 Determine the spherical and deviatoric stress tensors for the stress states given in Exercise 3.2.

3.17 For the stress state given in Example 3.1, determine the von Mises and octahedral stresses defined in Section 3.5.

3.18 For the case of *pure shear*, the stress matrix is given by

$$\sigma_{ij} = \begin{bmatrix} 0 & \tau & 0 \\ \tau & 0 & 0 \\ 0 & 0 & 0 \end{bmatrix}$$

where τ is a given constant. Determine the principal stresses and directions, and then using principal values compute the normal and shear stress on the octahedral plane. How could you determine N and S on the octahedral plane using the original non-principal stresses?

3.19 For the two-dimensional plane stress case (see Exercise 3.5), show that the relationship between the maximum shear stress and von Mises stress is given by $\tau_{max}^2 = \frac{1}{3}\left(\sigma_e^2 - \frac{I_1^2}{4}\right)$

3.20* For the stress state in Exercise 3.8, plot contours of the von Mises stress in the region $0 \le x \le L$, $-c \le y \le c$, with $L = 1$, $c = 0.1$, and $P = 1$.

3.21 Starting with two-dimensional stress transformation relation $(3.3.5)_1$, set $d\sigma'_x/d\theta = 0$, and thus show that the relation to determine the angle to the principal stress direction θ_p is given by $\tan 2\theta_p = 2\tau_{xy}/\sigma_x - \sigma_y$. Next explicitly develop relation (3.6.2).

3.22 Consider the equilibrium of a two-dimensional differential element in Cartesian coordinates, as shown in the following figure. Explicitly sum the forces and moments and develop the two-dimensional equilibrium equations

$$\frac{\partial \sigma_x}{\partial x} + \frac{\partial \tau_{yx}}{\partial y} + F_x = 0$$

$$\frac{\partial \tau_{xy}}{\partial x} + \frac{\partial \sigma_y}{\partial y} + F_y = 0$$

$$\tau_{xy} = \tau_{yx}$$

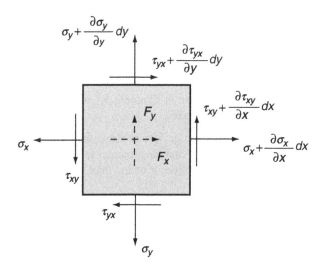

3.23 Consider the two-dimensional case described in Exercise 3.22 with no body forces. Show that equilibrium equations are identically satisfied if the stresses are expressed in the form

$$\sigma_x = \frac{\partial^2 \phi}{\partial y^2}, \; \sigma_y = \frac{\partial^2 \phi}{\partial x^2}, \; \tau_{xy} = -\frac{\partial^2 \phi}{\partial x \partial y}$$

where $\phi(x, y)$ is an arbitrary stress function. This stress representation will be used in Chapter 7 to establish a very useful solution scheme for two-dimensional problems.

3.24 Following similar procedures as in Exercise 3.22, sum the forces and moments on the two-dimensional differential element in polar coordinates (see figure), and explicitly develop the following two-dimensional equilibrium equations

$$\frac{\partial \sigma_r}{\partial r} + \frac{1}{r}\frac{\partial \tau_{\theta r}}{\partial \theta} + \frac{(\sigma_r - \sigma_\theta)}{r} + F_r = 0$$

$$\frac{\partial \tau_{r\theta}}{\partial r} + \frac{1}{r}\frac{\partial \sigma_\theta}{\partial \theta} + \frac{2\tau_{r\theta}}{r} + F_\theta = 0$$

$$\tau_{r\theta} = \tau_{\theta r}$$

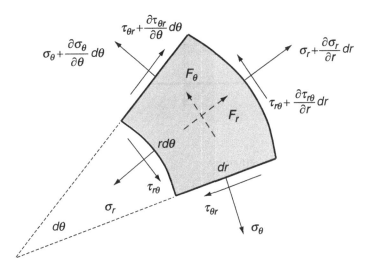

3.25 For a beam of circular cross-section, analysis from *elementary strength of materials theory* yields the following stresses

$$\sigma_x = -\frac{My}{I}, \ \tau_{xy} = \frac{V(R^2 - y^2)}{3I}, \ \sigma_y = \sigma_z = \tau_{xz} = \tau_{yz} = 0$$

where R is the section radius, $I = \pi R^4/4$, M is the bending moment, V is the shear force, and $dM/dx = V$. Assuming zero body forces, show that these stresses do not satisfy the equilibrium equations. This result is one of many that indicate the approximate nature of strength of materials theory.

3.26 A one-dimensional problem of a prismatic bar (see the following figure) loaded under its own weight can be modeled by the stress field $\sigma_x = \sigma_x(x)$, $\sigma_y = \sigma_z = \tau_{xy} = \tau_{yz} = \tau_{zx} = 0$, with body forces $F_x = \rho g$, $F_y = F_z = 0$, where ρ is the mass density and g is the local acceleration of gravity. Using the equilibrium equations, show that the nonzero stress will be given by $\sigma_x = \rho g(l - x)$, where l is the length of the bar.

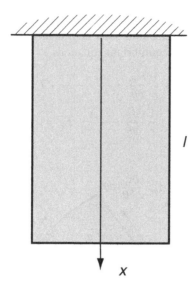

l

x

3.27 A *hydrostatic stress field* is specified by

$$\sigma_{ij} = -p\delta_{ij} = \begin{bmatrix} -p & 0 & 0 \\ 0 & -p & 0 \\ 0 & 0 & -p \end{bmatrix}$$

where $p = p\,(x_1, x_2, x_3)$ and may be called the pressure. Show that the equilibrium equations imply that the pressure must satisfy the relation $\nabla p = F$.

3.28 Verify the curvilinear cylindrical coordinate relations (3.8.5) and (3.8.6).

Material behavior—linear elastic solids

4

The previous two chapters establish elasticity field equations related to the kinematics of small deformation theory and the equilibrium of the associated internal stress field. Based on these physical concepts, six strain–displacement relations (2.2.5), six compatibility equations (2.6.2), and three equilibrium equations (3.7.5) were developed for the general three-dimensional case. Because moment equilibrium simply results in symmetry of the stress tensor, it is not normally included as a separate field equation set. Also, recall that the compatibility equations actually represent only three independent relations, and these equations are needed only to ensure that a given strain field will produce single-valued continuous displacements. Because the displacements are included in the general problem formulation, the solution normally gives continuous displacements, and the compatibility equations are not formally needed for the general system. Thus, excluding the compatibility relations, it is found that we have now developed nine field equations. The unknowns in these equations include three displacement components, six components of strain, and six stress components, yielding a total of 15 unknowns. Thus, the nine equations are not sufficient to solve for the 15 unknowns, and additional field equations are needed. This result should not be surprising since up to this point in our development we have not considered the material response.

We now wish to complete our general formulation by specializing to a particular material model that provides reasonable characterization of materials under small deformations. The model we will use is that of a linear elastic material, a name that categorizes the entire theory. This chapter presents the basics of the elastic model specializing the formulation for isotropic materials. Thermoelastic relations are briefly presented for later use in Chapter 12. Related theory for anisotropic media is developed in Chapter 11, and nonhomogeneous materials are examined in Chapter 14.

4.1 Material characterization

Relations that characterize the physical properties of materials are called *constitutive equations*. Because of the endless variety of materials and loadings, the study and development of constitutive equations is perhaps one of the most interesting and challenging fields in mechanics. Although continuum mechanics theory has established some principles for systematic development of constitutive equations (Sadd, 2019), many constitutive laws have been developed through empirical relations based on experimental evidence. Our interest here is limited to a special class of solid materials with loadings resulting from mechanical or thermal effects. The mechanical behavior of solids is normally defined by constitutive stress–strain relations. Commonly, these relations express the stress as a function of the strain, strain rate, strain history, temperature, and material properties. We choose a rather simple material model called the *elastic solid* that does not include rate or history effects. The model may be

Elasticity. https://doi.org/10.1016/B978-0-12-815987-3.00004-9

described as a deformable continuum that recovers its original configuration when the loadings causing the deformation are removed. Furthermore, we restrict the constitutive stress—strain law to be linear, thus leading to a *linear elastic solid*. Although these assumptions greatly simplify the model, linear elasticity predictions have shown good agreement with experimental data and have provided useful methods to conduct stress analysis. Many structural materials including metals, plastics, ceramics, wood, rock, concrete, and so forth exhibit linear elastic behavior under small deformations.

As mentioned, experimental testing is commonly employed in order to characterize the mechanical behavior of real materials. One such technique is the simple tension test in which a specially prepared cylindrical or flat stock sample is loaded axially in a testing machine. Strain is determined by the change in length between prescribed reference marks on the sample and is usually measured by a clip gage. Load data collected from a load cell is divided by the cross-sectional area in the test section to calculate the stress. Axial stress—strain data is recorded and plotted using standard experimental techniques. Typical qualitative data for three types of structural metals (mild steel, aluminum, cast iron) is shown in Fig. 4.1. It is observed that each material exhibits an initial stress—strain response for small deformation that is approximately linear. This is followed by a change to nonlinear behavior that can lead to large deformation, finally ending with sample failure.

For each material the initial linear response ends at a point normally referred to as the *proportional limit*. Another observation in this initial region is that if the loading is removed, the sample returns to its original shape and the strain disappears. This characteristic is the primary descriptor of elastic behavior. However, at some point on the stress—strain curve unloading does not bring the sample back to zero strain and some permanent plastic deformation results. The point at which this nonelastic behavior begins is called the *elastic limit*. Although some materials exhibit different elastic and proportional limits, many times these values are taken to be approximately the same. Another demarcation on the stress—strain curve is referred to as the *yield point*, defined by the location where large plastic deformation begins.

Because mild steel and aluminum are ductile materials, their stress—strain response indicates extensive plastic deformation, and during this period the sample dimensions will be changing. In particular the sample's cross-sectional area undergoes significant reduction, and the stress calculation using division by the original area will now be in error. This accounts for the reduction in the stress at

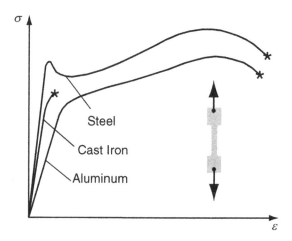

FIG. 4.1 Typical uniaxial stress—strain curves for three structural metals.

large strain. If we were to calculate the load divided by the true area, the *true stress* would continue to increase until failure. On the other hand, cast iron is known to be a brittle material, and thus its stress–strain response does not show large plastic deformation. For this material, very little nonelastic or nonlinear behavior is observed. It is therefore concluded from this and many other studies that a large variety of real materials exhibit linear elastic behavior under small deformations. This would lead to a linear constitutive model for the one-dimensional axial loading case given by the relation $\sigma = E\varepsilon$, where E is the slope of the uniaxial stress–strain curve. We now use this simple concept to develop the general three-dimensional forms of the linear elastic constitutive model.

4.2 Linear elastic materials—Hooke's law

Based on observations from the previous section, to construct a general three-dimensional constitutive law for linear elastic materials, we assume that each stress component is linearly related to each strain component

$$
\begin{aligned}
\sigma_x &= C_{11}e_x + C_{12}e_y + C_{13}e_z + 2C_{14}e_{xy} + 2C_{15}e_{yz} + 2C_{16}e_{zx} \\
\sigma_y &= C_{21}e_x + C_{22}e_y + C_{23}e_z + 2C_{24}e_{xy} + 2C_{25}e_{yz} + 2C_{26}e_{zx} \\
\sigma_z &= C_{31}e_x + C_{32}e_y + C_{33}e_z + 2C_{34}e_{xy} + 2C_{35}e_{yz} + 2C_{36}e_{zx} \\
\tau_{xy} &= C_{41}e_x + C_{42}e_y + C_{43}e_z + 2C_{44}e_{xy} + 2C_{45}e_{yz} + 2C_{46}e_{zx} \\
\tau_{yz} &= C_{51}e_x + C_{52}e_y + C_{53}e_z + 2C_{54}e_{xy} + 2C_{55}e_{yz} + 2C_{56}e_{zx} \\
\tau_{zx} &= C_{61}e_x + C_{62}e_y + C_{63}e_z + 2C_{64}e_{xy} + 2C_{65}e_{yz} + 2C_{66}e_{zx}
\end{aligned}
\tag{4.2.1}
$$

where the coefficients C_{ij} are material parameters and the factors of 2 arise because of the symmetry of the strain. Note that this relation could also be expressed by writing the strains as a linear function of the stress components. These relations can be cast into a matrix format as

$$
\begin{bmatrix} \sigma_x \\ \sigma_y \\ \sigma_z \\ \tau_{xy} \\ \tau_{yz} \\ \tau_{zx} \end{bmatrix}
=
\begin{bmatrix}
C_{11} & C_{12} & \cdot & \cdot & \cdot & C_{16} \\
C_{21} & \cdot & & \cdot & \cdot & \cdot \\
\cdot & \cdot & & \cdot & \cdot & \cdot \\
\cdot & \cdot & & \cdot & \cdot & \cdot \\
\cdot & \cdot & & \cdot & \cdot & \cdot \\
C_{61} & \cdot & & \cdot & \cdot & C_{66}
\end{bmatrix}
\begin{bmatrix} e_x \\ e_y \\ e_z \\ 2e_{xy} \\ 2e_{yz} \\ 2e_{zx} \end{bmatrix}
\tag{4.2.2}
$$

Relations (4.2.1) can also be expressed in standard tensor notation by writing

$$
\sigma_{ij} = C_{ijkl}e_{kl}
\tag{4.2.3}
$$

where C_{ijkl} is a *fourth-order elasticity tensor* whose components include all the material parameters necessary to characterize the material. Based on the symmetry of the stress and strain tensors, the elasticity tensor must have the following properties (see Exercise 4.2)

$$
\begin{aligned}
C_{ijkl} &= C_{jikl} \\
C_{ijkl} &= C_{ijlk}
\end{aligned}
\tag{4.2.4}
$$

In general, the fourth-order tensor C_{ijkl} has 81 components. However, relations (4.2.4) reduce the number of independent components to 36, and this provides the required match with form (4.2.1) or (4.2.2). Later in Chapter 6 we introduce the concept of strain energy, and this leads to the relation

$C_{ijkl} = C_{klij}$ or equivalently $C_{ij} = C_{ji}$, which provides further reduction to 21 independent elastic components. The components of C_{ijkl} or equivalently C_{ij} are called *elastic moduli* and have units of stress (force/area). In order to continue further, we must address the issues of material homogeneity and isotropy.

If the material is homogeneous, the elastic behavior does not vary spatially, and thus all elastic moduli are constant. For this case, the elasticity formulation is straightforward, leading to the development of many analytical solutions to problems of engineering interest. A homogeneous assumption is an appropriate model for most structural applications, and thus we primarily choose this particular case for subsequent formulation and problem solution. However, there are some important nonhomogeneous applications that warrant further discussion.

Studies in geomechanics have found that the material behavior of soil and rock commonly depends on distance below the earth's surface. In order to simulate particular geomechanics problems, researchers have used nonhomogeneous elastic models applied to semi-infinite domains. Typical applications have involved modeling the response of a semi-infinite soil mass under surface or subsurface loadings with variation in elastic moduli with depth (Poulos and Davis, 1974). Another more recent application involves the behavior of *functionally graded materials* (FGMs) (see Erdogan, 1995; Parameswaran and Shukla, 1999, 2002). FGMs are relatively a new class of engineered materials developed with spatially varying properties to suit particular applications. The graded composition of such materials is commonly established and controlled using powder metallurgy, chemical vapor deposition, or centrifugal casting. Typical analytical studies of these materials have assumed linear, exponential, and power-law variation in elastic moduli of the form

$$C_{ij}(x) = C_{ij}^o(1 + ax)$$
$$C_{ij}(x) = C_{ij}^o e^{ax} \qquad\qquad (4.2.5)$$
$$C_{ij}(x) = C_{ij}^o x^a$$

where C_{ij}^o and a are prescribed constants and x is the spatial coordinate. Further details on the formulation and solution of nonhomogeneous elasticity problems are given in Chapter 14.

Similar to homogeneity, another fundamental material property is *isotropy*. This property has to do with differences in material moduli with respect to orientation. For example, many materials including crystalline minerals, wood, and fiber-reinforced composites have different elastic moduli in different directions. Materials such as these are said to be *anisotropic*. Note that for most real anisotropic materials there exist particular directions where the properties are the same. These directions indicate *material symmetries*. However, for many engineering materials (most structural metals and many plastics), the orientation of crystalline and grain microstructure is distributed randomly so that macroscopic elastic properties are found to be essentially the same in all directions. Such materials with complete symmetry are called *isotropic*. As expected, an anisotropic model complicates the formulation and solution of problems. We therefore postpone development of such solutions until Chapter 11 and continue our current development under the assumption of isotropic material behavior. The interested reader can explore Sections 11.1 and 11.2 to see the complete development from anisotropy to isotropy in detail.

The tensorial form (4.2.3) provides a convenient way to establish the desired isotropic stress–strain relations. If we assume isotropic behavior, the elasticity tensor must be the same under all rotations of the coordinate system. Using the basic transformation properties from relation (1.5.1)$_5$, the fourth-order elasticity tensor must satisfy

$$C_{ijkl} = Q_{im}Q_{jn}Q_{kp}Q_{lq}C_{mnpq}$$

It can be shown (Sadd, 2019) that the most general form that satisfies this isotropy condition is given by

$$C_{ijkl} = \alpha\delta_{ij}\delta_{kl} + \beta\delta_{ik}\delta_{jl} + \gamma\delta_{il}\delta_{jk} \tag{4.2.6}$$

where α, β, and γ are arbitrary constants. Verification of the isotropic property of form (4.2.6) was given as Exercise 1.9. Using the general form (4.2.6) in the stress–strain relation (4.2.3) gives

$$\sigma_{ij} = \lambda e_{kk}\delta_{ij} + 2\mu e_{ij} \tag{4.2.7}$$

where we have relabeled particular constants using λ and μ. The elastic constant λ is called *Lamé's constant*, and μ is referred to as the *shear modulus* or *modulus of rigidity*. Some texts use the notation G for the shear modulus. Eq. (4.2.7) can be written out in individual scalar equations as

$$
\begin{aligned}
\sigma_x &= \lambda(e_x + e_y + e_z) + 2\mu e_x \\
\sigma_y &= \lambda(e_x + e_y + e_z) + 2\mu e_y \\
\sigma_z &= \lambda(e_x + e_y + e_z) + 2\mu e_z \\
\tau_{xy} &= 2\mu e_{xy} \\
\tau_{yz} &= 2\mu e_{yz} \\
\tau_{zx} &= 2\mu e_{zx}
\end{aligned}
\tag{4.2.8}
$$

Relations (4.2.7) or (4.2.8) are called the *generalized Hooke's law for linear isotropic elastic solids*. They are named after Robert Hooke, who in 1678 first proposed that the deformation of an elastic structure is proportional to the applied force. Notice the significant simplicity of the isotropic form when compared to the general stress–strain law originally given by (4.2.1). It should be noted that only *two independent elastic constants* are needed to describe the behavior of isotropic materials. As shown in Chapter 11, additional numbers of elastic moduli are needed in the corresponding relations for anisotropic materials.

Stress–strain relations (4.2.7) or (4.2.8) may be inverted to express the strain in terms of the stress. In order to do this it is convenient to use the index notation form (4.2.7) and set the two free indices the same (contraction process) to get

$$\sigma_{kk} = (3\lambda + 2\mu)e_{kk} \tag{4.2.9}$$

This relation can be solved for e_{kk} and substituted back into (4.2.7) to get

$$e_{ij} = \frac{1}{2\mu}\left(\sigma_{ij} - \frac{\lambda}{3\lambda + 2\mu}\sigma_{kk}\delta_{ij}\right)$$

which is more commonly written as

$$e_{ij} = \frac{1+\nu}{E}\sigma_{ij} - \frac{\nu}{E}\sigma_{kk}\delta_{ij} \tag{4.2.10}$$

where $E = \mu(3\lambda + 2\mu)/(\lambda + \mu)$ and is called *the modulus of elasticity* or *Young's modulus*, and $\nu = \lambda/[2(\lambda + \mu)]$ is referred to as *Poisson's ratio*. The index notation relation (4.2.10) may be written out in component (scalar) form, giving the six equations

$$e_x = \frac{1}{E}[\sigma_x - \nu(\sigma_y + \sigma_z)]$$

$$e_y = \frac{1}{E}[\sigma_y - \nu(\sigma_z + \sigma_x)]$$

$$e_z = \frac{1}{E}[\sigma_z - \nu(\sigma_x + \sigma_y)]$$

$$e_{xy} = \frac{1+\nu}{E}\tau_{xy} = \frac{1}{2\mu}\tau_{xy} \qquad (4.2.11)$$

$$e_{yz} = \frac{1+\nu}{E}\tau_{yz} = \frac{1}{2\mu}\tau_{yz}$$

$$e_{zx} = \frac{1+\nu}{E}\tau_{zx} = \frac{1}{2\mu}\tau_{zx}$$

Constitutive form (4.2.10) or (4.2.11) again illustrates that *only two elastic constants are needed to formulate Hooke's law for isotropic materials.* By using any of the isotropic forms of Hooke's law, it can be shown that the principal axes of stress coincide with the principal axes of strain (see Exercise 4.7). This result also holds for some but not all anisotropic materials.

4.3 Physical meaning of elastic moduli

For the isotropic case, the previously defined elastic moduli have simple physical meaning. These can be determined through investigation of particular states of stress commonly realized in laboratory materials testing as shown in Fig. 4.2.

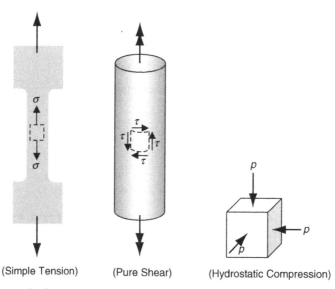

(Simple Tension)   (Pure Shear)   (Hydrostatic Compression)

FIG. 4.2 Special characterization states of stress.

4.3.1 Simple tension

Consider the simple tension test as discussed previously with a sample subjected to tension in the x direction (see Fig. 4.2). The state of stress is closely represented by the one-dimensional field

$$\sigma_{ij} = \begin{bmatrix} \sigma & 0 & 0 \\ 0 & 0 & 0 \\ 0 & 0 & 0 \end{bmatrix}$$

Using this in relation (4.2.10) gives a corresponding strain field

$$e_{ij} = \begin{bmatrix} \dfrac{\sigma}{E} & 0 & 0 \\ 0 & -\dfrac{\nu}{E}\sigma & 0 \\ 0 & 0 & -\dfrac{\nu}{E}\sigma \end{bmatrix}$$

Therefore, $E = \sigma/e_x$ and is simply the slope of the stress–strain curve, while $\nu = -e_y/e_x = -e_z/e_x$ is minus the ratio of the transverse strain to the axial strain. Standard measurement systems can easily collect axial stress and transverse and axial strain data, and thus through this one type of test both elastic constants can be determined for materials of interest.

4.3.2 Pure shear

If a thin-walled cylinder is subjected to torsional loading (as shown in Fig. 4.2), the state of stress on the surface of the cylindrical sample is given by

$$\sigma_{ij} = \begin{bmatrix} 0 & \tau & 0 \\ \tau & 0 & 0 \\ 0 & 0 & 0 \end{bmatrix}$$

Again, by using Hooke's law, the corresponding strain field becomes

$$e_{ij} = \begin{bmatrix} 0 & \tau/2\mu & 0 \\ \tau/2\mu & 0 & 0 \\ 0 & 0 & 0 \end{bmatrix}$$

and thus the shear modulus is given by $\mu = \tau/2e_{xy} = \tau/\gamma_{xy}$, and this modulus is simply the slope of the shear stress–shear strain curve.

4.3.3 Hydrostatic compression (or tension)

The final example is associated with the uniform compression (or tension) loading of a cubical specimen, as previously shown in Fig. 4.2. This type of test would be realizable if the sample was placed in a high-pressure compression chamber. The state of stress for this case is given by

$$\sigma_{ij} = \begin{bmatrix} -p & 0 & 0 \\ 0 & -p & 0 \\ 0 & 0 & -p \end{bmatrix} = -p\delta_{ij}$$

This is an isotropic state of stress and the strains follow from Hooke's law

$$
e_{ij} = \begin{bmatrix} -\dfrac{1-2\nu}{E}p & 0 & 0 \\[2ex] 0 & -\dfrac{1-2\nu}{E}p & 0 \\[2ex] 0 & 0 & -\dfrac{1-2\nu}{E}p \end{bmatrix}
$$

The dilatation that represents the change in material volume (see Exercise 2.11) is thus given by $\vartheta = e_{kk} = -3(1-2\nu)p/E$, which can be written as

$$p = -k\vartheta \tag{4.3.1}$$

where $k = E/[3(1-2\nu)]$ is called the *bulk modulus of elasticity*. This additional elastic constant represents the ratio of pressure to the dilatation, which could be referred to as the volumetric stiffness of the material. Notice that as Poisson's ratio approaches 0.5, the bulk modulus becomes unbounded; the material does not undergo any volumetric deformation and hence is incompressible.

Our discussion of elastic moduli for isotropic materials has led to the definition of five constants λ, μ, E, ν, and k. However, keep in mind that only two of these are needed to characterize the material. Although we have developed a few relationships between various moduli, many other such relations can also be found. In fact, it can be shown that all five elastic constants are interrelated, and if any two are given, the remaining three can be determined by using simple formulae. Results of these relations are conveniently summarized in Table 4.1. This table should be marked for future reference, because it will prove to be useful for calculations throughout the text.

Typical nominal values of elastic constants for particular engineering materials are given in Table 4.2. These moduli represent average values, and some variation will occur for specific materials. Further information and restrictions on elastic moduli require strain energy concepts, which are developed in Chapter 6.

Before concluding this section, we wish to discuss the forms of Hooke's law in curvilinear coordinates. Previous chapters have mentioned that cylindrical and spherical coordinates (see Figs. 1.5 and 1.6) are used in many applications for problem solution. Figs. 3.10 and 3.11 defined the stress components in each curvilinear system. In regard to these figures, it follows that the orthogonal curvilinear coordinate directions can be obtained from a base Cartesian system through a simple rotation of the coordinate frame. For isotropic materials, the elasticity tensor C_{ijkl} is the same in all coordinate frames, and thus *the structure of Hooke's law remains the same in any orthogonal curvilinear system.* Therefore, form (4.2.8) can be expressed in cylindrical and spherical coordinates as

$$
\begin{aligned}
\sigma_r &= \lambda(e_r + e_\theta + e_z) + 2\mu e_r & \sigma_R &= \lambda(e_R + e_\phi + e_\theta) + 2\mu e_R \\
\sigma_\theta &= \lambda(e_r + e_\theta + e_z) + 2\mu e_\theta & \sigma_\phi &= \lambda(e_R + e_\phi + e_\theta) + 2\mu e_\phi \\
\sigma_z &= \lambda(e_r + e_\theta + e_z) + 2\mu e_z & \sigma_\theta &= \lambda(e_R + e_\phi + e_\theta) + 2\mu e_\theta \\
\tau_{r\theta} &= 2\mu e_{r\theta} & \tau_{R\phi} &= 2\mu e_{R\phi} \\
\tau_{\theta z} &= 2\mu e_{\theta z} & \tau_{\phi\theta} &= 2\mu e_{\phi\theta} \\
\tau_{zr} &= 2\mu e_{zr} & \tau_{\theta R} &= 2\mu e_{\theta R}
\end{aligned}
\tag{4.3.2}
$$

Table 4.1 Relations among elastic constants.

	E	ν	k	μ	λ
E,ν	E	ν	$\dfrac{E}{3(1-2\nu)}$	$\dfrac{E}{2(1+\nu)}$	$\dfrac{E\nu}{(1+\nu)(1-2\nu)}$
E,k	E	$\dfrac{3k-E}{6k}$	k	$\dfrac{3kE}{9k-E}$	$\dfrac{3k(3k-E)}{9k-E}$
E,μ	E	$\dfrac{E-2\mu}{2\mu}$	$\dfrac{\mu E}{3(3\mu-E)}$	μ	$\dfrac{\mu(E-2\mu)}{3\mu-E}$
E,λ	E	$\dfrac{2\lambda}{E+\lambda+R}$	$\dfrac{E+3\lambda+R}{6}$	$\dfrac{E-3\lambda+R}{4}$	λ
ν,k	$3k(1-2\nu)$	ν	k	$\dfrac{3k(1-2\nu)}{2(1+\nu)}$	$\dfrac{3k\nu}{1+\nu}$
ν,μ	$2\mu(1+\nu)$	ν	$\dfrac{2\mu(1+\nu)}{3(1-2\nu)}$	μ	$\dfrac{2\mu\nu}{1-2\nu}$
ν,λ	$\dfrac{\lambda(1+\nu)(1-2\nu)}{\nu}$	ν	$\dfrac{\lambda(1+\nu)}{3\nu}$	$\dfrac{\lambda(1-2\nu)}{2\nu}$	λ
k,μ	$\dfrac{9k\mu}{3k+\mu}$	$\dfrac{3k-2\mu}{6k+2\mu}$	k	μ	$k-\dfrac{2}{3}\mu$
k,λ	$\dfrac{9k(k-\lambda)}{3k-\lambda}$	$\dfrac{\lambda}{3k-\lambda}$	k	$\dfrac{3}{2}(k-\lambda)$	λ
μ,λ	$\dfrac{\mu(3\lambda+2\mu)}{\lambda+\mu}$	$\dfrac{\lambda}{2(\lambda+\mu)}$	$\dfrac{3\lambda+2\mu}{3}$	μ	λ

$R=\sqrt{E^2+9\lambda^2+2E\lambda}$

Table 4.2 Typical values of elastic moduli for common engineering materials.

	E(GPa)	ν	μ(GPa)	λ(GPa)	k(GPa)	$\alpha(10^{-6}/^\circ\text{C})$
Aluminum	68.9	0.34	25.7	54.6	71.8	25.5
Concrete	27.6	0.20	11.5	7.7	15.3	11
Copper	89.6	0.34	33.4	71	93.3	18
Glass	68.9	0.25	27.6	27.6	45.9	8.8
Nylon	28.3	0.40	10.1	4.04	47.2	102
Rubber	0.0019	0.499	0.654×10^{-3}	0.326	0.326	200
Steel	207	0.29	80.2	111	164	13.5

The complete set of elasticity field equations in each of these coordinate systems is given in Appendix A.

4.4 Thermoelastic constitutive relations

It is well known that a temperature change in an unrestrained elastic solid produces deformation. Thus, a general strain field results from both mechanical and thermal effects. Within the context of linear small deformation theory, the total strain can be decomposed into the sum of mechanical and thermal components as

$$e_{ij} = e_{ij}^{(M)} + e_{ij}^{(T)} \tag{4.4.1}$$

If T_o is taken as the reference temperature and T as an arbitrary temperature, the thermal strains in an unrestrained solid can be written in the linear constitutive form

$$e_{ij}^{(T)} = \alpha_{ij}(T - T_o) \tag{4.4.2}$$

where α_{ij} is the *coefficient of thermal expansion tensor*. Notice that it is the temperature difference that creates thermal strain. If the material is taken as isotropic, then α_{ij} must be an isotropic second-order tensor, and (4.4.2) simplifies to

$$e_{ij}^{(T)} = \alpha(T - T_o)\delta_{ij} \tag{4.4.3}$$

where α is a material constant called the *coefficient of thermal expansion*. Table 4.2 provides typical values of this constant for some common materials. Notice that for isotropic materials, no shear strains are created by temperature change. By using (4.4.1), this result can be combined with the mechanical relation (4.2.10) to give

$$e_{ij} = \frac{1+\nu}{E}\sigma_{ij} - \frac{\nu}{E}\sigma_{kk}\delta_{ij} + \alpha(T - T_o)\delta_{ij} \tag{4.4.4}$$

The corresponding results for the stress in terms of strain can be written as

$$\sigma_{ij} = C_{ijkl}e_{kl} - \beta_{ij}(T - T_o) \tag{4.4.5}$$

where β_{ij} is a second-order tensor containing thermoelastic moduli. This result is sometimes referred to as the *Duhamel–Neumann thermoelastic constitutive law*. The isotropic case can be found by simply inverting relation (4.4.4) to get

$$\sigma_{ij} = \lambda e_{kk}\delta_{ij} + 2\mu e_{ij} - (3\lambda + 2\mu)\alpha(T - T_o)\delta_{ij} \tag{4.4.6}$$

Thermoelastic solutions are developed in Chapter 12, and the current study will now continue under the assumption of isothermal conditions.

Having developed the necessary six constitutive relations, the elasticity field equation system is now complete with 15 equations (strain–displacement, equilibrium, Hooke's law) for 15 unknowns (displacements, strains, stresses). Obviously, further simplification is necessary in order to solve specific problems of engineering interest, and these processes are the subject of the next chapter.

References

Erdogan F: Fracture mechanics of functionally graded materials, *Compos Eng* 5:753–770, 1995.
Parameswaran V, Shukla A: Crack-tip stress fields for dynamic fracture in functionally gradient materials, *Mech Mater* 31:579–596, 1999.
Parameswaran V, Shukla A: Asymptotic stress fields for stationary cracks along the gradient in functionally graded materials, *J Appl Mech* 69:240–243, 2002.
Poulos HG, Davis EH: *Elastic solutions for soil and rock mechanics*, New York, 1974, John Wiley.
Sadd MH: *Continuum mechanics modeling of material behavior*, 2019, Elsevier.

Exercises

4.1 Show that the components of the C_{ij} matrix in Eq. (4.2.2) are related to the components of C_{ijkl} by the relation

$$C_{ij} = \begin{bmatrix} C_{1111} & C_{1122} & C_{1133} & C_{1112} & C_{1123} & C_{1131} \\ C_{2211} & C_{2222} & C_{2233} & C_{2212} & C_{2223} & C_{2231} \\ C_{3311} & C_{3322} & C_{3333} & C_{3312} & C_{3323} & C_{3331} \\ C_{1211} & C_{1222} & C_{1233} & C_{1212} & C_{1223} & C_{1231} \\ C_{2311} & C_{2322} & C_{2333} & C_{2312} & C_{2323} & C_{2331} \\ C_{3111} & C_{3122} & C_{3133} & C_{3112} & C_{3123} & C_{3131} \end{bmatrix}$$

4.2 Explicitly justify the symmetry relations (4.2.4). Note that the first relation follows directly from the symmetry of the stress, while the second condition requires a simple expansion into the form $e_{kl} = \frac{1}{2}(e_{kl} + e_{lk})$ to arrive at the required conclusion.

4.3 Substituting the general isotropic fourth-order form (4.2.6) into (4.2.3), explicitly develop the stress–strain relation (4.2.7).

4.4 For isotropic materials, show that the fourth-order elasticity tensor can be expressed in the following forms

$$C_{ijkl} = \lambda \delta_{ij}\delta_{kl} + \mu(\delta_{il}\delta_{jk} + \delta_{ik}\delta_{jl})$$

$$C_{ijkl} = \mu(\delta_{il}\delta_{jk} + \delta_{ik}\delta_{jl}) + \left(k - \frac{2}{3}\mu\right)\delta_{ij}\delta_{kl}$$

$$C_{ijkl} = \frac{Ev}{(1+v)(1-2v)}\delta_{ij}\delta_{kl} + \frac{E}{2(1+v)}(\delta_{il}\delta_{jk} + \delta_{ik}\delta_{jl})$$

4.5 Following the steps outlined in the text, invert the form of Hooke's law given by (4.2.7) and develop form (4.2.10). Explicitly show that $E = \mu(3\lambda + 2\mu)/(\lambda + \mu)$ and $v = \lambda/[2(\lambda + \mu)]$.

4.6 Using the results of Exercise 4.5, show that $\mu = E/[2(1+v)]$ and $\lambda = Ev/[(1+v)(1-2v)]$.

4.7 For isotropic materials show that the principal axes of strain coincide with the principal axes of stress. Further, show that the principal stresses can be expressed in terms of the principal strains as $\sigma_i = 2\mu e_i + \lambda e_{kk}$.

4.8 A rosette strain gage (see Exercise 2.7) is mounted on the surface of a stress-free elastic solid at point O as shown in the following figure. The three gage readings give surface extensional strains $e_a = 300 \times 10^{-6}$, $e_b = 400 \times 10^{-6}$, $e_c = 100 \times 10^{-6}$. Assuming that the material is steel with nominal properties given by Table 4.2, determine all stress components at O for the given coordinate system.

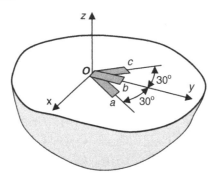

4.9 The displacements in an elastic material are given by

$$u = -\frac{M(1 - \nu^2)}{EI}xy, \quad v = \frac{M(1 + \nu)\nu}{2EI}y^2 + \frac{M(1 + \nu^2)}{2EI}\left(x^2 - \frac{l^2}{4}\right), \quad w = 0$$

where M, E, I, and l are constant parameters. Determine the corresponding strain and stress fields and show that this problem represents the pure bending of a rectangular beam in the x,y plane.

4.10 If the elastic constants E, k, and μ are required to be positive, show that Poisson's ratio must satisfy the inequality $-1 < \nu < \frac{1}{2}$. For most real materials it has been found that $0 < \nu < \frac{1}{2}$. Show that this more restrictive inequality in this problem implies that $\lambda > 0$.

4.11 Under the condition that E is positive and bounded, determine the elastic moduli λ, μ, and k for the special cases of Poisson's ratio: $\nu = 0, \frac{1}{4}, \frac{1}{2}$. Discuss the special circumstances for the case with $\nu = \frac{1}{2}$.

4.12* Further explore the behaviors of the elastic moduli in Exercise 4.11 by making non-dimensional plots of λ/E, μ/E and k/E for $0 \leq \nu \leq 0.5$.

4.13 Consider the case of incompressible elastic materials. For such materials, there will be a constraint on all deformations such that the change in volume must be zero, thus implying (see Exercise 2.11) that $e_{kk} = 0$. First show that, under this constraint, Poisson's ratio will become $\frac{1}{2}$ and the bulk modulus and Lamé's constant will become unbounded. Next show that the usual form of Hooke's law $\sigma_{ij} = \lambda e_{kk}\delta_{ij} + 2\mu e_{ij}$ will now contain an indeterminate term. For such cases, Hooke's law is commonly rewritten in the form $\sigma_{ij} = -p\delta_{ij} + 2\mu e_{ij}$, where

p is referred to as the *hydrostatic pressure*, which cannot be determined directly from the strain field but is normally found by solving the boundary-value problem. Finally justify that $p = -\sigma_{kk}/3$.

4.14 Consider the three deformation cases of simple tension, pure shear, and hydrostatic compression as discussed in Section 4.3. Using the nominal values from Table 4.2, calculate the resulting strains in each of these cases for:

> (a) Aluminum: with loadings $(\sigma = 150\ MPa,\ \tau = 75\ MPa,\ p = 500\ MPa)$
> (b) Steel: with loadings $(\sigma = 300\ MPa,\ \tau = 150\ MPa,\ p = 500\ MPa)$
> (c) Rubber: with loadings $(\sigma = 15\ MPa,\ \tau = 7\ MPa,\ p = 500\ MPa)$

Note that for aluminum and steel, these tensile and shear loadings are close to the yield values of the material.

4.15 Show that Hooke's law for an isotropic material may be expressed in terms of spherical and deviatoric tensors by the two relations

$$\tilde{\sigma}_{ij} = 3k\tilde{e}_{ij}, \quad \hat{\sigma}_{ij} = 2\mu\hat{e}_{ij}$$

4.16 A sample is subjected to a test under *plane stress* conditions (specified by $\sigma_z = \tau_{zx} = \tau_{zy} = 0$) using a special loading frame that maintains an in-plane loading constraint $\sigma_x = 2\sigma_y$. Determine the slope of the stress–strain response σ_x vs. e_x for this sample.

4.17* Using cylindrical coordinates, we wish to determine and compare the uniaxial stress–strain response of both an unconfined and confined isotropic homogeneous elastic cylindrical sample as shown. For both cases we will assume that the shear stresses and strains will vanish. For the unconfined sample, the stresses $\sigma_r = \sigma_\theta = 0$, while for the confined sample, the strains $e_r = e_\theta = 0$. Using Hooke's law in cylindrical coordinates (A.8), show that the uniaxial response for the unconfined case is given by the expected relation $\sigma_z = Ee_z$, while the corresponding relation for the confined situation is $\sigma_z = E^*e_z$, where $E^* = [(1 - v)E/(1 + v)(1 - 2v)]$. Finally make a plot of E^*/E versus v over the range $0 \leq v \leq 0.5$ and discuss the results.

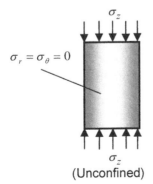
$\sigma_r = \sigma_\theta = 0$

σ_z

σ_z
(Unconfined)

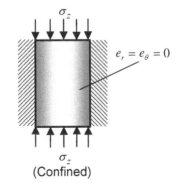
$e_r = e_\theta = 0$

σ_z

σ_z
(Confined)

4.18 A rectangular steel plate (thickness 4 mm) is subjected to a uniform biaxial stress field as shown in the following figure. Assuming all fields are uniform, determine changes in the dimensions of the plate under this loading.

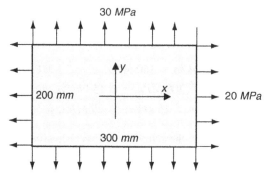

4.19 Redo Exercise 4.17 for the case where the vertical loading is 50 MPa in tension and the horizontal loading is 50 MPa in compression.

4.20 Consider the one-dimensional thermoelastic problem of a uniform bar constrained in the axial x direction but allowed to expand freely in the y and z directions, as shown in the following figure. Taking the reference temperature to be zero, show that the only nonzero stress and strain components are given by

$$\sigma_x = -E\alpha T$$

$$e_y = e_z = \alpha(1 + \nu)T$$

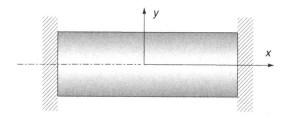

4.21 Verify that Hooke's law for isotropic thermoelastic materials can be expressed in the form

$$\sigma_x = \frac{E}{(1+\nu)(1-2\nu)}[(1-\nu)e_x + \nu(e_y + e_z)] - \frac{E}{1-2\nu}\alpha(T - T_o)$$

$$\sigma_y = \frac{E}{(1+\nu)(1-2\nu)}[(1-\nu)e_y + \nu(e_z + e_x)] - \frac{E}{1-2\nu}\alpha(T - T_o)$$

$$\sigma_z = \frac{E}{(1+\nu)(1-2\nu)}[(1-\nu)e_z + \nu(e_x + e_y)] - \frac{E}{1-2\nu}\alpha(T - T_o)$$

$$\tau_{xy} = \frac{E}{1+\nu}e_{xy}, \quad \tau_{yz} = \frac{E}{1+\nu}e_{yz}, \quad \tau_{zx} = \frac{E}{1+\nu}e_{zx}$$

Formulation and solution strategies

The previous chapters developed the basic field equations of elasticity theory. These results comprise a system of differential and algebraic relations among the stresses, strains, and displacements that express particular physics at all points *within* the body under investigation. In this chapter we now wish to complete the general formulation by first developing *boundary conditions* appropriate for use with the field equations. These conditions specify the physics that occur *on the boundary of the body*, and generally provide the loading inputs that physically create the interior stress, strain, and displacement fields. Although the field equations are the same for all problems, boundary conditions are different for each problem. Therefore, proper development of boundary conditions is essential for problem solution, and thus it is important to acquire a good understanding of such development procedures. Combining field equations with boundary conditions establishes the fundamental boundary-value problems of the theory. This eventually leads us into two different formulations: one in terms of displacements and the other in terms of stresses. Because boundary-value problems are difficult to solve, many different strategies have been developed to aid in problem solution. We review in a general way several of these strategies, and later chapters incorporate many of them into the solution of specific problems.

5.1 Review of field equations

We list here the basic field equations for linear isotropic elasticity before beginning our discussion on boundary conditions. Appendix A includes a more comprehensive listing of all field equations in Cartesian, cylindrical, and spherical coordinate systems. Because of its ease of use and compact properties, our formulation uses index notation.

Strain—displacement relations

$$e_{ij} = \frac{1}{2}\left(u_{i,j} + u_{j,i}\right) \tag{5.1.1}$$

Compatibility relations

$$e_{ij,kl} + e_{kl,ij} - e_{ik,jl} - e_{jl,ik} = 0 \tag{5.1.2}$$

Equilibrium equations

$$\sigma_{ij,j} + F_i = 0 \tag{5.1.3}$$

Elasticity. https://doi.org/10.1016/B978-0-12-815987-3.00005-0

Elastic constitutive law (Hooke's law)

$$\sigma_{ij} = \lambda e_{kk}\delta_{ij} + 2\mu e_{ij}$$

$$e_{ij} = \frac{1+\nu}{E}\sigma_{ij} - \frac{\nu}{E}\sigma_{kk}\delta_{ij}$$

(5.1.4)

As mentioned in Section 2.6, the compatibility relations ensure that the displacements are continuous and single-valued, and are necessary only when the strains are arbitrarily specified. If, however, the displacements are included in the problem formulation, the solution normally generates single-valued displacements and strain compatibility is automatically satisfied. Thus, in discussing the general system of equations of elasticity, the compatibility relations (5.1.2) are normally set aside, to be used only with the stress formulation that we discuss shortly. Therefore, the general system of elasticity field equations refers to the 15 relations (5.1.1), (5.1.3), and (5.1.4). It is convenient to define this entire system using a generalized operator notation as

$$\Im\{u_i, e_{ij}, \sigma_{ij}; \ \lambda, \mu, F_i\} = 0$$

(5.1.5)

This system involves 15 unknowns including three displacements u_i, six strains e_{ij}, and six stresses σ_{ij} to be determined. The terms after the semicolon indicate that the system is also dependent on two elastic material constants (for isotropic materials) and on the body force density, and these are to be given a priori with the problem formulation. It is reassuring that the number of equations matches the number of unknowns to be determined. However, for a general three-dimensional problem, this general system of equations is of such complexity that solutions using analytical methods are normally impossible and further simplification is required to solve problems of interest. Before proceeding with the development of such simplifications, it is useful to discuss typical boundary conditions connected with the elasticity model, and this leads us to the classification of the fundamental problems.

5.2 Boundary conditions and fundamental problem classifications

Similar to other field problems in engineering science (e.g., fluid mechanics, heat conduction, diffusion, electromagnetics), the solution of system (5.1.5) requires appropriate boundary conditions on the body under study. The common types of boundary conditions for elasticity applications normally include specification of how the body is being *supported* or *loaded*. This concept is mathematically formulated by specifying either the *displacements* or *tractions* at boundary points. Fig. 5.1 illustrates this general idea for three typical cases including tractions, displacements, and a *mixed case* for which tractions are specified on boundary S_t and displacements are given on the remaining portion S_u such that the total boundary is given by $S = S_t + S_u$.

Another type of mixed boundary condition can also occur. Although it is generally not possible to specify completely both the displacements and tractions at the *same* boundary point, it is possible to prescribe *part* of the displacement and *part* of the traction. Typically, this type of mixed condition involves the specification of a traction and displacement in two different orthogonal directions. A common example of this situation is shown in Fig. 5.2 for a case involving a surface of problem symmetry where the condition is one of a *rigid-smooth boundary* with zero normal displacement and zero tangential traction. Notice that in this example the body under study was subdivided along the symmetry line, thus creating a new boundary surface and resulting in a smaller region to analyze. Minimizing the size of the domain under study is commonly useful in computational modeling.

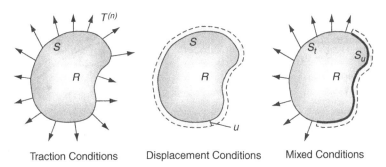

Traction Conditions   Displacement Conditions   Mixed Conditions

FIG. 5.1 Typical boundary conditions.

However, determining expected symmetry properties in the solution can often be useful to simplify analytical analyses as well.

Because boundary conditions play a very essential role in properly formulating and solving elasticity problems, it is important to acquire a clear understanding of their specification and use. Improper specification results in either no solution or a solution to a different problem than what was originally sought. Boundary conditions are normally specified using the coordinate system describing the problem, and thus particular components of the displacements and tractions are set equal to pre-scribed values. For displacement-type conditions, such a specification is straightforward, and a common example includes *fixed boundaries* where the displacements are to be zero. For traction boundary conditions, the specification can be a bit more complex.

Fig. 5.3 illustrates particular cases in which the boundaries coincide with Cartesian or polar co-ordinate surfaces. By using results from Section 3.2, the traction specification can be reduced to a stress specification. For the Cartesian example in which $y = $ constant, the normal traction becomes simply the stress component σ_y, while the tangential traction reduces to τ_{xy}. For this case, σ_x exists only

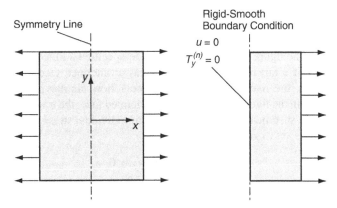

FIG. 5.2 Line of symmetry boundary condition.

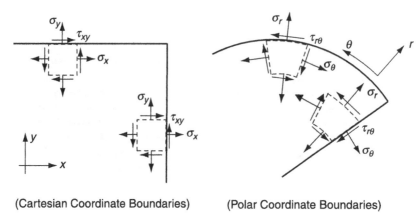

(Cartesian Coordinate Boundaries)   (Polar Coordinate Boundaries)

FIG. 5.3 Boundary stress components on coordinate surfaces.

inside the region; thus, this component of stress cannot be specified on the boundary surface $y = $ constant. A similar situation exists on the vertical boundary $x = $ constant, where the normal traction is now σ_x, the tangential traction is τ_{xy}, and the stress component σ_y exists inside the domain. Similar arguments can be made for polar coordinate boundary surfaces as shown. Drawing the appropriate element along the boundary as illustrated allows a clear visualization of the particular stress components that act *on* the surface in question. Such a sketch also allows determination of the positive directions of these boundary stresses, and this is useful to properly match with boundary loadings that might be prescribed. It is recommended that sketches similar to Fig. 5.3 be used to aid in the proper development of boundary conditions during problem formulation in later chapters.

Consider the pair of two-dimensional example problems with mixed conditions as shown in Fig. 5.4. For the rectangular plate problem, all four boundaries are coordinate surfaces, and this simplifies specification of particular boundary conditions. The fixed conditions on the left edge simply require that x and y displacement components vanish on $x = 0$, and this specification would not change even if this were not a coordinate surface. However, as per our previous discussion, the traction conditions on the other three boundaries simplify because they are coordinate surfaces. These simplifications are shown in the figure for each of the traction specified surfaces.

The second problem of a tapered cantilever beam has an inclined face that is not a coordinate surface. For this problem, the fixed end and top surface follow similar procedures as in the first example and are specified in the figure. However, on the inclined face, the traction is to be zero and this will not reduce to a simple specification of the vanishing of individual stress components. On this face each traction component is set to zero, giving the result

$$T_x^{(n)} = \sigma_x n_x + \tau_{xy} n_y = 0$$
$$T_y^{(n)} = \tau_{xy} n_x + \sigma_y n_y = 0$$

where n_x and n_y are the components of the unit normal vector to the inclined face. This is the more general type of specification, and it should be clearly noted that none of the individual stress

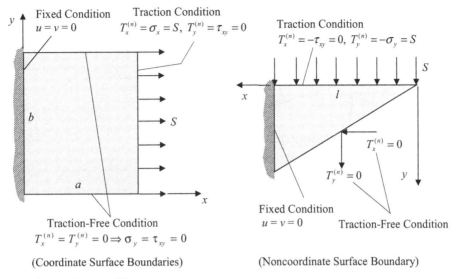

FIG. 5.4 Example boundary conditions.

components in the x,y system will vanish along this surface. It should also be pointed out for this problem that the unit normal vector components will be constants for all points on the inclined face. However, for curved boundaries the normal vector will change with surface position. Note that using strain–displacement relations in Hooke's law allows the stresses and hence tractions to be expressed in terms of displacement gradients. Thus, traction boundary conditions can actually be expressed in terms of displacements if so desired (see Exercise 5.4).

Another type of boundary condition formulation occurs for composite bodies that are composed of two or more pieces of different material with different elastic moduli. Examples of such situations are common and include many types of composite materials and structures as shown in Fig. 5.5. For such problems, the elasticity solution must be developed independently for each material body, thus

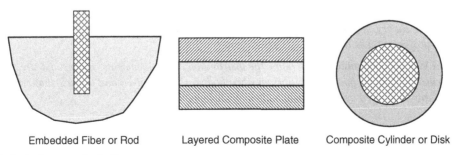

Embedded Fiber or Rod   Layered Composite Plate   Composite Cylinder or Disk

FIG. 5.5 Typical composite bodies.

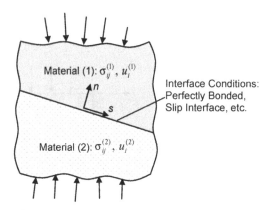

FIG. 5.6 Composite elastic continuum.

requiring specification of external boundary conditions and internal *interface conditions* that exist between each material phase (see Fig. 5.6). These interface conditions specify how the various composite pieces are joined together. Common simple examples include the *perfectly bonded interface* where both the displacements and tractions are continuous at the interface. With respect to the case shown in Fig. 5.6, a two-dimensional perfectly bonded model would specify the following interface conditions

$$\sigma_n^{(1)} = \sigma_n^{(2)}, \quad \tau_s^{(1)} = \tau_s^{(2)}$$
$$u_n^{(1)} = u_n^{(2)}, \quad u_s^{(1)} = u_s^{(2)}$$

Another common interface model is the *slip interface* that prescribes continuity of normal components of displacement and traction, vanishing of the tangential traction, and allows for a discontinuity in the tangential displacement. Other more complicated interface conditions have been proposed in the literature in an effort to model more sophisticated contact behavior between real material systems. Exercise 5.5 further explores interface formulations for some specific problems.

Although these examples provide some background on typical boundary conditions, many other types will be encountered throughout the text. Several exercises at the end of this chapter provide additional examples that will prove to be useful for students new to the elasticity formulation.

We are now in the position to formulate and classify the *three fundamental boundary-value problems in the theory of elasticity* that are related to solving the general system of field equations (5.1.5). Our presentation is limited to the static case.

Problem 1: Traction problem. *Determine the distribution of displacements, strains and stresses in the interior of an elastic body in equilibrium when body forces are given and the distribution of the tractions are prescribed over the surface of the body,*

$$T_i^{(n)}\left(x_i^{(s)}\right) = f_i\left(x_i^{(s)}\right) \tag{5.2.1}$$

where $x_i^{(s)}$ denotes boundary points and $f_i(x_i^{(s)})$ are the prescribed traction values.

Problem 2: Displacement problem. *Determine the distribution of displacements, strains and stresses in the interior of an elastic body in equilibrium when body forces are given and the distribution of the displacements are prescribed over the surface of the body,*

$$u_i\left(x_i^{(s)}\right) = g_i\left(x_i^{(s)}\right) \tag{5.2.2}$$

where $x_i^{(s)}$ denotes boundary points and $g_i(x_i^{(s)})$ are the prescribed displacement values.

Problem 3: Mixed problem. *Determine the distribution of displacements, strains and stresses in the interior of an elastic body in equilibrium when body forces are given and the distribution of the tractions are prescribed as per (5.2.1) over the surface S_t and the distribution of the displacements are prescribed as per (5.2.2) over the surface S_u of the body (see Fig. 5.1).*

As mentioned previously, the solution to any of these types of problems is formidable, and further reduction and simplification of (5.1.5) is required for the development of analytical solution methods. Based on the description of Problem 1 with only traction boundary conditions, it would appear to be desirable to express the fundamental system solely in terms of stress, that is, $\mathfrak{I}^{(t)}\{\sigma_{ij}; \lambda, \mu, F_i\}$, thereby reducing the number of unknowns in the system. Likewise for Problem 2, a displacement-only formulation of the form $\mathfrak{I}^{(u)}\{u_i; \lambda, \mu, F_i\}$ would appear to simplify the problem. We now pursue such specialized formulations and explicitly determine these reduced field equation systems.

5.3 Stress formulation

For the first fundamental problem in elasticity, the boundary conditions are to be given only in terms of the tractions or stress components. In order to develop solution methods for this case, it is very helpful to reformulate the general system (5.1.5) by eliminating the displacements and strains and thereby casting a new system solely in terms of the stresses. We now develop this reformulated system. By eliminating the displacements, we must now include the compatibility equations in the fundamental system of field equations. We start by using Hooke's law (5.1.4)$_2$ and eliminating the strains in the compatibility relations (5.1.2) to get

$$\sigma_{ij,kk} + \sigma_{kk,ij} - \sigma_{ik,jk} - \sigma_{jk,ik} =$$
$$\frac{\nu}{1+\nu}\left(\sigma_{mm,kk}\delta_{ij} + \sigma_{mm,ij}\delta_{kk} - \sigma_{mm,jk}\delta_{ik} - \sigma_{mm,ik}\delta_{jk}\right) \tag{5.3.1}$$

where we have used the arguments of Section 2.6, that the six meaningful compatibility relations are found by setting $k = l$ in (5.1.2). Although equations (5.3.1) represent the compatibility in terms of stress, a more useful result is found by incorporating the equilibrium equations into the system. Recall that from (5.1.3), $\sigma_{ij,j} = -F_i$, and also note that $\delta_{kk} = 3$. Substituting these results into (5.3.1) gives

$$\sigma_{ij,kk} + \frac{1}{1+\nu}\sigma_{kk,ij} = \frac{\nu}{1+\nu}\sigma_{mm,kk}\delta_{ij} - F_{i,j} - F_{j,i} \tag{5.3.2}$$

For the case $i = j$, relation (5.3.2) reduces to $\sigma_{ii,kk} = -\frac{1+\nu}{1-\nu}F_{i,i}$. Substituting this result back into (5.3.2) gives the desired relation

$$\sigma_{ij,kk} + \frac{1}{1+\nu}\sigma_{kk,ij} = -\frac{\nu}{1-\nu}\delta_{ij}F_{k,k} - F_{i,j} - F_{j,i} \tag{5.3.3}$$

This result gives the compatibility relations in terms of the stress; these relations are commonly called the *Beltrami—Michell compatibility equations*. For the case with no body forces, these relations can be expressed as the following six scalar equations

$$(1+v)\nabla^2\sigma_x + \frac{\partial^2}{\partial x^2}(\sigma_x + \sigma_y + \sigma_z) = 0$$

$$(1+v)\nabla^2\sigma_y + \frac{\partial^2}{\partial y^2}(\sigma_x + \sigma_y + \sigma_z) = 0$$

$$(1+v)\nabla^2\sigma_z + \frac{\partial^2}{\partial z^2}(\sigma_x + \sigma_y + \sigma_z) = 0$$

$$(1+v)\nabla^2\tau_{xy} + \frac{\partial^2}{\partial x\partial y}(\sigma_x + \sigma_y + \sigma_z) = 0$$

$$(1+v)\nabla^2\tau_{yz} + \frac{\partial^2}{\partial y\partial z}(\sigma_x + \sigma_y + \sigma_z) = 0$$

$$(1+v)\nabla^2\tau_{zx} + \frac{\partial^2}{\partial z\partial x}(\sigma_x + \sigma_y + \sigma_z) = 0$$

(5.3.4)

Recall that the six developed relations (5.3.3) or (5.3.4) actually represent three independent results as per our discussion in Section 2.6. Thus, combining these results with the three equilibrium equations (5.1.3) provides the necessary six relations to solve for the six unknown stress components for the general three-dimensional case.

This system constitutes the stress formulation for elasticity theory and is appropriate for use with traction boundary condition problems. Once the stresses have been determined, the strains may be found from Hooke's law (5.1.4)$_2$, and the displacements can be then be computed through integration of (5.1.1). As per our previous discussion in Section 2.2, such an integration process determines the displacements only up to an arbitrary rigid-body motion, and the displacements obtained are single-valued only if the region under study is simply connected.

The system of equations for the stress formulation is still rather complex, and analytical solutions are commonly determined for this case by making use of *stress functions*. This concept establishes a representation for the stresses that automatically satisfies the equilibrium equations. For the two-dimensional case, this concept represents the in-plane stresses in terms of a single function. The representation satisfies equilibrium, and the remaining compatibility equations yield a single partial differential equation (biharmonic equation) in terms of the stress function. Having reduced the system to a single equation we can employ many analytical methods to find solutions of interest. Further discussion on these techniques is presented in subsequent chapters.

5.4 Displacement formulation

We now wish to develop the reduced set of field equations solely in terms of the displacements. This system is referred to as the *displacement formulation* and is most useful when combined with displacement-only boundary conditions found in Problem 2 statement. This development is somewhat more straightforward than our previous stress formulation presentation. For this case, we wish to eliminate the strains and stresses from the fundamental system (5.1.5). This is easily accomplished by using the strain–displacement relations in Hooke's law to give

$$\sigma_{ij} = \lambda u_{k,k}\delta_{ij} + \mu\left(u_{i,j} + u_{j,i}\right) \tag{5.4.1}$$

which can be expressed as six scalar equations

$$\sigma_x = \lambda\left(\frac{\partial u}{\partial x} + \frac{\partial v}{\partial y} + \frac{\partial w}{\partial z}\right) + 2\mu\frac{\partial u}{\partial x}$$

$$\sigma_y = \lambda\left(\frac{\partial u}{\partial x} + \frac{\partial v}{\partial y} + \frac{\partial w}{\partial z}\right) + 2\mu\frac{\partial v}{\partial y}$$

$$\sigma_z = \lambda\left(\frac{\partial u}{\partial x} + \frac{\partial v}{\partial y} + \frac{\partial w}{\partial z}\right) + 2\mu\frac{\partial w}{\partial z} \tag{5.4.2}$$

$$\tau_{xy} = \mu\left(\frac{\partial u}{\partial y} + \frac{\partial v}{\partial x}\right), \quad \tau_{yz} = \mu\left(\frac{\partial v}{\partial z} + \frac{\partial w}{\partial y}\right), \quad \tau_{zx} = \mu\left(\frac{\partial w}{\partial x} + \frac{\partial u}{\partial z}\right)$$

Using these relations in the equilibrium equations gives the result

$$\mu u_{i,kk} + (\lambda + \mu)u_{k,ki} + F_i = 0 \tag{5.4.3}$$

which is the equilibrium equations in terms of the displacements and is referred to as *Navier's* or *Lamé's equations*. This system can be expressed in vector form as

$$\mu\nabla^2\boldsymbol{u} + (\lambda + \mu)\boldsymbol{\nabla}(\boldsymbol{\nabla}\cdot\boldsymbol{u}) + \boldsymbol{F} = 0 \tag{5.4.4}$$

or written out in terms of three scalar equations

$$\mu\nabla^2 u + (\lambda + \mu)\frac{\partial}{\partial x}\left(\frac{\partial u}{\partial x} + \frac{\partial v}{\partial y} + \frac{\partial w}{\partial z}\right) + F_x = 0$$

$$\mu\nabla^2 v + (\lambda + \mu)\frac{\partial}{\partial y}\left(\frac{\partial u}{\partial x} + \frac{\partial v}{\partial y} + \frac{\partial w}{\partial z}\right) + F_y = 0 \tag{5.4.5}$$

$$\mu\nabla^2 w + (\lambda + \mu)\frac{\partial}{\partial z}\left(\frac{\partial u}{\partial x} + \frac{\partial v}{\partial y} + \frac{\partial w}{\partial z}\right) + F_z = 0$$

where the Laplacian is given by $\nabla^2 = (\partial^2/\partial x^2) + (\partial^2/\partial y^2) + (\partial^2/\partial z^2)$. It should be noted that for the case with no body forces, Navier's equations are expressible in terms of a single elastic constant (Poisson's ratio)—see Exercise 5.10.

Navier's equations are the desired formulation for the displacement problem, and the system represents three equations for the three unknown displacement components. Similar to the stress formulation, this system is still difficult to solve, and additional mathematical techniques have been developed to further simplify these equations for problem solution. Common methods normally employ the use of *displacement potential functions*. It is shown in Chapter 13 that several such schemes can be developed that allow the displacement vector to be expressed in terms of particular potentials. These schemes generally simplify the problem by yielding uncoupled governing equations in terms of the displacement potentials. This then allows several analytical methods to be employed to solve problems of interest. Several of these techniques are discussed in later sections of the text.

To help acquire a general understanding of these results, a summary flow chart of the developed stress and displacement formulations is shown in Fig. 5.7. Note that for the stress formulation, the resulting system $\mathfrak{I}^{(t)}\{\sigma_{ij}; \lambda, \mu, F_i\}$ is actually dependent on only the single material constant Poisson's ratio, and thus it could be expressed as $\mathfrak{I}^{(t)}\{\sigma_{ij}; \nu, F_i\}$.

5.5 Principle of superposition

A very useful tool for the solution to many problems in engineering science is the principle of superposition. This technique applies to any problem that is governed by *linear equations*. Under the assumption of small deformations and linear elastic constitutive behavior, all elasticity field equations (see Fig. 5.7) are linear. Furthermore, the usual boundary conditions specified by relations (5.2.1) and (5.2.2) are also linear. Thus, under these conditions all governing equations are linear, and the superposition concept can be applied. It can be easily proved (see Chou and Pagano, 1967) that the general statement can be expressed as:

Principle of Superposition: For a given problem domain, if the state $\left\{\sigma_{ij}^{(1)}, e_{ij}^{(1)}, u_i^{(1)}\right\}$ *is a solution to the fundamental elasticity equations with prescribed body forces* $F_i^{(1)}$ *and surface tractions* $T_i^{(1)}$, *and the state* $\left\{\sigma_{ij}^{(2)}, e_{ij}^{(2)}, u_i^{(2)}\right\}$ *is a solution to the fundamental equations with prescribed body forces* $F_i^{(2)}$ *and surface tractions* $T_i^{(2)}$, *then the state* $\left\{\sigma_{ij}^{(1)} + \sigma_{ij}^{(2)}, e_{ij}^{(1)} + e_{ij}^{(2)}, u_i^{(1)} + u_i^{(2)}\right\}$ *will be a solution to the problem with body forces* $F_i^{(1)} + F_i^{(2)}$ *and surface tractions* $T_i^{(1)} + T_i^{(2)}$.

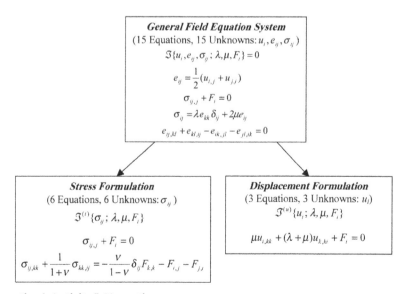

FIG. 5.7 Schematic of elasticity field equations.

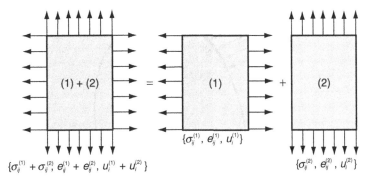

$$\{\sigma_{ij}^{(1)} + \sigma_{ij}^{(2)},\ e_{ij}^{(1)} + e_{ij}^{(2)},\ u_i^{(1)} + u_i^{(2)}\}$$

FIG. 5.8 Two-dimensional superposition example.

To see a more direct application of this principle, consider a simple two-dimensional case with no body forces as shown in Fig. 5.8. It can be observed that the solution to the more complicated biaxial loading case (1) + (2) is thus found by adding the two simpler problems. This is a common application of the superposition principle, and we make repeated use of it throughout the text. Thus, once the solutions to some simple problems are generated, we can combine these results to generate a solution to a more complicated case with similar geometry.

5.6 Saint–Venant's principle

Consider the set of three identical rectangular strips under compressive loadings as shown in Fig. 5.9. As indicated, the only difference between each problem is the loading. Because the total resultant load applied to each problem is identical (statically equivalent loadings), it is expected that the resulting stress, strain, and displacement fields near the bottom of each strip would be approximately the same.

This behavior can be generalized by considering an elastic solid with an arbitrary loading $T^{(n)}$ over a boundary portion S^*, as shown in Fig. 5.10. Based on experience from other field problems in

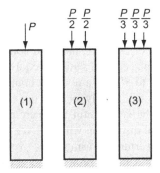

FIG. 5.9 Statically equivalent loadings.

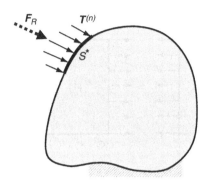

FIG. 5.10 Saint—Venant's principle.

engineering science, it seems logical that the particular boundary loading would produce detailed and characteristic effects only in the vicinity of S^*. In other words, we expect that at points far away from S^* the stresses generally depend more on the *resultant* F_R of the tractions rather than on the exact distribution. Thus, the *characteristic signature* of the generated stress, strain, and displacement fields from a given boundary loading tend to disappear as we move away from the boundary loading points. These concepts form the *principle of Saint—Venant*, which can be stated as follows:

> **Saint—Venant's Principle**: *The stress, strain, and displacement fields caused by two different statically equivalent force distributions on parts of the body far away from the loading points are approximately the same.*

This statement of the principle includes qualitative terms such as *far away* and *approximately the same*, and thus does not provide quantitative estimates of the differences between the two elastic fields in question. Quantitative results have been developed by von Mises (1945), Sternberg (1954), and Toupin (1965), while Horgan (1989) has presented a recent review of related work. Some of this work is summarized in Boresi and Chong (2010). Also Exercise 5.14 explores a specific problem involving Saint—Venant's principle.

If we restrict our solution to points away from the boundary loading, Saint—Venant's principle will allow us to change a given *pointwise* boundary condition to a simpler *statically equivalent resultant* statement and not affect the resulting solution. Some refer to this as substituting a *strong* boundary condition with a simpler *weak form*. For example, the fixed boundary condition in the tapered beam problem shown in Fig. 5.4 could be replaced by a statically equivalent resultant force system of a zero horizontal load, a vertical force equal to Sl, and a resultant counterclockwise moment of $Sl^2/2$. Such simplifications of boundary conditions will greatly increase our chances of finding analytical solutions to problems. This concept proves to be very useful for many beam and rod problems presented in Chapters 8 and 9, and some additional discussion will be given in the next section.

We close with a few comments of warning on using Saint—Venant's principle for structures that are not isotropic and homogeneous. Studies investigating elastic behaviors of anisotropic and inhomogeneous materials have found that for certain cases routine application of Saint—Venant's principle is not justified. For example, Horgan (1982) has shown that rates of decay of end effects in anisotropic materials were found to be much slower when compared to isotropic solids. Thus, for anisotropic materials the characteristic decay length over which end effects are significant can be several times larger than for isotropic solids. The example problems shown in Section 11.7 illustrate some of these behaviors. Similar findings have also been presented for nonhomogeneous materials; see, for example, Chan and Horgan (1998).

5.7 General solution strategies

Having completed our formulation and related solution principles, we now wish to examine some general solution strategies commonly used to solve elasticity problems. At this stage we categorize particular methods and outline only typical techniques that are commonly used. As we move further along in the text, many of these methods are developed in detail and are applied in specific problem solution. We first distinguish three general methods of solution called *direct*, *inverse*, and *semi-inverse*. Then we briefly discuss analytical, approximate, and numerical solution procedures.

5.7.1 Direct method

This method seeks to determine the solution by direct integration of the field equations (5.1.5) or equivalently the stress and/or displacement formulations given in Fig. 5.7. Boundary conditions are to be satisfied exactly. This method normally encounters significant mathematical difficulties, thus limiting its application to problems with simple geometry.

Example 5.1 Direct integration—stretching of prismatic bar under its own weight

As an example of a simple direct integration problem, consider the case of a uniform prismatic bar stretched by its own weight, as shown in Fig. 5.11. The body forces for this problem are $F_x = F_y = 0$, $F_z = -\rho g$, where ρ is the material mass density and g is the acceleration of gravity.

Assuming that on each cross-section we have uniform tension produced by the weight of the lower portion of the bar, the stress field would take the form

$$\sigma_x = \sigma_y = \tau_{xy} = \tau_{yz} = \tau_{zx} = 0 \qquad (5.7.1)$$

The equilibrium equations reduce to the simple result

$$\frac{\partial \sigma_z}{\partial z} = -F_z = \rho g \qquad (5.7.2)$$

This equation can be integrated directly, and applying the boundary condition $\sigma_z = 0$ at $z = 0$ gives the result $\sigma_z(z) = \rho g z$. Next, by using Hooke's law, the strains are easily calculated as

$$e_z = \frac{\rho g z}{E}, e_x = e_y = -\frac{\nu \rho g z}{E}$$
$$e_{xy} = e_{yz} = e_{xz} = 0 \qquad (5.7.3)$$

The displacements follow from integrating the strain–displacement relations (5.1.1), and for the case with boundary conditions of zero displacement and rotation at point A ($x = y = 0, z = l$), the final result is

$$u = -\frac{\nu \rho g x z}{E}, \quad v = -\frac{\nu \rho g y z}{E}$$
$$w = \frac{\rho g}{2E}\left[z^2 + \nu\left(x^2 + y^2\right) - l^2\right] \qquad (5.7.4)$$

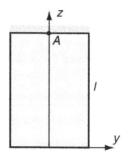

FIG. 5.11 Prismatic bar under self-weight.

5.7.2 Inverse method

For this technique, particular displacements or stresses are selected that satisfy the basic field equations. A search is then conducted to identify a specific problem that would be solved by this solution field. This amounts to determining appropriate problem geometry, boundary conditions, and body forces that would enable the solution to satisfy all conditions on the problem. Using this scheme it is sometimes difficult to construct solutions to a specific problem of practical interest.

Example 5.2 Inverse—pure beam bending

Consider the case of an elasticity problem under zero body forces with the following stress field

$$\sigma_x = Ay, \sigma_y = \sigma_z = \tau_{xy} = \tau_{yz} = \tau_{zx} = 0 \tag{5.7.5}$$

where A is a constant. It is easily shown that this simple linear stress field satisfies the equations of equilibrium and compatibility, and thus the field is a solution to an elasticity problem.

The question is, what problem would be solved by such a field? A common scheme to answer this question is to consider some trial domains and investigate the nature of the boundary conditions that would occur using the given stress field. Therefore, consider the two-dimensional rectangular domain shown in Fig. 5.12. Using the field (5.7.5), the tractions (stresses) on each boundary face give rise to zero loadings on the top and bottom and a linear distribution of normal stresses on the right and left sides as shown. Clearly, this type of boundary loading is related to a *pure bending problem*, whereby the loadings on the right and left sides produce no net force and only a pure bending moment.

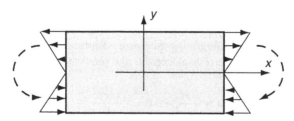

FIG. 5.12 Pure bending problem.

5.7.3 Semi-inverse method

In this scheme part of the displacement and/or stress field is specified, and the other remaining portion is determined from the fundamental field equations (normally using direct integration) and boundary conditions. It is often the case that constructing appropriate displacement and/or stress solution fields can be guided by approximate strength of materials theory. The usefulness of this approach is greatly enhanced by employing Saint–Venant's principle, whereby a complicated boundary condition can be replaced by a simpler statically equivalent distribution.

Example 5.3 Semi-inverse—torsion of prismatic bars

A simple semi-inverse example may be borrowed from the torsion problem that is discussed in detail in Chapter 9. Skipping for now the developmental details, we propose the following displacement field

$$u = -\alpha yz, \quad v = \alpha xz, \quad w = w(x, y) \tag{5.7.6}$$

where α is a constant. The assumed field specifies the x and y components of displacement, while the z component is left to be determined as a function of the indicated spatial variables. By using the strain–displacement relations and Hooke's law, the stress field corresponding to (5.7.6) is given by

$$\sigma_x = \sigma_y = \sigma_z = \tau_{xy} = 0$$

$$\tau_{xz} = \mu\left(\frac{\partial w}{\partial x} - \alpha y\right) \tag{5.7.7}$$

$$\tau_{yz} = \mu\left(\frac{\partial w}{\partial y} + \alpha x\right)$$

Using these stresses in the equations of equilibrium gives the following result

$$\frac{\partial^2 w}{\partial x^2} + \frac{\partial^2 w}{\partial y^2} = 0 \tag{5.7.8}$$

which is actually the form of Navier's equations for this case. This result represents a single equation (Laplace's equation) to determine the unknown part of the assumed solution form. It should be apparent that by assuming part of the solution field, the remaining equations to be solved are greatly simplified. A specific domain in the x, y plane along with appropriate boundary conditions is needed to complete the solution to a particular problem. Once this has been accomplished, the assumed solution form (5.7.6) has thus been shown to satisfy all the field equations of elasticity.

There are numerous mathematical techniques used to solve the elasticity field equations. Many techniques involve the development of *exact analytical solutions*, while others involve the construction of *approximate solution schemes*. A third procedure involves the establishment of *numerical solution methods*. We now briefly provide an overview of each of these techniques.

5.7.4 Analytical solution procedures

A variety of analytical solution methods are used to solve the elasticity field equations. The following sections outline some of the more common methods.

Power series method

For many two-dimensional elasticity problems, the stress formulation leads to the use of a stress function $\phi(x, y)$. It is shown that the entire set of field equations reduces to a single partial differential equation (biharmonic equation) in terms of this stress function. A general mathematical scheme to solve this equation is to look for solutions in terms of a power series in the independent variables, that is, $\phi(x, y) = \sum C_{mn} x^m y^n$ (see Neou, 1957). Use of the boundary conditions determines the coefficients and number of terms to be used in the series. This method is employed to develop two-dimensional solutions in Section 8.1.

Fourier method

A general scheme to solve a large variety of elasticity problems employs the Fourier method. This procedure is normally applied to the governing partial differential equations by using *separation of variables*, *superposition*, and *Fourier series* or *Fourier integral theory*. Although this is an ad hoc method, it has been shown to provide solutions to a large class of problems (see, for example, Pickett, 1944; Little, 1973). We make use of this scheme for two-dimensional problem solution in Chapter 8, for a torsion problem in Chapter 9, and for several three-dimensional solutions in Chapter 13.

Integral transform method

A very useful mathematical technique to solve partial differential equations is the use of integral transforms. By applying a particular linear integral transformation to the governing equations, certain differential forms can be simplified or eliminated, thus allowing simple solution for the unknown transformed variables. Through appropriate inverse transformation, the original unknowns are retrieved, giving the required solution. Typical transforms that have been successfully applied to elasticity problems include *Laplace*, *Fourier*, and *Hankel* transforms. We do not make specific use of this technique in the text, but example applications can be found in Sneddon (1978) and Sneddon and Lowengrub (1969).

Complex variable method

Several classes of problems in elasticity can be formulated in terms of functions of a complex variable. These include two-dimensional plane problems, the torsion problem, and some particular thermoelastic cases. The complex variable formulation is very powerful and useful because many solutions can be found that would be intractable by other techniques. Most of the original development of this method was done by a series of Russian elasticians and is summarized in the classic work of Muskhelishvili (1963). Chapter 10 formally develops this technique and employs the method to construct several solutions for plane isotropic elasticity problems. We also use the method in Chapter 11 to determine solutions of plane anisotropic problems and in Chapter 12 for some thermoelastic applications.

5.7.5 Approximate solution procedures

With the recognized difficulty in finding exact analytical solutions, considerable work has been done to develop approximate solutions to elasticity problems. Much of this work has been in the area of *variational methods*, which are related to energy theorems (see Chapter 6). The principal idea of this approach is the connection of the elasticity field equations with a variational problem of finding an extremum of a particular integral functional. One specific technique is outlined in the following section.

Ritz method

This scheme employs a set of approximating functions to solve elasticity problems by determining stationary values of a particular energy integral. The set of approximating functions is chosen to satisfy the boundary conditions on the problem, but only approximately make the energy integral take on an extremum. By including more terms in the approximating solution set, accuracy of the scheme is improved. Specific examples of this and related methods can be found in Reismann and Pawlik (1980), Reddy (1984), and Mura and Koya (1992). Because of the difficulty in finding proper approximating functions for problems of complex geometry, variational techniques have made only limited contributions to the solution of general problems. However, they have made very important applications in the finite element method.

5.7.6 Numerical solution procedures

Over the past several decades numerical methods have played a primary role in developing solutions to elasticity problems of complex geometry. Various schemes have been theoretically developed, and numerous private and commercial codes have been written for implementation on a variety of computer platforms. Several of the more important methods are briefly discussed here.

Finite difference method

The finite difference method (FDM) replaces derivatives in the governing field equations by *difference quotients*, which involve values of the solution at discrete mesh points in the domain under study. Repeated applications of this representation set up algebraic systems of equations in terms of unknown mesh point values. The method is a classical one, having been established over a century ago, and Timoshenko and Goodier (1970) provide some details on its applications in elasticity. The major difficulty with this scheme lies in the inaccuracies in dealing with regions of complex shape, although this problem can be eliminated through the use of coordinate transformation techniques.

Finite element method

The fundamental concept of the finite element method (FEM) lies in dividing the body under study into a finite number of pieces (subdomains) called *elements*. Particular assumptions are then made on the variation of the unknown dependent variable(s) across each element using so-called *interpolation* or *shape functions*. This approximated variation is quantified in terms of solution values at special locations within the element called *nodes*. Through this discretization process, the method sets up an algebraic system of equations for unknown nodal values that approximates the continuous solution. Because element size and shape are variable, the finite element method can handle problem domains of complicated shape, and thus it has become a very useful and practical tool (see Reddy, 2006). We briefly present an introduction to finite element methods in Chapter 16.

Boundary element method

The boundary element method (BEM) is based upon an *integral statement* of the governing equations of elasticity. The integral statement may be cast into a form that contains unknowns *only over the boundary* of the body domain. This boundary integral equation may then be solved by using concepts from the finite element method; that is, the boundary may be discretized into a number of elements and the interpolation approximation concept can then be applied. This method again produces an algebraic system of equations to solve for the unknown boundary nodal values, and the system is generally much smaller than that generated by FEM internal discretization. By avoiding interior discretization, the boundary element method has significant advantages over finite element schemes for infinite or very large domains and for cases in which only boundary information is sought (see Brebbia and Dominguez, 1992). A brief discussion of this technique is given in Chapter 16.

The previously mentioned numerical schemes are generally based on various discretization and approximation concepts, and ultimately require computers to carry out the very extensive computations necessary for problem solution. With the rapid development of computational methods involving *symbolic manipulation*, there has been a new trend in engineering analysis that has included elasticity. Using such computer tools, many analytical schemes and evaluations that had been previously abandoned because of their impractical algebraic complexity are now being evaluated using symbolic codes such as *Maple* or *Mathematica*. Such an approach is especially interesting, since it incorporates the fundamentals connected with exact closed-form analytical solutions and allows direct determination of solution details without having to resort to any further approximations or special limiting cases as was often done. Examples of such work can be found in Rand and Rovenski (2005), Constantinescu and Korsunsky (2007), and Barber (2010). Currently we will not pursue these solution techniques in the text.

Elasticity is a mature field and thus analytical solutions have been developed for a large number of problems. Kachanov, Shafiro, and Tsukrov (2003) have published an interesting compilation handbook of elasticity solutions collected from textbooks and journal articles.

5.8 Singular elasticity solutions

Before leaving this chapter, we wish to briefly discuss a somewhat unpleasant feature of our elasticity model. In later chapters we will find that some generated solutions will have particular singularities that result in unbounded and/or discontinuous stress and displacement predictions. These types of solutions normally come from problems that contain some form of singularity or discontinuity in the geometry and/or boundary conditions. Such problems commonly have geometries that include sharp corners, notches or cracks. Some cases include problems with various point or concentrated loadings, while other examples come from dislocation modeling where there is specification of particular displacement discontinuities. Fig. 5.13 illustrates some of these typical examples, and these particular cases will be formally analyzed in later chapters. Fortunately, much of these unbounded predications are generally localized to regions near the discontinuous boundary or loading as shown in shaded gray areas in the figure.

This type of solution behavior should not be surprising since our elasticity model is based on mathematical principles from continuum mechanics field theory. Such theories assume continuous distributions of mass and field variables at all length scales. While such assumptions allow the development of elegant theoretical models, they do not take into account real materials and real geometry which have microstructure. For example, actual materials are composed of atoms, molecules, grains, etc., and thus are not continuous at small length scales. Likewise, real structures do not have infinitely sharp corners, notches or cracks, and loadings can never be truly concentrated at a point. Furthermore we know that all materials have a finite stress or displacement limit that if exceeded will

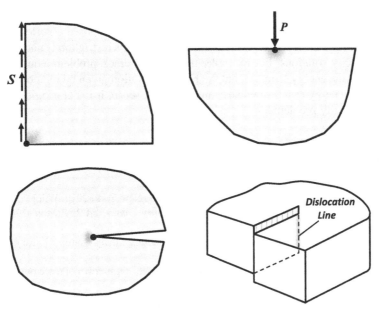

FIG. 5.13 Example domains containing corners, cracks, concentrated forces and dislocations. Each of these lead to some form of singularity in resulting stress or displacement fields in localized shaded regions.

result in non-elastic failure behavior. All of this implies that such unbounded elastic behaviors coming from these types of problems have to be viewed as approximations. There has been considerable research to develop *higher order theories* in an attempt to remove these unpleasant features of classical elasticity theory. We will further explore some of these in Chapter 15.

Some authors (Barber, 2010) have suggested an allowable singularity criterion based on stored strain energy (covered in detail in Chapter 6). The proposed criterion is that the *only acceptable singularities are those for which the total strain energy in a small region surrounding the singular point vanishes as the region's size shrinks to zero.* As will be shown, the total strain energy U_T in region V is given by the integral

$$U_T = \frac{1}{2} \iiint_V \sigma_{ij} e_{ij} dV \tag{5.8.1}$$

Now as an example, if the stress (and hence strain) is of the order r^n, then for a two-dimensional situation relation (5.8.1) can be written as

$$U_T = \frac{1}{2} \int_0^{2\pi} \int_0^r \sigma_{ij} e_{ij} r\, dr\, d\theta \sim K \int_0^r r^{2n+1} dr = K \frac{r^{2n+2}}{2n+2} , \quad (n \neq -1) \tag{5.8.2}$$

where K is a constant coming from terms that depend on elastic constants and the stress-strain distribution with respect to angle θ. We therefore conclude that the proposed criterion for an acceptable singularity would require $2n + 2 > 0 \Rightarrow n > -1$. Note that this restriction is not always satisfied by various exact elasticity solutions. For example, the singular crack problem example field given by (8.4.56) does satisfy the energy criterion, but the Flamant solution given by relation (8.4.34) does not. Hence although this energy criterion makes some physical sense, it is not a rigid rule when dealing with elasticity singularity problems.

References

Barber JR: *Elasticity*, ed 3, Dordrecht, 2010, Springer.

Boresi AP, Chong KP: *Elasticity in engineering mechanics*, ed 3, New York, 2010, Wiley.

Brebbia CA, Dominguez J: *Boundary elements: an introductory course*, ed 2, Boston, 1992, Computational Mechanics Publications.

Chan MM, Horgan CO: End effects in anti-plane shear for an inhomogeneous isotropic linear elastic semi-infinite strip, *J Elast* 51:227–242, 1998.

Chou PC, Pagano NJ: *Elasticity tensor, dyadic and engineering approaches*, Princeton, NJ, 1967, D. Van Nostrand.

Constantinescu A, Korsunsky A: *Elasticity with mathematica*, Cambridge, UK, 2007, Cambridge University Press.

Horgan CO: Saint-Venant end effects in composites, *J Compos Mater* 16:411–422, 1982.

Horgan CO: Recent developments concerning Saint-Venant's principle: an update, *Appl Mech Rev* 42:295–302, 1989.

Kachanov M, Shafiro B, Tsukrov I: *Handbook of elasticity solutions*, Dordrecht, Netherlands, 2003, Kluwer.

Little RW: *Elasticity*, Englewood Cliffs, NJ, 1973, Prentice Hall.

Mura T, Koya T: *Variational methods in mechanics*, New York, 1992, Oxford University Press.

Muskhelishvili NI: *Some basic problems of the theory of elasticity, trans*, Groningen, Netherlands, 1963, JRM Radok, P Noordhoff.

Neou CY: Direct method for determining Airy polynomial stress functions, *J Appl Mech* 24:387–390, 1957.

Pickett G: Application of the Fourier method to the solution of certain boundary problems in the theory of elasticity, *J Appl Mech* 11:176–182, 1944.

Rand O, Rovenski V: *Analytical methods in anisotropic elasticity*, Boston, 2005, Birkhauser.

Reddy JN: *Energy and variational methods in applied mechanics*, New York, 1984, Wiley.

Reddy JN: *An introduction to the finite element method*, New York, 2006, McGraw-Hill.

Reismann H, Pawlik PS: *Elasticity theory and applications*, New York, 1980, Wiley.

Sneddon IN: *Applications of integral transforms in the theory of elasticity*, New York, 1978, Springer-Verlag.

Sneddon IN, Lowengrub M: *Crack problems in the mathematical theory of elasticity*, New York, 1969, Wiley.

Sternberg E: On Saint-Venant's principle, *Quart J Appl Math* 11:393–402, 1954.

Timoshenko SP, Goodier JN: *Theory of elasticity*, New York, 1970, McGraw-Hill.

Toupin RA: Saint-Venant's principle, *Arch Ration Mech Anal* 18:83–96, 1965.

Von Mises R: On Saint-Venant's principle, *Bull Am Math Soc* 51:555–562, 1945.

Exercises

5.1 Using Cartesian coordinates, express all boundary conditions for each of the illustrated problems.

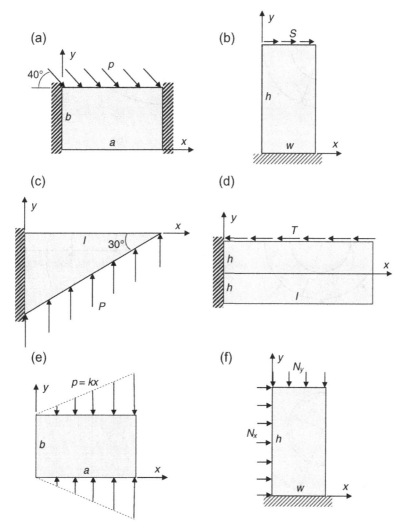

5.2 Using polar coordinates, express all boundary conditions for each of the illustrated problems.

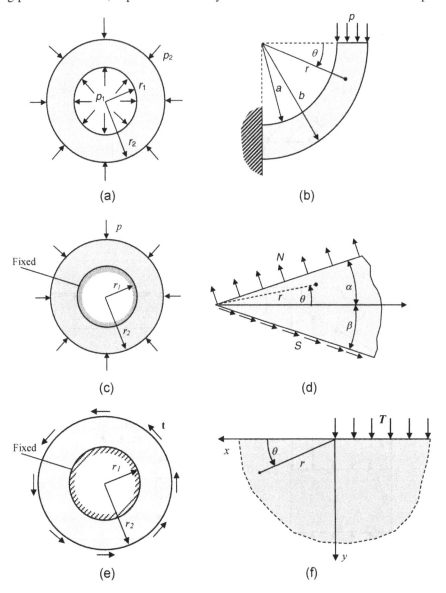

(a)

(b)

(c)

(d)

(e)

(f)

5.3 The tapered cantilever beam shown in the following figure is to have zero tractions on the bottom inclined surface. As discussed in the text (see Fig. 5.4), this may be specified by requiring $T_x^{(n)} = T_y^{(n)} = 0$. This condition can also be expressed in terms of components normal and tangential to the boundary surface as $T_n^{(n)} = T_s^{(n)} = 0$, thus implying that the normal and shearing stress on this surface should vanish. Show that these two specifications are equivalent.

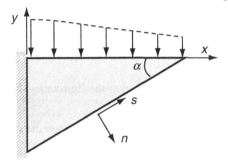

5.4 The following two-dimensional problems all have mixed boundary conditions involving both traction and displacement specifications. Using various field equations, formulate all boundary conditions for each problem solely in terms of displacements.

(a)   (b)

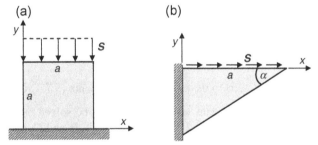

5.5 For problems involving composite bodies composed of two or more materials, the elasticity solution requires both boundary conditions and *interface conditions* between each material system. The solution process is then developed independently for each material by satisfying how the various material parts are held together. The composite bodies shown are composed of two different materials, (1) and (2), that are *perfectly bonded* thereby requiring displacement and stress continuity across the interface. Establish the required boundary and interface conditions for each problem.

(a)   (b)

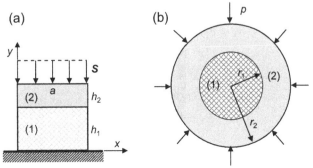

5.6 Solve Exercise 5.5 using slip interface conditions, where the normal components of displacement and traction are continuous, the tangential traction vanishes, and the tangential displacement can be discontinuous.

5.7 As mentioned in Section 5.6, Saint–Venant's principle will allow particular boundary conditions to be replaced by their statically equivalent resultant. For problems (b), (c), (d), and (f) in Exercise 5.1, the support boundaries that had fixed displacement conditions can be modified to specify the statically equivalent reaction force and moment loadings. For each case, determine these reaction loadings and then relate them to particular integrals of the tractions over the appropriate boundary.

5.8 Go through the details and explicitly develop the Beltrami–Michell compatibility equations (5.3.3).

5.9 For the displacement formulation, use relations (5.4.1) in the equilibrium equations and develop the Navier equations (5.4.3).

5.10 For the general displacement formulation with no body forces, show that Navier's equations (5.4.3) reduce to the form

$$u_{i,kk} + \frac{1}{1 - 2\nu} u_{k,ki} = 0$$

and thus the field equation formulation will now only depend on the single elastic constant, Poisson's ratio. For the case with only displacement boundary conditions, this fact would imply that the solution would also only depend on Poisson's ratio. Determine the reduction in this Navier equation for the case of $\nu = 0.5$.

5.11 Carry out the integration details to develop the displacements (5.7.4) in Example 5.1.

5.12 Using the inverse method, investigate which problem can be solved by the two-dimensional stress distribution $\sigma_x = Axy$, $\tau_{xy} = B + Cy^2$, $\sigma_y = 0$, where A, B, and C are constants. First show that the proposed stress field (with zero body force) satisfies the stress formulation field equations under the condition that $C = -A/2$. Note that for this two-dimensional plane stress case, the Beltrami–Michell compatibility equations reduce to the form given by (7.2.7). Next choose a rectangular domain $0 \le x \le l$ and $-h \le y \le h$, with $l \gg h$, and investigate the interior and boundary stresses. Finally use strength of materials theory to show that these stresses could represent the solution to a cantilever beam under end loading. Explicitly determine the required constants A, B, and C to solve the beam problem.

5.13 Show that the following stress components satisfy the equations of equilibrium with zero body forces, but are not the solution to a problem in elasticity

$$\sigma_x = c\left[y^2 + \nu\left(x^2 - y^2\right)\right]$$
$$\sigma_y = c\left[x^2 + \nu\left(y^2 - x^2\right)\right]$$
$$\sigma_z = c\nu\left(x^2 + y^2\right)$$
$$\tau_{xy} = -2c\nu xy$$
$$\tau_{yz} = \tau_{zx} = 0, \quad c = \text{constant} \ne 0$$

5.14* Consider the problem of a concentrated force acting normal to the free surface of a semi-infinite solid as shown in case (a) of the following figure. The two-dimensional stress field for this problem is given by equations (8.4.36) as

$$\sigma_x = -\frac{2Px^2y}{\pi\left(x^2 + y^2\right)^2}$$

$$\sigma_y = -\frac{2Py^3}{\pi\left(x^2 + y^2\right)^2}$$

$$\tau_{xy} = -\frac{2Pxy^2}{\pi\left(x^2 + y^2\right)^2}$$

Using this solution with the method of superposition, solve the problem with two concentrated forces as shown in case (b). Because problems (a) and (b) have the same resultant boundary loading, explicitly show that at distances far away from the loading points the stress fields for each case give approximately the same values. Explicitly plot and compare σ_y and τ_{xy} for each problem on the surface $y = 10a$ and $y = 100a$ (see Fig. 8.20).

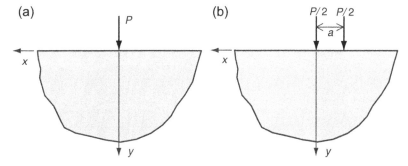

Strain energy and related principles

Before proceeding to the solution of specific elasticity problems, we wish to explore the associated concepts of work and energy. Boundary tractions and body forces will do work on an elastic solid, and this work will be stored inside the material in the form of strain energy. For the elastic case, removal of these loadings results in complete recovery of the stored energy within the body. Development of strain energy concepts can yield new and useful information not found by other methods. This study also leads to some new energy methods or principles that provide additional techniques to solve elasticity problems. In some sense these methods may be thought of as replacements of particular field equations that have been previously derived. For problems in structural mechanics involving rods, beams, plates, and shells, energy methods have proved to be very useful in developing the governing equations and associated boundary conditions. These schemes have also provided a method to generate approximate solutions to elasticity problems. More recently, particular energy and variational techniques have been used extensively in the development of finite and boundary element analysis. Our presentation here will only be a brief study on this extensive subject, and the interested reader is recommended to review Langhaar (1962), Washizu (1968), Reddy (1984), Mura and Koya (1992), or Fung and Tong (2001) for additional details and applications.

6.1 Strain energy

As mentioned, the work done by surface and body forces on an elastic solid is stored inside the body in the form of strain energy. For an idealized elastic body, this stored energy is completely recoverable when the solid is returned to its original unstrained configuration. In order to quantify this behavior, we now wish to determine the strain energy in terms of the resulting stress and strain fields within the elastic solid. Consider first the simple uniform uniaxial deformation case with no body forces, as shown in Fig. 6.1. The cubical element of dimensions dx, dy, dz is under the action of a uniform normal stress σ in the x direction as shown.

During this deformation process, we assume that the stress increases slowly from zero to σ_x, such that inertia effects can be neglected. The strain energy stored is equal to the net work done on the element, and this is given by the following equation

$$dU = \int_0^{\sigma_x} \sigma d\left(u + \frac{\partial u}{\partial x}dx\right)dydz - \int_0^{\sigma_x} \sigma du\, dydz = \int_0^{\sigma_x} \sigma d\left(\frac{\partial u}{\partial x}\right)dxdydz \qquad (6.1.1)$$

Using the strain displacement relations and Hooke's law

$$\frac{\partial u}{\partial x} = e_x = \frac{\sigma_x}{E}$$

Elasticity. https://doi.org/10.1016/B978-0-12-815987-3.00006-2

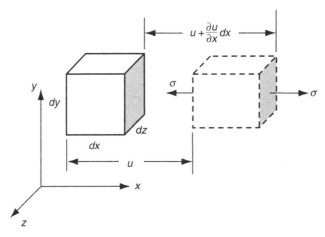

FIG. 6.1 Deformation under uniform uniaxial stress.

and thus (6.1.1) can be reduced to

$$dU = \int_0^{\sigma_x} \sigma \, \frac{d\sigma}{E} \, dxdydz = \frac{\sigma_x^2}{2E} \, dxdydz \tag{6.1.2}$$

The strain energy per unit volume, or *strain energy density*, is specified by

$$U = \frac{dU}{dxdydz} \tag{6.1.3}$$

and thus for this case we find

$$U = \frac{\sigma_x^2}{2E} = \frac{Ee_x^2}{2} = \frac{1}{2} \sigma_x e_x \tag{6.1.4}$$

This result can be interpreted from the stress–strain curve shown in Fig. 6.2. Because the material is linear elastic, the strain energy for the uniaxial case is simply the shaded area under the stress–strain curve.

We next investigate the strain energy caused by the action of uniform shear stress. Choosing the same cubical element as previously analyzed, consider the case under uniform τ_{xy} and τ_{yx} loading, as shown in Fig. 6.3. Following similar analyses, the strain energy is found to be

$$dU = \frac{1}{2} \tau_{xy} dydz \left(\frac{\partial v}{\partial x} dx \right) + \frac{1}{2} \tau_{yx} dxdz \left(\frac{\partial u}{\partial y} dy \right) = \frac{1}{2} \tau_{xy} \left(\frac{\partial u}{\partial y} + \frac{\partial v}{\partial x} \right) dxdydz \tag{6.1.5}$$

and thus the strain energy density can be expressed by

$$U = \frac{1}{2} \tau_{xy} \gamma_{xy} = \frac{\tau_{xy}^2}{2\mu} = \frac{\mu \gamma_{xy}^2}{2} \tag{6.1.6}$$

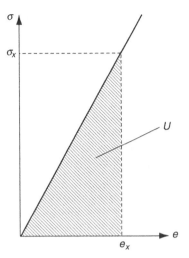

FIG. 6.2 Strain energy for uniaxial deformation.

Results from the previous two cases (6.1.4) and (6.1.6) indicate that the strain energy is not a linear function of the stresses or strains, and thus the principle of superposition cannot be directly applied to develop the strain energy for a multidimensional state of stress. However, from conservation of energy, the work done does not depend on the order of loading application, but only on the final magnitudes of

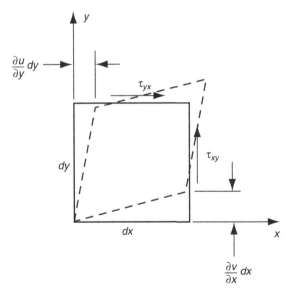

FIG. 6.3 Shear deformation.

the stresses and strains. This concept then allows normal and shear loadings to be applied one at a time and produces an additive total strain energy for a general three-dimensional state of stress and strain, as follows

$$U = \frac{1}{2}\left(\sigma_x e_x + \sigma_y e_y + \sigma_z e_z + \tau_{xy}\gamma_{xy} + \tau_{yz}\gamma_{yz} + \tau_{zx}\gamma_{zx}\right) = \frac{1}{2}\sigma_{ij}e_{ij} \tag{6.1.7}$$

Although the preceding results were developed for the case of uniform stress with no body forces, it can be shown (see Exercise 6.1) that identical results are found if body forces are included and the stresses are allowed to vary continuously. The total strain energy stored in an elastic solid occupying a region V is then given by the integral over the domain

$$U_T = \iiint_V U\,dx\,dy\,dz \tag{6.1.8}$$

Using Hooke's law, the stresses can be eliminated from relation (6.1.7) and the strain energy can be expressed solely in terms of strain. For the isotropic case, this result becomes

$$U(\mathbf{e}) = \frac{1}{2}\lambda e_{jj} e_{kk} + \mu e_{ij} e_{ij}$$

$$= \frac{1}{2}\lambda(e_x + e_y + e_z)^2 + \mu\left(e_x^2 + e_y^2 + e_z^2 + 2e_{xy}^2 + 2e_{yz}^2 + 2e_{zx}^2\right) \tag{6.1.9}$$

Likewise, the strains can be eliminated and the strain energy can be written in terms of stress

$$U(\boldsymbol{\sigma}) = \frac{1+\nu}{2E}\sigma_{ij}\sigma_{ij} - \frac{\nu}{2E}\sigma_{jj}\sigma_{kk}$$

$$= \frac{1+\nu}{2E}\left(\sigma_x^2 + \sigma_y^2 + \sigma_z^2 + 2\tau_{xy}^2 + 2\tau_{yz}^2 + 2\tau_{zx}^2\right) - \frac{\nu}{2E}(\sigma_x + \sigma_y + \sigma_z)^2 \tag{6.1.10}$$

After reviewing the various developed forms in terms of the stresses or strains, it is observed that the strain energy is a *positive definite quadratic form* with the property

$$U \geq 0 \tag{6.1.11}$$

for all values of σ_{ij} and e_{ij}, with equality only for the case with $\sigma_{ij} = 0$ or $e_{ij} = 0$. Actually, relation (6.1.11) is valid for all elastic materials, including both isotropic and anisotropic solids.

For the uniaxial deformation case, by using relation (6.1.4) the derivative of the strain energy in terms of strain yields

$$\frac{\partial U(\mathbf{e})}{\partial e_x} = \frac{\partial}{\partial e_x}\left(\frac{Ee_x^2}{2}\right) = Ee_x = \sigma_x$$

and likewise

$$\frac{\partial U(\boldsymbol{\sigma})}{\partial \sigma_x} = \frac{\partial}{\partial \sigma_x}\left(\frac{\sigma_x^2}{2E}\right) = \frac{\sigma_x}{E} = e_x$$

These specific uniaxial results can be generalized (see Exercise 6.4) for the three-dimensional case, giving the relations

$$\sigma_{ij} = \frac{\partial U(e)}{\partial e_{ij}}, \quad e_{ij} = \frac{\partial U(\sigma)}{\partial \sigma_{ij}} \tag{6.1.12}$$

These results are again true for all elastic materials (isotropic and anisotropic); see Langhaar (1962) or Boresi and Chang (2000) for a general derivation. Result (6.1.12) can also be developed directly from the continuum energy equation, see Sadd (2019). Thus, the strain energy function can be interpreted as playing a fundamental constitutive role in establishing general stress–strain relations for elastic materials. Such an approach in which the stresses are derivable from a strain energy function, that is, relation (6.1.12), is referred to as *hyperelasticity*. Note that this approach does not necessarily require that the relations between stress and strain be linear, and thus this scheme is commonly used in the development of constitutive relations for nonlinear elastic solids. Only linear relations given by Hooke's law (4.2.1) are incorporated in the text.

Using equations (6.1.12), the following symmetry relations can be developed (Exercise 6.6)

$$\frac{\partial \sigma_{ij}}{\partial e_{kl}} = \frac{\partial \sigma_{kl}}{\partial e_{ij}}$$

$$\frac{\partial e_{ij}}{\partial \sigma_{kl}} = \frac{\partial e_{kl}}{\partial \sigma_{ij}} \tag{6.1.13}$$

Going back to the general constitutive form $\sigma_{ij} = C_{ijkl} e_{kl}$, relations (6.1.13) can be used to develop the additional symmetry relations

$$C_{ijkl} = C_{klij} \tag{6.1.14}$$

Using constitutive form (4.2.2), result (6.1.14) implies that $C_{ij} = C_{ji}$, and thus *there are on 21 independent elastic constants for general anisotropic elastic materials.*

The strain energy in an elastic solid may be decomposed into two parts, one associated with *volumetric* change U_v and the other caused by *distortional* (change in shape) deformation U_d

$$U = U_v + U_d \tag{6.1.15}$$

The development of this decomposition is accomplished by using the definitions of spherical and deviatoric strain and stress tensors presented previously in Sections 2.5 and 3.5. For isotropic materials, the spherical stress produces only volumetric deformation, while the deviatoric stress causes only distortional changes. The volumetric strain energy is found by considering the spherical or hydrostatic components of stress and strain

$$U_v = \frac{1}{2} \tilde{\sigma}_{ij} \tilde{e}_{ij} = \frac{1}{6} \sigma_{jj} e_{kk} = \frac{1 - 2v}{6E} \sigma_{jj} \sigma_{kk} = \frac{1 - 2v}{6E} (\sigma_x + \sigma_y + \sigma_z)^2 \tag{6.1.16}$$

The distortional strain energy results from the deviatoric components or can be easily determined using relations (6.1.10), (6.1.15), and (6.1.16) to get

$$U_d = \frac{1}{12\mu} \left[(\sigma_x - \sigma_y)^2 + (\sigma_y - \sigma_z)^2 + (\sigma_z - \sigma_x)^2 + 6\left(\tau_{xy}^2 + \tau_{yz}^2 + \tau_{zx}^2\right) \right] \tag{6.1.17}$$

Particular failure theories of solids incorporate the strain energy of distortion by proposing that material failure or yielding will initiate when U_d reaches a critical value. It can be shown that the distortional strain energy is related to the octahedral shear stress previously discussed in Section 3.5 (see Exercise 6.12).

6.2 Uniqueness of the elasticity boundary-value problem

Although it would seem that the question of uniqueness of the elasticity boundary-value problem should have been covered in Chapter 5, the proof normally makes use of strain energy concepts and is therefore presented here. Consider the general mixed boundary-value problem in which tractions are specified over the boundary S_t and displacements are prescribed over the remaining part S_u. Assume that there are two different solutions $\left\{\sigma_{ij}^{(1)}, e_{ij}^{(1)}, u_i^{(1)}\right\}$ and $\left\{\sigma_{ij}^{(2)}, e_{ij}^{(2)}, u_i^{(2)}\right\}$ to the *same problem* with identical body forces and boundary conditions. Next define the *difference solution*

$$
\begin{aligned}
\sigma_{ij} &= \sigma_{ij}^{(1)} - \sigma_{ij}^{(2)} \\
e_{ij} &= e_{ij}^{(1)} - e_{ij}^{(2)} \\
u_i &= u_i^{(1)} - u_i^{(2)}
\end{aligned}
\tag{6.2.1}
$$

Because the solutions $\sigma_{ij}^{(1)}$ and $\sigma_{ij}^{(2)}$ have the same body force, the difference solution must satisfy the equilibrium equation

$$
\sigma_{ij,j} = 0
\tag{6.2.2}
$$

Likewise, the boundary conditions satisfied by the difference solution are given by

$$
\begin{aligned}
T_i^n &= \sigma_{ij}n_j = 0 \quad on \quad S_t \\
u_i &= 0 \quad on \quad S_u
\end{aligned}
\tag{6.2.3}
$$

Starting with the definition of strain energy, we may write

$$
\begin{aligned}
2\int_V U dV &= \int_V \sigma_{ij} e_{ij} dV = \int_V \sigma_{ij}\left(u_{i,j} - \omega_{ij}\right) dV \\
&= \int_V \sigma_{ij} u_{i,j} dV = \int_V \left(\sigma_{ij} u_i\right)_{,j} dV - \int_V \sigma_{ij,j} u_i dV \\
&= \int_S \sigma_{ij} n_j u_i dS - \int_V \sigma_{ij,j} u_i dV
\end{aligned}
\tag{6.2.4}
$$

where we have used the fact that $\sigma_{ij}\omega_{ij} = 0$ (symmetric times antisymmetric $= 0$) and have utilized the divergence theorem to convert the volume integral into a surface integral. Incorporating relations (6.2.2) and (6.2.3) and noting that the total surface $S = S_t + S_u$, (6.2.4) gives the result

$$
2\int_V U dV = 0
\tag{6.2.5}
$$

Relation (6.2.5) implies that U must vanish in the region V, and since the strain energy is a positive definite quadratic form, the associated strains and stresses also vanish; that is, $e_{ij} = \sigma_{ij} = 0$. If the strain field vanishes, then the corresponding displacements u_i can be at most rigid-body motion. However, if $u_i = 0$ on S_u, then the displacement field must vanish everywhere. Thus, we have shown that $\sigma_{ij}^{(1)} = \sigma_{ij}^{(2)}$, $e_{ij}^{(1)} = e_{ij}^{(2)}$, $u_i^{(1)} = u_i^{(2)}$ and therefore the problem solution is unique. Note that if tractions are prescribed over the entire boundary, then $u_i^{(1)}$ and $u_i^{(2)}$ may differ by rigid-body motion.

6.3 Bounds on the elastic constants

Strain energy concepts allow us to generate particular bounds on elastic constants. For the isotropic case, consider the following three stress states previously investigated in Section 4.3 (see Fig. 4.2).

6.3.1 Uniaxial tension

Uniform uniaxial deformation in the x direction is given by the stress state

$$\sigma_{ij} = \begin{bmatrix} \sigma & 0 & 0 \\ 0 & 0 & 0 \\ 0 & 0 & 0 \end{bmatrix} \tag{6.3.1}$$

For this case, the strain energy reduces to

$$U = \frac{1+\nu}{2E}\sigma^2 - \frac{\nu}{2E}\sigma^2 = \frac{\sigma^2}{2E} \tag{6.3.2}$$

Because the strain energy is positive definite, relation (6.3.2) implies that the modulus of elasticity must be positive

$$E > 0 \tag{6.3.3}$$

6.3.2 Simple shear

Consider next the case of uniform simple shear defined by the stress tensor

$$\sigma_{ij} = \begin{bmatrix} 0 & \tau & 0 \\ \tau & 0 & 0 \\ 0 & 0 & 0 \end{bmatrix} \tag{6.3.4}$$

The strain energy becomes

$$U = \frac{1+\nu}{2E}(2\tau^2) = \frac{\tau^2}{E}(1+\nu) \tag{6.3.5}$$

Again, invoking the positive definite property of the strain energy and using the previous result of $E > 0$ gives

$$1 + \nu > 0 \Rightarrow \nu > -1 \tag{6.3.6}$$

6.3.3 Hydrostatic compression

The final example is chosen as uniform hydrostatic compression specified by

$$\sigma_{ij} = \begin{bmatrix} -p & 0 & 0 \\ 0 & -p & 0 \\ 0 & 0 & -p \end{bmatrix} \tag{6.3.7}$$

Note that hydrostatic tension could also be used for this example. Evaluating the strain energy yields

$$U = \frac{1+\nu}{2E} 3p^2 - \frac{\nu}{2E}(-3p)^2 = \frac{3p^2}{2E}(1-2\nu) \tag{6.3.8}$$

Using the positive definite property with $E > 0$ gives the result

$$1 - 2\nu > 0 \Rightarrow \nu < \frac{1}{2} \tag{6.3.9}$$

Combining relations (6.3.6) and (6.3.9) places the following bounds on Poisson's ratio

$$-1 < \nu < \frac{1}{2} \tag{6.3.10}$$

Using relations between the elastic constants given in Table 4.1, the previous results also imply that

$$k > 0, \quad \mu > 0 \tag{6.3.11}$$

Experimental evidence indicates that most real materials have positive values of Poisson's ratio, and thus $0 < \nu < 1/2$. This further implies that $\lambda > 0$. Bounds on elastic moduli for the anisotropic case are more involved and are discussed in Chapter 11.

6.4 Related integral theorems

Within the context of linear elasticity, several integral relations based on work and energy can be developed. We now wish to investigate three particular results referred to as *Clapeyron's theorem*, *Betti's reciprocal theorem*, and *Somigliana's identity*.

6.4.1 Clapeyron's theorem

The strain energy of an elastic solid in static equilibrium is equal to one-half the work done by the external body forces F_i and surface tractions T_i^n

$$2 \int_V U dV = \int_S T_i^n u_i dS + \int_V F_i u_i dV \tag{6.4.1}$$

The proof of this theorem follows directly from results in relation (6.2.4).

6.4.2 Betti/Rayleigh reciprocal theorem

If an elastic body is subject to two body and surface force systems, then the work done by the first system of forces $\{T^{(1)}, F^{(1)}\}$ acting through the displacements $u^{(2)}$ of the second system is equal to the work done by the second system of forces $\{T^{(2)}, F^{(2)}\}$ acting through the displacements $u^{(1)}$ of the first system

$$\int_S T_i^{(1)} u_i^{(2)} dS + \int_V F_i^{(1)} u_i^{(2)} dV = \int_S T_i^{(2)} u_i^{(1)} dS + \int_V F_i^{(2)} u_i^{(1)} dV \tag{6.4.2}$$

The proof of this theorem starts by using the result from (6.2.4)

$$\int_V \sigma_{ij}^{(1)} e_{ij}^{(2)} dV = \int_S T_i^{(1)} u_i^{(2)} dS + \int_V F_i^{(1)} u_i^{(2)} dV$$

Interchanging states (1) and (2) gives

$$\int_V \sigma_{ij}^{(2)} e_{ij}^{(1)} dV = \int_S T_i^{(2)} u_i^{(1)} dS + \int_V F_i^{(2)} u_i^{(1)} dV$$

Now $\sigma_{ij}^{(1)} e_{ij}^{(2)} = C_{ijkl} e_{kl}^{(1)} e_{ij}^{(2)} = C_{klij} e_{kl}^{(1)} e_{ij}^{(2)} = C_{klij} e_{ij}^{(2)} e_{kl}^{(1)} = \sigma_{kl}^{(2)} e_{kl}^{(1)}$; therefore

$$\sigma_{ij}^{(1)} e_{ij}^{(2)} = \sigma_{ij}^{(2)} e_{ij}^{(1)} \tag{6.4.3}$$

Combing these results proves the theorem. In a very brief paper Truesdell (1963) presented a proof that Betti's reciprocal theorem provides a necessary and sufficient condition that the elastic material be hyperelastic. This reciprocal theorem can yield useful applications by special selection of the two systems. One such application follows.

6.4.3 Integral formulation of elasticity—Somigliana's identity

Using the reciprocal theorem (6.4.2), select the first system to be the desired solution to a particular problem $\{T, F, u\}$. The second system is chosen as the *fundamental solution* to the elasticity equations, and this corresponds to the solution of the displacement field at point x produced by a unit concentrated body force e located at point ξ. The fundamental solution is actually related to Kelvin's problem (concentrated force in an infinite domain) and is solved in Examples 13.1, 15.3, and 15.4. Using this concept, the displacement may be expressed as

$$u_i^{(2)}(x) = G_{ij}(x; \xi) e_j(\xi) \tag{6.4.4}$$

where G_{ij} represents the *displacement Green's function* to the elasticity equations. This function has been previously developed, and forms for both two- and three-dimensional domains have been given (Banerjee and Butterfield, 1981). The three-dimensional isotropic case, for example, is given by

$$G_{ij}(x, \xi) = \frac{1}{16\pi\mu(1-v)r} \left[(3-4v)\delta_{ij} + r_{,i} r_{,j}\right] \tag{6.4.5}$$

where $r_i = x_i - \xi_i$ and $r = |r|$. The stresses and tractions associated with this fundamental solution follow from the basic field equations and can be written in the form

$$\sigma_{ij}^{(2)} = T_{ijk}(x, \xi) e_k(\xi)$$

$$T_i^{(2)} = T_{ijk}(x, \xi) n_j e_k(\xi) \tag{6.4.6}$$

$$T_{ijk}(x, \xi) = \lambda G_{lk,l} \delta_{ij} + \mu(G_{ik,j} + G_{jk,i})$$

After some manipulation using these results in the reciprocal theorem (6.4.2) gives

$$c u_j(\xi) = \int_S [T_i(x) G_{ij}(x, \xi) - u_i T_{ikj}(x, \xi) n_k] dS + \int_V F_i G_{ij}(x, \xi) dV \tag{6.4.7}$$

where the coefficient c is given by

$$c = \begin{cases} 1, & \xi \ in \ V \\ \dfrac{1}{2}, & \xi \ on \ S \\ 0, & \xi \ outside \ V \end{cases}$$

Relation (6.4.7) is known as *Somigliana's identity* and represents an integral statement of the elasticity problem. This result is used in the development of *boundary integral equation* (BIE) *methods* in elasticity and leads to the computational technique of *boundary element methods* (BEM). A brief presentation of this numerical method is given in Chapter 16.

6.5 Principle of virtual work

Based on work and energy principles, several additional solution methods can be developed. These represent alternatives to the analytical methods based on differential equations outlined in Section 5.7. The *principle of virtual work* provides the foundation for many of these methods, and thus we begin our study by establishing this principle. The *virtual displacement* of a material point is a fictitious displacement such that the forces acting on the point remain unchanged. The work done by these forces during the virtual displacement is called the *virtual work*. For an object in static equilibrium, the virtual work is zero because the resultant force vanishes on every portion of an equilibrated body. The converse is also true that if the virtual work is zero, then the material point must be in equilibrium.

Let us introduce the following notational scheme. The virtual displacements of an elastic solid are denoted by $\delta u_i = \{\delta u, \delta v, \delta w\}$, and the corresponding virtual strains are then expressible as $\delta e_{ij} = 1/2(\delta u_{i,j} + \delta u_{j,i})$. Consider the standard elasticity boundary-value problem of a solid in equilibrium under the action of surface tractions over the boundary S_t with displacement conditions over the remaining boundary S_u (see Fig. 5.1). Now imagine that the body undergoes a virtual displacement δu_i from its equilibrium configuration. The virtual displacement is arbitrary except that it must not violate the *kinematic displacement boundary condition*, and thus $\delta u_i = 0$ on S_u.

The virtual work done by the surface and body forces can be written as

$$\delta W = \int_{S_t} T_i^n \delta u_i dS + \int_V F_i \delta u_i dV \tag{6.5.1}$$

Now, because the virtual displacement vanishes on S_u, the integration domain of the first integral can be changed to S. Following standard procedures, this surface integral can be changed to a volume integral and combined with the body force term. These steps are summarized as

$$
\begin{aligned}
\delta W &= \int_S T_i^n \delta u_i dS + \int_V F_i \delta u_i dV \\
&= \int_S \sigma_{ij} n_j \delta u_i dS + \int_V F_i \delta u_i dV \\
&= \int_V (\sigma_{ij} \delta u_i)_{,j} dV + \int_V F_i \delta u_i dV \\
&= \int_V (\sigma_{ij,j} \delta u_i + \sigma_{ij} \delta u_{i,j}) dV + \int_V F_i \delta u_i dV \\
&= \int_V (-F_i \delta u_i + \sigma_{ij} \delta e_{ij}) dV + \int_V F_i \delta u_i dV \\
&= \int_V \sigma_{ij} \delta e_{ij} dV
\end{aligned}
\tag{6.5.2}
$$

Now the last line in relation (6.5.2) is actually the virtual strain energy within the solid

$$\int_V \sigma_{ij}\delta e_{ij}dV = \int_V \left(\sigma_x\delta e_x + \sigma_y\delta e_y + \sigma_z\delta e_z + \tau_{xy}\delta\gamma_{xy} + \tau_{yz}\delta\gamma_{yz} + \tau_{zx}\delta\gamma_{zx}\right)dV \qquad (6.5.3)$$

Notice that the virtual strain energy does not contain the factor of 1/2 found in the general expression (6.1.7). This fact occurs because the stresses are constant during the virtual displacement.

Under the assumption of the existence of a strain energy function expressed in terms of the strains

$$\sigma_{ij} = \frac{\partial U(e)}{\partial e_{ij}} \qquad (6.5.4)$$

and thus relation (6.5.3) can be written as

$$\int_V \sigma_{ij}\delta e_{ij}dV = \int_V \frac{\partial U}{\partial e_{ij}}\delta e_{ij}dV = \delta\int_V UdV \qquad (6.5.5)$$

Because the external forces are unchanged during the virtual displacements and the region V is fixed, the operator δ can be placed before the integrals in (6.5.1). Combining this with relation (6.5.5) allows (6.5.2) to be written as

$$\delta\left(\int_V UdV - \int_{S_t} T_i^n u_i dS - \int_V F_i u_i dV\right) = \delta(U_T - W) = 0 \qquad (6.5.6)$$

This is one of the statements of the principle of virtual work for an elastic solid. The quantity $(U_T - W)$ actually represents the *total potential energy* of the elastic solid, and thus relation (6.5.6) states that the change in potential energy during a virtual displacement from equilibrium is zero. It should be noted that this principle is valid for all elastic materials including both linear and nonlinear stress–strain behavior. The principle of virtual work provides a convenient method for deriving equilibrium equations and associated boundary conditions for various special theories of elastic bodies, including rods, beams, plates, and shells. Several such examples are given by Reismann and Pawlik (1980). In fact, even the continuum equations previously developed can be re-established using this method.

To illustrate the process of using the principle of virtual work to re-derive the basic equilibrium equations and related boundary conditions for the general elasticity problem, we can start with relations (6.5.1) and (6.5.2) to write

$$\int_V \sigma_{ij}\delta e_{ij}dV - \int_S T_i^n \delta u_i dS - \int_V F_i\delta u_i dV = 0 \qquad (6.5.7)$$

The integrand of the first term can be reduced as

$$\sigma_{ij}\delta e_{ij} = \frac{1}{2}\sigma_{ij}\left(\delta u_{i,j} + \delta u_{j,i}\right) = \sigma_{ij}\delta u_{i,j} = \left(\sigma_{ij}\delta u_i\right)_{,j} - \sigma_{ij,j}\delta u_i$$

and thus (6.5.7) can be expressed as

$$\int_V \left[\left(\sigma_{ij}\delta u_i\right)_{,j} - \sigma_{ij,j}\delta u_i\right]dV - \int_S T_i^n \delta u_i dS - \int_V F_i\delta u_i dV = 0$$

$$\int_V \left(\sigma_{ij,j} + F_i\right)\delta u_i dV + \int_S \left(T_i^n - \sigma_{ij}n_j\right)\delta u_i dS = 0 \qquad (6.5.8)$$

where we have used the divergence theorem to convert the volume integral $\int_V (\sigma_{ij}\delta u_i)_{,j} dV$ to a surface integral over S. For arbitrary δu_i, relation (6.5.8) is satisfied if

$$\sigma_{ij,j} + F_i = 0 \in V$$
$$\text{and either} \tag{6.5.9}$$
$$\delta u_i = 0 \in S_u \quad \text{or} \quad T_i^n = \sigma_{ij}n_j \in S_t$$

Conditions on S_u are commonly referred to as *essential boundary conditions* while those on S_t are called *natural boundary conditions*. Thus, we have demonstrated that the principle of virtual work can be used to generate equilibrium equations and associated boundary conditions for the general elasticity problem.

6.6 Principles of minimum potential and complementary energy

We now wish to use the results of the previous section to develop principles of minimum potential and complementary energy. Denoting the potential energy by $\Pi = U_T - W$, the virtual work statement indicated that the variation in potential energy from an equilibrium configuration was zero. Another way this is commonly stated is that potential energy is *stationary* in an equilibrium configuration. Such a condition implies that the potential energy will take on a *local extremum* (maximum or minimum) value for this configuration. It can be shown (proof given by Sokolnikoff (1956); or Reismann and Pawlik (1980)) that the potential energy has a *local minimum* in the equilibrium configuration, and this leads to the following principle:

> **Principle of Minimum Potential Energy:** *Of all displacements satisfying the given boundary conditions of an elastic solid, those that satisfy the equilibrium equations make the potential energy a local minimum.*

An additional minimum principle can be developed by reversing the nature of the variation. Consider the variation of the stresses while holding the displacements constant. Let σ_{ij} be the actual stresses that satisfy the equilibrium equations and boundary conditions. Now consider a system of *stress variations* or *virtual stresses* $\delta\sigma_{ij}$ that also satisfies the stress boundary conditions (with $\delta T_i^n = \delta\sigma_{ij}n_j$ on S_t) and equilibrium equations with body force δF_i. In contrast to the previous development, we now investigate the *complementary virtual work*

$$\delta W^c = \int_S u_i \delta T_i^n dS + \int_V u_i \delta F_i dV \tag{6.6.1}$$

Employing the usual reduction steps as given in relations (6.5.2), the complementary virtual work statement is found to reduce to

$$\int_S u_i \delta T_i^n dS + \int_V u_i \delta F_i dV = \int_V e_{ij} \delta\sigma_{ij} dV \tag{6.6.2}$$

and the integral on the right-hand side is referred to as the *complementary virtual strain energy*.

Introducing the complementary strain energy density function U^c, which is taken as a function of the stresses

$$e_{ij} = \frac{\partial U^c(\boldsymbol{\sigma})}{\partial \sigma_{ij}} \tag{6.6.3}$$

Using this result, the right-hand side of Eq. (6.6.2) can be expressed as

$$\int_V e_{ij}\delta\sigma_{ij}dV = \int_V \frac{\partial U^c}{\partial \sigma_{ij}}\delta\sigma_{ij}dV = \delta\int_V U^c dV \qquad (6.6.4)$$

Because the displacements do not vary and the region V is fixed, the operator δ can be placed before the integrals in Eq. (6.6.1). Combining this with relation (6.6.4) allows (6.6.2) to be written as

$$\delta\left(\int_V U^c dV - \int_{S_t} u_i T_i^n dS - \int_V u_i F_i dV\right) = \delta\left(U_T^c - W^c\right) = 0 \qquad (6.6.5)$$

and thus the variation in total complementary energy $\Pi^c = U_T^c - W^c$ is also zero in an equilibrium configuration. As before, it can be shown that this extremum in the complementary energy corresponds to a local minimum, thus leading to the following principle:

> **Principle of Minimum Complementary Energy:** *Of all elastic stress states satisfying the given boundary conditions, those that satisfy the equilibrium equations make the complementary energy a local minimum.*

Each of the previously developed minimum principles used general constitutive relations, either (6.5.4) or (6.6.3), and thus both principles are valid for all elastic materials regardless of whether the stress−strain law is linear or nonlinear. Fundamentally, the strain energy is expressed in terms of strain, while the complementary energy is functionally written in terms of stress. As shown in Fig. 6.2, the strain energy for uniaxial deformation is equal to the area *under* the stress−strain curve, and thus $dU = \sigma\delta e$. On the other hand, the complementary energy may be expressed by $dU^c = e\delta\sigma$, and thus U^c will be the area *above* the uniaxial stress−strain curve, as shown in Fig. 6.4. For the uniaxial case with linear elastic behavior

$$U^c = \sigma_x e_x - U = \frac{\sigma_x^2}{E} - \frac{\sigma_x^2}{2E} = \frac{\sigma_x^2}{2E} = \frac{1}{2}\sigma_x e_x = U \qquad (6.6.6)$$

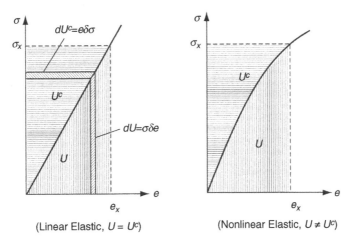

(Linear Elastic, $U = U^c$)   (Nonlinear Elastic, $U \neq U^c$)

FIG. 6.4 Strain and complementary energy for linear and nonlinear elastic materials.

This result is true for all deformations, and thus *for linear elastic materials the complementary energy is equal to the strain energy. Note also for this case*

$$\frac{\partial U^c}{\partial \sigma_x} = \frac{\partial}{\partial \sigma_x}\left(\frac{\sigma_x^2}{2E}\right) = \frac{\sigma_x}{E} = e_x \tag{6.6.7}$$

which verifies the general relation (6.6.3). For the nonlinear elastic case, as shown in Fig. 6.4, it is apparent that the strain energy and complementary energy will not be the same; that is, $U^c \neq U$. However, using the fact that $U^c = \sigma_x e_x - U$, it follows that

$$\frac{\partial U^c}{\partial \sigma_x} = e_x + \sigma_x\frac{\partial e_x}{\partial \sigma_x} - \frac{\partial U}{\partial e_x}\frac{\partial e_x}{\partial \sigma_x}$$

$$= e_x + \sigma_x\frac{\partial e_x}{\partial \sigma_x} - \sigma_x\frac{\partial e_x}{\partial \sigma_x} = e_x \tag{6.6.8}$$

which again verifies the general relation (6.6.3) for the nonlinear case.

Additional related principles can be developed, including Castigiliano's theorems and a mixed formulation called *Reissner's principle* (see Reismann and Pawlik, 1980 or Fung and Tong, 2001).

Example 6.1 Euler–Bernoulli beam theory

In order to demonstrate the utility of energy principles, consider an application dealing with the bending of an elastic beam, as shown in Fig. 6.5. The external distributed loading q will induce internal bending moments M and shear forces V at each section of the beam. According to classical Euler–Bernoulli theory, the bending stress σ_x and moment–curvature and moment–shear relations are given by

$$\sigma_x = -\frac{My}{I}, \quad M = EI\frac{d^2w}{dx^2}, \quad V = \frac{dM}{dx} \tag{6.6.9}$$

where $I = \iint_A y^2 dA$ is the area moment of inertia of the cross-section about the neutral axis, and w is the beam deflection (positive in y direction).

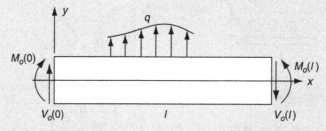

FIG. 6.5 Euler–Bernoulli beam geometry.

Considering only the strain energy caused by the bending stresses

$$U = \frac{\sigma_x^2}{2E} = \frac{M^2 y^2}{2EI^2} = \frac{E}{2}\left(\frac{d^2w}{dx^2}\right)^2 y^2$$

and thus the total strain energy in a beam of length l is

$$U_T = \int_0^l \left[\iint_A \frac{E}{2} \left(\frac{d^2w}{dx^2} \right) y^2 dA \right] dx = \int_0^l \frac{EI}{2} \left(\frac{d^2w}{dx^2} \right)^2 dx \qquad (6.6.10)$$

Now the work done by the external forces (tractions) includes contributions from the distributed loading q and the loadings at the ends $x = 0$ and l

$$W = \int_0^l qw dx - \left[V_o w - M_o \frac{dw}{dx} \right]_0^l \qquad (6.6.11)$$

Therefore, the total potential energy for this beam case is given by

$$\Pi = U_T - W = \int_0^l \left[\frac{EI}{2} \left(\frac{d^2w}{dx^2} \right)^2 - qw \right] dx + \left[V_o w - M_o \frac{dw}{dx} \right]_0^l \qquad (6.6.12)$$

The first variation of this quantity must vanish

$$\delta\Pi = \delta \int_0^l \left[\frac{EI}{2} \left(\frac{d^2w}{dx^2} \right)^2 - qw \right] dx + \delta \left[V_o w - M_o \frac{dw}{dx} \right]_0^l$$

$$= \int_0^l \delta \left[\frac{EI}{2} \left(\frac{d^2w}{dx^2} \right)^2 - qw \right] dx + \left[V_o \delta w - M_o \frac{d\delta w}{dx} \right]_0^l \qquad (6.6.13)$$

$$= \frac{EI}{2} \int_0^l 2 \frac{d^2w}{dx^2} \frac{d^2\delta w}{dx^2} dx - \int_0^l q\delta w dx + \left[V_o \delta w - M_o \frac{d\delta w}{dx} \right]_0^l = 0$$

Now the first integral term can be integrated by parts twice to get

$$\int_0^l \left(EI \frac{d^4w}{dx^4} - q \right) \delta w\, dx + \left[\frac{d\delta w}{dx} (M - M_o) - \delta w (V - V_o) \right]_0^l = 0 \qquad (6.6.14)$$

The integral and boundary terms must all vanish, thus implying

$$\int_0^l \left(EI \frac{d^4w}{dx^4} - q \right) \delta w dx = 0$$

$$V = V_o \text{ or } \delta w = 0, \quad x = 0, l \qquad (6.6.15)$$

$$M = M_o \text{ or } \delta \left(\frac{dw}{dx} \right) = 0, \quad x = 0, l$$

Continued

For this integral to vanish for all variations δw, the *fundamental lemma in the calculus of variations* implies that the integrand must be zero, giving

$$EI\frac{d^4 w}{dx^4} - q = 0 \qquad (6.6.16)$$

This result is simply the differential equilibrium equation for the beam, and thus the stationary value for the potential energy leads directly to the governing equilibrium equation in terms of displacement and the associated boundary conditions. Of course, this entire formulation is based on the simplifying assumptions found in Euler–Bernoulli beam theory, and resulting solutions would not match with the more exact theory of elasticity results.

6.7 Rayleigh–Ritz method

The previous beam example indicates a correspondence between the governing differential equation(s) and a variational problem of minimizing the potential energy of the system. Such a correspondence exists for many other types of problems in structural mechanics and elasticity. For problems of complicated shape or loading geometry, the solution to the governing differential equation cannot be found by analytical methods. For such cases, approximate solution schemes have been developed based on the variational form of the problem. Several such approximate schemes have been constructed, and one of the more important techniques is the *Rayleigh–Ritz method*.

This particular technique is based on the idea of constructing a series of trial approximating functions that satisfy the boundary conditions but not the differential equation(s). For the elasticity displacement formulation (Section 5.4), this concept would express the displacements in the form

$$u = u_o + a_1 u_1 + a_2 u_2 + a_3 u_3 + \ldots = u_o + \sum_{j=1}^{N} a_j u_j$$

$$v = v_o + b_1 v_1 + b_2 v_2 + b_3 v_3 + \ldots = v_o + \sum_{j=1}^{N} b_j v_j \qquad (6.7.1)$$

$$w = w_o + c_1 w_1 + c_2 w_2 + c_3 w_3 + \ldots = w_o + \sum_{j=1}^{N} c_j w_j$$

where the functions u_o, v_o, w_o are chosen to satisfy any nonhomogeneous displacement boundary conditions and u_j, v_j, w_j satisfy the corresponding homogeneous boundary conditions. Note that these forms are not required to satisfy the traction boundary conditions. Normally, these trial functions are chosen from some combination of elementary functions such as polynomial, trigonometric, or hyperbolic forms. The unknown constant coefficients a_j, b_j, c_j are to be determined so as to minimize the potential energy of the problem, thus approximately satisfying the variational formulation of the problem under study. Using this type of approximation, the total potential energy will thus be a function of these unknown coefficients

$$\Pi = \Pi(a_j, b_j, c_j) \qquad (6.7.2)$$

and the minimizing condition can be expressed as a series of expressions

$$\frac{\partial \Pi}{\partial a_j} = 0, \quad \frac{\partial \Pi}{\partial b_j} = 0, \quad \frac{\partial \Pi}{\partial c_j} = 0 \qquad (6.7.3)$$

This set forms a system of $3N$ algebraic equations that can be solved to obtain the parameters a_j, b_j, c_j. Under suitable conditions on the choice of trial functions (completeness property), the approximation will improve as the number of included terms is increased.

Commonly, this technique is applied to a reduced elasticity problem involving only one or two components of displacement typically found in rods, beams, plates, and shells. Once the approximate displacement solution is obtained, the strains and stresses can be calculated from the appropriate field equations. However, since the strains and stresses are determined through differentiation, the accuracy in these variables will in general not be as good as that obtained for the displacements themselves (see Exercises 6.18 and 6.19). In order to demonstrate the Ritz technique, consider again the Euler–Bernoulli beam problem from Example 6.1.

Example 6.2 Rayleigh–Ritz solution of a simply supported Euler–Bernoulli beam

Consider a simply supported Euler-Bernoulli beam of length l carrying a uniform loading q_o. This one-dimensional problem has displacement boundary conditions

$$w = 0 \text{ at } x = 0, l \qquad (6.7.4)$$

and tractions or moment conditions

$$EI\frac{d^2w}{dx^2} = 0 \text{ at } x = 0, l \qquad (6.7.5)$$

The Ritz approximation for this problem is of the form

$$w = w_o + \sum_{j=1}^{N} c_j w_j \qquad (6.7.6)$$

With no nonhomogeneous boundary conditions, $w_o = 0$. For this example, we choose a polynomial form for the trial solution. An appropriate choice that satisfies the homogeneous conditions (6.7.4) is $w_j = x^j(l - x)$. Note this form does not satisfy the traction conditions (6.7.5). Using the previously developed relation for the potential energy (6.6.12), we get

$$\Pi = \int_0^l \left[\frac{EI}{2}\left(\frac{d^2w}{dx^2}\right)^2 - q_o w \right] dx$$

$$= \int_0^l \left[\frac{EI}{2}\left(\sum_{j=1}^{N} c_j [j(j-1)lx^{j-2} - j(j+1)x^{j-1}] \right)^2 - q_o \sum_{j=1}^{N} c_j x^j (l-x) \right] dx \qquad (6.7.7)$$

Continued

Retaining only a two-term approximation ($N = 2$), the coefficients are found to be

$$c_1 = \frac{q_o l^2}{24EI}, \quad c_2 = 0$$

and this gives the following approximate solution

$$w = \frac{q_o l^2}{24EI} x(l - x) \tag{6.7.8}$$

Note that the approximate solution predicts a parabolic displacement distribution, while the exact solution to this problem is given by the cubic relation

$$w = \frac{q_o x}{24EI}\left(l^3 + x^3 - 2lx^2\right) \tag{6.7.9}$$

Actually, for this special case, the exact solution can be obtained from a Ritz scheme by including polynomials of degree three.

Other similar approximate techniques have been developed based on variational principles of complementary energy or the Reissner mixed formulation. A more generalized approximating scheme called the *weighted residual method* includes Ritz and several other techniques within the general approach. Although these approximate variational methods offer the potential to solve complex problems of engineering interest, they suffer a very important drawback involved with the selection of the approximating functions. Apart from the general properties the functions are required to satisfy, there exists no systematic procedure of constructing them. The selection process becomes more difficult when the domain is geometrically complex and/or the boundary conditions are complicated. Thus, these schemes have had limited success in solving such complicated problems. However, because these methods can easily provide approximate solutions over domains of simple shape with predetermined boundary conditions, they are ideally suited for *finite element* techniques, whereby a geometrically complex domain is divided into subdomains of simple geometry. Over each subdomain or element the governing differential equation may be formulated using variational methods, and the approximation functions can then be systematically generated for each typical element (Reddy, 2006). More details on these techniques are given in Chapter 16.

References

Banerjee PK, Butterfield R: *Boundary element methods in engineering science*, London, UK, 1981, McGraw-Hill.

Bharatha S, Levinson M: On physically nonlinear elasticity, *J Elast* 7:307–324, 1977.

Boresi AP, Chang KP: *Elasticity in engineering mechanics*, New York, 2000, Wiley.

Evans RJ, Pister KS: Constitutive equations for a class of nonlinear elastic solids, *Int J Solids Struct* 2:427–445, 1966.

Fung YC, Tong P: *Classical and computational solid mechanics*, Singapore, 2001, World Scientific.

Langhaar HL: *Energy methods in applied mechanics*, New York, 1962, Wiley.

Mura T, Koya T: *Variational methods in mechanics*, New York, 1992, Oxford University Press.

Reddy JN: *Energy and variational methods in applied mechanics*, New York, 1984, Wiley.

Reddy JN: *An introduction to the finite element method*, New York, 2006, McGraw-Hill.

Reismann H, Pawlik PS: *Elasticity theory and applications*, New York, 1980, Wiley.
Sadd MH: *Continuum Mechanics Modeling of Material Behavior*, Elsevier, 2019.
Sokolnikoff IS: *Mathematical theory of elasticity*, New York, 1956, McGraw-Hill.
Truesdell C: The meaning of Betti's reciprocal theorem, *J Res Natl Bureau Standards − B* 67B:85−86, 1963.
Washizu K: *Variational methods in elasticity and plasticity*, New York, 1968, Pergamon Press.

Exercises

6.1 The uniaxial deformation case as shown in Fig. 6.1 was used to determine the strain energy under uniform stress with zero body force. Determine this strain energy for the case in which the stress varies continuously as a function of x and also include the effect of a body force F_x. Neglecting higher-order terms, show that the result is the same as previously given by (6.1.4).

6.2 Since the strain energy has physical meaning that is independent of the choice of coordinate axes, it must be invariant to all coordinate transformations. Because U is a *quadratic form* in the strains or stresses, it cannot depend on the third invariants III_e or I_3, and so it must depend only on the first two invariants of the strain or stress tensors. Show that the strain energy can be written in the following forms

$$U = \left(\frac{1}{2}\lambda + \mu\right)I_e^2 - 2\mu II_e$$

$$= \frac{1}{2E}\left(I_1^2 - 2(1 + \nu)I_2\right)$$

6.3 Starting with the general expression (6.1.7), explicitly develop forms (6.1.9) and (6.1.10) for the strain energy density.

6.4 Differentiate the general three-dimensional isotropic strain energy form (6.1.9) to show that

$$\sigma_{ij} = \frac{\partial U(e)}{\partial e_{ij}}$$

6.5 For the isotropic case, express the strain energy function in terms of the principal strains, and then by direct differentiation show that

$$\sigma_1 = \frac{\partial U}{\partial e_1}, \quad \sigma_2 = \frac{\partial U}{\partial e_2}, \quad \sigma_3 = \frac{\partial U}{\partial e_3}$$

6.6 Using equations (6.1.12), develop the symmetry relations (6.1.13), and use these to prove the symmetry in the elasticity tensor $C_{ijkl} = C_{klij}$.

6.7 For the general anisotropic case with $\sigma_{ij} = C_{ijkl}e_{kl}$ assuming that $C_{ijkl} = C_{klij}$ show that $\sigma_{ij} = \partial U(e)/\partial e_{ij}$.

6.8 In light of Exercise 6.2, consider the formulation where the strain energy is assumed to be a function of the two invariants $U = U(I_e, II_e)$. Show that using the relation $\sigma_{ij} = \partial U/\partial e_{ij}$ and employing the chain rule, yields the expected constitutive law (4.2.7).

6.9 Consider the case of a *nonlinear elastic material*, and extend Exercise 6.8 for the case where the strain energy depends on *three invariants of the strain tensor* $U = U(I_{1e}, I_{2e}, I_{3e})$, where

$$I_{1e} = e_{kk}, \quad I_{2e} = \frac{1}{2} e_{km} e_{km}, \quad I_{3e} = \frac{1}{3} e_{km} e_{kn} e_{mn}$$

Note that this new choice of invariants does not change any fundamental aspects of the problem. Show that using the form $\sigma_{ij} = \partial U / \partial e_{ij}$ and employing the chain rule, yields the nonlinear relation $\sigma_{ij} = \phi_1 \delta_{ij} + \phi_2 e_{ij} + \phi_3 e_{ik} e_{jk}$ where $\phi_i = \phi_i(I_{je}) = \partial U / \partial I_{ie}$. This type of theory is often referred to as *physically nonlinear elasticity* in that the constitutive form retains the small deformation strain tensor but includes higher order nonlinear behavior in the constitutive law; see Evans and Pister (1966) and Bharatha and Levinson (1977).

6.10 Verify the decomposition of the strain energy into volumetric and deviatoric parts as given by equations (6.1.16) and (6.1.17).

6.11 Starting with relations (6.1.16) and (6.1.17), show that the volumetric and distortional strain energies can be expressed in terms of the invariants of the stress matrix as

$$U_v = \frac{1 - 2v}{6E} I_1^2$$

$$U_d = \frac{1}{6\mu} (I_1^2 - 3I_2)$$

Results from Exercise 6.10 may be helpful.

6.12 Show that the distortional strain energy given by (6.1.17) is related to the octahedral shear stress $(3.5.4)_2$ by the relation

$$U_d = \frac{3}{2} \frac{1 + v}{E} \tau_{oct}^2 = \frac{3}{4\mu} \tau_{oct}^2$$

Results from Exercise 3.5 may be helpful.

6.13 A two-dimensional state of *plane stress* in the x,y-plane is defined by the stress matrix

$$\sigma_{ij} = \begin{bmatrix} \sigma_x & \tau_{xy} & 0 \\ \tau_{xy} & \sigma_y & 0 \\ 0 & 0 & 0 \end{bmatrix}$$

Determine the strain energy density for this case in terms of these nonzero stress components.

6.14 The stress field for a beam of length $2l$ and depth $2c$ under end bending moments M (see Fig. 8.2) is given by

$$\sigma_x = -\frac{3M}{2c^3} y, \quad \sigma_y = \sigma_z = \tau_{xy} = \tau_{yz} = \tau_{zx} = 0$$

Determine the strain energy density and show that the total strain energy in the beam is given by

$$U_T = \frac{3M^2 l}{2Ec^3} = \frac{M^2 l}{EI}$$

where I is the section moment of inertia. Assume unit thickness in the z-direction.

6.15 The stress field for the torsion of a rod of circular cross-section is given by

$$\sigma_x = \sigma_y = \sigma_z = \tau_{xy} = 0, \tau_{xz} = -\mu\alpha y, \tau_{yz} = \mu\alpha x$$

where α is a constant and the z-axis coincides with the axis of the rod. Evaluate the strain energy density for this case, and determine the total strain energy in a rod with section radius R and length L.

6.16 From Chapter 9 using the Saint-Venant formulation, the stress field for the torsion of rod of general cross-section R can be expressed by

$$\sigma_x = \sigma_y = \sigma_z = \tau_{xy} = 0, \quad \tau_{xz} = \frac{\partial \phi}{\partial y}, \quad \tau_{yz} = -\frac{\partial \phi}{\partial x}$$

where $\phi = \phi(x, y)$ is the Prandtl stress function. Show that the total strain energy in a rod of length L is given by

$$U = \frac{L}{2\mu} \iint_R \left[\left(\frac{\partial \phi}{\partial x} \right)^2 + \left(\frac{\partial \phi}{\partial y} \right)^2 \right] dx dy$$

Next show that the total potential energy per unit length can be expressed by

$$\Pi = \iint_R \left\{ \frac{1}{2\mu} \left[\left(\frac{\partial \phi}{\partial x} \right) + \left(\frac{\partial \phi}{\partial y} \right)^2 \right] - 2\alpha\phi \right\} dx dy$$

where α is the constant angle of twist /length and reference should be made to relation (9.3.18).

6.17 Using the reciprocal theorem, choose the first state as $u_i^{(1)} = Ax_i$, $F_i^{(1)} = 0$, $T_i^{(1)} = 3kAn_i$, and take the second state as u_i, F_i, T_i to show that the change in volume of the body is given by

$$\Delta V = \int_V e_{ii} dV = \frac{1}{3k} \left\{ \int_V F_i x_i dV + \int_S T_i x_i dS \right\}$$

where A is an arbitrary constant and k is the bulk modulus.

6.18 Rework Example 6.2 using the trigonometric Ritz approximation $w_j = \sin \frac{j\pi x}{l}$. Develop a two-term approximate solution, and compare it with the displacement solution developed in the text. Also compare each of these approximations with the exact solution (6.7.9) at midspan $x = l/2$.

6.19 Using the bending formulae (6.6.9), compare the maximum bending stresses from the cases presented in Example 6.2 and Exercise 6.18. Numerically compare these results with the exact solution; see (6.7.9) at midspan $x = l/2$.

Two-dimensional formulation

Because of the complexity of the elasticity field equations, analytical closed-form solutions to fully three-dimensional problems are very difficult to accomplish. Thus, most solutions are developed for reduced problems that typically include axisymmetry or two-dimensionality to simplify particular aspects of the formulation and solution. We now wish to examine in detail the formulation of two-dimensional problems in elasticity. Our initial formulation will result in a boundary-value problem cast within a two-dimensional domain in the x,y-plane using Cartesian coordinates. This work will then be reformulated in polar coordinates to allow for the development of important solutions in that coordinate system. Because all real elastic structures are three-dimensional, the theories set forth here will be approximate models. The nature and accuracy of the approximation depends on problem and loading geometry. Although four different formulations are developed, the two basic theories of *plane strain* and *plane stress* represent the fundamental plane problem in elasticity. These two theories apply to significantly different types of two-dimensional bodies; however, their formulations yield very similar field equations. It will be shown that these two theories can be reduced to one governing equation in terms of a single unknown stress function. This reduction then allows many solutions to be generated to problems of engineering interest, and such solutions are presented in the following chapter. A detailed account of the history and development of plane elasticity theory has been given by Teodorescu (1964).

7.1 Plane strain

Consider an infinitely long cylindrical (prismatic) body shown in Fig. 7.1. If the body forces and tractions on the lateral boundaries are independent of the z-coordinate and have no z component, then the deformation field within the body can be taken in the reduced form

$$u = u(x, y), \quad v = v(x, y), \quad w = 0 \qquad (7.1.1)$$

This deformation is referred to as a state of *plane strain* in the x,y-plane. It should be obvious that for such a case all cross-sections R will have identical displacements, and thus the three-dimensional problem is reduced to a two-dimensional formulation in region R in the x,y-plane.

Using the strain−displacement relations (2.2.5), the strains corresponding to this plane problem become

$$e_x = \frac{\partial u}{\partial x}, \quad e_y = \frac{\partial v}{\partial y}, \quad e_{xy} = \frac{1}{2}\left(\frac{\partial u}{\partial y} + \frac{\partial v}{\partial x}\right) \qquad (7.1.2)$$

$$e_z = e_{xz} = e_{yz} = 0$$

Elasticity. https://doi.org/10.1016/B978-0-12-815987-3.00007-4

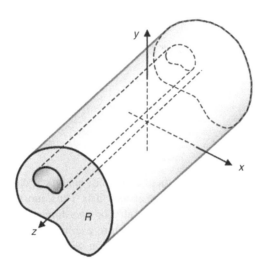

FIG. 7.1 Long cylindrical body representing plane strain conditions.

From the isotropic form of Hooke's law (4.2.8), the allowable stresses reduce to

$$
\begin{aligned}
\sigma_x &= \lambda(e_x + e_y) + 2\mu e_x \\
\sigma_y &= \lambda(e_x + e_y) + 2\mu e_y \\
\sigma_z &= \lambda(e_x + e_y) = v(\sigma_x + \sigma_y) \\
\tau_{xy} &= 2\mu e_{xy}, \quad \tau_{xz} = \tau_{yz} = 0
\end{aligned}
\tag{7.1.3}
$$

Note that the second expression for σ_z has used the first two relations of (7.1.3) to write σ_z in terms of the in-plane stress components. Thus, once σ_x and σ_y are determined, σ_z is easily found from Hooke's law. For this case, although $e_z = 0$, the corresponding normal stress σ_z will not in general vanish. It should be recognized that all strain and stress components will be functions of only the in-plane coordinates x and y.

For plane strain, the general equilibrium equations (3.7.5) reduce to

$$
\begin{aligned}
\frac{\partial \sigma_x}{\partial x} + \frac{\partial \tau_{xy}}{\partial y} + F_x &= 0 \\
\frac{\partial \tau_{xy}}{\partial x} + \frac{\partial \sigma_y}{\partial y} + F_y &= 0
\end{aligned}
\tag{7.1.4}
$$

where the third equation will vanish identically. Using relations (7.1.2) and (7.1.3), the equilibrium relations can be expressed in terms of displacement, yielding Navier equations

$$
\begin{aligned}
\mu \nabla^2 u + (\lambda + \mu) \frac{\partial}{\partial x}\left(\frac{\partial u}{\partial x} + \frac{\partial v}{\partial y}\right) + F_x &= 0 \\
\mu \nabla^2 v + (\lambda + \mu) \frac{\partial}{\partial y}\left(\frac{\partial u}{\partial x} + \frac{\partial v}{\partial y}\right) + F_y &= 0
\end{aligned}
\tag{7.1.5}
$$

where ∇^2 is the two-dimensional Laplacian operator $\nabla^2 = (\partial^2/\partial x^2) + (\partial^2/\partial y^2)$.

With regard to strain compatibility for plane strain, the Saint-Venant relations (2.6.2) reduce to

$$\frac{\partial^2 e_x}{\partial y^2} + \frac{\partial^2 e_y}{\partial x^2} = 2\frac{\partial^2 e_{xy}}{\partial x \partial y} \tag{7.1.6}$$

Expressing this relation in terms of stress gives the corresponding Beltrami−Michell equation

$$\nabla^2 (\sigma_x + \sigma_y) = -\frac{1}{1 - \nu} \left(\frac{\partial F_x}{\partial x} + \frac{\partial F_y}{\partial y} \right) \tag{7.1.7}$$

Thus, the plane strain problem is formulated in the two-dimensional region R with boundary S as shown in Fig. 7.2. The displacement formulation is given by relations (7.1.5) with boundary conditions

$$u = u_b(x, y), \quad v = v_b(x, y) \ \text{on} \ S \tag{7.1.8}$$

while the stress or traction formulation includes field relations (7.1.4) and (7.1.7) with boundary conditions

$$\begin{aligned} T_x^n = T_x^{(b)}(x, y) = \sigma_x^{(b)} n_x + \tau_{xy}^{(b)} n_y \\ T_y^n = T_y^{(b)}(x, y) = \tau_{xy}^{(b)} n_x + \sigma_y^{(b)} n_y \ \text{on} \ S \end{aligned} \tag{7.1.9}$$

Note that from our initial assumptions for plane strain, $T_z^n = 0$. The solution to the plane strain problem involves the determination of the in-plane displacements, strains, and stresses $\{u, v, e_x, e_y, e_{xy}, \sigma_x, \sigma_y, \tau_{xy}\}$ in R. The out-of-plane stress σ_z can be determined from the in-plane stresses by using relation $(7.1.3)_3$. This then completes our formulation of plane strain.

Before moving on to another case, let us consider the situation in which the cylindrical body in Fig. 7.1 is now of finite length. First consider the situation in which the body has *fixed and frictionless ends* at say $(z = \pm l)$. This case leads to end conditions

$$w(x, y, \pm l) = 0, \tau_{xz}(x, y, \pm l) = \tau_{yz}(x, y, \pm l) = 0$$

But these conditions are identically satisfied by the original plane strain formulation, and thus the original formulation also satisfies this finite length problem. Note that the restraining forces at the ends

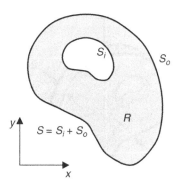

FIG. 7.2 Typical domain for the plane elasticity problem.

can be determined by integrating σ_z over the cross-section R. Although this specific problem has limited practical applications, the solution can be applied in an approximate sense for a long cylinder with any end conditions using Saint-Venant's principle.

If we wish to find the solution to a long but finite cylindrical body with no end tractions, a corrective solution must be added to the usual plane strain solution to remove the unwanted end loadings. Such a corrective solution must have zero tractions on the lateral sides of the body and prescribed end tractions equal but opposite to that obtained from the plane strain solution. Finding such a corrective solution to satisfy exact pointwise traction conditions on the ends is normally quite difficult. Commonly the Saint-Venant principle is invoked and the exact conditions are replaced by a simpler, statically equivalent distribution. Exercise 7.4 considers a specific problem of this type.

7.2 Plane stress

The second type of two-dimensional theory applies to domains bounded by two parallel planes separated by a distance that is small in comparison to other dimensions in the problem. Again, choosing the x, y-plane to describe the problem, the domain is bounded by two planes $z = \pm h$, as shown in Fig. 7.3. The theory further assumes that these planes are stress free, and thus $\sigma_z = \tau_{xz} = \tau_{yz} = 0$ on each face. Because the region is thin in the z direction, there can be little variation in these stress components through the thickness, and thus they will be approximately zero throughout the entire domain. Finally, because the region is thin in the z direction it can be argued that the other nonzero stress components will have little variation with z. These arguments can then be summarized by the stress state

$$\sigma_x = \sigma_x(x,y), \quad \sigma_y = \sigma_y(x,y), \quad \tau_{xy} = \tau_{xy}(x,y), \quad \sigma_z = \tau_{xz} = \tau_{yz} = 0 \qquad (7.2.1)$$

and this form constitutes a state of *plane stress* in an elastic solid. In order to maintain a stress field independent of z, there can be no body forces or surface tractions in the z direction. Furthermore, the nonzero body forces and tractions must be independent of z or distributed symmetrically about the midplane through the thickness, thus allowing average values to be used. Therefore, plane stress problems may be thought of as in-plane deformation of thin elastic plates.

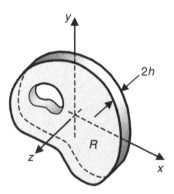

FIG. 7.3 Thin elastic plate representing plane stress conditions.

Using the simplified plane stress state, the corresponding strain field follows from Hooke's law

$$e_x = \frac{1}{E}(\sigma_x - \nu\sigma_y)$$

$$e_y = \frac{1}{E}(\sigma_y - \nu\sigma_x)$$

$$e_z = -\frac{\nu}{E}(\sigma_x + \sigma_y) = -\frac{\nu}{1-\nu}(e_x + e_y) \qquad (7.2.2)$$

$$e_{xy} = \frac{1+\nu}{E}\tau_{xy}, \quad e_{xz} = e_{yz} = 0$$

Similar to the previous plane strain theory, the second expression for e_z has used the first two relations of (7.2.2) to write the out-of-plane strain in terms of in-plane components. Note that although $e_z = 0$ for plane strain, it will not in general vanish for plane stress. It should be apparent from (7.2.2) that all strains will be independent of z. Relations (7.2.2) can be inverted to express the stresses in terms of the strains (see Exercise 7.6).

The strain–displacement equations for plane stress reduce to

$$e_x = \frac{\partial u}{\partial x}, \quad e_y = \frac{\partial v}{\partial y}, \quad e_z = \frac{\partial w}{\partial z}$$

$$e_{xy} = \frac{1}{2}\left(\frac{\partial u}{\partial y} + \frac{\partial v}{\partial x}\right)$$

$$e_{yz} = \frac{1}{2}\left(\frac{\partial v}{\partial z} + \frac{\partial w}{\partial y}\right) = 0 \qquad (7.2.3)$$

$$e_{xz} = \frac{1}{2}\left(\frac{\partial u}{\partial z} + \frac{\partial w}{\partial x}\right) = 0$$

The relations involving the three out-of-plane strains e_z, e_{xz}, and e_{yz} produce some unwanted results. For example, the last two relations of (7.2.3) imply that the in-plane displacements u and v are functions of z, thus making the theory three-dimensional. Likewise, the relation for e_z when viewed with Eq. (7.2.2)$_3$ implies that w is a linear function of z. Exercise 7.8 investigates these issues in more detail, and further discussion is given in Timoshenko and Goodier (1970), Article 98. Normally, these results are not used in the theory and this leads to an approximation in the formulation.

Under plane stress conditions, the equilibrium equations reduce to the identical form as in plane strain theory

$$\frac{\partial \sigma_x}{\partial x} + \frac{\partial \tau_{xy}}{\partial y} + F_x = 0$$

$$\frac{\partial \tau_{xy}}{\partial x} + \frac{\partial \sigma_y}{\partial y} + F_y = 0 \qquad (7.2.4)$$

where F_x and F_y are functions of x and y and $F_z = 0$. Expressing these equilibrium equations in terms of the displacements yields the Navier equations for plane stress

$$\mu\nabla^2 u + \frac{E}{2(1-v)}\frac{\partial}{\partial x}\left(\frac{\partial u}{\partial x} + \frac{\partial v}{\partial y}\right) + F_x = 0$$

$$\mu\nabla^2 v + \frac{E}{2(1-v)}\frac{\partial}{\partial y}\left(\frac{\partial u}{\partial x} + \frac{\partial v}{\partial y}\right) + F_y = 0$$

(7.2.5)

Notice that the corresponding system for plane strain (7.1.5) is similar *but not identical* to this plane stress result.

To develop the plane stress reduction in the compatibility relations (2.6.2), the three relations involving the out-of-plane strain component e_z are commonly neglected. This again brings out the approximate nature of the plane stress formulation. The neglected compatibility relations are further examined in Exercise 7.9. Under these conditions, the only remaining compatibility relation for plane stress is identical to that found in plane strain

$$\frac{\partial^2 e_x}{\partial y^2} + \frac{\partial^2 e_y}{\partial x^2} = 2\frac{\partial^2 e_{xy}}{\partial x \partial y}$$

(7.2.6)

Expressing this relation in terms of stress gives the corresponding Beltrami–Michell equation

$$\nabla^2(\sigma_x + \sigma_y) = -(1+v)\left(\frac{\partial F_x}{\partial x} + \frac{\partial F_y}{\partial y}\right)$$

(7.2.7)

Notice that this result is again similar but not identical to the corresponding plane strain relation.

Similar to plane strain, the plane stress problem is formulated in the two-dimensional region R with boundary S (see Fig. 7.2). The displacement formulation is specified by the governing Navier relations (7.2.5) with boundary conditions of the form given by Eq. (7.1.8). The stress or traction formulation includes the governing equations (7.2.4) and (7.2.7) with boundary conditions of the form (7.1.9). The solution to the plane stress problem then involves the determination of the in-plane displacements, strains, and stresses $\{u, v, e_x, e_y, e_{xy}, \sigma_x, \sigma_y, \tau_{xy}\}$ in R. The out-of-plane strain e_z can be determined from the in-plane strains by using relation (7.2.2)$_3$. This completes our formulation of plane stress.

In following the formulation developments of plane strain and plane stress, it should be apparent that although unfortunately the two theories do not have identical governing equations, many relations are quite similar. Note that each theory has identical equilibrium equations (7.1.4) and (7.2.4) and boundary condition specifications. Furthermore, each theory has similar Navier equations (7.1.5) and (7.2.5) and compatibility relations (7.1.7) and (7.2.7). Focusing attention on these similar relations, it is observed that the basic difference between these equations is simply some coefficients involving the elastic material constants. This leads to the idea that perhaps a simple change in elastic moduli would bring one set of relations into an exact match with the corresponding result from the other plane theory. This in fact is the case, and it is easily shown that through transformation of the elastic moduli E and v, as specified in Table 7.1, all plane stress problems can be transformed into the corresponding plane strain model and vice versa. Thus, solving one type of plane problem automatically gives the other solution through a simple transformation of elastic moduli in the final

Table 7.1 Elastic moduli transformation relations for conversion between plane stress and plane strain problems.

	E	ν
Plane stress to plane strain	$\dfrac{E}{1 - \nu^2}$	$\dfrac{\nu}{1 - \nu}$
Plane strain to plane stress	$\dfrac{E(1 + 2\nu)}{(1 + \nu)^2}$	$\dfrac{\nu}{1 + \nu}$

answer. It should be noted that for the particular value of Poisson's ratio $\nu = 0$, plane strain and plane stress solutions will be identical.

7.3 Generalized plane stress

Recall that the approximate nature of the plane stress formulation produced some inconsistencies, in particular out-of-plane behavior, and this resulted in some three-dimensional effects in which the in-plane displacements were functions of z. In order to avoid this situation, the more mathematically inclined elasticians have developed an alternate approach commonly referred to as *generalized plane stress*. This theory is based on *averaging* the field quantities through the thickness of the domain shown in Fig. 7.3. The averaging operator is defined by

$$\overline{\varphi}(x, y) = \frac{1}{2h} \int_{-h}^{h} \varphi(x, y, z) dz \tag{7.3.1}$$

and it is noted that this operation removes the z dependency from the function. We again assume that h is much smaller than other dimensions associated with the problem.

The tractions on surfaces $z = \pm h$ are again taken to be zero, while the loadings on the edge of the plate have no z component and are either independent of z or are symmetrically distributed through the thickness. Likewise, any body forces cannot have a z component and they must also be either independent of z or symmetrically distributed through the thickness. Under these assumptions, the out-of-plane displacement will be an odd function of z, implying $w(x,y,z) = -w(x,y,-z)$, and points on the middle plane will have no z displacement; that is, $w(x,y,0) = 0$. These conditions imply that the average value of w will be zero

$$\overline{w} = \frac{1}{2h} \int_{-h}^{h} w(x, y, z) dz = 0 \tag{7.3.2}$$

The assumed tractions on $z = \pm h$ can be expressed as

$$\sigma_z(x, y, \pm h) = \tau_{xz}(x, y, \pm h) = \tau_{yz}(x, y, \pm h) = 0 \tag{7.3.3}$$

The equilibrium equation in the z direction becomes

$$\frac{\partial \tau_{xz}}{\partial x} + \frac{\partial \tau_{yz}}{\partial y} + \frac{\partial \sigma_z}{\partial z} = 0$$

Evaluating this relation at $z = \pm h$ and using (7.3.3) drops the first two derivatives and gives

$$\frac{\partial \sigma_z(x, y, \pm h)}{\partial z} = 0$$

Thus, both σ_z and its normal derivative vanish at $z = \pm h$. A simple Taylor series expansion of σ_z through the thickness would then imply that this stress is of order h^2, and this further justifies the assumption that σ_z vanishes throughout the interior of the thin plate.

If we now take the average value of all remaining field equations, the resulting system is given by

$$\bar{u} = \bar{u}(x, y), \quad \bar{v} = \bar{v}(x, y), \quad \bar{w} = 0$$

$$\bar{\sigma}_z = \bar{\tau}_{xz} = \bar{\tau}_{yz} = 0$$

$$\bar{\sigma}_x = \lambda^* \left(\bar{e}_x + \bar{e}_y \right) + 2\mu \bar{e}_x$$

$$\bar{\sigma}_y = \lambda^* \left(\bar{e}_x + \bar{e}_y \right) + 2\mu \bar{e}_y \tag{7.3.4}$$

$$\bar{\tau}_{xy} = 2\mu \bar{e}_{xy}$$

$$\bar{e}_z = -\frac{\lambda}{\lambda + 2\mu} \left(\bar{e}_x + \bar{e}_y \right)$$

where $\lambda^* = \dfrac{2\lambda\mu}{(\lambda + 2\mu)}$. The equilibrium equations become

$$\frac{\partial \bar{\sigma}_x}{\partial x} + \frac{\partial \bar{\tau}_{xy}}{\partial y} + \bar{F}_x = 0$$

$$\frac{\partial \bar{\tau}_{xy}}{\partial x} + \frac{\partial \bar{\sigma}_y}{\partial y} + \bar{F}_y = 0 \tag{7.3.5}$$

and written in terms of displacements

$$\mu \nabla^2 \bar{u} + (\lambda^* + \mu) \frac{\partial}{\partial x} \left(\frac{\partial \bar{u}}{\partial x} + \frac{\partial \bar{v}}{\partial y} \right) + \bar{F}_x = 0$$

$$\mu \nabla^2 \bar{v} + (\lambda^* + \mu) \frac{\partial}{\partial y} \left(\frac{\partial \bar{u}}{\partial x} + \frac{\partial \bar{v}}{\partial y} \right) + \bar{F}_y = 0 \tag{7.3.6}$$

Note the coefficient $\lambda^* + \mu = E/2(1 - v)$. Finally, in terms of the averaged variables, all compatibility relations reduce to the single statement

$$\nabla^2 \left(\bar{\sigma}_x + \bar{\sigma}_y \right) = \frac{-2(\lambda^* + \mu)}{\lambda^* + 2\mu} \left(\frac{\partial \bar{F}_x}{\partial x} + \frac{\partial \bar{F}_y}{\partial y} \right) \tag{7.3.7}$$

and again the coefficient reduces as $2(\lambda^* + \mu)/(\lambda^* + 2\mu) = 1 + v$. It is then evident that generalized plane stress relations (7.3.4)–(7.3.7) in terms of the averaged values are the same as the original plane stress results in terms of the actual values.

The only advantage of pursuing the generalized plane stress formulation lies in the fact that all equations are satisfied *exactly* by these average variables, thereby eliminating the inconsistencies found in the previous plane stress formulation. However, this gain in rigor does not generally contribute much to applications, and thus only the plane strain and plane stress formulations from Sections 7.1 and 7.2 are normally used.

7.4 Antiplane strain

One additional plane theory of elasticity involves a formulation based on the existence of only out-of-plane deformation. This theory is sometimes used in geomechanics applications to model deformations of portions of the earth's interior. The formulation begins with the assumed displacement field

$$u = v = 0, \quad w = w(x, y) \tag{7.4.1}$$

For such a system of displacements, the strain field becomes

$$e_x = e_y = e_z = e_{xy} = 0$$
$$e_{xz} = \frac{1}{2} \frac{\partial w}{\partial x}, \quad e_{yz} = \frac{1}{2} \frac{\partial w}{\partial y} \tag{7.4.2}$$

and from Hooke's law the stresses reduce to

$$\sigma_x = \sigma_y = \sigma_z = \tau_{xy} = 0$$
$$\tau_{xz} = 2\mu e_{xz} = \mu \frac{\partial w}{\partial x}, \quad \tau_{yz} = 2\mu e_{yz} = \mu \frac{\partial w}{\partial y} \tag{7.4.3}$$

The equilibrium equations imply that

$$F_x = F_y = 0$$
$$\frac{\partial \tau_{xz}}{\partial x} + \frac{\partial \tau_{yz}}{\partial y} + F_z = 0 \tag{7.4.4}$$

and written out in terms of the single displacement component, the equilibrium statement becomes

$$\mu \nabla^2 w + F_z = 0 \tag{7.4.5}$$

where again ∇^2 is the two-dimensional Laplacian operator. It is observed that for zero body forces, the single displacement component satisfies Laplace's equation. Because many solution schemes can be applied to this equation, the displacement formulation provides a convenient method to solve this type of problem.

Similar to the other plane problems, antiplane strain is formulated in the two-dimensional region R with boundary S (see Fig. 7.2). The boundary conditions associated with the problem would take either the displacement form

$$w = w_b(x, y) \quad \text{on } S \tag{7.4.6}$$

or traction form:

$$T_z^n = T_z^{(b)}(x, y) = \tau_{xz}^{(b)} n_x + \tau_{yz}^{(b)} n_y$$

$$= \mu \left(\frac{\partial w}{\partial x} n_x + \frac{\partial w}{\partial y} n_y \right)^{(b)} \text{on } S \tag{7.4.7}$$

The solution to the antiplane strain problem then involves the determination of the out-of-plane displacement, strains, and stresses $\{w, e_{xz}, e_{yz}, \tau_{xz}, \tau_{yz}\}$ in R.

7.5 Airy stress function

Numerous solutions to plane strain and plane stress problems can be determined through the use of a particular stress function technique. The method employs the *Airy stress function* and will reduce the general formulation to a single governing equation in terms of a single unknown. The resulting governing equation is then solvable by several methods of applied mathematics, and thus many analytical solutions to problems of interest can be generated. The stress function formulation is based on the general idea of developing a representation for the stress field that satisfies equilibrium and yields a single governing equation from the compatibility statement.

The method is started by reviewing the equilibrium equations for the plane problem, either relations (7.1.4) or (7.2.4). For now, we retain the body forces but assume that they are derivable from a *potential function V* such that

$$F_x = -\frac{\partial V}{\partial x}, \quad F_y = -\frac{\partial V}{\partial y} \tag{7.5.1}$$

This assumption is not very restrictive because many body forces found in applications (e.g., gravity loading) fall into this category. Under form (7.5.1), the plane equilibrium equations can be written as

$$\frac{\partial(\sigma_x - V)}{\partial x} + \frac{\partial \tau_{xy}}{\partial y} = 0$$

$$\frac{\partial \tau_{xy}}{\partial x} + \frac{\partial(\sigma_y - V)}{\partial y} = 0 \tag{7.5.2}$$

It is observed that these equations will be identically satisfied by choosing a representation

$$\sigma_x = \frac{\partial^2 \phi}{\partial y^2} + V$$

$$\sigma_y = \frac{\partial^2 \phi}{\partial x^2} + V \tag{7.5.3}$$

$$\tau_{xy} = -\frac{\partial^2 \phi}{\partial x \partial y}$$

where $\phi = \phi(x, y)$ is an arbitrary form called the Airy stress function.

With equilibrium now satisfied, we focus attention on the remaining field equations in the stress formulation, that is, the compatibility relations in terms of stress. These equations were given by (7.1.7) for plane strain and (7.2.7) for plane stress, and it is noted that they differ only by the coefficient in front of the body force terms. Substituting the stress function form (7.5.3) into these compatibility relations gives the following pair

$$\frac{\partial^4 \phi}{\partial x^4} + 2\frac{\partial^4 \phi}{\partial x^2 \partial y^2} + \frac{\partial^4 \phi}{\partial y^4} = -\frac{1-2\nu}{1-\nu}\left(\frac{\partial^2 V}{\partial x^2} + \frac{\partial^2 V}{\partial y^2}\right) \cdots \text{plane strain}$$

$$\frac{\partial^4 \phi}{\partial x^4} + 2\frac{\partial^4 \phi}{\partial x^2 \partial y^2} + \frac{\partial^4 \phi}{\partial y^4} = -(1-\nu)\left(\frac{\partial^2 V}{\partial x^2} + \frac{\partial^2 V}{\partial y^2}\right) \cdots \text{plane stress}$$

(7.5.4)

which can also be written as

$$\nabla^4 \phi = -\frac{1-2\nu}{1-\nu}\nabla^2 V \cdots \text{plane strain}$$

$$\nabla^4 \phi = -(1-\nu)\nabla^2 V \cdots \text{plane stress}$$

(7.5.5)

The form $\nabla^4 = \nabla^2\nabla^2$ is called the *biharmonic operator*. If the body force vanishes or the potential function satisfies Laplace's equation $\nabla^2 V = 0$, then both the plane strain and plane stress forms reduce to

$$\frac{\partial^4 \phi}{\partial x^4} + 2\frac{\partial^4 \phi}{\partial x^2 \partial y^2} + \frac{\partial^4 \phi}{\partial y^4} = \nabla^4 \phi = 0$$

(7.5.6)

This relation is called the *biharmonic equation*, and its solutions are known as *biharmonic functions*. Thus, the plane problem of elasticity has been reduced to a single equation in terms of the Airy stress function ϕ. This function is to be determined in the two-dimensional region R bounded by the boundary S, as shown in Fig. 7.2. Appropriate boundary conditions over S are necessary to complete a solution. Using relations (7.5.3), traction boundary conditions would involve the specification of second derivatives of the stress function. However, this general traction condition can be reformulated to relate the resultant boundary loadings to first-order derivatives; see Section 11.5 or Sokolnikoff (1956) for details. An interesting review article by Meleshko (2003) gives a detailed historical overview of the formulation and solution methods for the two-dimensional biharmonic equation. Applications to specific boundary-value problems are demonstrated in the next chapter. Displacement boundary conditions require more development and are postponed until Chapter 10. Further general details on stress functions are given in Chapter 13.

It is interesting to observe that for the case of zero body forces, the governing Airy stress function equation (7.5.6) is the same for both plane strain and plane stress and is independent of elastic constants. Therefore, if the region is simply connected (see Fig. 2.9) and the boundary conditions specify only tractions, the stress fields for plane strain and plane stress will be identical and independent of elastic constants. Note, however, that the resulting strains and displacements calculated from these common stresses would not be the same for each plane theory. This occurs because plane strain and plane stress have different forms for Hooke's law and strain–displacement relations. Of course, because the two plane elasticity problems represent significantly different models, we would not expect that all parts of the solution field be identical. Problems with multiply connected regions or displacement boundary conditions bring additional displacement relations into the formulation, and thus we can no longer make the argument that the stress fields will be the same and remain independent of elastic moduli.

7.6 Polar coordinate formulation

Because we will make use of polar coordinates in the solution of many plane problems in elasticity, the previous governing equations will now be developed in this curvilinear system. Polar coordinates were originally presented in Fig. 1.8, and Example 1.5 developed the basic vector differential operations. For such a coordinate system, the solution to plane strain and plane stress problems involves the determination of the in-plane displacements, strains, and stresses $\{u_r, u_\theta, e_r, e_\theta, e_{r\theta}, \sigma_r, \sigma_\theta, \tau_{r\theta}\}$ in R subject to prescribed boundary conditions on S (see Fig. 7.2).

The polar coordinate form of the strain–displacement relations can be extracted from developments of Section 2.7 or results of Exercise 2.20. Dropping the z dependency in the cylindrical coordinate forms (2.7.3) directly gives the following desired results

$$e_r = \frac{\partial u_r}{\partial r}$$

$$e_\theta = \frac{1}{r}\left(u_r + \frac{\partial u_\theta}{\partial \theta}\right) \tag{7.6.1}$$

$$e_{r\theta} = \frac{1}{2}\left(\frac{1}{r}\frac{\partial u_r}{\partial \theta} + \frac{\partial u_\theta}{\partial r} - \frac{u_\theta}{r}\right)$$

These relations can also be developed using displacement and strain transformation laws (see Exercise 7.16). As per the discussion in Section 4.3, the basic form of Hooke's law will not change when moving to an orthogonal curvilinear system, and the cylindrical form given by relation (4.3.2) can be applied to the plane problem in polar coordinates. Thus, the original plane strain and plane stress forms for Hooke's law do not change other than a simple transformation of the subscripts from x and y to r and θ

Plane strain

$$\sigma_r = \lambda(e_r + e_\theta) + 2\mu e_r$$

$$\sigma_\theta = \lambda(e_r + e_\theta) + 2\mu e_\theta$$

$$\sigma_z = \lambda(e_r + e_\theta) = \nu(\sigma_r + \sigma_\theta)$$

$$\tau_{r\theta} = 2\mu e_{r\theta}, \quad \tau_{\theta z} = \tau_{rz} = 0$$

Plane stress

$$e_r = \frac{1}{E}(\sigma_r - \nu\sigma_\theta)$$

$$e_\theta = \frac{1}{E}(\sigma_\theta - \nu\sigma_r) \tag{7.6.2}$$

$$e_z = -\frac{\nu}{E}(\sigma_r + \sigma_\theta) = -\frac{\nu}{1-\nu}(e_r + e_\theta)$$

$$e_{r\theta} = \frac{1+\nu}{E}\tau_{r\theta}, \quad e_{\theta z} = e_{rz} = 0$$

Likewise, the results of Section 3.7 or Exercise 3.24 provide the appropriate forms for the equilibrium equations

$$\frac{\partial \sigma_r}{\partial r} + \frac{1}{r}\frac{\partial \tau_{r\theta}}{\partial \theta} + \frac{(\sigma_r - \sigma_\theta)}{r} + F_r = 0$$

$$\frac{\partial \tau_{r\theta}}{\partial r} + \frac{1}{r}\frac{\partial \sigma_\theta}{\partial \theta} + \frac{2\tau_{r\theta}}{r} + F_\theta = 0 \qquad (7.6.3)$$

Expressing the preceding relations in terms of displacements gives the following set of Navier equations

Plane strain

$$\mu\left(\nabla^2 u_r - \frac{2}{r^2}\frac{\partial u_\theta}{\partial \theta} - \frac{u_r}{r^2}\right) + (\lambda + \mu)\frac{\partial}{\partial r}\left(\frac{\partial u_r}{\partial r} + \frac{u_r}{r} + \frac{1}{r}\frac{\partial u_\theta}{\partial \theta}\right) + F_r = 0$$

$$\mu\left(\nabla^2 u_\theta + \frac{2}{r^2}\frac{\partial u_r}{\partial \theta} - \frac{u_\theta}{r^2}\right) + (\lambda + \mu)\frac{1}{r}\frac{\partial}{\partial \theta}\left(\frac{\partial u_r}{\partial r} + \frac{u_r}{r} + \frac{1}{r}\frac{\partial u_\theta}{\partial \theta}\right) + F_\theta = 0$$

Plane stress $\qquad (7.6.4)$

$$\mu\left(\nabla^2 u_r - \frac{2}{r^2}\frac{\partial u_\theta}{\partial \theta} - \frac{u_r}{r^2}\right) + \frac{E}{2(1-\nu)}\frac{\partial}{\partial r}\left(\frac{\partial u_r}{\partial r} + \frac{u_r}{r} + \frac{1}{r}\frac{\partial u_\theta}{\partial \theta}\right) + F_r = 0$$

$$\mu\left(\nabla^2 u_\theta + \frac{2}{r^2}\frac{\partial u_r}{\partial \theta} - \frac{u_\theta}{r^2}\right)_\theta + \frac{E}{2(1-\nu)}\frac{1}{r}\frac{\partial}{\partial \theta}\left(\frac{\partial u_r}{\partial r} + \frac{u_r}{r} + \frac{1}{r}\frac{\partial u_\theta}{\partial \theta}\right) + F_\theta = 0$$

where we have used results from Example 1.5, and the two-dimensional Laplacian is given by

$$\nabla^2 = \frac{\partial^2}{\partial r^2} + \frac{1}{r}\frac{\partial}{\partial r} + \frac{1}{r^2}\frac{\partial^2}{\partial \theta^2} \qquad (7.6.5)$$

Again using results from Example 1.5 and the fact that $\sigma_x + \sigma_y = \sigma_r + \sigma_\theta$, the compatibility equations (7.1.7) and (7.2.7) can be expressed as

$$\nabla^2(\sigma_r + \sigma_\theta) = -\frac{1}{1-\nu}\left(\frac{\partial F_r}{\partial r} + \frac{F_r}{r} + \frac{1}{r}\frac{\partial F_\theta}{\partial \theta}\right)...\text{plane strain}$$

$$\nabla^2(\sigma_r + \sigma_\theta) = -(1+\nu)\left(\frac{\partial F_r}{\partial r} + \frac{F_r}{r} + \frac{1}{r}\frac{\partial F_\theta}{\partial \theta}\right)...\text{plane stress} \qquad (7.6.6)$$

Relations (7.5.3) between the stress components and Airy function can be easily transformed to polar form using results from Exercise 3.3 and the chain rule to convert spatial derivatives. For the case of zero body forces, this yields

$$\sigma_r = \frac{1}{r}\frac{\partial \phi}{\partial r} + \frac{1}{r^2}\frac{\partial^2 \phi}{\partial \theta^2}$$

$$\sigma_\theta = \frac{\partial^2 \phi}{\partial r^2} \qquad (7.6.7)$$

$$\tau_{r\theta} = -\frac{\partial}{\partial r}\left(\frac{1}{r}\frac{\partial \phi}{\partial \theta}\right)$$

It can be verified that this form will satisfy the equilibrium equations (7.6.3) identically, and in the absence of body forces the compatibility relations (7.6.6) reduce to the biharmonic equation in polar coordinates

$$\nabla^4\phi = \left(\frac{\partial^2}{\partial r^2} + \frac{1}{r}\frac{\partial}{\partial r} + \frac{1}{r^2}\frac{\partial^2}{\partial\theta^2}\right)\left(\frac{\partial^2}{\partial r^2} + \frac{1}{r}\frac{\partial}{\partial r} + \frac{1}{r^2}\frac{\partial^2}{\partial\theta^2}\right)\phi = 0 \qquad (7.6.8)$$

Again, the plane problem is then formulated in terms of an Airy function $\phi(r, \theta)$ with a single governing biharmonic equation. Referring to Fig. 7.2, this function is to be determined in the two-dimensional region R bounded by the boundary S. Appropriate boundary conditions over S are necessary to complete a solution. Several example solutions in polar coordinates are given in the next chapter.

References

Meleshko VV: Selected topics in the history of the two-dimensional biharmonic problem, *Appl Mech Rev* 56: 33–85, 2003.
Sokolnikoff IS: *Mathematical theory of elasticity*, New York, 1956, McGraw-Hill.
Teodorescu PP: One hundred years of investigations in the plane problem of the theory of elasticity, *Appl Mech Rev* 17:175–186, 1964.
Timoshenko SP, Goodier JN: *Theory of elasticity*, New York, 1970, McGraw-Hill.

Exercises

7.1 Invert the plane strain form of Hooke's law (7.1.3) and express the strains in terms of the stresses as

$$e_x = \frac{1+\nu}{E}[(1-\nu)\sigma_x - \nu\sigma_y]$$

$$e_y = \frac{1+\nu}{E}[(1-\nu)\sigma_y - \nu\sigma_x]$$

$$e_{xy} = \frac{1+\nu}{E}\tau_{xy}$$

7.2 For the plane strain case, develop Navier equations (7.1.5) and the Beltrami–Michell compatibility relation (7.1.7).

7.3 Verify the following relations for the case of plane strain with constant body forces

$$\frac{\partial}{\partial y}\nabla^2 u = \frac{\partial}{\partial x}\nabla^2 v$$

$$\frac{\partial}{\partial x}\nabla^2 u = -\frac{\partial}{\partial y}\nabla^2 v$$

$$\nabla^4 u = \nabla^4 v = 0$$

7.4 At the end of Section 7.1, it was pointed out that the plane strain solution to a cylindrical body of finite length with zero end tractions could be found by adding a corrective solution to remove the unwanted end loadings being generated from the axial stress relation $\sigma_z = \nu(\sigma_x + \sigma_y)$. Using Saint-Venant's principle, show that such a corrective solution may be generated using a simple strength of materials approximation incorporating axial and bending stresses of the form $\sigma_z^{(c)} = Ax + By + C$, where A, B, and C are constants. Using principal centroidal x,y-axes, show how these constants could be determined.

7.5 In the absence of body forces, show that the following stresses

$$\sigma_x = kxy, \quad \sigma_y = kx, \quad \sigma_z = \nu kx(1 + y)$$

$$\tau_{xy} = -\frac{1}{2}ky^2, \quad \tau_{xz} = \tau_{yz} = 0, \quad k = \text{constant}$$

satisfy the plane strain stress formulation relations.

7.6 Invert the plane stress form of Hooke's law (7.2.2) and express the stresses in terms of the strain components

$$\sigma_x = \frac{E}{1 - \nu^2}(e_x + \nu e_y)$$

$$\sigma_y = \frac{E}{1 - \nu^2}(e_y + \nu e_x)$$

$$\tau_{xy} = \frac{E}{1 + \nu}e_{xy}$$

7.7 Using the results from Exercise 7.6, eliminate the stresses from the plane stress equilibrium equations and develop Navier equations (7.2.5). Also, formally establish the Beltrami–Michell equation (7.2.7).

7.8 For plane stress, investigate the unwanted three-dimensional results coming from integration of the strain–displacement relations involving the out-of-plane strains e_z, e_{xz}, and e_{yz}.

7.9 For the plane stress problem, show that the neglected nonzero compatibility relations involving the out-of-plane component e_z are

$$\frac{\partial^2 e_z}{\partial x^2} = 0, \quad \frac{\partial^2 e_z}{\partial y^2} = 0, \quad \frac{\partial^2 e_z}{\partial x \partial y} = 0$$

Next integrate these relations to show that the most general form for this component is given by

$$e_z = ax + by + c$$

where a, b, and c are arbitrary constants. In light of relation (7.2.2)$_3$, will this result for e_z be satisfied in general? Explain your reasoning.

7.10 Using the transformation results shown in Table 7.1, determine the required corresponding changes in Lamé's constant λ and the shear modulus μ.

7.11 Verify the validity of the transformation relations given in Table 7.1 by:
   (a) Transforming the plane strain equations (7.1.5) and (7.1.7) into the corresponding plane stress results.
   (b) Transforming the plane stress equations (7.2.5) and (7.2.7) into the corresponding plane strain results.

7.12 Verify the validity of the transformation relations given in Table 7.1 by:
   (a) Transforming the plane strain Hooke's law (7.1.3) into the corresponding plane stress results given in Exercise 7.6.
   (b) Transforming the plane stress Hooke's law (7.2.2) into the corresponding plane strain results given in Exercise 7.1.

7.13* For the pure bending problem shown in Example 8.2, the plane stress displacement field was determined and given by relations (8.1.22)$_2$ as

$$u = -\frac{Mxy}{EI}, \quad v = \frac{M}{2EI}\left[vy^2 + x^2 - l^2\right], \quad -l \le x \le l$$

Using the appropriate transformation relations from Table 7.1, determine the corresponding displacements for the plane strain case. Next develop a comparison plot for each case of the y-displacement along the x-axis ($y = 0$) with Poisson's ratio $v = 0.4$. Use dimensionless variables and plot $v(x, 0)/(Ml^2/EI)$ versus x/l. Which displacement is larger and what happens as Poisson's ratio goes to zero?

7.14* Consider the problem of a stress-free hole in an infinite domain under equal and uniform far-field loading T, as shown in Fig. 8.11. The plane strain radial displacement solution for this problem is found to be

$$u_r = \frac{T(1+v)}{E}\left[(1 - 2v)r + \frac{r_1^2}{r}\right]$$

where r_1 is the hole radius. Using the appropriate transformation relations from Table 7.1, determine the corresponding displacement for the plane stress case. Next develop a comparison plot for each case of the radial displacement versus radial distance r with Poisson's ratio $v = 0.4$. Use dimensionless variables and plot $u_r/(Tr_1/E)$ versus r/r_1 over the range $0 \le r/r_1 \le 10$. Which displacement is larger and what happens as Poisson's ratio goes to zero? Finally plot the dimensionless radial displacement on the hole boundary $r = r_1$ versus Poisson's ratio over the range $0 \le v \le 0.5$.

7.15 Explicitly develop the governing equations (7.5.4) in terms of the Airy function for plane strain and plane stress.

7.16 Derive the polar coordinate strain–displacement relations (7.6.1) by using the transformation equations

$$u = u_r \cos \theta - u_\theta \sin \theta$$

$$v = u_r \sin \theta + u_\theta \cos \theta$$

$$e_r = e_x \cos^2 \theta + e_y \sin^2 \theta + 2e_{xy} \sin \theta \cos \theta$$

$$e_\theta = e_x \sin^2 \theta + e_y \cos^2 \theta - 2e_{xy} \sin \theta \cos \theta$$

$$e_{r\theta} = -e_x \sin \theta \cos \theta + e_y \sin \theta \cos \theta + e_{xy}\left(\cos^2 \theta - \sin^2 \theta\right)$$

7.17 Using the polar strain–displacement relations (7.6.1), derive the strain–compatibility relation

$$\frac{\partial}{\partial r}\left(2r\frac{\partial e_{r\theta}}{\partial \theta} - r^2\frac{\partial e\theta}{\partial r}\right) + r\frac{\partial e_r}{\partial r} - \frac{\partial^2 e_r}{\partial \theta^2} = 0$$

7.18 For the axisymmetric polar case where all field functions depend only on the radial coordinate r, show that a strain compatibility statement can expressed as

$$e_r = \frac{d}{dr}(re_\theta)$$

while the shear strain-displacement relation becomes

$$e_{r\theta} = \frac{1}{2}r\frac{d}{dr}\left(\frac{u_\theta}{r}\right)$$

7.19 For the plane strain case, starting with the equilibrium equations (7.6.3), develop Navier equations (7.6.4)$_{1,2}$. Also verify the compatibility relation (7.6.6)$_1$.

7.20 For the plane stress case, starting with the equilibrium equations (7.6.3), develop Navier equations (7.6.4)$_{3,4}$. Also verify the compatibility relation (7.6.6)$_2$.

7.21 Using the chain rule and stress transformation theory, develop the stress–Airy function relations (7.6.7). Verify that this form satisfies equilibrium identically.

7.22 For rigid-body motion, the strains will vanish. Under these conditions, integrate the strain–displacement relations (7.6.1) to show that the most general form of a rigid-body motion displacement field in polar coordinates is given by

$$u_{r*} = a\sin\theta + b\cos\theta$$
$$u_{\theta*} = a\cos\theta - b\sin\theta + cr$$

where a, b, c are constants. Also show that this result is consistent with the Cartesian form given by relation (2.2.9).

7.23 Consider the two-dimensional plane stress field of the form $\sigma_r = \sigma_r(r, \theta)$, $\sigma_\theta = \tau_{r\theta} = 0$. This is commonly referred to as a *radial stress distribution*. For this case, first show that the equilibrium equations reduce to

$$\frac{\partial \sigma_r}{\partial r} + \frac{\sigma_r}{r} = \frac{1}{r}\frac{\partial}{\partial r}(r\sigma_r) = 0$$

Next integrate this result to get $\sigma_r = f(\theta)/r$ where $f(\theta)$ is an arbitrary function of θ. Finally using compatibility relation (7.6.6), show that the final form for the non-zero stress is given by

$$\sigma_r = \frac{1}{r}(A\sin\theta + B\cos\theta)$$

Note that this matches with the Flamant solution given in Section 8.4.7.

7.24 Consider the antiplane strain problem of a distributed loading F (per unit length) along the entire z-axis of an infinite medium. This will produce an axisymmetric deformation field. Using cylindrical coordinates, show that in the absence of body forces, governing equation (7.4.5) will reduce to

$$\frac{d^2 w}{dr^2} + \frac{1}{r}\frac{dw}{dr} = 0$$

Solve this equation to determine the form of the displacement and then show that the stresses are given by $\tau_{rz} = -F/2\pi r$, $\tau_{\theta z} = 0$.

Two-dimensional problem solution

The previous chapter developed the general formulation for the plane problem in elasticity. This formulation results in two types of in-plane problems—plane strain and plane stress. It was further shown that solution to each of these problem types could be conveniently handled using the Airy stress function approach. This scheme reduces the field equations to a single partial differential equation, and for the case of zero body forces, this result was the biharmonic equation. Thus, the plane elasticity problem was reduced to finding the solution to the biharmonic equation in a particular domain of interest. Such a solution must also satisfy the given boundary conditions associated with the problem under study. Several general solution techniques were briefly discussed in Section 5.7. These include the use of power series or polynomials and Fourier methods. An extensive review of two-dimensional biharmonic solutions has been given by Meleshko (2003). We now pursue the solution to several two-dimensional problems using these methods. Our formulation and solution are conducted using both Cartesian and polar coordinate systems. In many cases we use MATLAB® software to plot the stress and displacement field distributions in order to better understand the nature of the solution. Plane problems can also be solved using complex variable theory, and this powerful method is discussed in Chapter 10.

8.1 Cartesian coordinate solutions using polynomials

We begin the solution to plane elasticity problems with no body forces by considering problems formulated in Cartesian coordinates. When taking boundary conditions into account, this formulation is most useful for problems with rectangular domains. The method is based on the inverse solution concept where we assume a form of the solution to the biharmonic equation

$$\frac{\partial^4 \phi}{\partial x^4} + 2\frac{\partial^4 \phi}{\partial x^2 \partial y^2} + \frac{\partial^4 \phi}{\partial y^4} = 0 \tag{8.1.1}$$

and then try to determine which problem may be solved by this solution. The assumed solution form for the Airy stress function is taken to be a general polynomial of the in-plane coordinates, and this form can be conveniently expressed in the power series

$$\phi(x, y) = \sum_{m=0}^{\infty} \sum_{n=0}^{\infty} A_{mn} x^m y^n \tag{8.1.2}$$

where A_{mn} are constant coefficients to be determined. This representation was given by Neou (1957), who proposed a systematic scheme to solve such plane problems.

Elasticity. https://doi.org/10.1016/B978-0-12-815987-3.00008-6

163

Using the stress—stress function relations (7.5.3) with zero body forces

$$\sigma_x = \frac{\partial^2 \phi}{\partial y^2}, \quad \sigma_y = \frac{\partial^2 \phi}{\partial x^2}, \quad \tau_{xy} = -\frac{\partial^2 \phi}{\partial x \partial y} \tag{8.1.3}$$

Note that in the Airy function form the three lowest-order terms with $m + n \leq 1$ do not contribute to the stresses and therefore are dropped. It is observed that second-order terms produce a constant stress field, third-order terms give a linear distribution of stress, and so on for higher-order polynomials.

Terms with $m + n \leq 3$ automatically satisfy the biharmonic equation (8.1.1) for any choice of constants A_{mn}. However, for higher-order terms with $m + n > 3$, the constants A_{mn} must be related in order to have polynomial satisfy the governing equation. For example, the fourth-order polynomial terms $A_{40}x^4 + A_{22}x^2y^2 + A_{04}y^4$ will not satisfy the biharmonic equation unless $3A_{40} + A_{22} + 3A_{04} = 0$. This condition specifies one constant in terms of the other two, thus leaving two constants to be determined by the boundary conditions.

Considering the general case, substituting the series form (8.1.2) into the governing biharmonic equation (8.1.1) yields

$$\sum_{m=4}^{\infty} \sum_{n=0}^{\infty} m(m-1)(m-2)(m-3)A_{mn}x^{m-4}y^n$$

$$+2 \sum_{m=2}^{\infty} \sum_{n=2}^{\infty} m(m-1)n(n-1)A_{mn}x^{m-2}y^{n-2} \tag{8.1.4}$$

$$+ \sum_{m=0}^{\infty} \sum_{n=4}^{\infty} n(n-1)(n-2)(n-3)A_{mn}x^m y^{n-4} = 0$$

Collecting like powers of x and y, the preceding equation may be written as

$$\sum_{m=2}^{\infty} \sum_{n=2}^{\infty} \left[(m+2)(m+1)m(m-1)A_{m+2, n-2} + 2m(m-1)n(n-1)A_{mn} \right.$$
$$\left. + (n+2)(n+1)n(n-1)A_{m-2, n+2}\right] x^{m-2} y^{n-2} = 0 \tag{8.1.5}$$

Because this relation must be satisfied for all values of x and y, the coefficient in brackets must vanish, giving the result

$$(m+2)(m+1)m(m-1)A_{m+2, n-2} + 2m(m-1)n(n-1)A_{mn}$$
$$+(n+2)(n+1)n(n-1)A_{m-2, n+2} = 0 \tag{8.1.6}$$

For each m,n pair, (8.1.6) is the general relation that must be satisfied to ensure that the polynomial grouping is biharmonic. Note that the fourth-order case ($m = n = 2$) was discussed previously.

Because this method produces polynomial stress distributions, we would not expect the scheme to satisfy general boundary conditions. However, this limitation can be circumvented by modifying boundary conditions on the problem using the Saint—Venant principle. This is accomplished by replacing a complicated nonpolynomial boundary condition with a statically equivalent polynomial condition. The solution to the modified problem would then be accurate at points sufficiently far away from the boundary where adjustments were made. Normally, this method has applications to problems of rectangular shape in which one dimension is much larger than the other. This would include a variety of beam problems, and we shall now consider three such examples. Solutions to each of these problems are made under plane stress conditions. The corresponding plane strain solutions can easily be determined by using the simple change in elastic constants given in Table 7.1. Of course, for the case with zero body forces and traction boundary conditions, the stress fields will be identical in either theory.

Example 8.1 Uniaxial tension of a beam

As a simple example, consider the two-dimensional plane stress case of a long rectangular beam under uniform tension T at each end, as shown in Fig. 8.1. This problem could be considered the *Saint–Venant approximation* to the more general case with nonuniformly distributed tensile forces at the ends $x = \pm l$. For such an interpretation, the actual boundary conditions are replaced by the statically equivalent uniform distribution, and the solution to be developed will be valid at points away from these ends.

The boundary conditions on this problem may be written as

$$\sigma_x(\pm l, y) = T, \quad \sigma_y(x, \pm c) = 0$$
$$\tau_{xy}(\pm l, y) = \tau_{xy}(x, \pm c) = 0 \tag{8.1.7}$$

These conditions should be carefully verified by making reference to Fig. 5.3. Because the boundary conditions specify constant stresses on each of the beam's boundaries, we are motivated to try a second-order stress function of the form

$$\phi = A_{02}y^2 \tag{8.1.8}$$

and this gives the following constant stress field

$$\sigma_x = 2A_{02}, \quad \sigma_y = \tau_{xy} = 0 \tag{8.1.9}$$

The first boundary condition (8.1.7) implies that $A_{02} = T/2$ and all other boundary conditions are identically satisfied. Therefore, the stress field solution to this problem is given by

$$\sigma_x = T, \quad \sigma_y = \tau_{xy} = 0 \tag{8.1.10}$$

Next we wish to determine the displacement field associated with this stress distribution. This is accomplished by a standard procedural technique. First, the strain field is calculated using Hooke's

Continued

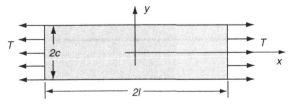

FIG. 8.1 Uniaxial tension problem.

law. Then the strain–displacement relations are used to determine various displacement gradients, and these expressions are integrated to find the individual displacements. Using this scheme, the in-plane displacement gradients are found to be

$$\frac{\partial u}{\partial x} = e_x = \frac{1}{E}(\sigma_x - \nu\sigma_y) = \frac{T}{E}$$

$$\frac{\partial v}{\partial y} = e_y = \frac{1}{E}(\sigma_y - \nu\sigma_x) = -\nu\frac{T}{E}$$

(8.1.11)

These results are easily integrated to get

$$u = \frac{T}{E}x + f(y)$$

$$v = -\nu\frac{T}{E}y + g(x)$$

(8.1.12)

where $f(y)$ and $g(x)$ are arbitrary functions of the indicated variable coming from the integration process (See Example 2.2). To complete the problem solution, these functions must be determined, and this is accomplished using the remaining Hooke's law and the strain–displacement relation for the shear stress and strain

$$\frac{\partial u}{\partial y} + \frac{\partial v}{\partial x} = 2e_{xy} = \frac{\tau_{xy}}{\mu} = 0 \Rightarrow f'(y) + g'(x) = 0$$

(8.1.13)

This result can be separated into two independent relations $g'(x) = -f'(y) = \text{constant}$ and integrated to get

$$f(y) = -\omega_o y + u_o$$

$$g(x) = \omega_o x + v_o$$

(8.1.14)

where ω_o, u_o, v_o are arbitrary constants of integration. The expressions given by relation (8.1.14) represent *rigid-body motion* terms where ω_o is the rotation about the z-axis and u_o and v_o are the translations in the x and y directions. Such terms will always result from the integration of the strain–displacement relations, and it is noted that they do not contribute to the strain or stress fields. Thus, *the displacements are determined from the strain field only up to an arbitrary rigid-body motion*. Additional displacement boundary conditions, referred to here as *fixity conditions*, are needed to explicitly determine these rigid-body motion terms. For two-dimensional problems, fixity conditions would require three independent statements commonly involving specification of the x- and y-displacements and rotation at a particular point. The choice of such conditions is normally made based on the expected deformation of the physical problem. For example, if we agree that the center of the beam does not move and the x-axis does not rotate, all rigid-body terms will vanish and $f = g = 0$.

Example 8.2 Pure bending of a beam

As a second plane stress example, consider the case of a straight beam subjected to end moments as shown in Fig. 8.2. The exact pointwise loading on the ends only involve the specification of shear stress while end normal stress distribution is specified by the statically equivalent effect of zero force and moment. Hence, the boundary conditions on this problem are written as

$$\sigma_y(x, \pm c) = 0, \quad \tau_{xy}(x, \pm c) = \tau_{xy}(\pm l, y) = 0$$

$$\int_{-c}^{c} \sigma_x(\pm l, y)dy = 0, \quad \int_{-c}^{c} \sigma_x(\pm l, y)ydy = -M \qquad (8.1.15)$$

Thus, the boundary conditions on the ends of the beam have been relaxed, and only the statically equivalent condition for σ_x will be satisfied. This fact leads to a solution that is not necessarily valid near the ends of the beam.

The choice of stress function is based on the fact that a third-order function will give rise to a linear stress field, and a particular linear boundary loading on the ends $x = \pm l$ will reduce to a pure moment. Based on these two concepts, we choose

$$\phi = A_{03}y^3 \qquad (8.1.16)$$

and the resulting stress field takes the form

$$\sigma_x = 6A_{03}y, \quad \sigma_y = \tau_{xy} = 0 \qquad (8.1.17)$$

This field automatically satisfies the boundary conditions on $y = \pm c$ and gives zero net forces at the ends of the beam. The remaining moment conditions at $x = \pm l$ are satisfied if $A_{03} = -M/4c^3$, and thus the stress field is determined as

$$\sigma_x = -\frac{3M}{2c^3}y, \quad \sigma_y = \tau_{xy} = 0 \qquad (8.1.18)$$

The displacements are again calculated in the same fashion as in the previous example. Assuming plane stress, Hooke's law will give the strain field, which is then substituted into the strain–displacement relations and integrated, yielding the result

$$\frac{\partial u}{\partial x} = -\frac{3M}{2Ec^3}y \Rightarrow u = -\frac{3M}{2Ec^3}xy + f(y)$$

$$\frac{\partial v}{\partial y} = v\frac{3M}{2Ec^3}y \Rightarrow v = \frac{3Mv}{2Ec^3}y^2 + g(x) \qquad (8.1.19)$$

Continued

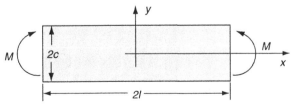

FIG. 8.2 Beam under end moments.

where f and g are arbitrary functions of integration. Using the shear stress–strain relations

$$\frac{\partial u}{\partial y}+\frac{\partial v}{\partial x}=0 \Rightarrow -\frac{3M}{2Ec^3}x+f'(y)+g'(x)=0 \tag{8.1.20}$$

this result can again be separated into two independent relations in x and y, and upon integration the arbitrary functions f and g are determined as

$$f(y)=-\omega_o y+u_o$$
$$g(x)=\frac{3M}{4Ec^3}x^2+\omega_o x+v_o \tag{8.1.21}$$

Again, rigid-body motion terms are brought out during the integration process. For this problem, the beam would normally be simply supported, and thus the fixity displacement boundary conditions could be specified as $v(\pm l,0)=0$ and $u(-l,0)=0$. This specification leads to determination of the rigid-body terms as $u_o=\omega_o=0$, $v_o=-3Ml^2/4Ec^3$.

We now wish to compare this elasticity solution with that developed by elementary strength of materials (often called mechanics of materials). Appendix D, Section D.3, conveniently provides a brief review of this undergraduate theory. Introducing the cross-sectional area moment of inertia $I=2c^3/3$ (assuming unit thickness), our stress and displacement field can be written as

$$\sigma_x=-\frac{M}{I}y, \quad \sigma_y=\tau_{xy}=0$$
$$u=-\frac{Mxy}{EI}, \quad v=\frac{M}{2EI}\left[\nu y^2+x^2-l^2\right] \tag{8.1.22}$$

Note that for this simple moment-loading case, we have verified the classic assumption from elementary beam theory *that plane sections remain plane*. Note, however, that this will not be the case for more complicated loadings. The elementary strength of materials solution is obtained using Euler–Bernoulli beam theory and gives the bending stress and deflection of the beam centerline as

$$\sigma_x=-\frac{M}{I}y, \quad \sigma_y=\tau_{xy}=0$$
$$v=v(x,0)=\frac{M}{2EI}\left[x^2-l^2\right] \tag{8.1.23}$$

Comparing these two solutions, it is observed that they are identical, with the exception of the x displacements. In general, however, the two theories will not match for other beam problems with more complicated loadings, and we investigate such a problem in the next example.

Example 8.3 Bending of a beam by uniform transverse loading

Our final example in this section is that of a beam carrying a uniformly distributed transverse loading w along its top surface, as shown in Fig. 8.3. Again, plane stress conditions are chosen, and we relax the boundary conditions on the ends and consider only statically equivalent effects. Exact pointwise boundary conditions will be specified on the top and bottom surfaces, while at the ends the resultant horizontal force and moment are set to zero and the resultant vertical force will be specified to satisfy overall equilibrium. Thus, the boundary conditions on this problem can be written as

$$\tau_{xy}(x, \pm c) = 0$$

$$\sigma_y(x, c) = 0$$

$$\sigma_y(x, -c) = -w$$

$$\int_{-c}^{c} \sigma_x(\pm l, y)dy = 0 \qquad (8.1.24)$$

$$\int_{-c}^{c} \sigma_x(\pm l, y)y\,dy = 0$$

$$\int_{-c}^{c} \tau_{xy}(\pm l, y)dy = \mp wl$$

Again, it is suggested that these conditions be verified, especially the last statement.

Using the polynomial solution format, we choose a trial Airy stress function including second-, third-, and fifth-order terms (a choice that has come from previous trial and error)

$$\phi = A_{20}x^2 + A_{21}x^2 y + A_{03}y^3 + A_{23}x^2 y^3 - \frac{A_{23}}{5}y^5 \qquad (8.1.25)$$

Continued

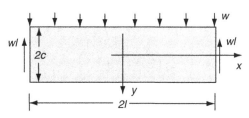

FIG. 8.3 Beam carrying uniformly transverse loading.

It is noted that the fifth-order term has been generated to satisfy the biharmonic equation. The resulting stress field from this stress function is given by

$$\sigma_x = 6A_{03}y + 6A_{23}\left(x^2 y - \frac{2}{3}y^3\right)$$

$$\sigma_y = 2A_{20} + 2A_{21}y + 2A_{23}y^3 \tag{8.1.26}$$

$$\tau_{xy} = -2A_{21}x - 6A_{23}xy^2$$

Applying the first three boundary conditions in the set (8.1.24) gives three equations among the unknown coefficients A_{20}, A_{21}, and A_{23}. Solving this system determines these constants, giving the result

$$A_{20} = -\frac{w}{4}, \quad A_{21} = \frac{3w}{8c}, \quad A_{23} = -\frac{w}{8c^3} \tag{8.1.27}$$

Using these results, it is found that the stress field will now also satisfy the fourth and sixth conditions in (8.1.24). The remaining condition of vanishing end moments gives the following

$$A_{03} = -A_{23}\left(l^2 - \frac{2}{5}c^2\right) = \frac{w}{8c}\left(\frac{l^2}{c^2} - \frac{2}{5}\right) \tag{8.1.28}$$

This completes determination of the four constants in the trial Airy stress function. The problem is now solved since the stress function satisfies the governing equation and all boundary conditions are also satisfied. The resulting stress field in now given by

$$\sigma_x = \frac{3w}{4c}\left(\frac{l^2}{c^2} - \frac{2}{5}\right)y - \frac{3w}{4c^3}\left(x^2 y - \frac{2}{3}y^3\right)$$

$$\sigma_y = -\frac{w}{2} + \frac{3w}{4c}y - \frac{w}{4c^3}y^3 \tag{8.1.29}$$

$$\tau_{xy} = -\frac{3w}{4c}x + \frac{3w}{4c^3}xy^2$$

We again wish to compare this elasticity solution with that developed by elementary strength of materials, and thus the elasticity stress field is rewritten in terms of the cross-sectional area moment of inertia $I = 2c^3/3$, as

$$\sigma_x = \frac{w}{2I}\left(l^2 - x^2\right)y + \frac{w}{I}\left(\frac{y^3}{3} - \frac{c^2 y}{5}\right)$$

$$\sigma_y = -\frac{w}{2I}\left(\frac{y^3}{3} - c^2 y + \frac{2}{3}c^3\right) \tag{8.1.30}$$

$$\tau_{xy} = -\frac{w}{2I}x\left(c^2 - y^2\right)$$

The corresponding results from strength of materials for this case (see Appendix D, Section D.3) are given by

$$\sigma_x = \frac{My}{I} = \frac{w}{2I}(l^2 - x^2)y$$

$$\sigma_y = 0 \tag{8.1.31}$$

$$\tau_{xy} = \frac{VQ}{It} = -\frac{w}{2I}x(c^2 - y^2)$$

where the bending moment $M = w(l^2 - x^2)/2$, the shear force $V = -wx$, the first moment of a sectioned cross-sectional area is $Q = (c^2 - y^2)/2$, and the thickness t is taken as unity.

Comparing the two theories, we see that the shear stresses are identical, while the two normal stresses are not. The two normal stress distributions are plotted in Figs. 8.4 and 8.5. The normalized bending stress σ_x for the case $x = 0$ is shown in Fig. 8.4. Note that the elementary theory predicts linear variation, while the elasticity solution indicates nonlinear behavior. The maximum difference between the two theories exists at the outer fibers (top and bottom) of the beam, and the actual difference in the stress values is simply $w/5$, a result independent of the beam dimensions. For most common beam problems where $l \gg c$, the bending stresses will be much greater than w, and thus the differences between elasticity theory and strength of materials will be relatively small. For example, the set of curves in Fig. 8.4 for $l/c = 4$ gives a maximum difference of about only 1%. Fig. 8.5 illustrates the behavior of the stress σ_y; the maximum difference between the two theories is given by w and this occurs at the top of the beam. Again, this difference will be negligibly small for most beam problems where $l \gg c$. These results are generally true for beam problems with other transverse loadings. That is, for the case with $l \gg c$, approximate bending stresses determined from strength of materials will generally closely match those developed from theory of elasticity.

Continued

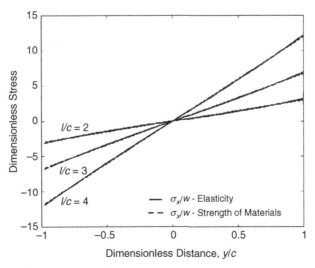

FIG. 8.4 Comparison of bending stress in the beam.

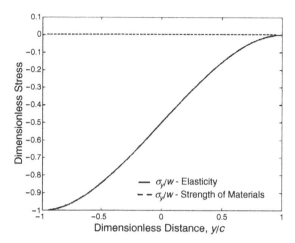

FIG. 8.5 Comparison of σ_y stress in the beam.

Next let us determine the displacement field for this problem. As in the previous examples, the displacements are developed through integration of the strain—displacement relations. Integrating the first two normal strain—displacement relations gives the result

$$u = \frac{w}{2EI}\left[\left(l^2 x - \frac{x^3}{3}\right)y + x\left(\frac{2y^3}{3} - \frac{2c^2 y}{5}\right) + vx\left(\frac{y^3}{3} - c^2 y + \frac{2c^3}{3}\right)\right] + f(y)$$

$$v = -\frac{w}{2EI}\left[\left(\frac{y^4}{12} - \frac{c^2 y^2}{2} + \frac{2c^3 y}{3}\right) + v(l^2 - x^2)\frac{y^2}{2} + v\left(\frac{y^4}{6} - \frac{c^2 y^2}{5}\right)\right] + g(x)$$

(8.1.32)

where $f(y)$ and $g(x)$ are arbitrary functions of integration. Using these results in the shear strain—displacement equation gives the relation

$$\frac{w}{2EI}\left[l^2 x - \frac{x^3}{3} + x\left(2y^2 - \frac{2c^2}{5}\right) + vx(y^2 - c^2)\right] + f'(y)$$

$$+ \frac{w}{2EI}vxy^2 + g'(x) = -\frac{w}{2\mu I}x(c^2 - y^2)$$

(8.1.33)

This result can again be rewritten in a separable form and integrated to determine the arbitrary functions

$$f(y) = \omega_o y + u_o$$

$$g(x) = \frac{w}{24EI}x^4 - \frac{w}{4EI}\left[l^2 - \left(\frac{8}{5} + v\right)c^2\right]x^2 - \omega_o x + v_o$$

(8.1.34)

Choosing the fixity conditions $u(0,y) = v(\pm l,0) = 0$, the rigid-body motion terms are found to be

$$u_o = \omega_o = 0, \quad v_o = \frac{5wl^4}{24EI}\left[1 + \frac{12}{5}\left(\frac{4}{5} + \frac{\nu}{2}\right)\frac{c^3}{l^2}\right] \tag{8.1.35}$$

Using these results, the final form of the displacements is given by

$$u = \frac{w}{2EI}\left[\left(l^2 x - \frac{x^3}{3}\right)y + x\left(\frac{2y^3}{3} - \frac{2c^2 y}{5}\right) + \nu x\left(\frac{y^3}{3} - c^2 y + \frac{2c^3}{3}\right)\right]$$

$$v = -\frac{w}{2EI}\left(\frac{y^4}{12} - \frac{c^2 y^2}{2} + \frac{2c^3 y}{3} + \nu\left[(l^2 - x^2)\frac{y^2}{2} + \frac{y^4}{6} - \frac{c^2 y^2}{5}\right] - \frac{x^4}{12} + \left[\frac{l^2}{2} + \left(\frac{4}{5} + \frac{\nu}{2}\right)c^2\right]x^2\right)$$

$$+ \frac{5wl^4}{24EI}\left[1 + \frac{12}{5}\left(\frac{4}{5} + \frac{\nu}{2}\right)\frac{c^2}{l^2}\right]$$

$$\tag{8.1.36}$$

The maximum deflection of the beam axis is given by

$$v(0,0) = v_{max} = \frac{5wl^4}{24EI}\left[1 + \frac{12}{5}\left(\frac{4}{5} + \frac{\nu}{2}\right)\frac{c^2}{l^2}\right] \tag{8.1.37}$$

while the corresponding value calculated from strength of materials is

$$v_{max} = \frac{5wl^4}{24EI} \tag{8.1.38}$$

The difference between the two theories given by relations (8.1.37) and (8.1.38) is specified by $\frac{wl^4}{2EI}\left(\frac{4}{5} + \frac{\nu}{2}\right)\frac{c^2}{l^2}$, and this term is caused by the presence of the shear force. For beams where $l \gg c$, this difference is very small. Thus, we again find that for long beams, strength of materials predictions match closely to theory of elasticity results. Note from Eq. (8.1.36) that the x component of displacement indicates that plane sections undergo nonlinear deformation and do not remain plane. It can also be shown that the Euler–Bernoulli relation $M = EI\frac{d^2 v(x,0)}{dx^2}$ used in strength of materials theory is not satisfied by this elasticity solution. Timoshenko and Goodier (1970) provide additional discussion on such differences.

Additional rectangular beam problems of this type with different support and loading conditions can be solved using various polynomial combinations. Several of these are given in the exercises. A general issue should be pointed out that for these types of problems the stresses coming from the Airy stress function will automatically be in equilibrium. Thus the entire structure will be in equilibrium. This implies that for particular problems, once boundary conditions are enforced, one boundary condition typically will be automatically satisfied since it was determined from overall equilibrium of the entire structure. This would commonly occur at say the fix end of cantilever beam problems – see Exercises 8.4–8.7. Barber's text (2010) provides some interesting applications and general strategies for polynomial Airy function problems using *Maple* and/or *Mathematica* software.

8.2 Cartesian coordinate solutions using Fourier methods

A more general solution scheme for the biharmonic equation may be found by using *Fourier methods*. Such techniques generally use *separation of variables* along with *Fourier series* or *Fourier integrals*. Use of this method began over a century ago, and the work of Pickett (1944), Timoshenko and Goodier (1970), and Little (1973) provides details on the technique.

In Cartesian coordinates, the method may be initiated by looking for an Airy stress function of the separable form

$$\phi(x, y) = X(x)Y(y) \tag{8.2.1}$$

Although the functions X and Y could be left somewhat general, the solution is obtained more directly if exponential forms are chosen as $X = e^{\alpha x}$, $Y = e^{\beta y}$. Substituting these results into the biharmonic equation (8.1.1) gives

$$\left(\alpha^4 + 2\alpha^2\beta^2 + \beta^4\right)e^{\alpha x}e^{\beta y} = 0$$

and this result implies that the term in parentheses must be zero, giving the *auxiliary* or *characteristic equation*

$$\left(\alpha^2 + \beta^2\right)^2 = 0 \tag{8.2.2}$$

The solution to this equation gives double roots of the form

$$\alpha = \pm i\beta \tag{8.2.3}$$

The general solution to the problem then includes the superposition of the zero root cases plus the general roots. For the zero root condition with $\beta = 0$, there is a fourfold multiplicity of the roots, yielding a general solution of the form

$$\phi_{\beta=0} = C_0 + C_1 x + C_2 x^2 + C_3 x^3 \tag{8.2.4}$$

while for the case with $\alpha = 0$, the solution is given by

$$\phi_{\alpha=0} = C_4 y + C_5 y^2 + C_6 y^3 + C_7 xy + C_8 x^2 y + C_9 xy^2 \tag{8.2.5}$$

Expressions (8.2.4) and (8.2.5) represent polynomial solution terms satisfying the biharmonic equation. For the general case given by Eq. (8.2.3), the solution becomes

$$\phi = e^{i\beta x}\left[Ae^{\beta y} + Be^{-\beta y} + Cye^{\beta y} + Dye^{-\beta y}\right]$$
$$+ e^{-i\beta x}\left[A'e^{\beta y} + B'e^{-\beta y} + C'ye^{\beta y} + D'ye^{-\beta y}\right] \tag{8.2.6}$$

The parameters C_i, A, B, C, D, A', B', C', and D' represent arbitrary constants to be determined from boundary conditions. The complete solution is found by the superposition of solutions (8.2.4), (8.2.5), and (8.2.6). Realizing that the final solution must be real, the exponentials are replaced by equivalent trigonometric and hyperbolic forms, thus giving

$$\begin{aligned}
\phi = {} & \sin \beta x[(A + C\beta y)\sinh \beta y + (B + D\beta y)\cosh \beta y] \\
& + \cos \beta x[(A' + C'\beta y)\sinh \beta y + (B' + D'\beta y)\cosh \beta y] \\
& + \sin \alpha y[(E + G\alpha x)\sinh \alpha x + (F + H\alpha x)\cosh \alpha x] \\
& + \cos \alpha y[(E' + G'\alpha x)\sinh \alpha x + (F' + H'\alpha x)\cosh \alpha x] \\
& + \phi_{\alpha=0} + \phi_{\beta=0}
\end{aligned} \tag{8.2.7}$$

Using this solution form along with superposition and Fourier series concepts, many problems with complex boundary loadings can be solved. Two particular problems are now presented.

Example 8.4 Beam subject to transverse sinusoidal loading

Consider the simply supported beam carrying a sinusoidal loading along its top edge as shown in Fig. 8.6.

The boundary conditions for this problem can be written as

$$\sigma_x(0,y) = \sigma_x(l,y) = 0$$
$$\tau_{xy}(x, \pm c) = 0$$
$$\sigma_y(x, -c) = 0$$
$$\sigma_y(x, c) = -q_o \sin(\pi x/l)$$
$$\int_{-c}^{c} \tau_{xy}(0,y)dy = -q_o l/\pi$$
$$\int_{-c}^{c} \tau_{xy}(l,y)dy = q_o l/\pi$$

(8.2.8)

Note that these conditions do not specify the pointwise distribution of shear stress on the ends of the beam, but rather stipulate the resultant condition based on overall problem equilibrium. Thus, we again are generating a solution valid away from the ends that would be most useful for the case where $l \gg c$. Because the vertical normal stress has a sinusoidal variation in x along $y = c$, an appropriate trial solution from the general case is

$$\phi = \sin \beta x[(A + C\beta y) \sinh \beta y + (B + D\beta y) \cosh \beta y]$$

(8.2.9)

The stresses from this trial form are

$$\sigma_x = \beta^2 \sin \beta x[A \sinh \beta y + C(\beta y \sinh \beta y + 2 \cosh \beta y) + B \cosh \beta y + D(\beta y \cosh \beta y + 2 \sinh \beta y)]$$
$$\sigma_y = -\beta^2 \sin \beta x[(A + C\beta y) \sinh \beta y + (B + D\beta y) \cosh \beta y]$$
$$\tau_{xy} = -\beta^2 \cos \beta x[A + \cosh \beta y + C(\beta y \cosh \beta y + \sinh \beta y) + B \sinh \beta y + D(\beta y \sinh \beta y + \cosh \beta y)]$$

(8.2.10)

Continued

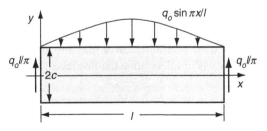

FIG. 8.6 Beam carrying sinusoidal transverse loading.

Condition $(8.2.8)_2$ implies that

$$[A \cosh \beta y + C(\beta y \cosh \beta y + \sinh \beta y) + B \sinh \beta y + D(\beta y \sinh \beta y + \cosh \beta y)]_{y=\pm c} = 0 \qquad (8.2.11)$$

This condition can be equivalently stated by requiring that the *even* and *odd* functions of y independently vanish at the boundary, thus giving the result

$$A \cosh \beta c + D(\beta c \sinh \beta c + \cosh \beta c) = 0$$
$$B \sinh \beta c + C(\beta c \cosh \beta c + \sinh \beta c) = 0 \qquad (8.2.12)$$

Solving for the constants A and B gives

$$A = -D(\beta c \tanh \beta c + 1)$$
$$B = -C(\beta c \coth \beta c + 1) \qquad (8.2.13)$$

and thus the vertical normal stress becomes

$$\sigma_y = - \beta^2 \sin \beta x \{D[\beta y \cosh \beta y - (\beta c \tanh \beta c + 1) \sinh \beta y]$$
$$+ C[\beta y \sinh \beta y - (\beta c \coth \beta c + 1) \cosh \beta y]\} \qquad (8.2.14)$$

Applying boundary condition $(8.2.8)_3$ to this result gives the relation between C and D

$$C = -\tanh \beta c \left[\frac{\beta c - \sinh \beta c \cosh \beta c}{\beta c + \sinh \beta c \cosh \beta c} \right] D \qquad (8.2.15)$$

while condition $(8.2.8)_4$ gives

$$q_o \sin \frac{\pi x}{l} = 2\beta^2 \sin \beta x \left[\frac{\beta c - \sinh \beta c \cosh \beta c}{\cosh \beta c} \right] D \qquad (8.2.16)$$

In order for relation (8.2.16) to be true for all x, $\beta = \pi/l$, and so the constant D is thus determined as

$$D = \frac{q_o \cosh \dfrac{\pi c}{l}}{2\dfrac{\pi^2}{l^2} \left[\dfrac{\pi c}{l} - \sinh \dfrac{\pi c}{l} \cosh \dfrac{\pi c}{l} \right]} \qquad (8.2.17)$$

This result can be substituted into (8.2.15) to give the remaining constant C

$$C = \frac{-q_o \sinh \dfrac{\pi c}{l}}{2\dfrac{\pi^2}{l^2} \left[\dfrac{\pi c}{l} + \sinh \dfrac{\pi c}{l} \cosh \dfrac{\pi c}{l} \right]} \qquad (8.2.18)$$

Using these results, the remaining boundary conditions $(8.2.8)_1$ and $(8.2.8)_{5,6}$ will now be satisfied. Thus, we have completed the determination of the stress field for this problem. Following our usual solution steps, we now wish to determine the displacements, and these are again developed through integration of the plane stress strain–displacement relations. Skipping the details, the final results are given by

$$u = -\frac{\beta}{E}\cos\beta x\{A(1+v)\sinh\beta y + B(1+v)\cosh\beta y$$

$$+ C[(1+v)\beta y\sinh\beta y + 2\cosh\beta y]$$

$$+ D[(1+v)\beta y\cosh\beta y + 2\sinh\beta y]\} - \omega_o y + u_o$$

$$(8.2.19)$$

$$v = \frac{-\beta}{E}\sin\beta x\{A(1+v)\cosh\beta y + B(1+v)\sinh\beta y$$

$$+ C[(1+v)\beta y\cosh\beta y - (1+v)\sinh\beta y]$$

$$+ D[(1+v)\beta y\sinh\beta y - (1-v)\cosh\beta y]\} + \omega_o y + v_o$$

To model a simply supported beam, we choose displacement fixity conditions as

$$u(0,0) = v(0,0) = v(l,0) = 0 \qquad (8.2.20)$$

These conditions determine the rigid-body terms, giving the result

$$\omega_o = v_o = 0, \quad u_o = \frac{\beta}{E}[B(1+v) + 2C] \qquad (8.2.21)$$

To compare with strength of materials theory, the vertical centerline displacement is determined. Using $(8.2.19)_2$ and $(8.2.13)_1$, the deflection of the beam axis reduces to

$$v(x,0) = \frac{D\beta}{E}\sin\beta x[2 + (1+v)\beta c\tanh\beta c] \qquad (8.2.22)$$

For the case $l \gg c$, $D \approx -3q_o l^5/4c^3\pi^5$, and so the previous relation becomes

$$v(x,0) = -\frac{3q_o l^4}{2c^3\pi^4 E}\sin\frac{\pi x}{l}\left[1 + \frac{1+v}{2}\frac{\pi c}{l}\tanh\frac{\pi c}{l}\right] \qquad (8.2.23)$$

The corresponding deflection from strength of materials theory is given by

$$v(x,0) = -\frac{3q_o l^4}{2c^3\pi^4 E}\sin\frac{\pi x}{l} \qquad (8.2.24)$$

Considering again the case $l \gg c$, the second term in brackets in relation (8.2.23) can be neglected, and thus the elasticity result matches with that found from strength of materials.

8.2.1 Applications involving Fourier series

More sophisticated applications of the Fourier solution method commonly incorporate Fourier series theory. This is normally done by using superposition of solution forms to enable more general boundary conditions to be satisfied. For example, in the previous problem the solution was obtained for a single sinusoidal loading. However, this solution form could be used to generate a series of solutions with sinusoidal loadings having different periods; that is, $\beta = \beta_n = n\pi/l$, $(n = 1, 2, 3, \cdots)$. Invoking the principle of superposition, we can form a linear combination of these sinusoidal solutions, thus leading to a Fourier series representation for a general transverse boundary loading on the beam.

In order to use such a technique, we shall briefly review some basic concepts of Fourier series theory. Further details may be found in Kreyszig (2011) or Churchill (1963). A function $f(x)$ periodic with period $2l$ can be represented on the interval $(-l, l)$ by the *Fourier trigonometric series*

$$f(x) = \frac{1}{2}a_o + \sum_{n=1}^{\infty} \left(a_n \cos \frac{n\pi x}{l} + b_n \sin \frac{n\pi x}{l} \right) \tag{8.2.25}$$

where

$$a_n = \frac{1}{l} \int_{-l}^{l} f(\xi) \cos \frac{n\pi\xi}{l} \, d\xi, \quad n = 0, 1, 2, \cdots$$

$$b_n = \frac{1}{l} \int_{-l}^{l} f(\xi) \sin \frac{n\pi\xi}{l} \, d\xi, \quad n = 1, 2, 3, \cdots \tag{8.2.26}$$

This representation simplifies for some special cases that often arise in applications. For example, if $f(x)$ is an *even function*, $f(x) = f(-x)$, then representation (8.2.25) reduces to the *Fourier cosine series*

$$f(x) = \frac{1}{2}a_o + \sum_{n=1}^{\infty} a_n \cos \frac{n\pi x}{l}$$

$$a_n = \frac{2}{l} \int_{0}^{l} f(\xi) \cos \frac{n\pi\xi}{l} \, d\xi, \quad n = 0, 1, 2, \cdots \tag{8.2.27}$$

on the interval $(0, l)$. If $f(x)$ is an *odd function*, $f(x) = -f(-x)$, then representation (8.2.25) reduces to the *Fourier sine series*

$$f(x) = \sum_{n=1}^{\infty} b_n \sin \frac{n\pi x}{l}$$

$$b_n = \frac{2}{l} \int_{0}^{l} f(\xi) \sin \frac{n\pi\xi}{l} \, d\xi, \quad n = 1, 2, 3, \cdots \tag{8.2.28}$$

on interval $(0,l)$. We now develop the solution to a specific elasticity problem using these tools.

Example 8.5 Rectangular domain with arbitrary boundary loading

Consider again a rectangular domain with arbitrary compressive boundary loading on the top and bottom of the body, as shown in Fig. 8.7. Although a more general boundary loading could be considered on all four sides, the present case will sufficiently demonstrate the use of Fourier series theory for problem solution. For this problem, dimensions a and b are to be of the same order, and thus we cannot use the Saint–Venant principle to develop an approximate solution valid away from a particular boundary. Thus, the solution is developed using the exact pointwise boundary conditions

$$\sigma_x(\pm a, y) = 0$$
$$\tau_{xy}(\pm a, y) = 0$$
$$\tau_{xy}(x, \pm b) = 0$$
$$\sigma_y(x, \pm b) = -p(x)$$

(8.2.29)

To ease the solution details, we shall assume that the boundary loading $p(x)$ is an even function; that is, $p(x) = p(-x)$. Normal stresses are expected to be symmetric about the x- and y-axes, and this leads to a proposed stress function of the form

$$\phi = \sum_{n=1}^{\infty} \cos \beta_n x [B_n \cosh \beta_n y + C_n \beta_n y \sinh \beta_n y]$$

$$+ \sum_{m=1}^{\infty} \cos \alpha_m y [F_m \cosh \alpha_m x + G_m \alpha_m x \sinh \alpha_m x] + C_0 x^2$$

(8.2.30)

The stresses derived from this Airy stress function become

$$\sigma_x = \sum_{n=1}^{\infty} \beta_n^2 \cos \beta_n x [B_n \cosh \beta_n y + C_n(\beta_n y \sinh \beta_n y + 2 \cosh \beta_n y)]$$

$$- \sum_{m=1}^{\infty} \alpha_m^2 \cos \alpha_m y [F_m \cosh \alpha_m x + G_m \alpha_m x \sinh \alpha_m x]$$

$$\sigma_y = - \sum_{n=1}^{\infty} \beta_n^2 \cos \beta_n x [B_n \cosh \beta_n y + C_n \beta_n y \sinh \beta_n y] + 2C_0$$

(8.2.31)

$$+ \sum_{m=1}^{\infty} \alpha_m^2 \cos \alpha_m y [F_m \cosh \alpha_m x + G_m(\alpha_m x \sinh \alpha_m x + 2 \cosh \alpha_m x)]$$

$$\tau_{xy} = \sum_{n=1}^{\infty} \beta_n^2 \sin \beta_n x [B_n \sinh \beta_n y + C_n(\beta_n y \cosh \beta_n y + \sinh \beta_n y)]$$

$$+ \sum_{m=1}^{\infty} \alpha_m^2 \sin \alpha_m y [F_m \sinh \alpha_m x + G_m(\alpha_m x \cosh \alpha_m x + \sinh \alpha_m x)]$$

To satisfy the homogeneous boundary conditions $(8.2.29)_{1,2,3}$,

$$\alpha_m = m\pi/b \text{ and } \beta_n = n\pi/a$$

Condition $(8.2.29)_2$ implies that

$$F_m = -G_m(1 + \alpha_m a \coth \alpha_m a)$$

(8.2.32)

Continued

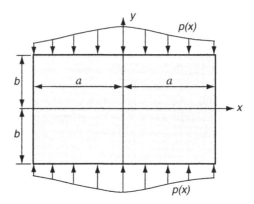

FIG. 8.7 General boundary loading on a rectangular elastic plate.

while $(8.3.29)_3$ gives

$$B_n = -C_n(1 + \beta_n b \coth \beta_n b) \tag{8.2.33}$$

Boundary condition $(8.2.29)_1$ gives

$$\sum_{n=1}^{\infty} \beta_n^2 \cos \beta_n a [B_n \cosh \beta_n y + C_n(\beta_n y \sinh \beta_n y + 2 \cosh \beta_n y)]$$

$$= \sum_{m=1}^{\infty} \alpha_m^2 \cos \alpha_m y [F_m \cosh \alpha_m a + G_m \alpha_m a \sinh \alpha_m a]$$

which can be written as

$$\sum_{m=1}^{\infty} A_m \cos \alpha_m y = \sum_{n=1}^{\infty} \beta_n^2 (-1)^{n+1} [B_n \cosh \beta_n y + C_n(\beta_n y \sinh \beta_n y + 2 \cosh \beta_n y)] \tag{8.2.34}$$

where

$$A_m = \frac{\alpha_m^2}{\sinh \alpha_m a} (\alpha_m a + \sinh \alpha_m a \cosh \alpha_m a) G_m \tag{8.2.35}$$

The expression given by (8.2.34) can be recognized as the Fourier cosine series for the terms on the right-hand side of the equation. Thus, using Fourier series theory from relations (8.2.27), the coefficients may be expressed as

$$A_m = \frac{2}{b} \sum_{n=1}^{\infty} \beta_n^2 (-1)^{n+1} \int_0^b [B_n \cosh \beta_n \xi + C_n(\beta_n \xi \sinh \beta_n \xi + 2 \cosh \beta_n \xi)] \cos \alpha_m \xi \, d\xi \tag{8.2.36}$$

Carrying out the integrals and using (8.2.35) gives

$$G_m = \frac{-4 \sinh \alpha_m a}{b(\alpha_m a + \sinh \alpha_m a \cosh \alpha_m a)} \sum_{n=1}^{\infty} C_n \frac{\beta_n^3 (-1)^{m+n} \sinh \beta_n b}{(\alpha_m^2 + \beta_n^2)^2} \tag{8.2.37}$$

The final boundary condition $(8.2.29)_4$ involves the specification of the nonzero loading $p(x)$ and implies

$$-\sum_{n=1}^{\infty} \beta_n^2 \cos \beta_n x [B_n \cosh \beta_n b + C_n \beta_n b \sinh \beta_n b] + 2C_0$$

$$+\sum_{m=1}^{\infty} \alpha_m^2 \cos \alpha_m b [F_m \cosh \alpha_m x + G_m(\alpha_m x \sinh \alpha_m x + 2 \cosh \alpha_m x)] = -p(x)$$

and this can be written in more compact form as

$$\sum_{n=0}^{\infty} A_n^* \cos \beta_n x = -p(x)$$

$$+\sum_{m=1}^{\infty} \alpha_m^2 (-1)^{m+1} [F_m \cosh \alpha_m x + G_m(\alpha_m x \sinh \alpha_m x + 2 \cosh \alpha_m x)]$$

(8.2.38)

where

$$A_n^* = \frac{\beta_n^2}{\sinh \beta_n b} (\beta_n b + \sinh \beta_n b \cosh \beta_n b) C_n$$

$$A_0^* = 2C_0$$

(8.2.39)

As before, (8.2.38) is a Fourier cosine series form, and so the series coefficients A_n^* can be easily determined from the theory given in relations (8.2.27). This then determines the coefficients C_n to be

$$C_n = \frac{-4 \sinh \beta_n b}{a(\beta_n b + \sinh \beta_n b \cosh \beta_n b)} \sum_{m=1}^{\infty} G_m \frac{\alpha_m^3 (-1)^{m+n} \sinh \alpha_m a}{(\alpha_m^2 + \beta_n^2)^2}$$

$$-\frac{2 \sinh \beta_n b}{a\beta_n^2(\beta_n b + \sinh \beta_n b \cosh \beta_n b)} \int_0^a p(\xi) \cos \beta_n \xi d\xi, \quad n = 1, 2, 3, \cdots$$

(8.2.40)

$$C_0 = -\frac{1}{2a} \int_0^a p(\xi) d\xi$$

The rather formidable systems of equations given by (8.2.37) and $(8.2.40)_1$ can be written in compact form as

$$G_m + \sum_{n=1}^{\infty} R_{mn} C_n = 0$$

$$C_n + \sum_{m=1}^{\infty} S_{nm} G_m = T_n$$

(8.2.41)

with appropriate definitions of R_{mn}, S_{nm}, and T_n. The system (8.2.41) then represents a doubly infinite set of equations in the doubly infinite set of unknowns C_n and G_m. An approximate solution

Continued

may be found by truncating the system to a finite number of equations, which can be solved for the remaining unknowns. Improved accuracy in the solution is achieved by including more equations in the truncated system. Thus, all unknown coefficients in the solution (8.2.30) are now determined, and the problem solution is completed. Little (1973) provides additional details on this solution. Obviously this solution requires an extensive amount of analysis and numerical work to determine specific stress or displacement values.

8.3 General solutions in polar coordinates

As discussed in Section 7.6, the geometry of many two-dimensional problems requires the use of polar coordinates to develop a solution. We now wish to explore the general solutions to such problems using the field equations developed in polar coordinates.

8.3.1 General Michell solution

Employing the Airy stress function approach, the governing biharmonic equation was given by

$$\nabla^4 \phi = \left(\frac{\partial^2}{\partial r^2} + \frac{1}{r}\frac{\partial}{\partial r} + \frac{1}{r^2}\frac{\partial^2}{\partial \theta^2} \right)^2 \phi = 0 \tag{8.3.1}$$

We shall first look for a general solution to this equation by assuming a separable form $\phi(r,\theta) = f(r)e^{b\theta}$, where b is a parameter to be determined. Substituting this form into the biharmonic equation and canceling the common $e^{b\theta}$ term yields

$$f'''' + \frac{2}{r}f''' - \frac{1 - 2b^2}{r^2}f'' + \frac{1 - 2b^2}{r^3}f' + \frac{b^2(4 + b^2)}{r^4}f = 0 \tag{8.3.2}$$

To solve this equation, we make the change of variable $r = e^\xi$, and this will transform (8.3.2) into the differential equation with constant coefficients

$$f'''' - 4f''' + (4 + 2b^2)f'' - 4b^2 f' + b^2(4 + b^2)f = 0 \tag{8.3.3}$$

where primes now denote $d/d\xi$. The solution to this equation is found by employing the usual scheme of substituting in $f = e^{a\xi}$, and this generates the following characteristic equation

$$(a^2 + b^2)(a^2 - 4a + 4 + b^2) = 0 \tag{8.3.4}$$

The roots to this equation may be written as

$$a = \pm ib, \quad a = 2 \pm ib$$

or   (8.3.5)

$$b = \pm ia, \quad b = \pm i(a - 2)$$

We shall consider only periodic solutions in θ, and these are obtained by choosing $b = in$, where n is an integer. Note this choice also implies that a is an integer. For particular values of n, repeated roots occur, and these require special consideration in the development of the solution. Details of the complete solution have been given by Little (1973), although the original development is credited to Michell (1899). The final form (commonly called the Michell solution) can be written as

$$\phi = a_0 + a_1 \log r + a_2 r^2 + a_3 r^2 \log r$$

$$+ \left(a_4 + a_5 \log r + a_6 r^2 + a_7 r^2 \log r\right)\theta$$

$$+ \left(a_{11}r + a_{12}r \log r + \frac{a_{13}}{r} + a_{14}r^3 + a_{15}r\theta + a_{16}r\theta \log r\right)\cos\theta$$

$$+ \left(b_{11}r + b_{12}r \log r + \frac{b_{13}}{r} + b_{14}r^3 + b_{15}r\theta + b_{16}r\theta \log r\right)\sin\theta \qquad (8.3.6)$$

$$+ \sum_{n=2}^{\infty}\left(a_{n1}r^n + a_{n2}r^{2+n} + a_{n3}r^{-n} + a_{n4}r^{2-n}\right)\cos n\theta$$

$$+ \sum_{n=2}^{\infty}\left(b_{n1}r^n + b_{n2}r^{2+n} + b_{n3}r^{-n} + b_{n4}r^{2-n}\right)\sin n\theta$$

where a_n, a_{nm}, and b_{nm} are constants to be determined. Note that this general solution is restricted to the periodic case, which has the most practical applications because it allows the Fourier method to be applied to handle general boundary conditions. Various pieces of the general solution (8.3.6) will now be used in many of the upcoming example problems. Since any/all pieces of (8.3.6) will automatically satisfy the governing equation, we will only have to satisfy the boundary conditions on the problem under study to determine the solution. As was discussed in Section 8.1, an Airy stress function of the form $\phi = A_{00} + A_{10}x + A_{01}y$ will produce no stress. Therefore terms in the Michell solution $\phi = a_0 + a_{11}r \cos\theta + b_{11}r \sin\theta$ will also produce no stress and should be dropped.

8.3.2 Axisymmetric solution

For the *axisymmetric case*, field quantities are independent of the angular coordinate, and this can be accomplished by choosing an Airy function solution from (8.3.6) by dropping all θ-terms, giving

$$\phi = a_1 \log r + a_2 r^2 + a_3 r^2 \log r \qquad (8.3.7)$$

Using relations (7.6.7), the resulting stresses for this case are

$$\sigma_r = 2a_3 \log r + \frac{a_1}{r^2} + a_3 + 2a_2$$

$$\sigma_\theta = 2a_3 \log r - \frac{a_1}{r^2} + 3a_3 + 2a_2 \qquad (8.3.8)$$

$$\tau_{r\theta} = 0$$

The displacements corresponding to these stresses can be determined by the usual methods of integrating the strain−displacement relations. For the case of plane stress, the result is

$$u_r = \frac{1}{E}\left[-\frac{(1+\nu)}{r}a_1 + 2(1-\nu)a_3 r \log r - (1+\nu)a_3 r + 2a_2(1-\nu)r\right]$$

$$+ A \sin\theta + B \cos\theta \qquad (8.3.9)$$

$$u_\theta = \frac{4r\theta}{E}a_3 + A \cos\theta - B \sin\theta + Cr$$

where A, B, and C are arbitrary constants associated with the rigid-body motion terms (see Exercise 7.22).

Plane strain results follow by simple change of elastic constants as per Table 7.1. If the body includes the origin, then a_3 and a_1 must be set to zero for the stresses to remain finite, and thus the stress field would be constant. Also note that the a_3 term in the tangential displacement relation leads to multivalued behavior if the domain geometry is such that the origin can be encircled by any contour lying entirely in the body. Exercise 8.24 is concerned with a particular problem that requires such multivalued behavior in the tangential displacement.

It should be pointed out that not all axisymmetric stress fields come from the Airy stress function given by (8.3.7). Reviewing the general form (8.3.6) indicates that the $a_4\theta$ term generates the stress field $\sigma_r = \sigma_\theta = 0$, $\tau_{r\theta} = a_4/r^2$, which is also axisymmetric. This solution could be used to solve problems with shear fields that produce tangential displacements that are independent of θ. Exercise 8.15 demonstrates such an example.

It has been previously pointed out that for multiply connected regions, the compatibility equations are not sufficient to guarantee single-valued displacements. With this in mind, we can investigate the displacement solution directly from the Navier equations. With zero body forces, Navier equations (5.4.4) or (7.6.4) for the axisymmetric case $\boldsymbol{u} = u_r(r)\boldsymbol{e}_r$ reduce to

$$\frac{d^2 u_r}{dr^2} + \frac{1}{r}\frac{du_r}{dr} - \frac{1}{r^2}u_r = 0 \tag{8.3.10}$$

The solution to this equation is given by

$$u_r = C_1 r + C_2 \frac{1}{r} \tag{8.3.11}$$

where C_1 and C_2 are constants. Notice this solution form is not the same as that given by (8.3.9), because we have a priori assumed that $u_\theta = 0$. Furthermore, the stresses corresponding to displacement solution (8.3.11) do not contain the logarithmic terms given in relations (8.3.8). Thus, these terms are not consistent with single-valued displacements.

8.4 Example polar coordinate solutions

With the general solution forms determined, we now explore the solution to several specific problems of engineering interest, including cases with both axisymmetric and general geometries.

Example 8.6 Thick-walled cylinder under uniform boundary pressure

The first example to be investigated involves a hollow thick-walled cylinder under the action of uniform internal and external pressure loadings, as shown in Fig. 8.8. We shall assume that the cylinder is long and this problem may be modeled under plane strain conditions.

Using the stress solution (8.3.8) without the log terms, the nonzero stresses are given by the form

$$\sigma_r = \frac{A}{r^2} + B$$

$$\sigma_\theta = -\frac{A}{r^2} + B \tag{8.4.1}$$

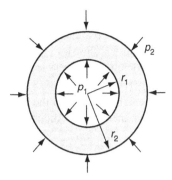

FIG. 8.8 Thick-walled cylinder problem.

Applying the boundary conditions $\sigma_r(r_1) = -p_1$, $\sigma_r(r_2) = -p_2$ creates two equations for the two unknown constants A and B. Solving for these constants gives the result

$$A = \frac{r_1^2 r_2^2 (p_2 - p_1)}{r_2^2 - r_1^2}$$

$$B = \frac{r_1^2 p_1 - r_2^2 p_2}{r_2^2 - r_1^2}$$

(8.4.2)

Substituting these values back into relation (8.4.1) gives the final result for the stress field

$$\sigma_r = \frac{r_1^2 r_2^2 (p_2 - p_1)}{r_2^2 - r_1^2} \frac{1}{r^2} + \frac{r_1^2 p_1 - r_2^2 p_2}{r_2^2 - r_1^2}$$

$$\sigma_\theta = -\frac{r_1^2 r_2^2 (p_2 - p_1)}{r_2^2 - r_1^2} \frac{1}{r^2} + \frac{r_1^2 p_1 - r_2^2 p_2}{r_2^2 - r_1^2}$$

(8.4.3)

From plane strain theory, the out-of-plane longitudinal stress is given by

$$\sigma_z = \nu(\sigma_r - \sigma_\theta) = 2\nu \frac{r_1^2 p_1 - r_2^2 p_2}{r_2^2 - r_1^2}$$

(8.4.4)

Using the strain–displacement relations (7.6.1) and Hooke's law (7.6.2), the radial displacement is easily determined as

$$u_r = \frac{1+\nu}{E} r \left[(1 - 2\nu)B - \frac{A}{r^2} \right]$$

$$= \frac{1+\nu}{E} \left[-\frac{r_1^2 r_2^2 (p_2 - p_1)}{r_2^2 - r_1^2} \frac{1}{r} + (1 - 2\nu)\frac{r_1^2 p_1 - r_2^2 p_2}{r_2^2 - r_1^2} r \right]$$

(8.4.5)

Reviewing this solution, it is noted that for the traction boundary-value problem with no body forces, the stress field does not depend on the elastic constants. However, the resulting displacements do depend on both E and ν.

Continued

For the case of only internal pressure ($p_2 = 0$ and $p_1 = p$) with $r_1/r_2 = 0.5$, the nondimensional stress distribution through the wall thickness is shown in Fig. 8.9 (using MATALB code C.4). The radial stress decays from $-p$ to zero, while the hoop stress is always positive with a maximum value at the inner radius $(\sigma_\theta)_{max} = (r_1^2 + r_2^2)/(r_2^2 - r_1^2)p = (5/3)p$.

For the case of a thin-walled tube, it can be shown that the hoop stress reduces to the well-known relation found from strength of materials theory (see Appendix D, Section D.5)

$$\sigma_\theta \approx \frac{p r_o}{t} \qquad (8.4.6)$$

where $t = r_2 - r_1$ is the thickness and $r_0 = (r_1 + r_2)/2$ is the mean radius.

The general solution to this example can be used to generate the solution to other problems through appropriate limiting processes. Two such cases are now presented.

8.4.1 Pressurized hole in an infinite medium

Consider the problem of a hole under uniform pressure in an infinite medium, as shown in Fig. 8.10. The solution to this problem can be easily determined from the general case of Example 8.6 by choosing $p_2 = 0$ and $r_2 \to \infty$. Taking these limits in relations (8.4.3) and (8.4.4) gives

$$\sigma_r = -p_1\frac{r_1^2}{r^2}, \quad \sigma_\theta = p_1\frac{r_1^2}{r^2}, \quad \sigma_z = 0 \qquad (8.4.7)$$

and the displacement field follows from (8.4.5)

$$u_r = \frac{1+\nu}{E}\frac{p_1 r_1^2}{r} \qquad (8.4.8)$$

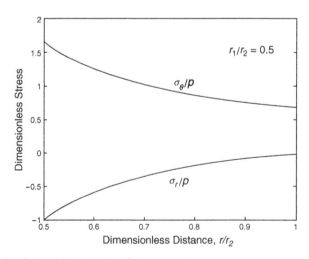

FIG. 8.9 Stress distribution in the thick-walled cylinder example.

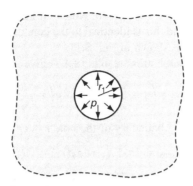

FIG. 8.10 Pressurized hole in an infinite medium.

Although both the stress and displacement fields decrease to zero as $r \rightarrow \infty$, there is a fundamental difference in their rate of decay. The stress field decays at a higher rate, of order $O(1/r^2)$, while the displacement field behaves as $O(1/r)$. Because stresses are proportional to displacement gradients, this behavior is to be expected.

8.4.2 Stress-free hole in an infinite medium under equal biaxial loading at infinity

Another example that can be generated from the general thick-walled cylinder problem is that of a stress-free hole in an unbounded medium with equal and uniform tensile loadings in the horizontal and vertical directions, as shown in Fig. 8.11. This particular case can be found from the general solution

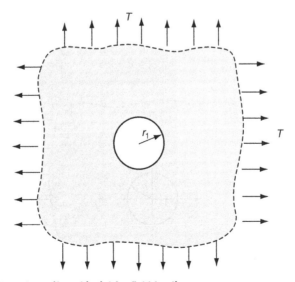

FIG. 8.11 Stress-free hole under uniform biaxial far-field loading.

by letting $r_2 \to \infty$ and taking $p_2 = -T$ and $p_1 = 0$. Note that the far-field stress in this problem is a hydrostatic state with $\sigma_x = \sigma_y = T$ and this is identical to the condition $\sigma_r = \sigma_\theta = T$. Thus, our limiting case matches with the far conditions shown in Fig. 8.11.

Under these conditions, the general stress results (8.4.3) give

$$\sigma_r = T\left(1 - \frac{r_1^2}{r^2}\right), \quad \sigma_\theta = T\left(1 + \frac{r_1^2}{r^2}\right) \tag{8.4.9}$$

The maximum stress σ_θ occurs at the boundary of the hole $r = r_1$ and is given by

$$\sigma_{max} = (\sigma_\theta)_{max} = \sigma_\theta(r_1) = 2T \tag{8.4.10}$$

and thus the *stress concentration factor* σ_{max}/T for a stress-free circular hole in an infinite medium under uniform all-around tension is 2. This result is of course true for plane strain or plane stress. Because of their importance, we shall next study several other stress concentration problems involving a stress-free hole under different far-field loading.

Example 8.7 Infinite medium with a stress-free hole under uniform far-field tension loading

Consider now an infinite medium with a circular stress-free hole subjected to a uniform far-field tension in a single direction, as shown in Fig. 8.12. Note that this problem will not be axisymmetric, and it requires particular θ-dependent terms from the general Michell solution.

The boundary conditions on this problem are

$$\sigma_r(a, \theta) = \tau_{r\theta}(a, \theta) = 0$$

$$\sigma_r(\infty, \theta) = \frac{T}{2}(1 + \cos 2\theta)$$

$$\sigma_\theta(\infty, \theta) = \frac{T}{2}(1 - \cos 2\theta) \tag{8.4.11}$$

$$\tau_{r\theta}(\infty, \theta) = -\frac{T}{2}\sin 2\theta$$

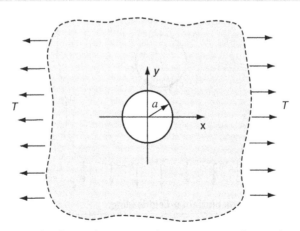

FIG. 8.12 Stress-free hole in an infinite medium under uniform far-field tension loading.

where the far-field conditions have been determined using the transformation laws established in Exercise 3.3 (or see Appendix B).

We start the solution to this example by considering the state of stress in the medium if there were no hole. This stress field is simply $\sigma_x = T$, $\sigma_y = \tau_{xy} = 0$ and can be derived from the Airy stress function

$$\phi = \frac{1}{2}Ty^2 = \frac{T}{2}r^2 \sin^2\theta = \frac{T}{4}r^2(1 - \cos 2\theta)$$

The presence of the hole acts to disturb this uniform field. We expect that this disturbance will be local in nature; the disturbed field will decay to zero as we move far away from the hole. Based on this, we choose a trial solution that includes the axisymmetric and $\cos 2\theta$ terms from the general Michell solution (8.3.6)

$$\begin{aligned}\phi = {}& a_1 \log r + a_2 r^2 + a_3 r^2 \log r \\ & + \left(a_{21} r^2 + a_{22} r^4 + a_{23} r^{-2} + a_{24}\right) \cos 2\theta\end{aligned} \tag{8.4.12}$$

The stresses corresponding to this Airy function are

$$\sigma_r = a_3(1 + 2\log r) + 2a_2 + \frac{a_1}{r^2} - \left(2a_{21} + \frac{6a_{23}}{r^4} + \frac{4a_{24}}{r^2}\right)\cos 2\theta$$

$$\sigma_\theta = a_3(3 + 2\log r) + 2a_2 - \frac{a_1}{r^2} + \left(2a_{21} + 12a_{22}r^4 + \frac{6a_{23}}{r^4}\right)\cos 2\theta \tag{8.4.13}$$

$$\tau_{r\theta} = \left(2a_{21} + 6a_{22}r^2 - \frac{6a_{23}}{r^4} - \frac{2a_{24}}{r^2}\right)\sin 2\theta$$

For finite stresses at infinity, we must take $a_3 = a_{22} = 0$. Applying the five boundary conditions in (8.4.11) gives

$$2a_2 + \frac{a_1}{a^2} = 0$$

$$2a_{21} + \frac{6a_{23}}{a^4} + \frac{4a_{24}}{a^2} = 0$$

$$2a_{21} - \frac{6a_{23}}{a^4} - \frac{2a_{24}}{a^2} = 0 \tag{8.4.14}$$

$$2a_{21} = -\frac{T}{2}$$

$$2a_2 = \frac{T}{2}$$

This system is easily solved for the constants, giving

$$a_1 = -\frac{a^2 T}{2}, \quad a_2 = \frac{T}{4}, \quad a_{21} = -\frac{T}{4}, \quad a_{23} = -\frac{a^4 T}{4}, \quad a_{24} = \frac{a^2 T}{2}$$

Substituting these values back into (8.4.13) gives the stress field

Continued

$$\sigma_r = \frac{T}{2}\left(1 - \frac{a^2}{r^2}\right) + \frac{T}{2}\left(1 + \frac{3a^4}{r^4} - \frac{4a^2}{r^2}\right)\cos 2\theta$$

$$\sigma_\theta = \frac{T}{2}\left(1 + \frac{a^2}{r^2}\right) - \frac{T}{2}\left(1 + \frac{3a^4}{r^4}\right)\cos 2\theta \tag{8.4.15}$$

$$\tau_{r\theta} = -\frac{T}{2}\left(1 - \frac{3a^4}{r^4} + \frac{2a^2}{r^2}\right)\sin 2\theta$$

The strain and displacement field can then be determined using the standard procedures used previously.

The hoop stress variation around the boundary of the hole is given by

$$\sigma_\theta(a, \theta) = T(1 - 2\cos 2\theta) \tag{8.4.16}$$

and this is shown in the polar plot in Fig. 8.13 (using MATLAB code C.5). This distribution indicates that the stress actually vanishes at $\theta = 30°$ and leads to a maximum value at $\theta = 90°$

$$\sigma_{\max} = \sigma_\theta(a, \pm\pi/2) = 3T \tag{8.4.17}$$

Therefore, the stress concentration factor for this problem is 3, a result that is higher than that found in the previous example shown in Fig. 8.11 for uniform tension in two orthogonal directions. This illustrates an interesting, nonintuitive point that additional vertical loading to the problem of Fig. 8.12 actually *reduces* the stress concentration.

The effects of the hole in perturbing the uniform stress field can be shown by plotting the variation of the stress with radial distance. Considering the case of the hoop stress at an angle $\pi/2$, Fig. 8.14 shows the distribution of $\sigma_\theta(r,\pi/2)/T$ versus nondimensional radial distance r/a. It is seen that the stress concentration around the hole is highly localized and decays very rapidly, essentially disappearing when $r > 5a$.

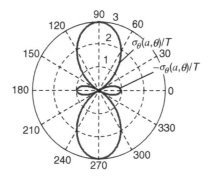

FIG. 8.13 Variation of hoop stress around hole boundary in Example 8.7.

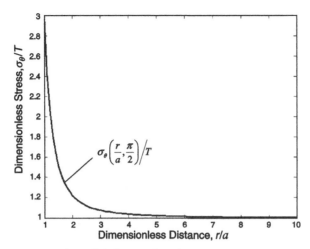

FIG. 8.14 Variation in hoop stress with radial distance from hole.

8.4.3 Biaxial and shear loading cases

Another interesting stress concentration problem is shown in Fig. 8.15. For this case, the far-field stress is biaxial, with tension in the horizontal and compression in the vertical. This far-field loading is equivalent to a pure shear loading on planes rotated 45° as shown in case (b). Thus, the solution to this case would apply to either the biaxial or shear-loading problems as shown.

The biaxial problem solution can be easily found from the original solution (8.4.15). This is done by adding to the original state another stress field with loading replaced by $-T$ and having coordinate axes rotated 90°. Details of this process are left as an exercise, and the final result is given by

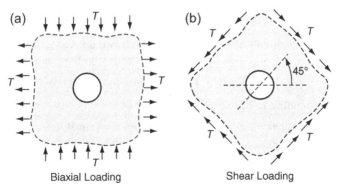

Biaxial Loading   Shear Loading

FIG. 8.15 Stress-free hole under biaxial and shear loading.

$$\sigma_r = T\left(1 + \frac{3a^4}{r^4} - \frac{4a^2}{r^2}\right)\cos 2\theta$$

$$\sigma_\theta = -T\left(1 + \frac{3a^4}{r^4}\right)\cos 2\theta \qquad (8.4.18)$$

$$\tau_{r\theta} = -T\left(1 - \frac{3a^4}{r^4} + \frac{2a^2}{r^2}\right)\sin 2\theta$$

The maximum stress is found to be the hoop stress on the boundary of the hole given by

$$\sigma_\theta(a,0) = \sigma_\theta(a,\pi) = -4T, \quad \sigma_\theta(a,\pi/2) = \sigma_\theta(a,3\pi/2) = 4T$$

and thus the stress concentration factor for this case is 4. It is interesting to compare this case with our previous examples shown in Figs. 8.11 and 8.12. The equal biaxial tension in Fig. 8.11 gives a stress concentration factor of 2, while the uniaxial far-field loading in Fig. 8.12 produces a factor of 3. It therefore appears that the equal but opposite biaxial loadings in Fig. 8.15A enhance the local stress field, thus giving the highest concentration effect of 4.

Other loading cases of stress concentration around a stress-free hole in an infinite medium can be developed by these techniques. The problem of determining such stress distributions for the case where the hole is in a medium of *finite size* poses a much more difficult boundary-value problem that would generally require Fourier methods using a series solution; see Little (1973). However, as shown in the stress plot in Fig. 8.14, the localized concentration effects decay rapidly. Thus, these infinite domain solutions could be used as a good approximation to finite size problems with boundaries located greater than about five hole radii away from the origin. Numerical techniques employing finite and boundary element methods are applied to these stress concentration problems in Chapter 16 (see Examples 16.2 and 16.5).

Example 8.8 Wedge and semi-infinite domain problems

In this example, we shall develop the solution to several problems involving the wedge domain shown in Fig. 8.16. The two boundaries are defined by the lines $\theta = \alpha$ and $\theta = \beta$. By making special choices for angles α and β and the boundary loadings on each face, many different problems can be generated.

Using the general Michell solution (8.3.6), we first choose an Airy stress function to include terms that are bounded at the origin and give uniform stresses on the boundaries

$$\phi = r^2(a_2 + a_6\theta + a_{21}\cos 2\theta + b_{21}\sin 2\theta) \qquad (8.4.19)$$

The stresses corresponding to this solution are given by

$$\sigma_r = 2a_2 + 2a_6\theta - 2a_{21}\cos 2\theta - 2b_{21}\sin 2\theta$$
$$\sigma_\theta = 2a_2 + 2a_6\theta + 2a_{21}\cos 2\theta + 2b_{21}\sin 2\theta \qquad (8.4.20)$$
$$\tau_{r\theta} = -a_6 - 2b_{21}\cos 2\theta + 2a_{21}\sin 2\theta$$

Note that this general stress field is independent of the radial coordinate.

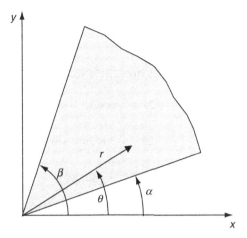

FIG. 8.16 Wedge domain geometry.

8.4.4 Quarter-plane example

Consider the specific case of a quarter-plane ($\alpha = 0$ and $\beta = \pi/2$) as shown in Fig. 8.17. The problem has a uniform shear loading along one boundary (y-axis) and no loading on the other boundary.

The boundary conditions on this problem can thus be stated as

$$\sigma_\theta(r,0) = \tau_{r\theta}(r,0) = 0$$
$$\sigma_\theta(r,\pi/2) = 0, \quad \tau_{r\theta}(r,\pi/2) = S$$

(8.4.21)

Using the general stress solution (8.4.20), these boundary conditions give the following four equations

$$2a_2 + 2a_{21} = 0$$
$$-a_6 - 2b_{21} = 0$$
$$2a_2 - 2a_{21} + a_6\pi = 0$$
$$-a_6 + 2b_{21} = S$$

(8.4.22)

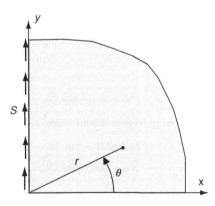

FIG. 8.17 Quarter-plane example.

These are easily solved for the unknown constants, giving

$$a_2 = \frac{S\pi}{8}, \quad a_6 = -\frac{S}{2}, \quad a_{21} = -\frac{S\pi}{8}, \quad b_{21} = \frac{S}{4} \tag{8.4.23}$$

Back-substituting these results determines the stress field solution

$$\sigma_r = \frac{S}{2}\left(\frac{\pi}{2} - 2\theta + \frac{\pi}{2}\cos 2\theta - \sin 2\theta\right)$$

$$\sigma_\theta = \frac{S}{2}\left(\frac{\pi}{2} - 2\theta - \frac{\pi}{2}\cos 2\theta + \sin 2\theta\right) \tag{8.4.24}$$

$$\tau_{r\theta} = \frac{S}{2}\left(1 - \cos 2\theta - \frac{\pi}{2}\sin 2\theta\right)$$

It has been pointed out that this problem has an apparent inconsistency in the shear stress component at the origin—that is, $\tau_{xy} \neq \tau_{yx}$ at $x = y = 0$. To further investigate this, let us reformulate the problem in Cartesian coordinates. The stress function can be expressed as

$$\phi = S\left(\frac{\pi(x^2 + y^2)}{8} - \frac{(x^2 + y^2)}{2}\tan^{-1}\frac{y}{x} - \frac{\pi}{8}(x^2 - y^2) + \frac{xy}{2}\right) \tag{8.4.25}$$

The shear stress is then given by

$$\tau_{xy} = -\frac{\partial^2\phi}{\partial x\partial y} = \frac{-Sy^2}{x^2 + y^2} \tag{8.4.26}$$

Excluding the origin, this expression tends to zero for $y \to 0$ and to $-S$ for $x \to 0$, and thus has the proper limiting behavior for $r \neq 0$. However, it has been shown by Barber (2010) that the stress gradients in the tangential direction are of order $O(r^{-1})$.

8.4.5 Half-space examples

Let us next consider the solution to several half-space examples with a domain specified by $\alpha = 0$ and $\beta = \pi$. We shall investigate examples with uniform loadings over portions of the boundary surface and also cases with concentrated forces.

8.4.6 Half-space under uniform normal stress over $x \le 0$

The problem of a half-space with uniform normal stress over the negative x-axis is shown in Fig. 8.18. For the particular angles of α and β that create the half-space domain, the general Airy stress function solution form (8.4.19) can be reduced to

$$\phi = a_6 r^2\theta + b_{21}r^2 \sin 2\theta \tag{8.4.27}$$

The hoop and shear stresses corresponding to this function are

$$\sigma_\theta = 2a_6\theta + 2b_{21} \sin 2\theta$$

$$\tau_{r\theta} = -a_6 - 2b_{21} \cos 2\theta \tag{8.4.28}$$

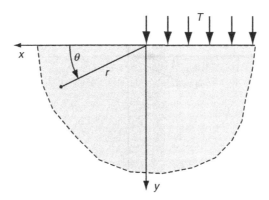

FIG. 8.18 Half-space under uniform loading over half of the free surface.

Applying boundary conditions $\sigma_\theta(r,0) = \tau_{r\theta}(r,0) = \tau_{r\theta}(r,\pi) = 0$, $\sigma_\theta(r,\pi) = -T$ determines constants $a_6 = -T/2\pi$, $b_{21} = T/4\pi$. Thus, the stress field solution is now determined as

$$\sigma_r = -\frac{T}{2\pi}(\sin 2\theta + 2\theta)$$

$$\sigma_\theta = \frac{T}{2\pi}(\sin 2\theta - 2\theta) \tag{8.4.29}$$

$$\tau_{r\theta} = \frac{T}{2\pi}(1 - \cos 2\theta)$$

It is again noted that this field depends only on the angular coordinate. Because of the discontinuity of the boundary loading, there is a lack of continuity of the stress at the origin. This can be seen by considering the behavior of the Cartesian shear stress component. Using the transformation relations in Appendix B, the Cartesian shear stress for this problem is found to be

$$\tau_{xy} = -\frac{T}{2\pi}(1 - \cos 2\theta) \tag{8.4.30}$$

Along the positive x-axes ($\theta = 0$) $\tau_{xy} = 0$, while on the y-axes ($\theta = \pi/2$) $\tau_{xy} = -T/\pi$. Thus, as we approach the origin along these two different paths, the values will not coincide.

8.4.7 Half-space under concentrated surface force system (Flamant problem)

As another half-space example, consider the case of a concentrated force system acting at the origin, as shown in Fig. 8.19. This example is commonly called the *Flamant problem*, one of the classical problems in elasticity.

Specifying boundary conditions for such problems with only concentrated loadings requires some modification of our previous methods. For this example, the tractions on any semicircular arc C enclosing the origin must balance the applied concentrated loadings. Because the area of such an arc is proportional to the radius r, the stresses must be of order $1/r$ to allow such an equilibrium statement to hold on any radius. The appropriate terms in the general Michell solution (8.3.6) that will give stresses of order $1/r$ are specified by

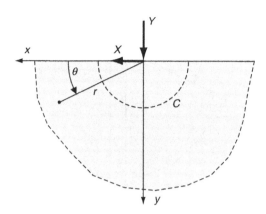

FIG. 8.19 Flamant problem.

$$\phi = (a_{12}r \log r + a_{15}r\theta) \cos \theta + (b_{12}r \log r + b_{15}r\theta)\sin \theta \qquad (8.4.31)$$

The stresses resulting from this stress function are

$$\sigma_r = \frac{1}{r}[(a_{12} + 2b_{15}) \cos \theta + (b_{12} - 2a_{15}) \sin \theta]$$

$$\sigma_\theta = \frac{1}{r}[a_{12} \cos \theta + b_{12} \sin \theta] \qquad (8.4.32)$$

$$\tau_{r\theta} = \frac{1}{r}[a_{12} \sin \theta - b_{12} \cos \theta]$$

With zero normal and shear stresses on $\theta = 0$ and π, $a_{12} = b_{12} = 0$, and thus $\sigma_\theta = \tau_{r\theta} = 0$ everywhere. Therefore, the state of stress is sometimes called a *radial distribution* (see Exercise 7.23). Note that this result is also true for the more general case of a wedge domain with arbitrary angles of α and β (see Exercise 8.30). To determine the remaining constants a_{15} and b_{15}, we apply the equilibrium statement that the summation of the tractions over the semicircular arc C of radius a must balance the applied loadings

$$X = -\int_0^\pi \sigma_r(a, \theta)a \cos\theta d\theta = -\pi b_{15}$$

$$\qquad (8.4.33)$$

$$Y = -\int_0^\pi \sigma_r(a, \theta)a \sin\theta d\theta = \pi a_{15}$$

Thus, the constants are determined as $a_{15} = Y/\pi$ and $b_{15} = -X/\pi$, and the stress field is now given by

$$\sigma_r = -\frac{2}{\pi r}[X \cos\theta + Y \sin\theta]$$

$$\qquad (8.4.34)$$

$$\sigma_\theta = \tau_{r\theta} = 0$$

As expected, the stress field is singular at the origin directly under the point loading. However, what is not expected is the result that σ_θ and $\tau_{r\theta}$ vanish even for the case of tangential loading.

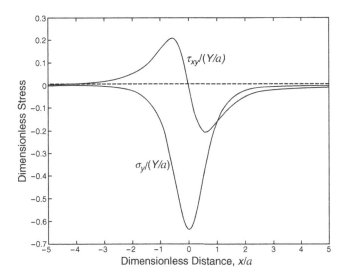

FIG. 8.20 Normal and shear stress distributions below the free surface for the Flamant problem.

To investigate this problem further, we will restrict the case to only normal loading and set $X = 0$. For this loading, the stresses are

$$\sigma_r = -\frac{2Y}{\pi r} \sin \theta$$

$$\sigma_\theta = \tau_{r\theta} = 0$$

(8.4.35)

The Cartesian components corresponding to this stress field are determined using the transformation relations, given in Appendix B or directly from Exercise 3.4. The results are found to be

$$\sigma_x = \sigma_r \cos^2\theta = -\frac{2Yx^2y}{\pi\left(x^2 + y^2\right)^2}$$

$$\sigma_y = \sigma_r \sin^2\theta = -\frac{2Yy^3}{\pi\left(x^2 + y^2\right)^2}$$

(8.4.36)

$$\tau_{xy} = \sigma_r \sin\theta\cos\theta = -\frac{2Yxy^2}{\pi\left(x^2 + y^2\right)^2}$$

The distribution of the normal and shearing stresses on a horizontal line located a distance a below the free surface of the half-space is shown in Fig. 8.20. The maximum normal stress directly under the load is given by $|\sigma_y| = 2Y/\pi a$. It is observed that the effects of the concentrated loading are highly localized, and the stresses are vanishingly small for distances where $x > 5a$. Stress contours of σ_r are shown in Fig. 8.21. From solution (8.4.35), lines of constant radial stress are circles tangent to the half-space surface at the loading point.

We now wish to determine the displacements for the normal concentrated force problem. Assuming *plane stress* conditions, Hooke's law and the strain–displacement relations give

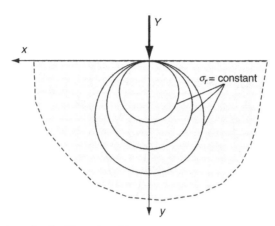

FIG. 8.21 Radial stress contours for the Flamant problem.

$$e_r = \frac{\partial u_r}{\partial r} = \frac{1}{E}(\sigma_r - \nu\sigma_\theta) = -\frac{2Y}{\pi E r}\sin\theta$$

$$e_\theta = \frac{u_r}{r} + \frac{1}{r}\frac{\partial u_\theta}{\partial \theta} = \frac{1}{E}(\sigma_\theta - \nu\sigma_r) = \frac{2\nu Y}{\pi E r}\sin\theta \tag{8.4.37}$$

$$2e_{r\theta} = \frac{1}{r}\frac{\partial u_r}{\partial \theta} + \frac{\partial u_\theta}{\partial r} - \frac{u_\theta}{r} = \frac{1}{\mu}\tau_{r\theta} = 0$$

Integrating $(8.4.37)_1$ yields the radial displacement

$$u_r = -\frac{2Y}{\pi E}\sin\theta \ \log r + f(\theta) \tag{8.4.38}$$

where f is an arbitrary function of the angular coordinate.

Substituting (8.4.38) into $(8.4.37)_2$ allows separation of the derivative of the tangential displacement component

$$\frac{\partial u_\theta}{\partial \theta} = \frac{2\nu Y}{\pi E}\sin\theta + \frac{2Y}{\pi E}\sin\theta \log r - f(\theta)$$

Integrating this equation gives

$$u_\theta = -\frac{2\nu Y}{\pi E}\cos\theta - \frac{2Y}{\pi E}\cos\theta \log r - \int f(\theta)d\theta + g(r) \tag{8.4.39}$$

where $g(r)$ is an arbitrary function of the indicated variable. Determination of the arbitrary functions f and g is accomplished by substituting Eqs. (8.4.38) and (8.4.39) into $(8.4.37)_3$. Similar to our previous Cartesian examples, the resulting equation can be separated into the two relations

$$g(r) - rg'(r) = K$$

$$-(1-\nu)\frac{2Y}{\pi E}\cos\theta + f'(\theta) + \int f(\theta)d\theta = K \tag{8.4.40}$$

where K is an arbitrary constant. The solutions to this system are

$$g(r) = Cr + K$$

$$f(\theta) = \frac{(1-\nu)Y}{\pi E} \theta \cos\theta + A \sin\theta + B \cos\theta \qquad (8.4.41)$$

where A, B and C are constants of integration, and K must vanish to satisfy $(8.4.37)_3$

Collecting these results, the displacements thus can be written as

$$u_r = \frac{(1-\nu)Y}{\pi E} \theta \cos\theta - \frac{2Y}{\pi E} \log r \sin\theta + A \sin\theta + B \cos\theta$$

$$u_\theta = -\frac{(1-\nu)Y}{\pi E} \theta \sin\theta - \frac{2Y}{\pi E} \log r \cos\theta - \frac{(1+\nu)Y}{\pi E} \cos\theta \qquad (8.4.42)$$

$$+ A \cos\theta - B \sin\theta + Cr$$

The terms involving the constants A, B, and C represent rigid-body motion (see Exercise 7.22). These terms can be set to any arbitrary value without affecting the stress distribution. Rather than setting them all to zero, they will be selected to satisfy the expected symmetry condition that the horizontal displacements along the y-axis should be zero. This condition can be expressed by $u_\theta(r,\pi/2) = 0$, and this relation requires

$$C = 0, \quad B = -\frac{(1-\nu)Y}{2E}$$

The vertical rigid-body motion may be taken as zero, thus implying that $A = 0$. Values for all constants are now determined, and the final result for the displacement field is

$$u_r = \frac{Y}{\pi E}\left[(1-\nu)\left(\theta - \frac{\pi}{2}\right)\cos\theta - 2 \log r \sin\theta\right]$$

$$u_\theta = \frac{Y}{\pi E}\left[-(1-\nu)\left(\theta - \frac{\pi}{2}\right)\sin\theta - 2 \log r \cos\theta - (1+\nu)\cos\theta\right] \qquad (8.4.43)$$

It should be pointed out that these results contain unbounded logarithmic terms that would lead to unrealistic predictions at infinity. This unpleasant situation is a result of the two-dimensional model, and is certainly expected since the radial stress distribution was required to be $O(1/r)$. The corresponding three-dimensional problem (Boussinesq's problem) is solved in Chapter 13. The resulting 3-D displacement field does not have logarithmic terms and is bounded at infinity; see equations (13.4.16).

The radial displacement along the free surface is given by

$$u_r(r,0) = u_r(r, \pi) = -\frac{Y}{2E}(1-\nu) \qquad (8.4.44)$$

Since $(1-\nu) > 0$, we see an unexpected result in that the horizontal displacement of all points on the half-space surface move an equal amount toward the loading point. The tangential displacement component on the surface is given by

$$u_\theta(r,0) = -u_\theta(r, \pi) = -\frac{Y}{\pi E}[(1+\nu) + 2 \log r] \qquad (8.4.45)$$

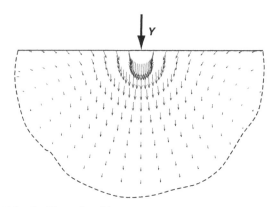

FIG. 8.22 Displacement field for the Flamant problem.

which as expected is singular at the origin under the point loading. Again, the corresponding three-dimensional solution in Chapter 13 predicts quite different surface displacements. For example, the three-dimensional result corresponding to Eq. (8.4.45) gives a vertical displacement of order $O(1/r)$. A MATLAB$^{®}$ vector distribution plot of the general displacement field resulting from solution (8.4.43) is shown in Fig. 8.22 (using code C.6). The total displacement vectors are illustrated using suitable units for the near-field case ($0 < r < 0.5$) with a Poisson's ratio of 0.3, and $Y/E = 1$. The field pattern would significantly change for $r > 1$.

Some authors (for example, Timoshenko and Goodier, 1970) have tried to remove the unpleasant logarithmic effects by invoking a somewhat arbitrary condition $u_r(r_0, \pi/2) = 0$, where r_0 is some arbitrary distance from the loading point. This condition may be used to determine the vertical rigid-body term, thus determining the constant $A = (2Y/\pi E)\log r_0$. Under this condition, the displacement solution can then be written as

$$u_r = \frac{Y}{\pi E}\left[(1-\nu)\left(\theta - \frac{\pi}{2}\right)\cos\theta + 2\log\left(\frac{r_0}{r}\right)\sin\theta\right]$$

$$u_\theta = \frac{Y}{\pi E}\left[-(1-\nu)\left(\theta - \frac{\pi}{2}\right)\sin\theta + 2\log\left(\frac{r_0}{r}\right)\cos\theta - (1+\nu)\ \cos\theta\right]$$

(8.4.46)

8.4.8 Half-space under a surface concentrated moment

Other half-space problems with concentrated loadings can be generated from the previous single-force solution. For example, the concentrated moment problem can be found from the superposition of two equal but opposite forces separated by a distance d, as shown in Fig. 8.23. The limit is taken with $d \to 0$ but with the product $Pd \to M$. Details are left as an exercise, and the final resulting stress field is given by

$$\sigma_r = -\frac{4M}{\pi r^2}\sin\theta\cos\theta$$

$$\sigma_\theta = 0$$

(8.4.47)

$$\tau_{r\theta} = -\frac{2M}{\pi r^2}\sin^2\theta$$

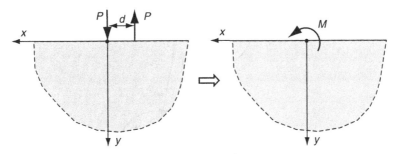

FIG. 8.23 Half-space with a concentrated surface moment loading.

8.4.9 Half-space under uniform normal loading over $-a \geq x \geq a$

As a final half-space example, consider the case of a uniform normal loading acting over a finite portion $(-a \geq x \geq a)$ of the free surface, as shown in Fig. 8.24. This problem can be solved by using the superposition of the single normal force solution developed previously. Using the Cartesian stress solution (8.4.36) for the single-force problem

$$\sigma_x = \sigma_r \cos^2 \theta = -\frac{2Y}{\pi r} \sin\theta \cos^2 \theta$$

$$\sigma_y = \sigma_r \sin^2 \theta = -\frac{2Y}{\pi r} \sin^3 \theta \qquad (8.4.48)$$

$$\tau_{xy} = \sigma_r \sin\theta \cos\theta = -\frac{2Y}{\pi r} \sin^2 \theta \cos\theta$$

For the distributed loading case, a differential load acting on the free surface length dx may be expressed by $dY = pdx$. Using the geometry in Fig. 8.25, $dx = rd\theta/\sin\theta$, and thus the differential loading is given by $dY = prd\theta/\sin\theta$.

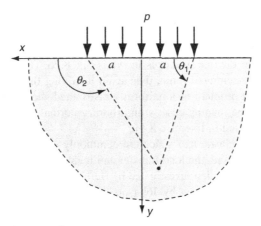

FIG. 8.24 Half-space under uniform loading over $-a \geq x \geq a$.

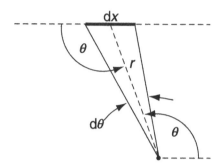

FIG. 8.25 Surface geometry for the distributed loading example.

Using the differential loading in relations (8.4.48) gives the differential stress field

$$d\sigma_x = -\frac{2p}{\pi} \cos^2 \theta \, d\theta$$

$$d\sigma_y = -\frac{2p}{\pi} \sin^2 \theta \, d\theta \qquad (8.4.49)$$

$$d\tau_{xy} = -\frac{2p}{\pi} \sin\theta \cos\theta \, d\theta$$

Integrating this result over the entire load distribution gives the total stress field

$$\sigma_x = -\frac{2p}{\pi} \int_{\theta_1}^{\theta_2} \cos^2\theta \, d\theta = -\frac{p}{2\pi}[2(\theta_2 - \theta_1) + (\sin 2\theta_2 - \sin 2\theta_1)]$$

$$\sigma_y = -\frac{2p}{\pi} \int_{\theta_1}^{\theta_2} \sin^2\theta \, d\theta = -\frac{p}{2\pi}[2(\theta_2 - \theta_1) - (\sin 2\theta_2 - \sin 2\theta_1)] \qquad (8.4.50)$$

$$\tau_{xy} = -\frac{2p}{\pi} \int_{\theta_1}^{\theta_2} \sin\theta \cos\theta \, d\theta = \frac{p}{2\pi}[\cos 2\theta_2 - \cos 2\theta_1]$$

with θ_1 and θ_2 defined in Fig. 8.24. The distribution of the normal and shearing stresses on a horizontal line located a distance a below the free surface is shown in Fig. 8.26. This distribution is similar to that in Fig. 8.20 for the single concentrated force, thus again justifying the Saint–Venant principle. The solution of the corresponding problem of a uniformly distributed shear loading is given in Exercise 8.40. The more general surface loading case, with arbitrary normal and shear loading over the free surface $(-a \leq s \leq a)$, is included in Exercise 8.41.

Distributed loadings on an elastic half-space are commonly used to simulate *contact mechanics* problems, which are concerned with the load transfer and local stress distribution in elastic bodies in contact. Problems of this type were first investigated by Hertz (1882), and numerous studies have been conducted over the last century (see text by Johnson, 1985). Because interest in these problems is

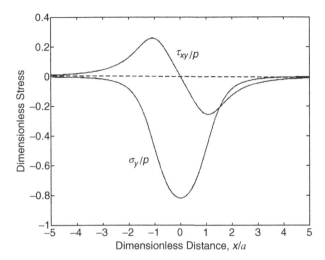

FIG. 8.26 Normal and shear stress distributions for the distributed loading example.

normally restricted to near-field behavior, boundary dimensions and curvatures can often be neglected and a distributed loading on a half-space can provide an estimate of the local stress distribution. Of course, the simple uniform normal load distribution in Fig. 8.24 would only provide an approximation to the actual nonuniform loading generated by bodies in contact.

The high local stresses commonly generated in such problems have been found to cause material failure under repeated loading conditions found in rotating wheels, gears, and bearings. Because failure of ductile materials can be related to the maximum shear stress, consider the behavior of τ_{max} under the loading in Fig. 8.24. Along the y-axis below the loading, $\tau_{xy} = 0$, thus the x- and y-axes are principal at these points and the maximum shear stress is given by $\tau_{max} = 1/2|\sigma_x - \sigma_y|$. A plot of this stress versus depth below the surface is shown in Fig. 8.27. It is interesting to observe that τ_{max} takes on a maximum value of p/π below the free surface at $y = a$, and thus initial material failure is expected to start at this subsurface location. The corresponding stress distribution for the concentrated loading problem of Fig. 8.21 is also shown. In contrast, it is seen that the concentrated loading produces monotonically decreasing behavior from the singular value directly under the load.

To compare these theoretical results with actual material behavior, photoelastic contact examples (taken from Johnson, 1985) are shown in Fig. 8.28. The figure illustrates and compares near-field photoelastic fringe patterns in a rectangular plate with four different contact loadings. The photoelastic model (plate) is made of a transparent material that exhibits *isochromatic fringe patterns* when viewed under polarized light. As mentioned in Section 3.6, these fringes represent lines of *constant maximum shear stress* and can be used to determine the nature of the local stress field. Under point loading, the maximum stress appears to be located directly under the load, while for the uniform distributed loading case the maximum contour occurs at a small distance below the contact surface. These results provide qualitative agreement with the theoretical predictions shown in Fig. 8.27, and Exercise 8.39 involves the development and plotting of the maximum shear stress contours.

Fig. 8.28 also shows surface loading from a flat punch and circular cylinder. The flat punch loading generates high local stresses at the edges of the punch, and this is caused by the singularity of the

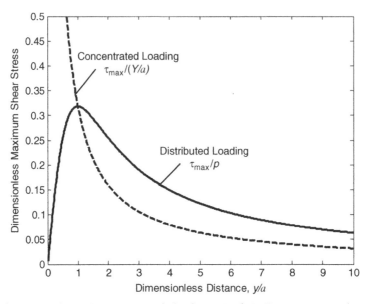

FIG. 8.27 Comparison of maximum shear stress variation in an elastic half-space under point and distributed surface loadings.

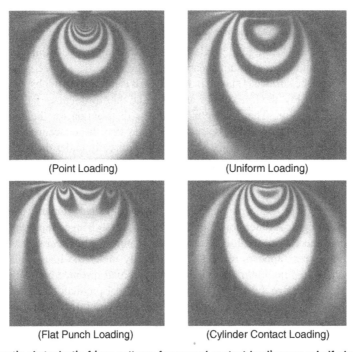

(Point Loading)   (Uniform Loading)

(Flat Punch Loading)   (Cylinder Contact Loading)

FIG. 8.28 Isochromatic photoelastic fringe patterns for several contact loadings on a half-plane.

(Taken from Contact Mechanics by KL Johnson, reprinted with the permission of Cambridge University Press.)

loading at these two points. Although the stress fields of the cylinder and uniform loading cases look similar, the detailed stresses will not be the same. The cylinder case will create a *nonuniform* contact loading profile that decreases to zero at the ends of the contact area. We will explore in more detail some of these contact mechanics solution behaviors in Section 8.5. Other distributed loading problems can be solved in a similar superposition fashion, and the solution to several cases have been given by Timoshenko and Goodier (1970), Poulos and Davis (1974), and Kachanov et al. (2003).

8.4.10 Notch and crack problems

Consider the original wedge problem shown in Fig. 8.16 for the case where angle α is small and β is $2\pi - \alpha$. This case generates a thin notch in an infinite medium, as shown in Fig. 8.29. We pursue the case where $\alpha \approx 0$, and thus the notch becomes a crack. The boundary surfaces of the notch are taken to be stress free and thus the problem involves only far-field loadings.

Starting with the Michell solution (8.3.6), the Airy stress function is chosen from the generalized form

$$\phi = r^\lambda [A \sin \lambda\theta + B \cos \lambda\theta + C \sin(\lambda - 2)\theta + D \cos(\lambda - 2)\theta] \tag{8.4.51}$$

where we are now allowing λ to be a noninteger. The boundary stresses corresponding to this stress function are

$$\sigma_\theta = \lambda(\lambda - 1)r^{\lambda-2}[A \sin \lambda\theta + B \cos \lambda\theta + C \sin(\lambda - 2)\theta + D \cos(\lambda - 2)\theta]$$
$$\tau_{r\theta} = -(\lambda - 1)r^{\lambda-2}[A\lambda \cos \lambda\theta - B\lambda \sin \lambda\theta + C(\lambda - 2)\cos(\lambda - 2)\theta - D(\lambda - 2)\sin(\lambda - 2)\theta]$$
$$\tag{8.4.52}$$

The stress-free boundary conditions at $\theta = \alpha \approx 0$ give

$$B + D = 0$$
$$\lambda A + (\lambda - 2)C = 0 \tag{8.4.53}$$

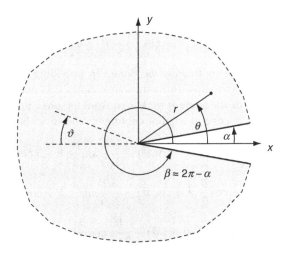

FIG. 8.29 Crack problem geometry.

while the identical conditions at $\theta = \beta = 2\pi - \alpha \approx 2\pi$ produce

$$\left[\sin 2\pi(\lambda - 2) - \frac{\lambda - 2}{\lambda}\sin 2\pi\lambda\right]C$$

$$+[\cos 2\pi(\lambda - 2) - \cos 2\pi\lambda]D = 0 \qquad (8.4.54)$$

$$[(\lambda - 2)\cos 2\pi(\lambda - 2) - (\lambda - 2)\cos 2\pi\lambda]C$$

$$-[(\lambda - 2)\sin 2\pi(\lambda - 2) - \lambda \sin 2\pi\lambda]D = 0$$

where we have used relations (8.4.53) to reduce the form of (8.4.54). These relations represent a system of four homogeneous equations for the four unknowns A, B, C, and D. Thus, the determinant of the coefficient matrix must vanish, and this gives the result

$$\sin 2\pi(\lambda - 1) = 0$$

This relation implies that $2\pi(\lambda - 1) = n\pi$, with $n = 0, 1, 2, \ldots$, thus giving

$$\lambda = \frac{n}{2} + 1, \quad n = 0, 1, 2, \ldots \qquad (8.4.55)$$

Near the tip of the notch $r \to 0$ and the stresses will be of order $O(r^{\lambda-2})$, while the displacements are $O(r^{\lambda-1})$. At this location the displacements are expected to be finite, thus implying $\lambda > 1$, while the stresses are expected to be singular, requiring $\lambda < 2$. Therefore, we find that the allowable range for λ is given by $1 < \lambda < 2$. In light of relation (8.4.55), we need only consider the case with $n = 1$, giving $\lambda = 3/2$. Thus, for the crack problem the local stresses around the crack tip will be $O(1/\sqrt{r})$ and the displacements $O(\sqrt{r})$.

Using these results, the stress field can then be written as

$$\sigma_r = -\frac{3}{4}\frac{1}{\sqrt{r}}\left[A\left(\sin\frac{3}{2}\theta + 5\sin\frac{\theta}{2}\right) + B\left(\cos\frac{3}{2}\theta + \frac{5}{3}\cos\frac{\theta}{2}\right)\right]$$

$$\sigma_\theta = \frac{3}{4}\frac{1}{\sqrt{r}}\left[A\left(\sin\frac{3}{2}\theta - 3\sin\frac{\theta}{2}\right) + B\left(\cos\frac{3}{2}\theta - \cos\frac{\theta}{2}\right)\right] \qquad (8.4.56)$$

$$\tau_{r\theta} = -\frac{3}{4}\frac{1}{\sqrt{r}}\left[A\left(\cos\frac{3}{2}\theta - \cos\frac{\theta}{2}\right) - B\left(\sin\frac{3}{2}\theta - \frac{1}{3}\sin\frac{\theta}{2}\right)\right]$$

Such relations play an important role in *fracture mechanics* by providing information on the nature of the singular state of stress near crack tips. In fracture mechanics it is normally more convenient to express the stress field in terms of the angle ϑ measured from the direction of crack propagation, as shown in Fig. 8.29. Making the change in angular coordinate, the stress field now becomes

$$\sigma_r = -\frac{3}{2}\frac{A}{\sqrt{r}}\cos\frac{\vartheta}{2}(3 - \cos\vartheta) - \frac{B}{2\sqrt{r}}\sin\frac{\vartheta}{2}(1 - 3\cos\vartheta)$$

$$\sigma_\theta = -\frac{3}{2}\frac{A}{\sqrt{r}}\cos\frac{\vartheta}{2}(1 + \cos\vartheta) - \frac{3B}{2\sqrt{r}}\sin\frac{\vartheta}{2}(1 + \cos\vartheta) \qquad (8.4.57)$$

$$\tau_{r\theta} = \frac{3}{2}\frac{A}{\sqrt{r}}\sin\frac{\vartheta}{2}(1 + \cos\vartheta) + \frac{B}{2\sqrt{r}}\cos\frac{\vartheta}{2}(1 - 3\cos\vartheta)$$

The remaining constants A and B are determined from the far-field boundary conditions. However, an important observation is that the *form* of this crack-tip stress field is not dependent on such boundary conditions. With respect to the angle ϑ, it is noted that terms with the A coefficient include *symmetric* normal stresses, while the remaining terms containing B have *antisymmetric* behavior. The symmetric terms are normally referred to as *opening* or *mode I* behavior, while antisymmetric terms correspond to *shear* or *mode II*. Constants A and B can be related to the *stress intensity factors* commonly used in fracture mechanics studies. Further information on stress analysis around cracks can be found in the classic monograph by Tada et al. (2000). Other cases of notch problems with different geometry and boundary conditions have been presented by Little (1973). Additional stress analysis around cracks is investigated in Chapters 10–12 using the powerful method of complex variable theory.

Example 8.9 Curved beam problems

We shall now investigate the solution to some curved beam problems defined by an annular region cut by two radial lines. Similar to the previous beam examples, we use resultant force boundary conditions at the ends and exact pointwise specifications along the lateral curved boundaries. Comparisons with strength of materials predictions are made for specific cases.

8.4.11 Pure bending example

The first example is the simple case of a curved beam loaded by end moments, as shown in Fig. 8.30. The solution to such a problem is independent of the angular coordinate. As usual, we satisfy the pointwise boundary conditions on the sides of the beam but address only the resultant effects at each cross-sectional end. Thus, the boundary conditions on this problem are formulated as

$$\sigma_r(a) = \sigma_r(b) = 0$$
$$\tau_{r\theta}(a) = \tau_{r\theta}(b) = 0$$
$$\int_a^b \sigma_\theta dr = 0 \tag{8.4.58}$$
$$\int_a^b \sigma_\theta r dr = -M$$

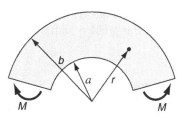

FIG. 8.30 Curved beam with end moments.

Using the general axisymmetric stress solution (8.3.8) in boundary relations (8.4.58) gives

$$2A \log a + \frac{C}{a^2} + A + 2B = 0$$

$$2A \log b + \frac{C}{b^2} + A + 2B = 0$$

(8.4.59)

$$b \left(2A \log b + \frac{C}{b^2} + A + 2B \right) - a \left(2A \log a + \frac{C}{a^2} + A + 2B \right) = 0$$

$$-C \log \left(\frac{b}{a} \right) + A \left(b^2 \log b - a^2 \log a \right) + B \left(b^2 - a^2 \right) = -M$$

Because the third equation is a linear combination of the first two, only three of these four relations are independent. Solving these equations for the three constants gives

$$A = -\frac{2M}{N} \left(b^2 - a^2 \right)$$

$$B = \frac{M}{N} \left[b^2 - a^2 + 2 \left(b^2 \log b - a^2 \log a \right) \right]$$

(8.4.60)

$$C = -\frac{4M}{N} a^2 b^2 \log \left(\frac{b}{a} \right)$$

where

$$N = \left(b^2 - a^2 \right)^2 - 4a^2 b^2 \left[\log \left(\frac{b}{a} \right) \right]^2$$

The stresses thus become

$$\sigma_r = -\frac{4M}{N} \left[\frac{a^2 b^2}{r^2} \log \left(\frac{b}{a} \right) + b^2 \log \left(\frac{r}{b} \right) + a^2 \log \left(\frac{a}{r} \right) \right]$$

$$\sigma_\theta = -\frac{4M}{N} \left[-\frac{a^2 b^2}{r^2} \log \left(\frac{b}{a} \right) + b^2 \log \left(\frac{r}{b} \right) + a^2 \log \left(\frac{a}{r} \right) + b^2 - a^2 \right]$$

(8.4.61)

$$\tau_{r\theta} = 0$$

The bending stress distribution σ_θ through the beam thickness is shown in Fig. 8.31 for the case of $b/a = 4$. Also shown in the figure is the corresponding result from strength of materials theory (see Appendix D, Section D.4, and Exercise 8.46). Both theories predict nonlinear stress distributions with maximum values on the inner fibers. For this problem, differences between elasticity and strength of materials predictions are very small.

8.4.12 Curved cantilever under end loading

Consider the curved cantilever beam carrying an end loading, as shown in Fig. 8.32. For this problem, the stress field depends on the angular coordinate. The boundary conditions require zero stress on

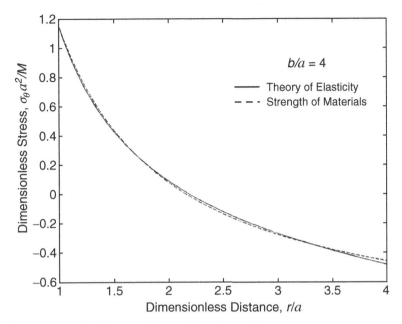

FIG. 8.31 Bending stress results of a curved beam with end moments.

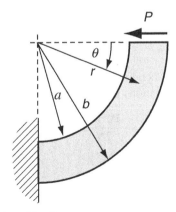

FIG. 8.32 Curved cantilever beam with end loading.

$r = a$ and b and a resultant shear load on the end $\theta = 0$. These conditions are thus formulated as follows

$$\sigma_r(a, \theta) = \sigma_r(b, \theta) = 0$$

$$\tau_{r\theta}(a, \theta) = \tau_{r\theta}(b, \theta) = 0$$

$$\int_a^b \tau_{r\theta}(r, 0)dr = P$$

$$\int_a^b \sigma_\theta(r, 0)dr = \int_a^b \sigma_\theta(r, 0)rdr = 0$$

(8.4.62)

$$\int_a^b \sigma_\theta(r, \pi/2)dr = -P$$

$$\int_a^b \sigma_\theta(r, \pi/2)rdr = 0$$

$$\int_a^b \tau_{r\theta}(r, \pi/2)dr = 0$$

Based on the required angular dependence of the stress field, the Airy stress function for this problem is selected from the general Michell solution (8.3.6), including only terms with $\sin\theta$ dependence

$$\phi = \left(Ar^3 + \frac{B}{r} + Cr + Dr \log r \right) \sin\theta$$

(8.4.63)

This form gives the following stresses

$$\sigma_r = \left(2Ar - \frac{2B}{r^3} + \frac{D}{r} \right) \sin\theta$$

$$\sigma_\theta = \left(6Ar + \frac{2B}{r^3} + \frac{D}{r} \right) \sin\theta$$

(8.4.64)

$$\tau_{r\theta} = -\left(2Ar - \frac{2B}{r^3} + \frac{D}{r} \right) \cos\theta$$

Using these results in the boundary condition relations (8.4.62) generates three equations for the unknown constants. Solving these equations gives the results

$$A = \frac{P}{2N}, \quad B = -\frac{Pa^2b^2}{2N}, \quad D = -\frac{P}{N}(a^2 + b^2)$$

where

$$N = a^2 - b^2 + (a^2 + b^2) \log\left(\frac{b}{a}\right)$$

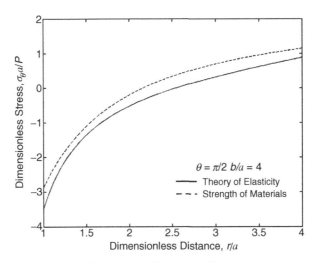

FIG. 8.33 Stress results of a curved cantilever beam with end loading.

Substituting these forms into (8.4.64) gives the stress field solution

$$\sigma_r = \frac{P}{N}\left(r + \frac{a^2 b^2}{r^3} - \frac{a^2 + b^2}{r}\right)\sin\theta$$

$$\sigma_\theta = \frac{P}{N}\left(3r - \frac{a^2 b^2}{r^3} - \frac{a^2 + b^2}{r}\right)\sin\theta \qquad (8.4.65)$$

$$\tau_{r\theta} = -\frac{P}{N}\left(r + \frac{a^2 b^2}{r^3} - \frac{a^2 + b^2}{r}\right)\cos\theta$$

These elasticity results can again be compared with the corresponding predictions from strength of materials. Fig. 8.33 illustrates the comparison of the hoop stress component through the beam thickness at $\theta = \pi/2$ (fixed end) for the case of $b/a = 4$. As in the previous example, results from the two theories are similar, but for this case differences are more sizable. Other problems of end-loaded cantilever beams can be solved using similar methods (see Exercise 8.49).

Example 8.10 Disk under diametrical compression

Let us now investigate the solution to the plane problem shown in Fig. 8.34 of a circular disk or cylinder loaded by equal but opposite concentrated forces along a given diameter. This particular problem is of special interest since this geometry is used in standard testing (ASTM D-4123, 1987) of *bituminous* and other *brittle* materials such as concrete, asphalt, rock, and ceramics. Normally referred to as the *Brazilian* or *indirect tension test*, the sample and loading geometry create a tension zone along the loaded diameter, thus allowing determination of the tensile strength of the specimen

Continued

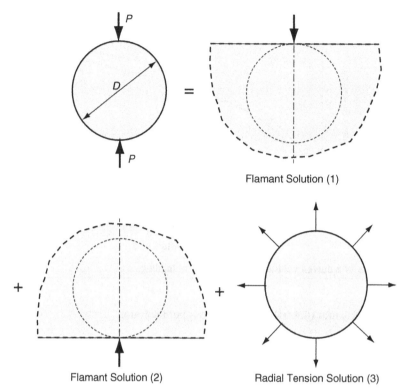

FIG. 8.34 Disk under diametrical compression—superposition solution.

material. Standard direct tension testing on such brittle materials has led to difficulty in establishing a failure region in the sample's central interior away from the gripping locations (see the simple tension sample geometry in Fig. 4.2).

This problem can be solved by more than one method, but perhaps the most interesting technique employs a clever superposition scheme, as shown in Fig. 8.34. The method uses superposition of three particular stress fields, including two Flamant solutions along with a uniform radial tension loading. As will be shown, the Flamant solutions provide the required singular behaviors at the top and bottom of an imaginary disk within each half-space, while the radial loading removes the resulting boundary tractions on the disk that were created by the two point loadings.

To combine the two Flamant solutions, it is more convenient to redefine the angular coordinate as shown in Fig. 8.35. Using the previous results from Eq. (8.4.36), the stress fields for each Flamant solution can be written as

$$\sigma_x^{(1)} = -\frac{2P}{\pi r_1}\cos\theta_1\sin^2\theta_1, \quad \sigma_x^{(2)} = -\frac{2P}{\pi r_2}\cos\theta_2\sin^2\theta_2$$

$$\sigma_y^{(1)} = -\frac{2P}{\pi r_1}\cos^3\theta_1, \qquad \sigma_y^{(2)} = -\frac{2P}{\pi r_2}\cos^3\theta_2 \qquad (8.4.66)$$

$$\tau_{xy}^{(1)} = -\frac{2P}{\pi r_1}\cos^2\theta_1\sin\theta_1, \quad \tau_{xy}^{(2)} = -\frac{2P}{\pi r_2}\cos^2\theta_2\sin\theta_2$$

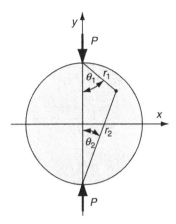

FIG. 8.35 Disk under diametrical compression.

From the general solution (8.4.35), each Flamant solution produces only a constant radial stress of $\sigma_r = -2P/\pi D$ on the circular boundary of the disk (see Fig. 8.21). The resultant boundary traction from the two combined Flamant loadings is found to be normal to the disk surface with a magnitude given by

$$T_n = -\sqrt{\left(\sigma_r^{(1)}\cos\theta_1\right)^2 + \left(\sigma_r^{(2)}\sin\theta_1\right)^2}$$

$$= -\sqrt{\left(-\frac{2P}{\pi D}\cos\theta_1\right)^2 + \left(-\frac{2P}{\pi D}\sin\theta_1\right)^2} = -\frac{2P}{\pi D}$$

(8.4.67)

Thus, the final superposition of a uniformly loaded disk with the opposite tractions of (8.4.67) removes the boundary forces and yields the solution to the desired problem. The uniformly loaded disk problem creates a simple hydrostatic state of stress given by

$$\sigma_x^{(3)} = \sigma_y^{(3)} = \frac{2P}{\pi D}, \quad \tau_{xy}^{(3)} = 0$$

(8.4.68)

Applying the superposition of states (1), (2), and (3), relations (8.4.66) and (8.4.68) are added, giving the final stress field solution

$$\sigma_x = -\frac{2P}{\pi}\left[\frac{(R-y)x^2}{r_1^4} + \frac{(R+y)x^2}{r_2^4} - \frac{1}{D}\right]$$

$$\sigma_y = -\frac{2P}{\pi}\left[\frac{(R-y)^3}{r_1^4} + \frac{(R+y)^3}{r_2^4} - \frac{1}{D}\right]$$

(8.4.69)

$$\tau_{xy} = \frac{2P}{\pi}\left[\frac{(R-y)^2 x}{r_1^4} - \frac{(R+y)^2 x}{r_2^4}\right]$$

Continued

where $R = D/2$ and $r_{1,2} = \sqrt{x^2 + (R \mp y)^2}$. On the x-axis ($y = 0$) these results simplify to give

$$\sigma_x(x, 0) = \frac{2P}{\pi D} \left[\frac{D^2 - 4x^2}{D^2 + 4x^2} \right]^2$$

$$\sigma_y(x, 0) = -\frac{2P}{\pi D} \left[\frac{4D^4}{(D^2 + 4x^2)^2} - 1 \right] \tag{8.4.70}$$

$$\tau_{xy}(x, 0) = 0$$

while on the y-axis ($x = 0$) the stresses are

$$\sigma_x(0, y) = \frac{2P}{\pi D}$$

$$\sigma_y(0, y) = -\frac{2P}{\pi} \left[\frac{2}{D - 2y} + \frac{2}{D + 2y} - \frac{1}{D} \right] \tag{8.4.71}$$

$$\tau_{xy}(0, y) = 0$$

Thus, along the loaded diameter ($x = 0$), the body will have a *uniform tensile stress* of $\sigma_x = 2P/\pi D$, and this result is the primary basis of using the geometry for indirect tension testing. Knowing the sample size and failure (fracture) loading, the simple stress relation allows the determination of the failing tensile stress or material strength. Plots of the stress distribution along the x-axis ($y = 0$) are left as an exercise. Additional applications of this problem can be found in models of granular materials in which particles are simulated by circular disks loaded by several contact forces (see Exercise 8.51).

The maximum shearing stresses in the disk can be calculated by the relation

$$\tau_{max} = \sqrt{\left(\frac{\sigma_x - \sigma_y}{2} \right)^2 + \tau_{xy}^2} \tag{8.4.72}$$

Using the stress results (8.4.69) in this relation, the τ_{max} distribution may be determined, and these results are illustrated in Fig. 8.36. The theoretical maximum shear stress contours are plotted using MATLAB®. The corresponding photoelastic results are also shown in the figure. In general, the theoretical contours match quite well with the experimental results except for the regions near the loading points at the top and bottom of the disk. This lack of correspondence is caused by the fact that the photoelastic isochromatics were generated with a loading distributed over a small but finite contact area, and thus the maximum shear stress occurs slightly below the contact surface, as per earlier discussions of Fig. 8.28. A numerical analysis of this problem using the finite element method is developed in Chapter 16 (see Example 16.3 and Fig. 16.6).

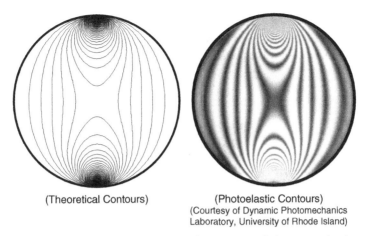

(Theoretical Contours)   (Photoelastic Contours)
(Courtesy of Dynamic Photomechanics
Laboratory, University of Rhode Island)

FIG. 8.36 Maximum shear stress contours and corresponding photoelastic isochromatics for a disk under diametrical compression.

Example 8.11 Rotating disk problem

As a final example in this section, consider the problem of a thin uniform circular disk subject to constant rotation ω, as shown in Fig. 8.37. The rotational motion generates centrifugal acceleration on each particle of the disk, and this then becomes the source of external loading for the problem. No other additional external loadings are considered.

It is convenient to handle the centrifugal force loading by relating it to a body force density through the disk. For the case of constant angular velocity, the body force is only in the radial direction given by

$$F_r = \rho \omega^2 r \tag{8.4.73}$$

where ρ is the material mass density. This problem is axisymmetric, and thus the equilibrium equations reduce to

$$\frac{d\sigma_r}{dr} + \frac{\sigma_r - \sigma_\theta}{r} + \rho \omega^2 r = 0 \tag{8.4.74}$$

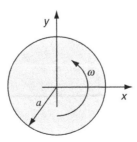

FIG. 8.37 Rotating circular disk.

The solution can be efficiently handled by using a special stress function that automatically satisfies the equilibrium equation. The particular stress–stress function relation with this property is given by

$$\sigma_r = \varphi/r$$
$$\sigma_\theta = \frac{d\varphi}{dr} + \rho\omega^2 r^2 \qquad (8.4.75)$$

where $\varphi = \varphi(r)$ is the stress function.

As usual, the governing equation for the stress function is determined from the compatibility statement. For this axisymmetric case, the displacement field is of the form $u_r = u_r(r)$ and $u_\theta = 0$. Therefore, the strain field is given by

$$e_r = \frac{du_r}{dr}, \quad e_\theta = \frac{u_r}{r}, \quad e_{r\theta} = 0$$

Eliminating u_r from these equations develops the simple compatibility statement

$$\frac{d}{dr}(re_\theta) - e_r = 0 \qquad (8.4.76)$$

Recall that the more general polar coordinate case was given as Exercise 7.17. Using Hooke's law for plane stress, the strains are given by

$$e_r = \frac{1}{E}(\sigma_r - \nu\sigma_\theta) = \frac{1}{E}\left(\frac{\varphi}{r} - \nu\frac{d\varphi}{dr} - \nu\rho\omega^2 r^2\right)$$
$$e_\theta = \frac{1}{E}(\sigma_\theta - \nu\sigma_r) = \frac{1}{E}\left(\frac{d\varphi}{dr} + \rho\omega^2 r^2 - \nu\frac{\varphi}{r}\right) \qquad (8.4.77)$$

Using this result in the compatibility relation (8.4.76) generates the desired governing equation

$$\frac{d^2\varphi}{dr^2} + \frac{1}{r}\frac{d\varphi}{dr} - \frac{\varphi}{r^2} + (3+\nu)\rho\omega^2 r = 0$$

which can be written as

$$\frac{d}{dr}\left(\frac{1}{r}\frac{d}{dr}(r\varphi)\right) = -(3+\nu)\rho\omega^2 r \qquad (8.4.78)$$

This equation is easily integrated, giving the result

$$\varphi = -\frac{(3+\nu)}{8}\rho\omega^2 r^3 + \frac{1}{2}C_1 r + C_2\frac{1}{r} \qquad (8.4.79)$$

where C_1 and C_2 are constants. The stresses corresponding to this solution are

$$\sigma_r = -\frac{(3+\nu)}{8}\rho\omega^2 r^2 + \frac{C_1}{2} + \frac{C_2}{r^2}$$
$$\sigma_\theta = -\frac{1+3\nu}{8}\rho\omega^2 r^2 + \frac{C_1}{2} - \frac{C_2}{r^2} \qquad (8.4.80)$$

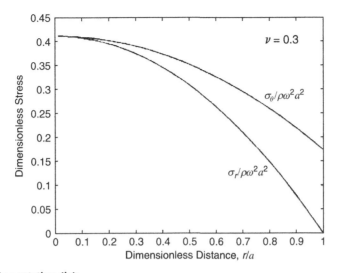

FIG. 8.38 Stresses in a rotating disk.

For a solid disk, the stresses must be bounded at the origin and so $C_2 = 0$. The condition that the disk is stress free at $r = a$ gives the remaining constant $C_1 = (3 + \nu)\rho\omega^2 a^2/4$. The final form of the stress field is then

$$\sigma_r = \frac{3 + \nu}{8}\rho\omega^2(a^2 - r^2)$$

$$\sigma_\theta = \frac{\rho\omega^2}{8}\left[(3 + \nu)a^2 - (1 + 3\nu)r^2\right]$$

(8.4.81)

The stress distribution within the disk is shown in Fig. 8.38 for the case $\nu = 0.3$. Notice that even though the body force is largest at the disk's outer boundary, the maximum stress occurs at the center of the disk where $F_r = 0$. The maximum stress is given by

$$\sigma_{max} = \sigma_r(0) = \sigma_\theta(0) = \frac{3 + \nu}{8}\rho\omega^2 a^2$$

For an *annular disk* with $a < r < b$, the maximum stress occurs on the inner boundary, and for the case of a very small inner hole with $a << b$, the maximum stress is approximately twice that of the solid disk (see Exercise 8.53).

The solution to this problem could also be obtained by formulation in terms of the radial displacement, thus generating the Navier equation, which can be easily integrated. The corresponding plane strain solution for a rotating cylinder is found from these results through the usual simple change in elastic constants, by letting $\nu \to \nu/(1 - \nu)$.

8.5 Simple plane contact problems

As mentioned back in Sections 8.4.7 and 8.4.9, contact mechanics is the study of stress and deformation of two or more solid bodies that touch each other at one or more points. Such studies have very important engineering applications in wheel–rail contact, mechanical gears, bearings and other linkages, metal working, tribology, etc. Because of this importance we now will look at the problem in more detail. In general, when two elastic solids are pressed together a contact surface is generated. Predicting details of the contact area can be very challenging as its behavior will depend on the geometries of the bodies and loadings, elastic moduli, and interfacial frictional characteristics. The general problem actually becomes nonlinear, since the size and shape of the contact surface depends on the stress solution to the problem. In this section we will explore only a couple of simple contact problems involving frictionless indentation of an elastic half-space from rigid indenters of particular shape. Much of the theory needed to explore such problems basically comes from the Flamant solution previously presented in Sections 8.4.7 and 8.4.9. Hertz (1882) was the first to study these problems, and numerous other investigations of increasing complexity have been conducted since. The texts by Johnson (1985) and Jaeger (2005) are especially useful sources to view more details and recent work, while Asaro and Lubarda (2006) and Barber (2010) also provide further material on the topic. The handbook by Kachanov et al. (2003) gives a concise set of solution results for many contact problems.

The general class of problems to be presented is illustrated in Fig. 8.39 and represents displacement-based contact. Such problems require specification of the displacement under the indenter as a boundary condition. Going back to the Flamant problem results in Section 8.4.7, the stresses were given by

$$\sigma_r = -\frac{2}{\pi r}[X \cos \theta + Y \sin \theta]$$

$$\sigma_\theta = \tau_{r\theta} = 0$$

(8.5.1)

where X and Y are the tangential and normal force components shown in Fig. 8.19. The surface displacements (plane stress case) for the normal loading were given by relations (8.4.44) and (8.4.45) while the corresponding displacements for the tangential load case can be determined from Exercise (8.34), assuming no vertical displacements of the y-axis and no rotation. Summarizing these results gives

Normal loading	Tangential loading

$$u_r(r,0) = u_r(r,\pi) = -\frac{Y}{2E}(1-\nu) \qquad u_r(r,0) = -u_r(r,\pi) = -\frac{2X}{\pi E}\log r + B$$

$$u_\theta(r,0) = -u_\theta(r,\pi) = -\frac{Y}{\pi E}[(1+\nu) + 2\log r] \quad u_\theta(r,0) = u_\theta(r,\pi) = \frac{(1-\nu)X}{2E}$$

(8.5.2)

where B is a constant coming from rigid-body motion.

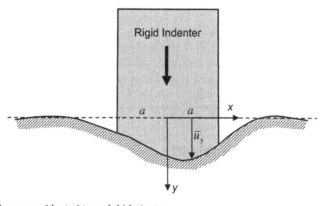

FIG. 8.39 Elastic half-space subjected to a rigid indenter.

We now wish to apply these results for a half-space that is subjected to a distributed surface displacement determined from the shape and frictional characteristics of the indenter (see Fig. 8.39). Following the general superposition schemes used in Section 8.4.9 and Exercise 8.41, we will transfer to Cartesian coordinates and use Flamant solution results (8.5.2). Both the stress and displacement fields can then be determined by integral superposition using the Flamant solution as a Green's function. The shift to Cartesian coordinates requires writing the log r as $\mathrm{sgn}(x)\log|x|$. Thus, for the general case with both normal and shear distributions $p(x)$ and $t(x)$ over the free surface $-a \leq x \leq a$, the surface displacements $\bar{u}_x$ and $\bar{u}_y$ can be expressed by

$$\bar{u}_x = -\frac{1-\nu}{2E}\left[\int_{-a}^{x}p(s)ds - \int_{x}^{a}p(s)ds\right] - \frac{2}{\pi E}\int_{-a}^{a}t(s)\log|x-s|ds + a_1$$

$$\bar{u}_y = -\frac{2}{\pi E}\int_{-a}^{a}p(s)\log|x-s|ds + \frac{1-\nu}{2E}\left[\int_{-a}^{x}t(s)ds - \int_{x}^{a}t(s)ds\right] + a_2$$

(8.5.3)

Note that we have separated the integration range to properly handle the sign switch inherent in u_r and u_θ relations (8.5.2); see Johnson (1985) for details. The constants a_1 and a_2 correspond to undetermined rigid-body motion terms.

Restricting the problem to frictionless indenters, the shear loading distribution $t(x)$ will vanish, and we eliminate the undetermined constants by differentiating (8.5.3) to determine the displacement gradients as

$$\frac{d\bar{u}_x}{dx} = -\frac{1-\nu}{E}p(x)$$

$$\frac{d\bar{u}_y}{dx} = -\frac{2}{\pi E}\int_{-a}^{a}\frac{p(s)}{x-s}ds$$

(8.5.4)

Since the integral in (8.5.4) is singular at $x = s$, it is interpreted as the usual Cauchy Principal Value sense, i.e.

$$\int_{-a}^{a}\frac{p(s)}{x-s}ds = \lim_{\varepsilon\to 0}\left[\int_{-a}^{x-\varepsilon}\frac{p(s)}{x-s}ds + \int_{x+\varepsilon}^{a}\frac{p(s)}{x-s}ds\right]$$

(8.5.5)

Now if we consider a flat rigid indenter as shown in Fig. 8.40, $\bar{u}_y = u_y^o = $ constant, and thus (8.5.4)$_2$ gives the simple singular integral relation

$$\int_{-a}^{a}\frac{p(s)}{x-s}ds = 0$$

(8.5.6)

The solution to this equation for the normal load distribution has been given by Johnson (1985) as

$$p(x) = \frac{P}{\pi\sqrt{a^2 - x^2}}$$

(8.5.7)

where P is the total load applied by the rigid indenter. Note that contact load distribution is singular at the edges of the indenter $x = \pm a$.

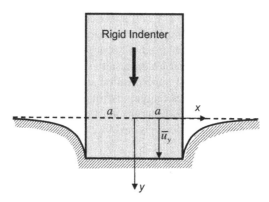

FIG. 8.40 Elastic half-space subjected to a flat rigid indenter.

With the loading now determined, the surface displacements follow from relations (8.5.3)

$$\bar{u}_x = -\frac{1-v}{\pi E}P\sin^{-1}(x/a), \quad x < |a|$$

$$\bar{u}_y = -\frac{2}{\pi E}\log\left[\frac{x}{a} + \left(\frac{x^2}{a^2} - 1\right)^{1/2}\right] + \bar{u}_y^o, \quad x > |a| \tag{8.5.8}$$

Note that for Poisson's ratio in the usual range $-1 \leq v \leq 1/2$, horizontal surface displacements under the indenter move toward the center. Realistically this motion would be opposed by friction, which was originally neglected.

The individual stress components for this case follow from the results of Exercise 8.41

$$\sigma_x = -\frac{2y}{\pi}\int_{-a}^{a}\frac{p(s)(x-s)^2}{\left[(x-s)^2 + y^2\right]^2}\,ds = -\frac{2Py}{\pi^2}\int_{-a}^{a}\frac{(x-s)^2}{\sqrt{a^2 - s^2}\left[(x-s)^2 + y^2\right]^2}\,ds$$

$$\sigma_y = -\frac{2y^3}{\pi}\int_{-a}^{a}\frac{p(s)}{\left[(x-s)^2 + y^2\right]^2}\,ds = -\frac{2Py^3}{\pi^2}\int_{-a}^{a}\frac{1}{\sqrt{a^2 - s^2}\left[(x-s)^2 + y^2\right]^2}\,ds \tag{8.5.9}$$

$$\tau_{xy} = -\frac{2y^2}{\pi}\int_{-a}^{a}\frac{p(s)(x-s)}{\left[(x-s)^2 + y^2\right]^2}\,ds = -\frac{2Py^2}{\pi^2}\int_{-a}^{a}\frac{(x-s)}{\sqrt{a^2 - s^2}\left[(x-s)^2 + y^2\right]^2}\,ds$$

Analytical evaluation of these integrals appears to be formidable, and thus we employ numerical integration using MATLAB software (see code C.7 in Appendix C) to extract stress values. Results of the maximum shearing stress contours below the indenter are illustrated in Fig. 8.41. Since the surface load distribution (8.5.7) is singular at $x = \pm a$, the stress is also unbounded at these points. These results should be compared with the photoelastic τ_{max} contours shown in Fig. 8.28 for the flat punch loading. Comparisons with the other cases shown in this figure qualitatively illustrate differences in stresses coming from different contact loading situations.

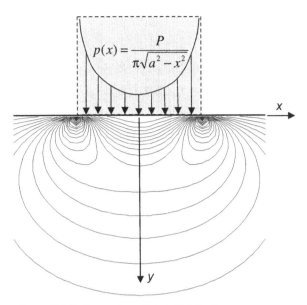

FIG. 8.41 Maximum shear stress distribution under a flat rigid indenter.

Next we consider a similar frictionless contact problem but with a rigid cylindrical indenter, as shown in Fig. 8.42. For this case the prescribed vertical surface displacement $\bar{u}_y$ with respect to an arbitrary datum is proportional to $-x^2/2R$, and thus relation $(8.5.4)_2$ can be written as

$$\int_{-a}^{a} \frac{p(s)}{x-s}\,ds = \frac{\pi E}{2R}x \tag{8.5.10}$$

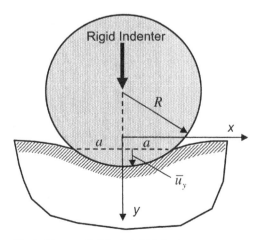

FIG. 8.42 Elastic half-space subjected to a cylindrical rigid indenter.

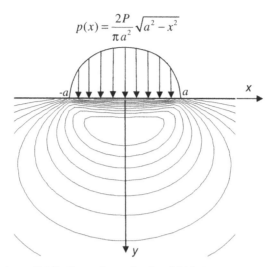

FIG. 8.43 Maximum shear stress distribution under a circular rigid indenter.

Johnson (1985) again gives the solution to this integral equation for the normal load distribution as

$$p(x) = \frac{2P}{\pi a^2} \sqrt{a^2 - x^2} \qquad (8.5.11)$$

where P is the total load applied by the indenter. This result is the classic elliptical distribution that vanishes at $x = \pm a$ (see Fig. 8.43). For this problem the contact semi-width a is related to the total applied load, and the expression for the plane stress case is given by

$$a^2 = \frac{4PR}{\pi E} \qquad (8.5.12)$$

With the contact load distribution now determined, the stresses in the elastic half-space can be determined employing the scheme previously used for the flat punch shown in equations (8.5.9). The individual stress components for this case can thus be expressed by

$$\sigma_x = -\frac{2y}{\pi} \int_{-a}^{a} \frac{p(s)(x-s)^2}{\left[(x-s)^2 + y^2\right]^2} \, ds = -\frac{4Py}{\pi^2 a^2} \int_{-a}^{a} \frac{\sqrt{a^2 - s^2}(x-s)^2}{\left[(x-s)^2 + y^2\right]^2} \, ds$$

$$\sigma_y = -\frac{2y^3}{\pi} \int_{-a}^{a} \frac{p(s)}{\left[(x-s)^2 + y^2\right]^2} \, ds = -\frac{4Py^3}{\pi^2 a^2} \int_{-a}^{a} \frac{\sqrt{a^2 - s^2}}{\left[(x-s)^2 + y^2\right]^2} \, ds \qquad (8.5.13)$$

$$\tau_{xy} = -\frac{2y^2}{\pi} \int_{-a}^{a} \frac{p(s)(x-s)}{\left[(x-s)^2 + y^2\right]^2} \, ds = -\frac{4Py^2}{\pi^2 a^2} \int_{-a}^{a} \frac{\sqrt{a^2 - s^2}(x-s)}{\left[(x-s)^2 + y^2\right]^2} \, ds$$

As before, numerical methods (MATLAB®) are used to evaluate these integrals, and maximum shearing stress contours below the indenter are illustrated in Fig. 8.43. Differing from the previous flat punch case shown in Fig. 8.41, the circular indenter produces no singular contact loading, and thus the half-space stresses are all bounded. These results can again be compared with the photoelastic τ_{max} contours shown in Fig. 8.28 for the cylinder contact loading case.

Similar methods can be applied to solve the more general case where each of the two bodies in contact is treated as linear elastic. For such cases, classical Hertz theory assumes that at the contact point each nonconforming surface has a particular radius of curvature, thus allowing a similar representation for the surface displacements as used in the previous cylindrical indenter case. Each radius of curvature is assumed to be sufficiently large so that the Flamant half-space solution can be employed as before. More details on such contact problems can be found in references quoted at the beginning of this section.

References

American Society for Testing and Materials (ASTM): *D 4123: standard test method for indirect tension test for resilient modulus of bituminous mixtures*, 1987.

Asaro RJ, Lubarda VA: *Mechanics of solids and materials*, New York, 2006, Cambridge University Press.

Barber JR: *Elasticity*, Dordrecht, The Netherlands, 2010, Springer.

Churchill RV: *Fourier series and boundary value problems*, New York, 1963, McGraw-Hill.

Hertz H: Über die Berührung fester elastischer Körper, *J f d reine u angewandte, Math* 92:156−171, 1882 (For English translation, see Miscellaneous Papers by H. Hertz, eds. Jones and Schott, Macmillan, London, 1896.).

Jaeger J: *New solutions in contact mechanics*, Southampton, UK, 2005, WIT Press.

Johnson KL: *Contact mechanics*, London, 1985, Cambridge University Press.

Kachanov M, Shafiro B, Tsukrov I: *Handbook of elasticity solutions*, Dordrecht, The Netherlands, 2003, Kluwer Academic Press.

Kreyszig E: *Advanced engineering mathematics*, ed 10, New York, 2011, Wiley.

Little RW: *Elasticity*, Englewood Cliffs, NJ, 1973, Prentice Hall.

Meleshko VV: Selected topics in the history of the two-dimensional biharmonic problem, *Appl Mech Rev* 56, 2003.

Michell JH: On the direct determination of stress in an elastic solid with application to the theory of plates, *Proc Lond Math Soc* 31:100−124, 1899.

Neou CY: Direct method for determining Airy polynomial stress functions, *J Appl Mech* 24:387−390, 1957.

Pickett G: Application of the Fourier method to the solution of certain boundary problems in the theory of elasticity, *J Appl Mech* 11:176−182, 1944.

Poulos HG, Davis EH: *Elastic solutions for soil and rock mechanics*, New York, 1974, Wiley.

Tada H, Paris PC, Irwin GR: *The stress analysis of cracks handbook*, ed 3, New York, 2000, American Society of Mechanical Engineers.

Timoshenko SP, Goodier JN: *Theory of elasticity*, New York, 1970, McGraw-Hill.

Exercises

8.1 Explicitly show that the fourth-order polynomial Airy stress function

$$A_{40}x^4 + A_{22}x^2y^2 + A_{04}y^4$$

will not satisfy the biharmonic equation unless $3A_{40} + A_{22} + 3A_{04} = 0$.

8.2 Show that the Airy function

$$\phi = \frac{3P}{4c}\left(xy - \frac{xy^3}{3c^2}\right) + \frac{N}{4c}y^2$$

solves the following cantilever beam problem, as shown in the following figure. As usual for such problems, boundary conditions at the ends ($x = 0$ and L) should be formulated only in terms of the *resultant force system*, while at $y = \pm c$ the exact *pointwise* specification should be used. For the case with $N = 0$, compare the elasticity stress field with the corresponding results from strength of materials theory. *Answer*:

$$\sigma_x = -\frac{3Pxy}{2c^3} + \frac{N}{2c}, \quad \sigma_y = 0, \quad \tau_{xy} = -\frac{3P}{4c}\left(1 - \frac{y^2}{c^2}\right)$$

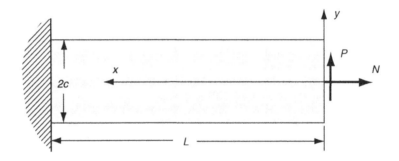

8.3 Determine the displacement field for the beam problem in Exercise 8.2. To determine the rigid-body motion terms, choose fixity conditions

$$u(L,0) = v(L,0) = \frac{\partial v(L,0)}{\partial x} = 0$$

Note that with our approximate Saint—Venant solution, we cannot ensure pointwise conditions all along the built-in end. Finally, for the special case with $N = 0$, compare the elasticity displacement field with the corresponding results from mechanics of materials theory (see Appendix D). *Answer*:

$$N = 0: \quad v_{elasticity}(x,0) = \frac{P}{4Ec^3}\left(x^3 - 3L^2x + 2L^3\right) = v_{MOM}(x)$$

8.4 The solution to the illustrated two-dimensional cantilever beam problem is proposed using the Airy stress function $\phi = C_1 x^2 + C_2 x^2 y + C_3 y^3 + C_4 y^5 + C_5 x^2 y^3$, where C_i are constants. First determine requirements on the constants so that ϕ satisfies the governing equation. Next find the values of the remaining constants by applying exact pointwise boundary conditions on the top and bottom of the beam and integrated resultant boundary conditions on the ends $x = 0$ and $x = L$.

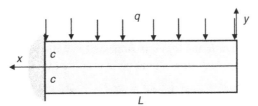

8.5 Verify that the Airy stress function

$$\phi = \frac{s}{4}\left(xy + \frac{ly^2}{c} + \frac{ly^3}{c^2} - \frac{xy^2}{c} - \frac{xy^3}{c^2}\right)$$

solves the problem of a cantilever beam loaded by uniform shear along its bottom edge as shown. Use pointwise boundary conditions on $y = \pm c$ and only resultant effects at ends $x = 0$ and l. Note, however, you should be able to show that σ_x vanishes at $x = l$.

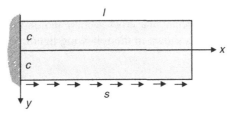

8.6 The following stress function

$$\phi = C_1 xy + C_2\frac{x^3}{6} + C_3\frac{x^3 y}{6} + C_4\frac{xy^3}{6} + C_5\frac{x^3 y^3}{9} + C_6\frac{xy^5}{20}$$

is proposed to solve the problem of a cantilever beam carrying uniformly varying loading as shown in the following figure. Explicitly verify that this stress function will satisfy all conditions on the problem and determine each of the constants C_i and the resulting stress field. Use resultant force boundary conditions at the beam ends. *Answers:*

$$C_1 = -\frac{pc}{40L}, \quad C_2 = -\frac{p}{2L}, \quad C_3 = -\frac{3p}{4Lc}, \quad C_4 = \frac{3p}{10Lc}, \quad C_5 = \frac{3p}{8Lc^3}$$

$$C_6 = -\frac{p}{2Lc^3}, \quad \sigma_x = \frac{pxy}{20Lc^3}\left(5x^2 - 10y^2 + 6c^2\right)$$

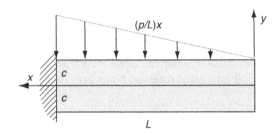

8.7 The cantilever beam shown in the figure is subjected to a distributed shear stress $\tau_o x/l$ on the upper face. The following Airy stress function is proposed to solve this problem

$$\phi = c_1 y^2 + c_2 y^3 + c_3 y^4 + c_4 y^5 + c_5 x^2 + c_6 x^2 y + c_7 x^2 y^2 + c_8 x^2 y^3$$

Determine the constants c_i and find the stress distribution in the beam. Use resultant force boundary conditions at the ends. (*Answer:* $c_1 = \tau_o c/12l$, $c_2 = \tau_o/20l$, $c_3 = -\tau_o/24cl$, ...)

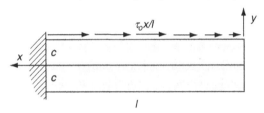

8.8* A triangular plate of narrow rectangular cross-section and uniform thickness is loaded uniformly along its top edge as shown in the following figure. Verify that the Airy stress function

$$\phi = \frac{p \cot \alpha}{2(1 - \alpha \cot \alpha)} \left[-x^2 \tan \alpha + xy + \left(x^2 + y^2\right)\left(\alpha - \tan^{-1}\frac{y}{x}\right) \right]$$

solves this plane problem. For the particular case of $\alpha = 30°$, explicitly calculate the normal and shear stress distribution over a typical cross-section AB and make comparison plots (MATLAB® recommended) of your results with those from elementary strength of materials. *Answer:*

$$\sigma_x = 2K\left[\alpha - \tan^{-1}\frac{y}{x} - \frac{xy}{x^2 + y^2}\right], \quad \sigma_y = 2K\left[\alpha - \tan\alpha - \tan^{-1}\frac{y}{x} + \frac{xy}{x^2 + y^2}\right]$$

$$\tau_{xy} = -2K\frac{y^2}{x^2 + y^2}, \quad K = \frac{p \cot \alpha}{2(1 - \alpha \cot \alpha)}$$

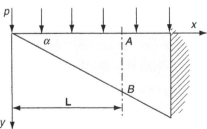

8.9* Redo Exercise 8.8* using polar coordinates.

8.10 A triangular plate of narrow rectangular cross-section and uniform thickness carries a uniformly varying loading along its top edge as shown. Verify that the Airy stress function

$$\phi = r^3 [a_{14} \cos \theta + b_{14} \sin \theta + a_{31} \cos 3\theta + b_{31} \sin 3\theta]$$

solves this plane problem.

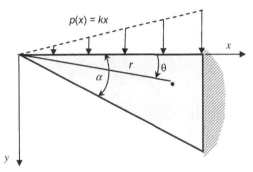

8.11* For the pure beam bending problem solved in Example 8.2, calculate and plot the in-plane displacement field given by relation $(8.1.22)_2$. Use the vector distribution plotting scheme illustrated in Appendix C, Example C.5. Your solution should look like the following figure.

8.12* For the beam problem in Example 8.3, the boundary conditions required that the resultant normal force vanish at each end $(x = \pm l)$. Show, however, that the normal stress on each end is not zero, and plot its distribution over $-c < y < c$.

8.13* Explicitly determine the bending stress σ_x for the problem in Example 8.4. For the case $l/c = 3$, plot this stress distribution through the beam thickness at $x = l/2$, and compare with strength of materials theory. For long beams $(l \gg c)$, show that the elasticity results approach the strength of materials predictions.

8.14 Develop the general displacement solution (8.3.9) for the axisymmetric case.

8.15 Consider the axisymmetric problem of an annular disk with a fixed inner radius and loaded with uniform shear stress τ over the outer radius. Using the Airy stress function term $a_4\theta$, show that the stress and displacement solution for this problem is given by

$$\sigma_r = \sigma_\theta = 0, \quad \tau_{r\theta} = \tau \frac{r_2^2}{r^2}$$

$$u_r = 0, \quad u_\theta = \frac{1+\nu}{E} \tau r_2^2 \left(\frac{r}{r_1^2} - \frac{1}{r} \right)$$

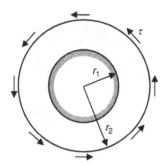

8.16 Under the conditions of polar axisymmetry, verify that the Navier equations (5.4.4) reduce to relation (8.3.10). Refer to Example 1.5 to evaluate vector terms in (5.4.4) properly. Next show that the general solution to this Cauchy–Euler differential equation is given by (8.3.11). Finally, use this solution to determine the stresses and show that they will not contain the logarithmic terms given in the general solution (8.3.8).

8.17 For the axisymmetric problem of Example 8.6, explicitly develop the displacement solution given by relation (8.4.5).

8.18 Consider the annular ring loaded with a sinusoidal distributed pressure as illustrated. Show that this problem can be solved using the Airy function

$$\phi = a_1 \log r + a_2 r^2 + \left(a_{21} r^2 + a_{22} r^4 + a_{23} r^{-2} + a_{24} \right) \cos 2\theta$$

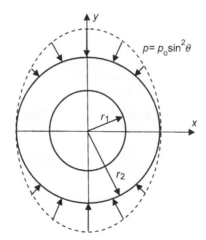

8.19 Through a *shrink-fit* process, a rigid solid cylinder of radius $r_1 + \delta$ is to be inserted into the hollow cylinder of inner radius r_1 and outer radius r_2 (as shown in the following figure). This process creates a displacement boundary condition $u_r(r_1) = \delta$. The outer surface of the hollow cylinder is to remain stress free. Assuming plane strain conditions, determine the resulting stress field within the cylinder ($r_1 < r < r_2$).

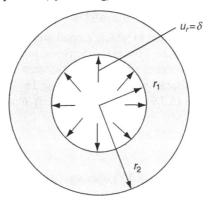

8.20 A long composite cylinder is subjected to the external pressure loading as shown in the following figure. Assuming idealized perfect bonding between the two materials, the normal stress and displacement will be continuous across the interface $r = r_1$ (see Section 5.2). Under these conditions, determine the stress and displacement fields in each material.

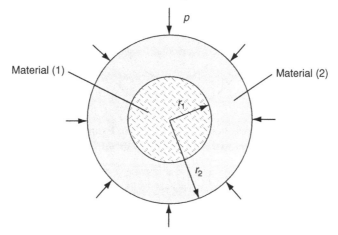

8.21[*] Numerically generate and plot the fields of stress (σ_r, σ_θ) and displacement (u_r) within the composite cylinder of Exercise 8.20 for the specific case with material (1) = steel and material (2) = aluminum. Use Table 4.2 for elastic moduli values. Explore and discuss the continuity issues for these field quantities at the interface $r = r_1$.

8.22 Resolve Exercise 8.20 for the case where material (1) is rigid and material (2) is elastic with modulus E and Poisson's ratio ν.

8.23 For the case of a thin-walled tube under internal pressure, verify that the general solution for the hoop stress $(8.4.3)_2$ will reduce to the strength of materials relation (see Appendix D, Section D.5)

$$\sigma_\theta \approx \frac{p r_o}{t}$$

where t is the wall thickness and r_o is the mean radius.

8.24 Consider the *cut-and-weld* problem in which a small wedge of angle α is removed from an annular ring as shown in the figure. The ring is then to be joined back together (welded) at the cut section. This operation produces an axisymmetric stress field, but the problem will contain a cyclic tangential displacement condition $u_\theta(r,2\pi) - u_\theta(r,0) = \alpha r$. First using the general plane stress solution $(8.3.9)_2$, drop the rigid-body motion terms and show that the constant a_3 is given by

$$a_3 = \frac{\alpha E}{8\pi}$$

Next use the general solution form (8.3.8) with zero boundary tractions on the inner and outer radii of the ring and determine the constants a_1 and a_2 and complete the stress field solution.

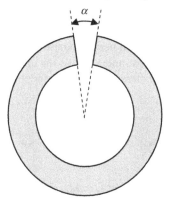

8.25 Using superposition of the stress field (8.4.15) given in Example 8.7, show that the problem of equal biaxial tension loading on a stress-free hole as shown in the figure is given by Eq. (8.4.9).

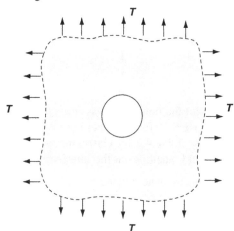

8.26[*] Using superposition of the stress field (8.4.15), develop solution (8.4.18) for the equal but opposite biaxial loading on a stress-free hole shown in Fig. 8.15A. Also justify that this solution will solve the shear loading case shown in Fig. 8.15B. Construct a polar plot (similar to Fig. 8.13) of $\sigma_\theta(a,\theta)/T$ for this case.

8.27 An elastic circular plug of radius a with properties E_1 and ν_1 is perfectly bonded and embedded in an infinite elastic medium with properties E_2 and ν_2. The composite is loaded with a uniform far-field biaxial stress T as shown. Using the results from Exercise 8.20, determine the stress and displacement solutions in each material. Explicitly explore the stresses on the boundary of the plug ($r = a$) and determine if any stress concentration will exist.

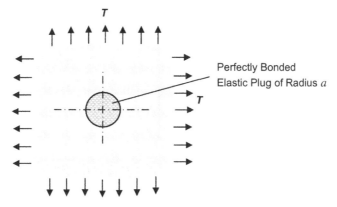

8.28 An infinite elastic medium contains a perfectly bonded rigid plug and is loaded with uniform far-field biaxial stress T as shown. Using the results from Exercise 8.22, determine the stress and displacement solution. Explicitly explore the stresses on the boundary of the plug ($r = a$) and determine if any stress concentration will exist.

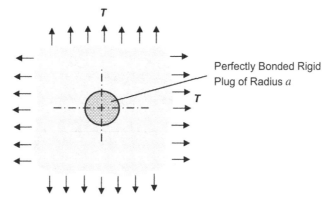

8.29 Show that the stress function

$$\phi = \frac{\tau_o r^2}{\pi} \left[\sin^2 \theta \log r + \theta \sin\theta \cos\theta - \sin^2 \theta \right]$$

gives the solution to the problem of an elastic half-space loaded by a uniformly distributed shear over the free surface $(x \le 0)$, as shown in the figure. Identify locations where the stresses are singular.

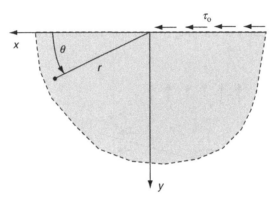

8.30 Show that the Flamant solution given by Eqs. (8.4.31) and (8.4.32) can also be used to solve the more general wedge problem as shown.

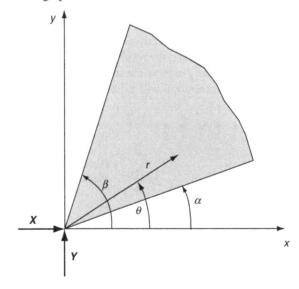

8.31 Consider the quarter plane domain problem shown in Fig. 8.17 but this time with a uniformly distributed normal loading N along the y-axis in the x-direction. Use the proposed general Airy function (8.4.19), and apply four proper boundary conditions to determine the unknown constants. Develop the stress solution and similar to Example 8.4.4, and explore the conditions at the corner $(x = y = 0)$.

8.32 The in-plane rotation in polar coordinates is given by relation $\omega_z = \frac{1}{2}\left(\frac{1}{r}\frac{\partial}{\partial r}(ru_\theta) - \frac{1}{r}\frac{\partial u_r}{\partial \theta}\right)$. Using this form with the Flamant displacement solution (8.4.43), show that $\omega_z = -\frac{2Y}{\pi E}\frac{\cos\theta}{r}$, and thus conclude that ω_z is an even function of θ that vanishes along the y-axis and is unbounded at the origin.

8.33 For the Flamant problem with only normal loading, explore the strain energy in a semi-circular area of radius R centered at the loading point. Refer to the discussion in Section 5.8, and use relation (6.1.10) to explicitly determine the strain energy. Comment on whether the expression is singular or not as $R \to 0$.

8.34 Show that plane stress displacements for the Flamant problem in Section 8.4.7 under only tangential force X are given by

$$u_r = -\frac{(1-v)X}{\pi E}\theta \sin\theta - \frac{2X}{\pi E}\log r \cos\theta + A \sin\theta + B \cos\theta$$

$$u_\theta = \frac{(1-v)X}{\pi E}\theta \cos\theta + \frac{2X}{\pi E}\log r \sin\theta + \frac{(1+v)X}{\pi E}\sin\theta$$
$$+A \cos\theta - B \sin\theta + Cr + K$$

8.35 Determine the stress field solution (8.4.47) for the problem of a half-space under a concentrated surface moment as shown in Fig. 8.23. It is recommended to use the superposition and limiting process as illustrated in the figure. This solution can be formally developed using either Cartesian or polar coordinate stress components. However, a simple and elegant solution can be found by noting that the superposition and limiting process yields the stress function solution $\phi_M = -d\partial\phi/\partial x$, where ϕ is the solution to the Flamant problem shown in Fig. 8.21.

8.36 Show that the problem of a half-space carrying a concentrated surface moment (see Fig. 8.23) can also be solved using the Airy function form $\phi = a_4\theta + b_{24}\sin 2\theta$.

8.37* Consider an elastic half-space loaded over its entire free surface with a sinusoidal pressure as shown. Using a portion of the general solution (8.2.6), show that an Airy stress function of the form $\phi = \cos \beta x[Be^{-\beta y} + Dye^{-\beta y}]$ will satisfy the appropriate boundary and far-field conditions and thus solve this problem. Next determine the x-location where the horizontal shear stress is a maximum and plot the distribution of this component versus depth into the medium. Such solutions have been used in geomechanics problems to simulate seafloor loading from surface water waves.

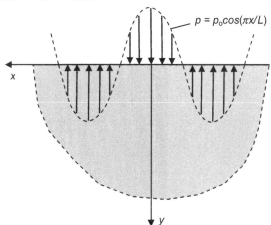

8.38 Working in polar coordinates, show that an Airy stress function of the form
$\phi = a_2 r^2 + a_{21} r^2 \cos 2\theta$ (where a_2 and a_{21} are constants) solves the illustrated problem of a shear loaded wedge problem. Are there any points in the body that exhibit singular behavior?

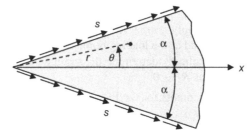

8.39* For the problem of a half-space under uniform normal loading as shown in Fig. 8.24, show that the maximum shear stress can be expressed by

$$\tau_{max} = \frac{p}{\pi} \sin(\theta_1 - \theta_2)$$

Plot the distribution of lines of constant maximum shear stress, and compare the results with the photoelastic fringes shown in Fig. 8.28. These results, along with several other loading cases, have been given by Poulos and Davis (1974).

8.40 Following a similar solution procedure as used in Section 8.4.9, show that the solution for a half-space carrying a uniformly distributed shear loading t is given by

$$\sigma_x = \frac{t}{2\pi}[4\log(\sin\theta_1/\sin\theta_2) - \cos2\theta_2 + \cos2\theta_1)]$$

$$\sigma_y = \frac{t}{2\pi}[\cos2\theta_2 - \cos2\theta_1)]$$

$$\tau_{xy} = -\frac{t}{2\pi}[2(\theta_2 - \theta_1) + \sin2\theta_2 - \sin2\theta_1]$$

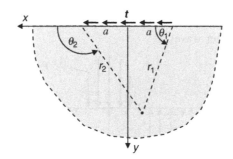

8.41 Generalize the integral superposition methods used in the examples shown in Section 8.4.9 and Exercise 8.40. In particular, show that the stress solution for a half-space carrying general normal and shear distributions $p(x)$ and $t(x)$ over the free surface $-a \le x \le a$ is given by

$$\sigma_x = -\frac{2y}{\pi} \int_{-a}^{a} \frac{p(s)(x-s)^2}{\left[(x-s)^2 + y^2\right]^2} ds - \frac{2}{\pi} \int_{-a}^{a} \frac{t(s)(x-s)^3}{\left[(x-s)^2 + y^2\right]^2} ds$$

$$\sigma_y = -\frac{2y^3}{\pi} \int_{-a}^{a} \frac{p(s)}{\left[(x-s)^2 + y^2\right]^2} ds - \frac{2y^2}{\pi} \int_{-a}^{a} \frac{t(s)(x-s)}{\left[(x-s)^2 + y^2\right]^2} ds$$

$$\tau_{xy} = -\frac{2y^2}{\pi} \int_{-a}^{a} \frac{p(s)(x-s)}{\left[(x-s)^2 + y^2\right]^2} ds - \frac{2y}{\pi} \int_{-a}^{a} \frac{t(s)(x-s)^2}{\left[(x-s)^2 + y^2\right]^2} ds$$

8.42 Using the formulation and boundary condition results of the thin notch crack problem shown in Fig. 8.29, explicitly develop the stress components given by relations (8.4.56) and (8.4.57).

8.43* Photoelastic studies of the stress distribution around the tip of a crack have produced the isochromatic fringe pattern (opening mode I case) as shown in the figure. Using the solution given in (8.4.57), show that maximum shear stresses for each mode case are given by

$$(\tau_{max})_I = \frac{3A}{2\sqrt{r}} \sin \vartheta, \quad (\tau_{max})_{II} = \frac{B}{2\sqrt{r}} \sqrt{(1 + 3 \cos^2 \vartheta)}$$

Next, plot contours of constant maximum shear stress for modes I and II. In plotting each case, normalize τ_{max} by the coefficient A or B. For the mode I case, theoretical contours should compare with the following photoelastic picture.

(Courtesy of Dynamic Photomechanics
Laboratory, University of Rhode Island)

8.44 Consider the crack problem shown for the *antiplane strain* case with $u = v = 0$, $w = w(x,y)$. From Section 7.4, the governing equation for the unknown displacement component with zero body force was given by Laplace's equation, which in polar coordinates reads.

$$\nabla^2 w = \frac{\partial^2 w}{\partial r^2} + \frac{1}{r}\frac{\partial w}{\partial r} + \frac{1}{r^2}\frac{\partial^2 w}{\partial \theta^2} = 0$$

Use a separation of variables scheme with $w = r^\lambda f(\theta)$, where λ is a parameter to be determined and $f(\theta)$ is expected to be an odd function. Show that using this solution form in the governing equation gives the result $w = Ar^\lambda \sin\lambda\theta$, where A is a constant. Next determine the polar coordinate stress components, and following similar methods as in Section 8.4.10, show that the boundary condition of zero stress on the crack surfaces gives $\lambda = n/2$, where $n = 1, 3, 5, \ldots$ Finally, using the arguments of finite displacements but singular stresses at the crack tip, show that $0 < \lambda < 1$ and thus conclude that the displacement and stress near $r \approx 0$ must be of the form

$$w = A\sqrt{r}\sin\frac{\theta}{2}, \quad \tau_{z\theta} = \frac{\mu A}{2\sqrt{r}}\cos\frac{\theta}{2}, \quad \tau_{zr} = \frac{\mu A}{2\sqrt{r}}\sin\frac{\theta}{2}$$

Note that the order of stress singularity, $O(r^{-1/2})$, is identical to our previous study.

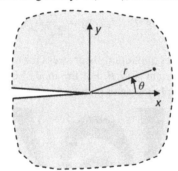

This particular crack deformation case is normally referred to as mode III in fracture mechanics literature.

8.45* For the antiplane stain crack problem solved in Exercise 8.44, plot contours of the displacement and stress fields.

8.46 Using results from Exercises 3.19 and 8.43, show that the mode I crack tip von Mises stress is given by

$$\sigma_e^2 = \frac{A^2}{4r}\left(27\sin^2\vartheta + 36\cos^2\left(\frac{\vartheta}{2}\right)\right).$$

Next solve this equation for the radial coordinate and make a polar plot for a typical von Mises stress contour (σ_e = constant) about the crack tip. Result should look like the following figure, and represents the *shape of the plastic zone size for the plane stress mode I crack problem.*

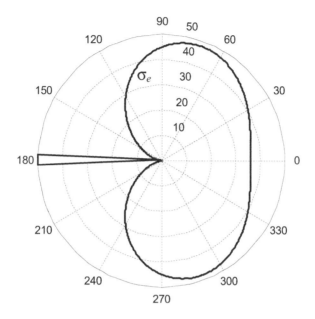

8.47 For the mode III crack problem in Exercise 8.44, explore the strain energy in a circular area of radius R centered at the crack tip. Refer to the discussion in Section 5.8, and use relation (6.1.10) to explicitly determine the strain energy. Comment on whether the expression is singular or not as $R \to 0$.

8.48* Using strength of materials theory (see Appendix D), the bending stress σ_θ for curved beams is given by $\sigma_\theta = -M(r - B)/[rA(R - B)]$, where $A = b - a$, $B = (b - a)/\log(b/a)$, $R = (a + b)/2$. For the problem shown in Fig. 8.30, compare and plot the strength of materials and elasticity predictions for the cases of $b/a = 2$ and 4. Follow the nondimensional plotting scheme used in Fig. 8.31.

8.49 Show that the curved beam problem with given end loadings can be solved by superimposing the solution from the Airy function $\phi = [Ar^3 + (B/r) + Cr + Dr \log r] \cos\theta$ with the pure bending solution (8.4.61).

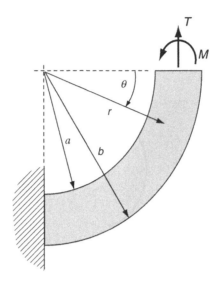

8.50* For the disk under diametrical compression (Fig. 8.35), plot the distribution of the two normal stresses σ_x and σ_y along the horizontal diameter ($y = 0$, $-R < x < R$).

8.51* The behavior of granular materials has often been studied using photoelastic models of circular particles as shown in the following figure. This provides the full-field distribution of local contact load transfer through the model assembly. Particles in such models are commonly loaded through multiple contacts with neighboring grains, and the particular example particle shown has four contact loads. Assuming the loadings are in-line and along two perpendicular diameters, use superposition of the solution given in Example 8.10 to determine the stress field within the model particle. Make a comparison plot of the distribution of normal stress along a loaded diameter with the corresponding results from Example 8.10.

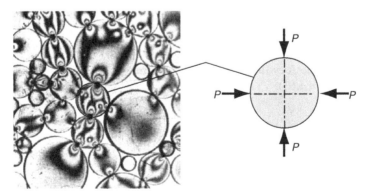

(Courtesy of Dynamic Photomechanics Laboratory, University of Rhode Island)

8.52 Consider an extension of Exercise 8.51 in which we wish to explore the stress distribution in a circular disk under an increasing number of boundary loadings. First show that for case (a) with four loadings, the center stresses are given by $\sigma_x = \sigma_y = -4P/\pi D$, $\tau_{xy} = 0$, where D is the disk diameter. Next for case (b) with eight loadings, show that the center stresses become $\sigma_x = \sigma_y = -8P/\pi D$, $\tau_{xy} = 0$. Hence conclude that for a general case with N loadings ($N = 4, 8, 16, \ldots$), the center stresses can be expressed by $\sigma_x = \sigma_y = -NP/\pi D$, $\tau_{xy} = 0$. Thus as $N \to \infty$, the boundary loading becomes uniformly distributed as shown in case (c), and center stresses are then given by $\sigma_x = \sigma_y = -p$, $\tau_{xy} = 0$ where $p = NP/\pi D$, which can be found from solution (8.4.3).

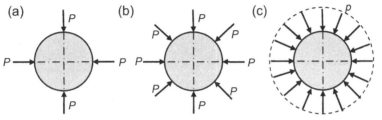

(a)   Four Boundary Loadings   (b)   Eight Boundary Loadings   (c)   Distributed Boundary Loading

8.53 Solve the rotating disk problem of Example 8.11 for the case of an annular disk with inner radius a and outer radius b being stress free. Explicitly show that for the case $b \gg a$, the maximum stress is approximately twice that of the solid disk.

8.54 Using relation (8.5.4)$_1$ along with Hooke's law for plane stress, show that the surface stresses under a frictionless indenter are given by the hydrostatic state $\bar{\sigma}_x = \bar{\sigma}_y = -p(x)$. This situation would tend to restrict the surface layer to deform plastically under such loading.

8.55* The example MATLAB code C.7 numerically integrated the integrals in solution (8.5.9) for the case of the flat rigid punch problem. Modify this example code to handle the case of the cylindrical punch given by solution (8.5.13) and thus generate the τ_{max} contours shown in Fig. 8.43.

Extension, torsion, and flexure of elastic cylinders

9

This chapter investigates particular solutions to the problem of cylindrical bars subjected to forces acting on the end planes. The general problem is illustrated in Fig. 9.1, where an elastic cylindrical bar with arbitrary cross-section R and lateral surface S carries general resultant end loadings of force $\boldsymbol{P}$ and moment $\boldsymbol{M}$. The lateral surface is taken to be free of external loading. The cylindrical body is a prismatic bar, and the constant cross-section may be solid or contain one or more holes. Considering the components of the general loading leads to a definition of four problem types including *extension*, *torsion*, *bending*, and *flexure*. These problems are inherently three-dimensional, and thus analytical solutions cannot be generally determined. In an attempt to obtain an approximate solution in central portions of the bar, Saint-Venant presumed that the character of the elastic field in this location would depend only in a secondary way on the exact distribution of tractions on the ends of the cylinder and that the principal effects are caused by the *force resultants* on the ends (Saint-Venant's principle). As such, he relaxed the original problem by no longer requiring the solution to satisfy pointwise traction conditions on the ends, but rather seeking one that had the same resultant loading. This approach is

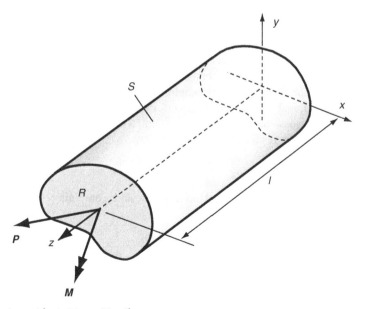

FIG. 9.1 Prismatic bar subjected to end loadings.

Elasticity. https://doi.org/10.1016/B978-0-12-815987-3.00009-8

similar to our previous two-dimensional studies of beam problems in Chapter 8. Under these condi-tions, the solution is not unique but provides reasonable results away from the ends of the cylinder.

9.1 General formulation

Formulation and solution of the extension, torsion, bending, and flexure problems are normally made using the *semi-inverse method*, as previously discussed in Section 5.7. Recall this method assumes a portion of the solution field and then determines the remaining unknowns by requiring that all fundamental field equations be satisfied. For a prismatic bar with zero body forces and under only end loadings as shown in Fig. 9.1, it is reasonable to assume that

$$\sigma_x = \sigma_y = \tau_{xy} = 0 \tag{9.1.1}$$

Note that this enforces zero tractions on the lateral surface S. Under these conditions, the equilibrium equations (3.6.5) and stress compatibility equations (5.3.4) give

$$\frac{\partial \tau_{xz}}{\partial z} = \frac{\partial \tau_{yz}}{\partial z} = 0$$

$$\frac{\partial^2 \sigma_z}{\partial x^2} = \frac{\partial^2 \sigma_z}{\partial y^2} = \frac{\partial^2 \sigma_z}{\partial z^2} = \frac{\partial^2 \sigma_z}{\partial x \partial y} = 0 \tag{9.1.2}$$

Thus, τ_{xz} and τ_{yz} must be independent of z, and σ_z must be a bilinear form in x,y,z such that $\sigma_z = C_1 x + C_2 y + C_3 z + C_4 xz + C_5 yz + C_6$, where C_i are arbitrary constants (see Exercise 9.1). For the extension, bending, and torsion problems, it can be further argued that σ_z must be independent of z. We now investigate the formulation and solution of extension, torsion, and flexure problems.

9.2 Extension formulation

Consider first the case of an axial resultant end loading $P = P_z e_3$ and $M = 0$. It is further assumed that the extensional loading P_z is applied at the centroid of the cross-section R so as not to produce any bending effects. Invoking the Saint-Venant principle, the exact end tractions can be replaced by a statically equivalent system, and this is taken as a uniform loading over the end section. Under these conditions, it is reasonable to assume that the stress σ_z is uniform over any cross-section throughout the solid, and this yields the simple results

$$\sigma_z = \frac{P_z}{A}, \tau_{xz} = \tau_{yz} = 0 \tag{9.2.1}$$

Using stress results (9.1.1) and (9.2.1) in Hooke's law and combining them with the strain− displacement relations gives

$$\frac{\partial u}{\partial x} = -\frac{\nu P_z}{AE}, \frac{\partial v}{\partial y} = -\frac{\nu P_z}{AE}, \frac{\partial w}{\partial z} = \frac{P_z}{AE}$$

$$\frac{\partial u}{\partial y} + \frac{\partial v}{\partial x} = 0, \frac{\partial v}{\partial z} + \frac{\partial w}{\partial y} = 0, \frac{\partial w}{\partial x} + \frac{\partial u}{\partial z} = 0$$

Integrating these results and dropping the rigid-body motion terms such that the displacements vanish at the origin yields

$$u = -\frac{\nu P_z}{AE}x, \quad v = -\frac{\nu P_z}{AE}y, \quad w = \frac{P_z}{AE}z \qquad (9.2.2)$$

These results then satisfy all elasticity field equations and complete the problem solution.

An additional extension example of a prismatic bar under uniform axial body force has been presented previously in Example 5.1. This problem was defined in Fig. 5.11 and corresponds to the deformation of a bar under its own weight. The problem includes no applied end tractions, and the deformation is driven by a uniformly distributed axial body force $F_z = -\rho g$. Relations for the stresses, strains, and displacements are given in the example.

9.3 Torsion formulation

For the general problem shown in Fig. 9.1, we next investigate the case of a torsional end loading $P = 0$ and $M = Te_3$. Formulation of this problem began at the end of the eighteenth century, and a very comprehensive review of analytical, approximate, and experimental solutions has been given by Higgins (1942, 1943, 1945). Studies on the torsional deformation of cylinders of circular cross-section have found the following:

- Each section rotates as a rigid body about the center axis.
- For small deformation theory, the amount of rotation is a linear function of the axial coordinate.
- Because of symmetry, circular cross-sections remain plane after deformation.

Guided by these observations, it is logical to assume the following for general cross-sections:

- The projection of each section on the x,y-plane rotates as a rigid body about the central axis.
- The amount of projected section rotation is a linear function of the axial coordinate.
- Plane cross-sections do not remain plane after deformation, thus leading to a warping displacement.

In order to quantify these deformation assumptions, consider the typical cross-section shown in Fig. 9.2. For convenience, the origin of the coordinate system is located at point O called the *center of twist*, which is defined by the location where $u = v = 0$. The location of this point depends on the shape of the section; however, the general problem formulation does not depend on the choice of coordinate origin (see Exercise 9.3). Under torque T, the displacement of a generic point P in the x,y-plane will move to location P' as shown. Line OP then rotates through a small angle β, and thus the arc length $PP' = r\beta$. This distance may be represented by a straight line normal to OP. The in-plane or projected displacements can thus be determined as

$$\begin{aligned} u &= -r\beta \sin\theta = -\beta y \\ v &= r\beta \cos\theta = \beta x \end{aligned} \qquad (9.3.1)$$

Using the assumption that the section rotation is a linear function of the axial coordinate, we can assume that the cylinder is fixed at $z = 0$ and take

$$\beta = \alpha z \qquad (9.3.2)$$

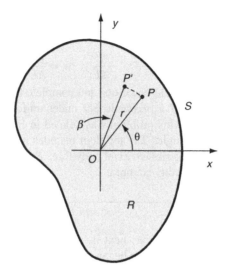

FIG. 9.2 In-plane displacements for the torsion problem.

where the parameter α is the *angle of twist* per *unit length*. The out-of-plane, warping displacement is assumed to be a function of only the in-plane coordinates and is left as an unknown to be determined. Collecting these results together, the displacements for the torsion problem can thus be written as

$$u = -\alpha yz$$
$$v = \alpha xz \qquad\qquad (9.3.3)$$
$$w = w(x, y)$$

This then establishes a semi-inverse scheme whereby requiring these displacements to satisfy all governing field equations generates a much simplified problem that can be solved for many particular cross-sectional shapes. We now proceed with the details of both a stress (stress function) and displacement formulation.

9.3.1 Stress–stress function formulation

The stress formulation leads to the use of a stress function similar to the scheme used in the plane problem discussed in Section 7.5. Using the displacement form (9.3.3), the strain–displacement relations give the following strain field

$$e_x = e_y = e_z = e_{xy} = 0$$
$$e_{xz} = \frac{1}{2}\left(\frac{\partial w}{\partial x} - \alpha y\right) \qquad\qquad (9.3.4)$$
$$e_{yz} = \frac{1}{2}\left(\frac{\partial w}{\partial y} + \alpha x\right)$$

The corresponding stresses follow from Hooke's law

$$\sigma_x = \sigma_y = \sigma_z = \tau_{xy} = 0$$

$$\tau_{xz} = \mu\left(\frac{\partial w}{\partial x} - \alpha y\right)$$

$$\tau_{yz} = \mu\left(\frac{\partial w}{\partial y} + \alpha x\right)$$

(9.3.5)

Note the strain and stress fields are functions only of x and y.

For this case, with zero body forces, the equilibrium equations reduce to

$$\frac{\partial \tau_{xz}}{\partial x} + \frac{\partial \tau_{yz}}{\partial y} = 0$$

(9.3.6)

Rather than using the general Beltrami–Michell compatibility equations, it is more direct to develop a special compatibility relation for this particular problem. This is easily done by simply differentiating (9.3.5)$_2$ with respect to y and (9.3.5)$_3$ with respect to x and subtracting the results to get

$$\frac{\partial \tau_{xz}}{\partial y} - \frac{\partial \tau_{yz}}{\partial x} = -2\mu\alpha$$

(9.3.7)

This represents an independent relation among the stresses developed under the continuity conditions of $w(x,y)$.

Relations (9.3.6) and (9.3.7) constitute the governing equations for the stress formulation. The coupled system pair can be reduced by introducing a stress function approach. For this case, the stresses are represented in terms of the *Prandtl stress function* $\phi = \phi(x,y)$ by

$$\tau_{xz} = \frac{\partial \phi}{\partial y}, \quad \tau_{yz} = -\frac{\partial \phi}{\partial x}$$

(9.3.8)

Note that here we are using the same notation for the stress function as used for the Airy function in the previous chapter that dealt with plane elasticity. Since the problem types are completely different, there should be little confusion. The equilibrium equations are then identically satisfied and the compatibility relation gives

$$\nabla^2 \phi = \frac{\partial^2 \phi}{\partial x^2} + \frac{\partial^2 \phi}{\partial y^2} = -2\mu\alpha$$

(9.3.9)

This single relation is then the governing equation for the problem and (9.3.9) is a *Poisson equation* that is amenable to several analytical solution techniques.

To complete the stress formulation we now must address the boundary conditions on the problem. As previously mentioned, the lateral surface of the cylinder S (see Fig. 9.1) is to be free of tractions, and thus

$$T_x^n = \sigma_x n_x + \tau_{yx} n_y + \tau_{zx} n_z = 0$$

$$T_y^n = \tau_{xy} n_x + \sigma_y n_y + \tau_{zy} n_z = 0$$

$$T_z^n = \tau_{xz} n_x + \tau_{yz} n_y + \sigma_z n_z = 0$$

(9.3.10)

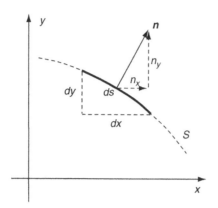

FIG. 9.3 Differential surface element.

The first two relations are identically satisfied because $\sigma_x = \sigma_y = \tau_{xy} = n_z = 0$ on S. To investigate the third relation, consider the surface element shown in Fig. 9.3. The components of the unit normal vector can be expressed as

$$n_x = \frac{dy}{ds} = \frac{dx}{dn}, \quad n_y = -\frac{dx}{ds} = \frac{dy}{dn} \tag{9.3.11}$$

Using this result along with (9.3.8) in (9.3.10)$_3$ gives

$$\frac{\partial \phi}{\partial x} \frac{dx}{ds} + \frac{\partial \phi}{\partial y} \frac{dy}{ds} = 0$$

which can be written as

$$\frac{d\phi}{ds} = 0, \quad \text{on} \quad S \tag{9.3.12}$$

This result indicates that the stress function ϕ must be a *constant* on the cross-section boundary. Because the value of this constant is not specified (at least for simply connected sections), we may choose any convenient value and this is normally taken to be zero.

Next consider the boundary conditions on the ends of the cylinder. On this boundary, components of the unit normal become $n_x = n_y = 0$, $n_z = \pm 1$, and thus the tractions simplify to

$$T_x^n = \pm \tau_{xz}$$
$$T_y^n = \pm \tau_{yz} \tag{9.3.13}$$
$$T_z^n = 0$$

Recall that we are only interested in satisfying the resultant end-loading conditions, and thus the resultant force should vanish while the moment should reduce to a pure torque T about the z-axis. These conditions are specified by

$$P_x = \iint_R T_x^n\,dxdy = 0$$
$$P_y = \iint_R T_y^n\,dxdy = 0$$
$$P_z = \iint_R T_z^n\,dxdy = 0$$
$$M_x = \iint_R yT_z^n\,dxdy = 0 \qquad (9.3.14)$$
$$M_y = \iint_R -xT_z^n\,dxdy = 0$$
$$M_z = \iint_R \left(xT_y^n - yT_x^n\right)dxdy = T$$

With $T_z^n = 0$, conditions $(9.3.14)_{3-5}$ are automatically satisfied. Considering the first condition in set (9.3.14), the x component of the resultant force on the ends may be written as

$$\iint_R T_x^n\,dxdy = \pm\iint_R \tau_{xz}\,dxdy = \pm\iint_R \frac{\partial\phi}{\partial y}\,dxdy \qquad (9.3.15)$$

Using Green's theorem (1.8.11), $\iint_R \frac{\partial\phi}{\partial y}\,dxdy = \oint_S \phi n_y\,ds$, and because ϕ vanishes on boundary S, the integral is zero and the resultant force P_x vanishes. Similar arguments can be used to show that the resultant force P_y will vanish. The final end condition $(9.3.14)_6$ involving the resultant torque can be expressed as

$$T = \iint_R \left(xT_y^n - yT_x^n\right)dxdy = -\iint_R \left(x\frac{\partial\phi}{\partial x} + y\frac{\partial\phi}{\partial y}\right)dxdy \qquad (9.3.16)$$

Again using results from Green's theorem

$$\iint_R x\frac{\partial\phi}{\partial x}\,dxdy = \iint_R \frac{\partial}{\partial x}(x\phi)\,dxdy - \iint_R \phi\,dxdy$$
$$= \oint_S x\phi n_x\,ds - \iint_R \phi\,dxdy$$
$$\iint_R y\frac{\partial\phi}{\partial y}\,dxdy = \iint_R \frac{\partial}{\partial y}(y\phi)\,dxdy - \iint_R \phi\,dxdy \qquad (9.3.17)$$
$$= \oint_S y\phi n_y\,ds - \iint_R \phi\,dxdy$$

Because ϕ is zero on S, the boundary integrals in (9.3.17) will vanish and relation (9.3.16) simplifies to

$$T = 2 \iint_R \phi \, dx \, dy \tag{9.3.18}$$

We have now shown that the assumed displacement form (9.3.3) produces a stress field that when represented by the Prandtl stress function relation (9.3.8) yields a governing Poisson equation (9.3.9) with the condition that the stress function vanishes on the boundary of the cross-section. All resultant boundary conditions on the ends of the cylinder are satisfied by the representation, and the overall torque is related to the stress function through relation (9.3.18). This then concludes the stress formulation of the torsion problem for simply connected sections.

9.3.2 Displacement formulation

The displacement formulation starts by expressing the equilibrium equation in terms of the warping displacement w. Using (9.3.5) in (9.3.6) gives

$$\frac{\partial^2 w}{\partial x^2} + \frac{\partial^2 w}{\partial y^2} = 0 \tag{9.3.19}$$

and thus the displacement component satisfies Laplace's equation in the cross-section R. The associated boundary condition on the lateral side S is given by $(9.3.10)_3$, and expressing this in terms of the warping displacement gives

$$\left(\frac{\partial w}{\partial x} - y\alpha \right) n_x + \left(\frac{\partial w}{\partial y} + x\alpha \right) n_y = 0 \tag{9.3.20}$$

Using relations (9.3.11), this result can be rewritten as

$$\frac{\partial w}{\partial x} \frac{dx}{dn} + \frac{\partial w}{\partial y} \frac{dy}{dn} = \alpha \left(x \frac{dx}{ds} + y \frac{dy}{ds} \right)$$

$$\frac{dw}{dn} = \frac{\alpha}{2} \frac{d}{ds} \left(x^2 + y^2 \right) \tag{9.3.21}$$

It can again be shown that the boundary conditions on the ends specified by equations $(9.3.14)_{1-5}$ will all be satisfied, and the resultant torque condition $(9.3.14)_6$ will give

$$T = \mu \iint_R \left(\alpha \left(x^2 + y^2 \right) + x \frac{\partial w}{\partial y} - y \frac{\partial w}{\partial x} \right) dx \, dy \tag{9.3.22}$$

This result is commonly written as

$$T = \alpha J \tag{9.3.23}$$

where J is called the *torsional rigidity* and is given by

$$J = \mu \iint_R \left(x^2 + y^2 + \frac{x}{\alpha} \frac{\partial w}{\partial y} - \frac{y}{\alpha} \frac{\partial w}{\partial x} \right) dx \, dy \tag{9.3.24}$$

This completes the displacement formulation for the torsion problem.

Comparing the two formulations, it is observed that the stress function approach results in a governing equation of the Poisson type (9.3.9) with a very simple boundary condition requiring only that the stress function be constant or vanish. On the other hand, the displacement formulation gives a somewhat simpler Laplace governing equation (9.3.19), but the boundary specification is expressed in terms of the normal derivative. An additional approach involving formulation in terms of a *conjugate function* (see Exercise 9.4) creates yet another scheme that yields a Laplace governing equation with a somewhat simpler boundary condition involving specification of the unknown itself. The boundary-value problems created by these approaches generally fall into the area of applied mathematics called *potential theory* (Kellogg, 1969). As such, many mathematical techniques have been developed to solve such problems, including potential theory, complex variables, Fourier methods, and some specialized simple schemes based on the boundary equation. In this presentation we only consider two solution schemes, one using the boundary equation and the other using Fourier methods. Before moving on to these solutions, we wish to establish briefly the necessary modifications to the formulations for cylinders with hollow sections. We shall also explore an analogous (membrane) problem that provides some useful information and interpretation for development of approximate solutions to the torsion problem.

9.3.3 Multiply connected cross-sections

We now wish to develop some additional relations necessary to solve the torsion of hollow cylinders with multiply connected cross-sections (see definitions in Section 2.6). Fig. 9.4 illustrates a typical section of this type with a single hole, and we shall establish theory capable of handling any number of holes. It is assumed that the original boundary conditions of zero tractions on all lateral surfaces apply to the external boundary S_o and all internal boundaries $S_1, \ldots$ Therefore, as before, condition $(9.3.10)_3$

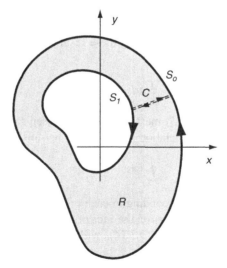

FIG. 9.4 Multiply connected cross-section.

would imply that the stress function is a constant and the displacement is specified as per (9.3.20) or (9.3.21) on each boundary S_i, $i = 0, 1, \ldots$

$$\phi = \phi_i \quad \text{on} \quad S_i$$

$$\frac{dw}{dn} = \alpha(yn_x - xn_y) \quad \text{on} \quad S_i \tag{9.3.25}$$

where ϕ_i are constants. These conditions imply that the stress function and warping displacement can be determined up to an arbitrary constant on each boundary S_i. With regard to the stress function, the value of ϕ_i may be arbitrarily chosen only on one boundary, and commonly this value is taken as zero on the outer boundary S_o similar to the simply connected case.

For multiply connected sections, the constant values of the stress function on each of the interior boundaries are determined by requiring that the displacement w be single-valued. Considering the doubly connected example shown in Fig. 9.4, the displacement will be single-valued if

$$\oint_{S_1} dw(x, y) = 0 \tag{9.3.26}$$

This integral can be written as

$$\oint_{S_1} dw(x, y) = \oint_{S_1} \left(\frac{\partial w}{\partial x} dx + \frac{\partial w}{\partial y} dy \right)$$

$$= \frac{1}{\mu} \oint_{S_1} (\tau_{xz} dx + \tau_{yz} dy) - \alpha \oint_{S_1} (x dy - y dx) \tag{9.3.27}$$

Now $\tau_{xz} dx + \tau_{yz} dy = \tau ds$, where τ is the resultant shear stress. Using Green's theorem (1.8.10)

$$\oint_{S_1} (x dy - y dx) = \iint_{A_1} \left(\frac{\partial x}{\partial x} + \frac{\partial y}{\partial y} \right) dx dy = 2 \iint_{A_1} dx dy = 2A_1 \tag{9.3.28}$$

where A_1 is the area enclosed by S_1. Combining these results, the single-valued condition (9.3.26) implies that

$$\oint_{S_1} \tau ds = 2\mu\alpha A_1 \tag{9.3.29}$$

The value of ϕ_1 on the inner boundary S_1 must therefore be chosen so that (9.3.29) is satisfied. If the cross-section has more than one hole, relation (9.3.29) must be satisfied for each; that is

$$\oint_{S_k} \tau ds = 2\mu\alpha A_k \tag{9.3.30}$$

where $k = 1, 2, 3, \ldots$ is the index corresponding to each of the interior holes.

It can be shown that boundary conditions on the ends of the cylinder given by (9.3.14)$_{1-5}$ will all be satisfied, and the resultant torque condition (9.3.14)$_6$ will give

$$T = 2 \iint_R \phi dx dy + 2\phi_1 A_1 \tag{9.3.31}$$

For the case with N holes, this relation becomes

$$T = 2 \iint_R \phi \, dx \, dy + \sum_{k=1}^{N} 2\phi_k A_k \tag{9.3.32}$$

Justifying these developments for multiply connected sections requires contour integration in a cut domain following the segments S_o, C, S_1, as shown in Fig. 9.4.

9.3.4 Membrane analogy

It was originally discovered by Prandtl in 1903 that the equations of the stress function formulation (9.3.9), (9.3.12), and (9.3.18) are identical with those governing the static deflection of an elastic membrane under uniform pressure. This fact then creates an *analogy* between the two problems and enables particular features from the membrane problem to be used to aid in solution of the torsion problem. Use of this analogy is generally limited to providing insight into qualitative features and to aid in developing approximate solutions.

Consider a thin elastic membrane stretched over a frame with shape S that encloses region R in the x,y-plane, as shown in Fig. 9.5A. The membrane is stretched with uniform tension N and is subjected to a uniform pressure p, which produces a transverse membrane deflection $z(x,y)$. For small deformation theory, it is assumed that the pressure loading will not alter the membrane tension. The governing membrane displacement equation is developed by applying equilibrium to a differential element shown in Fig. 9.5B. A side view of this element along the y-axis shown in Fig. 9.5C

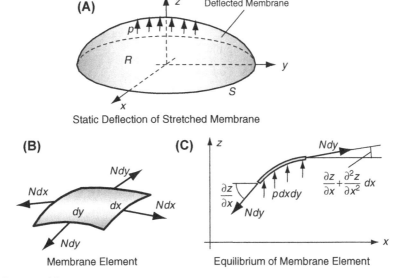

FIG. 9.5 Membrane problem.

illustrates the tension forces on each edge and the pressure loading. Summing forces in the z direction and including the tension forces in both x and y directions gives

$$Ndy\left(\frac{\partial z}{\partial x} + \frac{\partial^2 z}{\partial x^2}dx\right) - Ndy\left(\frac{\partial z}{\partial x}\right) + Ndx\left(\frac{\partial z}{\partial y} + \frac{\partial^2 z}{\partial y^2}dy\right) - Ndx\left(\frac{\partial z}{\partial y}\right) + pdxdy = 0$$

and this result simplifies to

$$\frac{\partial^2 z}{\partial x^2} + \frac{\partial^2 z}{\partial y^2} = -\frac{p}{N} \tag{9.3.33}$$

Because the membrane is stretched over the boundary S in the x,y-plane, the boundary condition for deflection is expressed by

$$z = 0 \text{ on } S \tag{9.3.34}$$

The volume enclosed by the deflected membrane and the x,y-plane is given by

$$V = \iint_R z dx dy \tag{9.3.35}$$

The analogy can now be recognized because relations (9.3.33)–(9.3.35) match the corresponding results from the torsion formulation providing $\phi = z$, $p/N = 2\mu\alpha$, $T = 2V$.

In order to extract some useful information from this analogy, consider first the relationship between the shear stress and stress function

$$\tau_{xz} = \frac{\partial \phi}{\partial y} = \frac{\partial z}{\partial y}$$
$$\tau_{yz} = -\frac{\partial \phi}{\partial x} = -\frac{\partial z}{\partial x} \tag{9.3.36}$$

A *contour line* on the membrane is defined as $z = $ constant (see Fig. 9.6). Using the analogy, such a contour is also a line of constant ϕ, and along the contour

$$\frac{\partial z}{\partial s} = \frac{\partial \phi}{\partial s} = \frac{\partial \phi}{\partial x}\frac{dx}{ds} + \frac{\partial \phi}{\partial y}\frac{dy}{ds} = 0 \tag{9.3.37}$$

$$= \tau_{yz}n_y + \tau_{xz}n_x = \tau_{zn}$$

where τ_{zn} is the component of shear stress normal to the contour line. Thus, the component τ_{zn} is zero along a contour line and the resultant shear stress must be tangent to the contour. This resultant shear stress is given by

$$\tau = \tau_{zt} = -\tau_{xz}n_y + \tau_{yz}n_x = -\tau_{xz}\frac{dy}{dn} + \tau_{yz}\frac{dx}{dn}$$
$$= -\left(\frac{\partial \phi}{\partial y}\frac{dy}{dn} + \frac{\partial \phi}{\partial x}\frac{dx}{dn}\right) = -\frac{d\phi}{dn} = -\frac{dz}{dn} \tag{9.3.38}$$

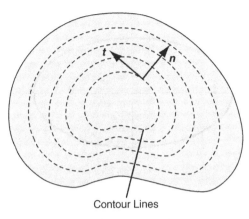

Contour Lines

FIG. 9.6 Contour lines for the torsion-membrane analogy.

Reviewing the previous findings related to the membrane analogy, the following concepts can be concluded. The shear stress at any point in the cross-section is given by the negative of the slope of the membrane in the direction normal to the contour line through the point. The maximum shear stress appears to be located on the boundary where the largest slope of the membrane occurs. Actually, this result can be explicitly proven (see Exercise 9.5). The torque T is given as twice the volume under the membrane. Using these membrane visualizations, a useful qualitative picture of the stress function distribution can be determined and approximate solutions can be constructed (see Exercise 9.8). However, it should be realized that trying to make slope measurements of an actual pressurized membrane would not provide an accurate method to determine the stresses in a bar under torsion.

We now explore the solution to several torsion problems using boundary equation schemes, Fourier methods, and membrane analogy techniques. These methods provide solutions to sections of simple geometry. More complicated sections can be solved using complex variable theory; see Sokolnikoff (1956) for a brief presentation of these techniques.

9.4 Torsion solutions derived from boundary equation

For simply connected sections, the stress function formulation requires that the function satisfy Poisson equation (9.3.9) and vanish on boundary S. If the boundary is expressed by the relation $f(x,y) = 0$, this suggests a possible simple solution scheme of expressing the stress function in terms of the boundary equation $\phi = Kf(x,y)$ where K is an arbitrary constant.

This form satisfies the boundary condition on S, and for some simple geometric shapes it will also satisfy the governing equation (9.3.9) with an appropriate choice of K. Unfortunately, this is not a general solution method but rather an ad hoc scheme that works only for special cross-sections of simple geometry. Nevertheless, it provides several solutions to problems of interest, and we now investigate some particular solutions using this scheme.

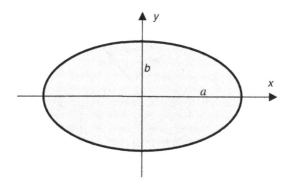

FIG. 9.7 Elliptical cross-section.

Example 9.1 Elliptical section

The first example of this solution method is that of an elliptical cross-section as shown in Fig. 9.7. The boundary equation has the usual form

$$\frac{x^2}{a^2} + \frac{y^2}{b^2} = 1 \tag{9.4.1}$$

where a and b are the semi major and minor axes as shown.

Using the boundary equation scheme, we look for a stress function of the form

$$\phi = K\left(\frac{x^2}{a^2} + \frac{y^2}{b^2} - 1\right) \tag{9.4.2}$$

This stress function satisfies the boundary condition by vanishing on S, and this form will also satisfy the governing equation (9.3.9) if the constant is chosen as

$$K = -\frac{a^2 b^2 \mu \alpha}{a^2 + b^2} \tag{9.4.3}$$

Because both the governing equation and the boundary conditions are satisfied, we have found the solution to the torsion of the elliptical section.

The load-carrying torque follows from relation (9.3.18)

$$T = -\frac{2a^2 b^2 \mu \alpha}{a^2 + b^2}\left(\frac{1}{a^2}\iint_R x^2 dxdy + \frac{1}{b^2}\iint_R y^2 dxdy - \iint_R dxdy\right) \tag{9.4.4}$$

The integrals in this expression have the following simple meaning and evaluation

$$A = \text{Area of section} = \iint_R dxdy = \pi ab$$

$$I_x = \text{Moment of inertia about } x\text{-axis} = \iint_R y^2 dxdy = \frac{\pi}{4}ab^3 \tag{9.4.5}$$

$$I_y = \text{Moment of inertia about } y\text{-axis} = \iint_R x^2 dxdy = \frac{\pi}{4}ba^3$$

Substituting these results back into (9.4.4) yields

$$T = \frac{\pi a^3 b^3 \mu \alpha}{a^2 + b^2}$$ (9.4.6)

which can be cast in the form to determine the angle of twist in terms of the applied loading

$$\alpha = \frac{T(a^2 + b^2)}{\pi a^3 b^3 \mu}$$ (9.4.7)

The shear stresses resulting from this solution are given by

$$\tau_{xz} = -\frac{2a^2 \mu \alpha}{a^2 + b^2} y = -\frac{2Ty}{\pi a b^3}$$

$$\tau_{yz} = \frac{2b^2 \mu \alpha}{a^2 + b^2} x = \frac{2Tx}{\pi b a^3}$$ (9.4.8)

Intuition from strength of materials theory would suggest that the maximum stress should occur at the boundary point most removed from the section's center; that is, at $x = \pm a$ and $y = 0$ (assuming $a > b$). However, the membrane analogy would argue for a boundary point closest to the center of the section where the membrane slope would be the greatest. Evaluating equations (9.4.8), we find that the resultant shear stress becomes

$$\tau = \sqrt{\tau_{xz}^2 + \tau_{yz}^2} = \frac{2T}{\pi a b} \sqrt{\frac{x^2}{a^4} + \frac{y^2}{b^4}}$$ (9.4.9)

For the case $a > b$, the maximum value of τ occurs at $x = 0$ and $y = \pm b$ and is given by

$$\tau_{max} = \frac{2T}{\pi a b^2}$$ (9.4.10)

This result then corresponds to arguments from the membrane analogy and thus differs from strength of materials suggestions. Contour lines of the stress function are shown in Fig. 9.8, and it is observed that the maximum slope of the stress function (membrane) occurs at $x = 0$ and $y = \pm b$ (on the top and bottom of the section).

Using the stress relations (9.4.8) in (9.3.5) yields a system that can be integrated to determine the displacement field

$$w = \frac{T(b^2 - a^2)}{\pi a^3 b^3 \mu} xy$$ (9.4.11)

Contour lines of this displacement field are represented by hyperbolas in the x,y-plane and are shown in Fig. 9.8 for the case of a positive counterclockwise torque applied to the section. These plots were generated using MATLAB Code C.8. Solid lines correspond to positive values of w, indicating that points move out of the section in the positive z direction, while dotted lines indicate negative values of displacement. A three-dimensional plot of the warping displacement surface is also shown in Fig. 9.8 (using MATLAB code C.9), illustrating the positive and negative behavior of the w displacement. Along each of the coordinate axes the displacement is zero, and for the special case with $a = b$ (circular section), the warping displacement vanishes everywhere. If the ends of the elliptical cylinder are restrained, normal stresses σ_z are generated as a result of the torsion.

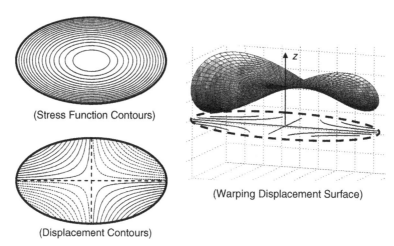

(Stress Function Contours)

(Warping Displacement Surface)

(Displacement Contours)

FIG. 9.8 Stress function and warping displacement contours and warping displacement surface for the elliptical section.

Example 9.2 Equilateral triangular section

Consider next the torsion of a cylinder with equilateral triangular section, as shown in Fig. 9.9. Following our boundary equation solution scheme, we look for a stress function of the form

$$\phi = K\left(x - \sqrt{3}\,y + 2a\right)\left(x + \sqrt{3}\,y + 2a\right)(x - a) \qquad (9.4.12)$$

where we have simply used a product form of each boundary line equation. In this fashion, the stress function vanishes on each side of the triangular section. It is found that this function satisfies the governing equation (9.3.9) if the constant is taken as

$$K = -\frac{\mu\alpha}{6a} \qquad (9.4.13)$$

All conditions on the problem are now satisfied, and we have thus determined the solution for the equilateral triangular case. The torque may be calculated through a lengthy integration using relation (9.3.18), giving the result

$$T = \frac{27}{5\sqrt{3}}\mu\alpha a^4 = \frac{3}{5}\mu\alpha I_p \qquad (9.4.14)$$

where $I_p = 3\sqrt{3}a^4$ is the polar moment of inertia of the cross-section about the centroid.
The stresses follow from relations (9.3.8)

$$\tau_{xz} = \frac{\mu\alpha}{a}(x - a)y$$

$$\tau_{yz} = \frac{\mu\alpha}{2a}\left(x^2 + 2ax - y^2\right) \qquad (9.4.15)$$

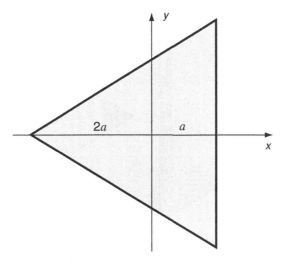

FIG. 9.9 Equilateral triangular cross-section.

Note that the component τ_{xz} vanishes along the edge $x = a$ as required by the problem boundary conditions, and this can also be argued by the membrane analogy. This component also vanishes along the x-axis. The maximum stress always occurs on the boundary, and the section symmetry implies that each boundary side has an identical resultant stress distribution. Therefore, we can choose one particular side to investigate and determine the maximum resultant shear stress. For convenience, we choose side $x = a$, and because $\tau_{xz} = 0$ on this edge, the resultant stress is given by

$$\tau = \tau_{yz}(a, y) = \frac{\mu\alpha}{2a}\left(3a^2 - y^2\right) \tag{9.4.16}$$

The maximum value of this expression gives

$$\tau_{\max} = \tau_{yz}(a, 0) = \frac{3}{2}\mu\alpha a = \frac{5\sqrt{3}T}{18a^3} \tag{9.4.17}$$

Contours of the stress function are shown in Fig. 9.10, and by using the membrane analogy it is evident that the maximum stress occurs at the midpoint of each boundary side.

The warping displacement again follows from integrating relations (9.3.5)

$$w = \frac{\alpha}{6a}y\left(3x^2 - y^2\right) \tag{9.4.18}$$

Contour lines of this displacement field are shown in Fig. 9.10 for the case of a positive counterclockwise torque applied to the section. Again, solid lines correspond to positive values, while dotted lines indicate negative displacements.

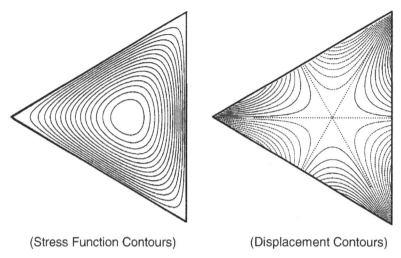

(Stress Function Contours)   (Displacement Contours)

FIG. 9.10 Stress function and warping displacement contours for the equilateral triangular section.

Example 9.3 Higher-order boundary polynomials

As a final example of the boundary equation scheme, consider the more general case of a section with a polynomial boundary equation. The trial stress function is taken of the form

$$\phi = K(a^2 - x^2 + cy^2)(a^2 + cx^2 - y^2) \qquad (9.4.19)$$

where K, a, and c are constants to be determined. The terms in parentheses can be rewritten as

$$x = \pm\sqrt{a^2 + cy^2}, \quad y = \pm\sqrt{a^2 + cx^2}$$

and these represent pairs of curves shown in Fig. 9.11 that can be interpreted as bounding a closed region R as shown. This region is taken as the cylinder section for the torsion problem. As before, this stress function vanishes on the boundary, and it satisfies the governing equation (9.3.9) if $c = 3 - \sqrt{8}$ and $K = -\mu\alpha/[4a^2(1 - \sqrt{2})]$. The stresses and displacements can be calculated using the previous procedures (see Exercise 9.16). Timoshenko and Goodier (1970) discuss additional examples of this type of problem.

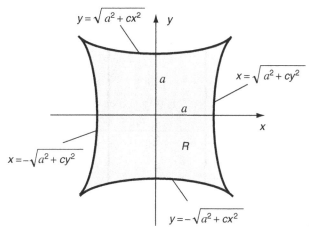

FIG. 9.11 Polynomial boundary example.

9.5 Torsion solutions using Fourier methods

Previously introduced in Section 8.2 for plane problems, Fourier methods also provide a useful technique for solving the torsion problem. Using separation of variables and Fourier series theory, solutions can be developed to particular problems formulated either in terms of the stress or displacement function. We now pursue one such case in Cartesian coordinates involving the torsion of a rectangular section.

Example 9.4 Rectangular section

We now wish to develop the solution to the torsion of a cylinder with rectangular section shown in Fig. 9.12. Trying the previous scheme of products of the boundary lines does not create a stress function that can satisfy the governing equation (see Exercise 9.17). Thus, we must resort to a more fundamental solution technique, and the Fourier method is ideally suited for this problem. We develop the solution using the stress function formulation, but a similar solution can also be determined using the displacement formulation.

The solution to governing equation (9.3.9) can be written as the sum of a general solution to the homogeneous Laplace equation plus a particular solution to the nonhomogeneous form; that is, $\phi = \phi_h + \phi_p$. A convenient particular solution can be chosen as

$$\phi_p(x, y) = \mu\alpha(a^2 - x^2) \tag{9.5.1}$$

Note that this choice of a parabolic form can be motivated using the membrane analogy for the case of a thin rectangle with $a \ll b$ (see Exercise 9.8). We discuss more on this limiting case at

---

Continued

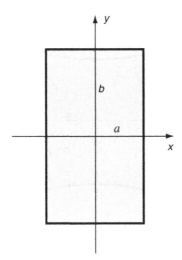

FIG. 9.12 Rectangular section example.

the end of the problem solution. Using this form, the homogeneous solution must then satisfy the following conditions

$$\nabla^2 \phi_h = 0$$
$$\phi_h(\pm a, y) = 0$$
$$\phi_h(x, \pm b) = -\mu\alpha(a^2 - x^2)$$

(9.5.2)

and this ensures that the combined stress function ϕ satisfies the general formulation conditions. Standard separation of variables methods are used to generate the homogeneous solution by looking for solutions of the form

$$\phi_h(x, y) = X(x)Y(y)$$

(9.5.3)

Substituting this form into $(9.5.2)_1$ allows the variables to be separated into the following pair of differential relations

$$X''(x) + \lambda^2 X(x) = 0$$
$$Y''(y) - \lambda^2 Y(y) = 0$$

(9.5.4)

where λ is the separation constant. The solution to (9.5.4) is given by

$$X(x) = A \sin \lambda x + B \cos \lambda x$$
$$Y(y) = C \sinh \lambda y + D \cosh \lambda y$$

(9.5.5)

where A, B, C, D are constants. Because of the given problem symmetry, we can immediately argue that the solution should be an even function of x and y, and thus the odd function terms must be dropped by taking $A = C = 0$. In order to satisfy condition $(9.5.2)_2$, the separation constant must be given by $\lambda = n\pi/2a$, $n = 1, 3, 5, \ldots$ Combining these results, the homogeneous solution can then be expressed by

$$\phi_h(x, y) = \sum_{n=1}^{\infty} B_n \cos\frac{n\pi x}{2a}\cosh\frac{n\pi y}{2a} \tag{9.5.6}$$

where we use the superposition of all solution forms and the coefficient B_n has absorbed the product term BD.

The final boundary condition $(9.5.2)_3$ yields the result

$$-\mu\alpha(a^2 - x^2) = \sum_{n=1}^{\infty} B_n^* \cos\frac{n\pi x}{2a} \tag{9.5.7}$$

where $B_n^* = B_n \cosh(n\pi b/2a)$. Equation (9.5.7) is recognized as the Fourier cosine series for the expression on the left-hand side. Using relations (8.2.27), Fourier series theory provides a simple scheme to determine the series coefficients, giving the result

$$B_n^* = -\frac{2\mu\alpha}{a}\int_0^a (a^2 - \xi^2)\cos\frac{n\pi\xi}{2a}d\xi \tag{9.5.8}$$

Evaluating this integral, the original coefficient can then be expressed as

$$B_n = -\frac{32\mu\alpha a^2(-1)^{(n-1)/2}}{n^3\pi^3\cosh\dfrac{n\pi b}{2a}} \tag{9.5.9}$$

The stress function has now been determined, and combining the previous results gives

$$\phi = \mu\alpha(a^2 - x^2) - \frac{32\mu\alpha a^2}{\pi^3}\sum_{n=1,3,5,\ldots}^{\infty}\frac{(-1)^{(n-1)/2}}{n^3\cosh\dfrac{n\pi b}{2a}}\cos\frac{n\pi x}{2a}\cosh\frac{n\pi y}{2a} \tag{9.5.10}$$

The stresses follow from relation (9.3.8)

$$\tau_{xz} = \frac{\partial\phi}{\partial y} = \frac{16\mu\alpha a}{\pi^2}\sum_{n=1,3,5,\ldots}^{\infty}\frac{(-1)^{(n-1)/2}}{n^2\cosh\dfrac{n\pi b}{2a}}\cos\frac{n\pi x}{2a}\sinh\frac{n\pi y}{2a}$$

$$\tau_{yz} = -\frac{\partial\phi}{\partial x} = 2\mu\alpha x - \frac{16\mu\alpha a}{\pi^2}\sum_{n=1,3,5,\ldots}^{\infty}\frac{(-1)^{(n-1)/2}}{n^2\cosh\dfrac{n\pi b}{2a}}\sin\frac{n\pi x}{2a}\cosh\frac{n\pi y}{2a} \tag{9.5.11}$$

and using (9.3.18), the torque is given by

$$T = \frac{16\mu\alpha a^3 b}{3} - \frac{1024\mu\alpha a^4}{\pi^5}\sum_{n=1,3,5,\ldots}^{\infty}\frac{1}{n^5}\tanh\frac{n\pi b}{2a} \tag{9.5.12}$$

Using our experience from the previous examples or from the membrane analogy, the maximum stress will occur on the boundary at the midpoint of the longest side. Under the assumption that $a < b$, these points are located at $x = \pm a$ and $y = 0$, and thus

Continued

$$\tau_{\max} = \tau_{yz}(a,0) = 2\mu\alpha a - \frac{16\mu\alpha a}{\pi^2} \sum_{n=1,3,5,\dots}^{\infty} \frac{1}{n^5 \cosh\dfrac{n\pi b}{2a}} \qquad (9.5.13)$$

Fig. 9.13 illustrates the stress function contours for this case, and it is observed that the maximum stresses occur at the midpoint of each of the longest boundary sides. For the square section case ($a = b$), the maximum stresses would occur at the midpoint of each side.

Again, the displacement field follows from integrating relations (9.3.5), giving the result

$$w = \alpha xy - \frac{32\alpha a^2}{\pi^3} \sum_{n=1,3,5,\dots}^{\infty} \frac{(-1)^{(n-1)/2}}{n^3 \cosh\dfrac{n\pi b}{2a}} \sin\frac{n\pi x}{2a} \sinh\frac{n\pi y}{2a} \qquad (9.5.14)$$

Contour lines of this displacement field are shown in Fig. 9.13 for three sections with different aspect ratios. Again, solid lines correspond to positive displacements, while dotted lines indicate

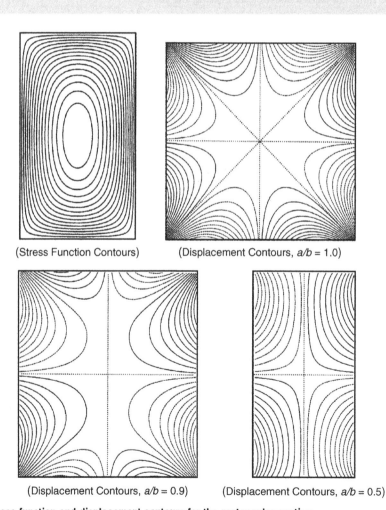

(Stress Function Contours)   (Displacement Contours, a/b = 1.0)

(Displacement Contours, a/b = 0.9)   (Displacement Contours, a/b = 0.5)

FIG. 9.13 Stress function and displacement contours for the rectangular section.

negative values. The square section case with $a/b = 1$ produces a displacement pattern with eight zones of symmetry. As the aspect ratio a/b is reduced, four of the displacement patterns disappear and the resulting displacement contours for $a/b = 0.5$ look similar to that from the elliptical section case shown in Fig. 9.8.

We now investigate these results for the special case of a *very thin rectangle* with $a << b$. Under the conditions of $b/a >> 1$, $\cosh(n\pi b/2a) \to \infty$ and $\tanh(n\pi b/2a) \to 1$, and we therefore find that the stress function, maximum shear stress, and torque relations reduce to

$$\phi = \mu\alpha(a^2 - x^2)$$

$$\tau_{max} = 2\mu\alpha a \qquad\qquad (9.5.15)$$

$$T = \frac{16}{3}\mu\alpha a^3 b$$

For this limiting case, it is observed that the stress function reduces to a parabolic distribution, and this would be predictable from the membrane analogy (see Exercise 9.8). These results can be applied to the torsion of sections composed of a number of thin rectangles such as the example shown in Fig. 9.14 with three rectangles. Note that these shapes can approximate many common structural beams with angle, channel, and I sections. Neglecting the local regions where the rectangles are joined and the free short edges, it can be assumed that the membrane has the parabolic distribution given by $(9.5.15)_1$ over each rectangle. Stress function contours (from a numerical solution) shown in Fig. 9.14 justify these assumptions. Thus, the load-carrying torque for such a composite section is given by

$$T = \frac{16}{3}\mu\alpha \sum_{i=1}^{N} a_i^3 b_i \qquad\qquad (9.5.16)$$

where a_i and b_i ($b_i >> a_i$) are the dimensions of the various N rectangles. Neglecting the high localized stresses at the re-entrant corners, the maximum shear stress can be estimated by using relation $(9.5.15)_2$ for the narrowest rectangle.

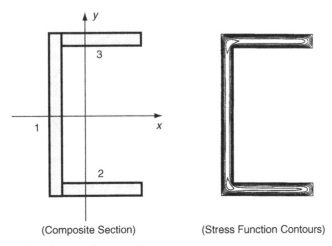

(Composite Section)   (Stress Function Contours)

FIG. 9.14 Composite section of three thin rectangles.

9.6 Torsion of cylinders with hollow sections

Section 9.3 developed the basic formulation for the torsion of hollow cylinders with multiply connected cross-sections. It was found that the stress function must be constant on all section boundaries. Although ϕ could be arbitrarily chosen as zero on the outer boundary, on each interior surface it is required to be a different constant determined by relation (9.3.30), a requirement that ensures single-valued displacements. Under such a formulation, analytical solutions of these problems are difficult to develop and only a few closed-form solutions exist. Complex variable theory using conformal mapping has provided some of these solutions, and Sokolnikoff (1956) provides references to a few specific cases. Rather than trying to pursue these details, we shall only present a couple of simple solutions in order to demonstrate some basic features of such problems.

Example 9.5 Hollow elliptical section

Consider the torsion of a bar with a hollow elliptical section as shown in Fig. 9.15. The inner boundary is simply a scaled ellipse similar to that of the outer boundary. Using the solid section solution from Example 9.1, it can be shown that the contour lines, or lines of constant shear stress, coincide with such a scaled concentric ellipse (see Exercise 9.10). The shear stress will then be tangent to the inner boundary contour and no stress will then act on the lateral surface of a cylinder with inner ellipse section. Therefore, the solution to the hollow section can be found by simply removing the inner core from the solid solution developed in Example 9.1, and this results in the same stress distribution in the remaining material.

Thus, the stress function solution for the hollow case is given by

$$\phi = -\frac{a^2 b^2 \mu \alpha}{a^2 + b^2} \left(\frac{x^2}{a^2} + \frac{y^2}{b^2} - 1 \right) \tag{9.6.1}$$

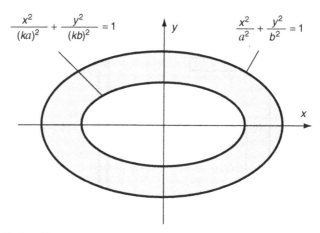

$$\frac{x^2}{(ka)^2} + \frac{y^2}{(kb)^2} = 1 \qquad \frac{x^2}{a^2} + \frac{y^2}{b^2} = 1$$

FIG. 9.15 Hollow elliptical section.

and this form satisfies the governing equation, boundary conditions, and the multiply connected condition (9.3.30). The constant value of the stress function on the inner boundary is found to be

$$\phi_i = -\frac{a^2 b^2 \mu \alpha}{a^2 + b^2} \left(k^2 - 1\right) \tag{9.6.2}$$

In order to determine the load-carrying capacity, the torque relation for the solid section (9.4.6) must be reduced by subtracting the load carried by the removed inner cylinder. This gives the result

$$T = \frac{\pi a^3 b^3 \mu \alpha}{a^2 + b^2} - \frac{\pi (ka)^3 (kb)^3 \mu \alpha}{(ka)^2 + (kb)^2}$$

$$= \frac{\pi \mu \alpha}{a^2 + b^2} a^3 b^3 \left(1 - k^4\right) \tag{9.6.3}$$

and this relation can also be determined from equation (9.3.31). As mentioned, the stress distribution in the hollow cylinder will be the same as that found in the corresponding material of the solid section; see relations (9.4.8). For the case $a > b$, the maximum stress still occurs at $x = 0$ and $y = \pm b$ and is given by

$$\tau_{\max} = \frac{2T}{\pi a b^2} \frac{1}{1 - k^4} \tag{9.6.4}$$

This solution scheme could be applied to other cross-sections whose inner boundary coincides with a contour line of the corresponding solid section problem.

Example 9.6 Hollow thin-walled sections

The torsion of hollow thin-walled cylinders can be effectively handled using an approximate solution based on the membrane analogy. Consider the general thin-walled tube shown in Fig. 9.16. We assume that thickness t is small, although not necessarily constant. A general section aa is taken through the tube wall at AB, and the expected membrane shape is shown. From our previous theory, the membrane (stress function) will be zero at the outer boundary (point B) and equal to a nonzero constant, say ϕ_o, on the inner boundary (point A). Because the thickness is small there will be little variation in the membrane slope, and thus shape BC can be approximated by a straight line. Because the membrane slope equals the resultant shear stress, we can write

$$\tau = \frac{\phi_o}{t} \tag{9.6.5}$$

The load-carrying relation (9.3.31) gives

$$T = 2 \iint_R \phi \, dx \, dy + 2 \phi_o A_i \tag{9.6.6}$$

Continued

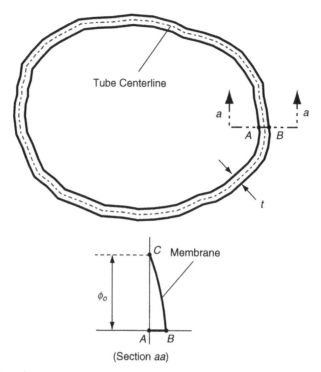

FIG. 9.16 Thin-walled section.

where A_i is the area enclosed by the inner boundary. Using our assumption that the membrane slope is constant over the section and neglecting variation in the wall thickness the integral over the cross-section R can be approximated by $A\phi_o/2$, where A is the section area. This allows the torque relation to be expressed by

$$T = 2\left(A\frac{\phi_o}{2}\right) + 2\phi_o A_i = 2\phi_o A_c \qquad (9.6.7)$$

where A_c is the area enclosed by the section's centerline. Combining relations (9.6.5) and (9.6.7) gives

$$\tau = \frac{T}{2A_c t} \qquad (9.6.8)$$

The angle of twist is determined using relation (9.3.29) with constant wall thickness

$$\oint_{S_c} \tau ds = 2\mu\alpha A_c \Rightarrow$$

$$\alpha = \frac{TS_c}{4A_c^2 \mu t} \qquad (9.6.9)$$

where S_c is the length of the centerline of the tube section. These results provide reasonable estimates of the stress, torque capacity, and angle of twist for thin-walled tubes under torsion. However, if the tube has sharp corners such as those found in square or rectangular sections, considerable stress concentration normally exists at these re-entrant locations. Timoshenko and Goodier (1970) provide additional details on calculating these stress concentration effects.

9.7 Torsion of circular shafts of variable diameter

The previous discussion on the torsion problem was limited to bars of constant section. We now wish to investigate the case of variable section, and in order to limit problem complexity we consider only circular cross-sections, as shown in Fig. 9.17. Cylindrical coordinates are the logical choice to formulate this type of problem, and the governing field equations have been previously given by (2.7.3), (3.8.6), and (4.3.2), or see Appendix A. Guided by studies on uniform circular cylinders, we assume that $u_r = u_z = 0$, and $u_\theta = u_\theta(r, z)$. For this semi-inverse scheme, it will be shown that the solution based on these assumptions satisfies all governing elasticity field equations, and therefore represents the true solution.

Under these assumptions, strain and stress fields are then determined as

$$e_r = e_\theta = e_z = e_{rz} = 0$$

$$e_{r\theta} = \frac{1}{2}\left(\frac{\partial u_\theta}{\partial r} - \frac{u_\theta}{r}\right), \quad e_{\theta z} = \frac{1}{2}\frac{\partial u_\theta}{\partial z} \qquad (9.7.1)$$

$$\sigma_r = \sigma_\theta = \sigma_z = \tau_{rz} = 0$$

$$\tau_{r\theta} = \mu\left(\frac{\partial u_\theta}{\partial r} - \frac{u_\theta}{r}\right), \quad \tau_{\theta z} = \mu\frac{\partial u_\theta}{\partial z} \qquad (9.7.2)$$

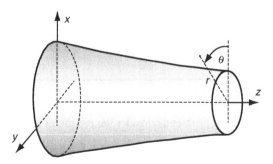

FIG. 9.17 Shaft of a variable circular section.

Using these stress results in the equilibrium equations with no body forces gives one nonvanishing relation

$$\frac{\partial}{\partial r}\left[r^3\frac{\partial}{\partial r}\left(\frac{u_\theta}{r}\right)\right] + \frac{\partial}{\partial z}\left[r^3\frac{\partial}{\partial z}\left(\frac{u_\theta}{r}\right)\right] = 0 \tag{9.7.3}$$

This particular form suggests attempting a stress function approach, and introducing the function Ψ, such that

$$\frac{\partial \Psi}{\partial z} = -r^3\frac{\partial}{\partial r}\left(\frac{u_\theta}{r}\right) = -\frac{r^2}{\mu}\tau_{r\theta}$$

$$\frac{\partial \Psi}{\partial r} = r^3\frac{\partial}{\partial z}\left(\frac{u_\theta}{r}\right) = \frac{r^2}{\mu}\tau_{\theta z} \tag{9.7.4}$$

satisfies the equilibrium equation identically. Differentiating relations (9.7.4) to eliminate u_θ generates the compatibility relation

$$\frac{\partial^2 \Psi}{\partial r^2} - \frac{3}{r}\frac{\partial \Psi}{\partial r} + \frac{\partial^2 \Psi}{\partial z^2} = 0 \tag{9.7.5}$$

The lateral sides of the bar are again taken to be traction free, and thus the boundary conditions are expressed as

$$\tau_{r\theta}n_r + \tau_{\theta z}n_z = 0 \tag{9.7.6}$$

As before $n_r = \dfrac{dz}{ds}$ and $n_z = -\dfrac{dr}{ds}$, and incorporating (9.7.4), boundary condition (9.7.6) becomes

$$\frac{\mu}{r^2}\left(\frac{\partial \Psi}{\partial r}\frac{dr}{ds} + \frac{\partial \Psi}{\partial z}\frac{dz}{ds}\right) = 0 \Rightarrow \frac{d\Psi}{ds} = 0 \tag{9.7.7}$$

and so, as before, the stress function must be a constant on the boundary of the section.

The load-carrying torque is given by

$$T = \int_0^{2\pi}\int_0^{R(z)}\tau_{\theta z}r^2\,dr\,d\theta = 2\pi\mu\int_0^{R(z)}\frac{\partial \Psi}{\partial r}\,dr \tag{9.7.8}$$

$$= 2\pi\mu[\Psi(R(z),z) - \Psi(0,z)]$$

where $R(z)$ is the variable radius of the section.

Example 9.7 Conical shaft

As an example of a variable section problem, consider the torsion of a conical shaft with cone angle φ, as shown in Fig. 9.18. We again have selected a problem whose boundary shape will help generate the solution. On the lateral sides of the conical boundary, $z/\sqrt{r^2 + z^2} = \cos\varphi$, which is a constant. Thus, any function of this ratio will satisfy the boundary condition (9.7.7). It can be shown that a linear combination of this ratio with its cube can be constructed to satisfy the governing equation (9.7.5), and the solution for the stress function is then given by

$$\Psi = C\left(\frac{z}{\sqrt{r^2 + z^2}} - \frac{1}{3}\frac{z^3}{(r^2 + z^2)^{3/2}}\right) \tag{9.7.9}$$

where the constant C has been determined to satisfy the load-carrying relation (9.7.8)

$$C = -\frac{T}{2\pi\mu\left(\frac{2}{3} - \cos\varphi + \frac{1}{3}\cos^3\varphi\right)} \tag{9.7.10}$$

The stresses follow from relations (9.7.4)

$$\tau_{r\theta} = -\frac{C\mu r^2}{(r^2 + z^2)^{5/2}}$$

$$\tau_{\theta z} = -\frac{C\mu rz}{(r^2 + z^2)^{5/2}} \tag{9.7.11}$$

and the displacement u_θ can be determined by integrating equations (9.7.2) to get

$$u_\theta = -\frac{Cr}{3(r^2 + z^2)^{3/2}} + \omega r \tag{9.7.12}$$

where ωr is the rigid-body rotation term about the z-axis and ω can be determined by specifying the shaft rotation at a specific z location. Additional examples of such problems are discussed in Timoshenko and Goodier (1970) and Sokolnikoff (1956).

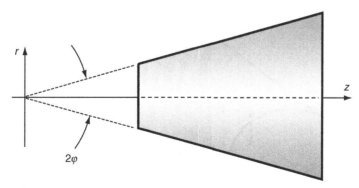

FIG. 9.18 Conical shaft geometry.

Before leaving the torsion problem, it should be mentioned that this problem can also be easily formulated and solved using the numerical finite element method. Chapter 16 discusses this important numerical scheme and provides a series of such solutions in Example 16.4 and Fig. 16.7. These examples illustrate the power and usefulness of this numerical method to solve problems with complicated geometry that could not be easily solved using analytical means.

9.8 Flexure formulation

We now investigate a final case of deformation of elastic cylinders under end loadings by considering the flexure of elastic beams subject to transverse end forces, as shown in Fig. 9.19. The problem geometry is formulated as a cantilever beam of arbitrary section with a fixed end at $z = 0$ and transverse end loadings P_x and P_y at $z = l$. Following our usual procedure, the problem is to be solved in the Saint-Venant sense, in that only the resultant end loadings P_x and P_y are used to formulate the boundary conditions at $z = l$.

From our general formulation in Section 9.1, $\sigma_x = \sigma_y = \tau_{xy} = 0$. The other three nonzero stresses will be determined to satisfy the equilibrium and compatibility relations and all associated boundary conditions. From our earlier work, the equilibrium and compatibility relations resulted in equations (9.1.2) from which it was argued that τ_{xz} and τ_{yz} were independent of z, and σ_z was a bilinear form in x,y,z (see Exercise 9.1). Motivated from strength of materials theory, we choose the arbitrary form for σ_z as follows

$$\sigma_z = (Bx + Cy)(l - z) \tag{9.8.1}$$

where B and C are constants.

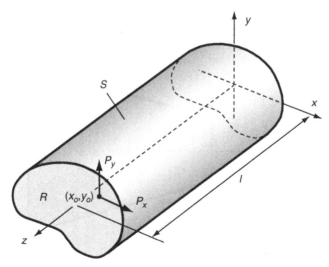

FIG. 9.19 Flexure problem geometry.

Using this result in the remaining equilibrium equation in the z direction gives

$$\frac{\partial \tau_{xz}}{\partial x} + \frac{\partial \tau_{yz}}{\partial y} - (Bx + Cy) = 0$$

which can be written in the form

$$\frac{\partial}{\partial x}\left[\tau_{xz} - \frac{1}{2}Bx^2\right] + \frac{\partial}{\partial y}\left[\tau_{yz} - \frac{1}{2}Cy^2\right] = 0 \qquad (9.8.2)$$

This equilibrium statement motivates the introduction of another stress function $F(x,y)$, such that

$$\tau_{xz} = \frac{\partial F}{\partial y} + \frac{1}{2}Bx^2$$

$$(9.8.3)$$

$$\tau_{yz} = -\frac{\partial F}{\partial x} + \frac{1}{2}Cy^2$$

This form then satisfies equilibrium identically, and using it in the remaining compatibility relations gives the results

$$\frac{\partial}{\partial y}\left(\nabla^2 F\right) + \frac{\nu B}{1 + \nu} = 0$$

$$(9.8.4)$$

$$-\frac{\partial}{\partial x}\left(\nabla^2 F\right) + \frac{\nu C}{1 + \nu} = 0$$

This system can be integrated to get

$$\nabla^2 F = \frac{\nu}{1 + \nu}(Cx - By) + k \qquad (9.8.5)$$

where k is a constant of integration. In order to determine this constant, consider the rotation about the z-axis. From the general relation (2.1.9), $\omega_z = [(\partial v/\partial x) - (\partial u/\partial y)]/2$, differentiating with respect to z and using Hooke's law and our previous results gives

$$\frac{\partial \omega_z}{\partial z} = \frac{1}{2}\left(\frac{\partial^2 v}{\partial x \partial z} - \frac{\partial^2 u}{\partial y \partial z}\right)$$

$$= \frac{1}{2\mu}\left(\frac{\partial \tau_{yz}}{\partial x} - \frac{\partial \tau_{xz}}{\partial y}\right) \qquad (9.8.6)$$

$$= -\frac{1}{2\mu}\nabla^2 F = -\frac{1}{2\mu}\left[\frac{\nu}{1 + \nu}(Cx - By) + k\right]$$

From the torsion formulation, the angle of twist per unit length was specified by the parameter α, and selecting the section origin ($x = y = 0$) at the center of twist, relation (9.8.6) then implies that $k = -2\mu\alpha$. Thus, the governing equation (9.8.5) can be written as

$$\nabla^2 F = \frac{\nu}{1 + \nu}(Cx - By) - 2\mu\alpha \qquad (9.8.7)$$

The zero loading boundary condition on the lateral surface S is expressed by

$$\tau_{xz} n_x + \tau_{yz} n_y = 0$$

and using the stress function definition, this can be written as

$$\frac{\partial F}{\partial x}\frac{dx}{dy} + \frac{\partial F}{\partial y}\frac{dy}{ds} + \frac{1}{2}\left(Bx^2\frac{dy}{ds} - Cy^2\frac{dx}{ds}\right) = 0 \quad \text{or} \quad \frac{dF}{ds} = -\frac{1}{2}\left(Bx^2\frac{dy}{ds} - Cy^2\frac{dx}{ds}\right) \tag{9.8.8}$$

It is convenient to separate the stress function F into a *torsional part* ϕ and a *flexural part* ψ, such that

$$F(x,y) = \phi(x,y) + \psi(x,y) \tag{9.8.9}$$

where the torsional part is formulated by

$$\nabla^2 \phi = -2\mu\alpha \text{ in } R$$

$$\frac{d\phi}{ds} = 0 \text{ on } S \tag{9.8.10}$$

while the flexural portion satisfies

$$\nabla^2 \psi = \frac{\nu}{1+\nu}(Cx - By) \text{ in } R$$

$$\frac{d\psi}{dx} = -\frac{1}{2}\left(Bx^2\frac{dy}{ds} - Cy^2\frac{dx}{ds}\right) \text{ on } S \tag{9.8.11}$$

Because we have already investigated the torsional part of this problem in the preceding sections, we now pursue only the flexural portion. The general solution to $(9.8.11)_1$ may be expressed as the sum of a particular solution plus a harmonic function

$$\psi(x,y) = f(x,y) + \frac{1}{6}\frac{\nu}{1+\nu}(Cx^3 - By^3) \tag{9.8.12}$$

where f is a harmonic function satisfying $\nabla^2 f = 0$. The boundary conditions on end $z = l$ can be stated as

$$\iint_R \tau_{xz}dxdy = P_x$$

$$\iint_R \tau_{yz}dxdy = P_y \tag{9.8.13}$$

$$\iint_R [x\tau_{yz} - y\tau_{xz}]dxdy = x_o P_y - y_o P_x$$

Using the first relation of this set gives

$$\iint_R \left[\frac{\partial}{\partial y}(\phi + \psi) + \frac{1}{2}Bx^2\right]dxdy = P_x \tag{9.8.14}$$

but from the torsion formulation $\iint_R \dfrac{\partial \phi}{\partial y} \, dxdy = 0$, and so (9.8.14) can be written as

$$\iint_R \left[\frac{\partial}{\partial x} \left(x \frac{\partial \psi}{\partial y} \right) - \frac{\partial}{\partial y} \left(x \frac{\partial \psi}{\partial x} \right) \right] dxdy + \iint_R \frac{1}{2} Bx^2 \, dxdy = P_x \tag{9.8.15}$$

Using Green's theorem and the boundary relation $(9.8.11)_2$, the first integral can be expressed as

$$\iint_R \left[\frac{\partial}{\partial x} \left(x \frac{\partial \psi}{\partial y} \right) - \frac{\partial}{\partial y} \left(x \frac{\partial \psi}{\partial x} \right) \right] dxdy = - \iint_R \left[\frac{3}{2} Bx^2 + Cxy \right] dxdy$$

and thus equation (9.8.15) reduces to

$$BI_y + CI_{xy} = -P_x \tag{9.8.16}$$

In a similar manner, boundary condition $(9.8.13)_2$ gives

$$BI_{xy} + CI_x = -P_y \tag{9.8.17}$$

The expressions I_x, I_y, and I_{xy} are the area moments of inertia of section R

$$I_x = \iint_R y^2 \, dxdy, \quad I_y = - \iint_R x^2 \, dxdy, \quad I_{xy} = \iint_R xy \, dxdy \tag{9.8.18}$$

Relations (9.8.16) and (9.8.17) can be solved for the constants B and C

$$B = -\frac{P_x I_x - P_y I_{xy}}{I_x I_y - I_{xy}^2}$$

$$C = -\frac{P_y I_y - P_x I_{xy}}{I_x I_y - I_{xy}^2} \tag{9.8.19}$$

The final boundary condition $(9.8.13)_3$ can be expressed as

$$-\iint_R \left[x \frac{\partial \phi}{\partial x} + y \frac{\partial \phi}{\partial y} \right] dxdy - \iint_R \left[x \frac{\partial \psi}{\partial x} + y \frac{\partial \psi}{\partial y} \right] dxdy$$

$$+ \iint_R \frac{1}{2} (Cxy^2 - Bx^2 y) dxdy = x_o P_y - y_o P_x \tag{9.8.20}$$

From the torsion formulation

$$-\iint_R \left[x \frac{\partial \phi}{\partial x} + y \frac{\partial \phi}{\partial y} \right] dxdy = 2 \iint_R \phi \, dxdy = T = \alpha J$$

so (9.8.20) becomes

$$\alpha J + \iint_R \left(\frac{1}{2} (Cxy^2 - Bx^2 y) - \left(x \frac{\partial \psi}{\partial x} + y \frac{\partial \psi}{\partial y} \right) \right) dxdy = x_o P_y - y_o P \tag{9.8.21}$$

Once the flexural stress function ψ is known, (9.8.21) will provide a relation to determine the angle of twist α.

Relation (9.8.21) can also be used to determine the location (x_o, y_o) for no induced torsional rotation, a point commonly called the *shear center* or *center of flexure*. Choosing $\alpha = 0$, this equation can be independently used for the two cases of $(P_x = 0, P_y \neq 0)$ and $(P_x \neq 0, P_y = 0)$ to generate two equations for locations x_o and y_o. If the x-axis is an axis of symmetry, then $y_o = 0$; likewise, if the y-axis is one of symmetry, then $x_o = 0$. For a section with two perpendicular axes of symmetry, the location (x_o, y_o) lies at the intersection of these two axes, which is at the *centroid* of the section. However, in general the shear center does not coincide with the section's centroid and need not even lie within the section.

9.9 Flexure problems without twist

Because we have previously studied examples of the torsion problem, we shall now develop flexure solutions to problems that do not include twist. The two examples to be investigated include simple symmetric cross-sections with single end loadings along an axis of symmetry.

Example 9.8 Circular section

Consider the flexure of an elastic beam of circular section, as shown in Fig. 9.20. The end loading $(P_x = 0, P_y = P)$ passes through the center of the section, which coincides with the centroid and center of twist. Thus, for this problem there will be no torsion ($\alpha = 0$), and so $\phi = 0$ and $F = \psi$. It is convenient to use polar coordinates for this problem, and the governing equation $(9.8.11)_1$ can then be written as

$$\nabla^2 \psi = -\frac{\nu}{1+\nu}\frac{P}{I_x} r \cos \theta \tag{9.9.1}$$

while the boundary condition $(9.8.11)_2$ becomes

$$\frac{1}{a}\frac{\partial \psi}{\partial \theta} = \frac{1}{2}\frac{P}{I_x}a^2\sin^3\theta \text{ on } r = a \tag{9.9.2}$$

The solution to (9.9.1) can then be taken as

$$\psi = \frac{P}{I_x}\left[f - \frac{1}{6}\frac{\nu}{1+\nu}r^3\cos^3\theta\right] \tag{9.9.3}$$

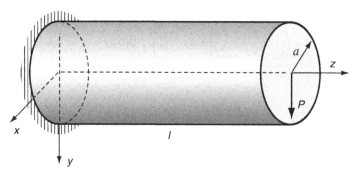

FIG. 9.20 **Flexure of a beam of circular section.**

Using trigonometric identities, relations (9.9.3) and (9.9.2) can be rewritten as

$$\psi = \frac{P}{I_x}\left[f - \frac{1}{24}\frac{v}{1+v}r^3(\cos 3\theta + 3\cos\theta)\right]$$

$$\frac{\partial \psi}{\partial \theta} = \frac{1}{8}\frac{P}{I_x}a^3(-\sin 3\theta + 3\sin\theta) \text{ on } r=a$$

(9.9.4)

Based on the previous relations, we look for solutions for the harmonic function in the form $f = \sum_n A_n r^n \cos n\theta$ and consider the two terms

$$f = A_1 r \cos\theta + A_3 r^3 \cos 3\theta$$

(9.9.5)

Combining (9.9.4) and (9.9.5) yields

$$\psi = \frac{P}{I_x}\left[\left[A_1 r - \frac{vr^3}{8(1+v)}\right]\cos\theta + \left[A_3 - \frac{v}{24(1+v)}\right]r^3\cos 3\theta\right]$$

(9.9.6)

Boundary condition (9.9.4)$_2$ yields two relations to determine the constants A_1 and A_3

$$A_1 = -\frac{3+2v}{8(1+v)}a^2$$

$$A_3 = \frac{1+2v}{24(1+v)}$$

(9.9.7)

and back-substituting this result into (9.9.6) gives the final form of the stress function

$$\psi = \frac{P}{I_x}\left[-\frac{3+2v}{8(1+v)}a^2x - \frac{1+2v}{8(1+v)}xy^2 + \frac{1-2v}{24(1+v)}x^3\right]$$

(9.9.8)

The stresses corresponding to this solution become

$$\tau_{xz} = -\frac{P}{4I_x}\frac{1+2v}{1+v}xy$$

$$\tau_{yz} = \frac{P}{I_x}\frac{3+2v}{8(1+v)}\left[a^2 - y^2 - \frac{1-2v}{3+2v}x^2\right]$$

$$\sigma_z = -\frac{P}{I_x}y(l-z)$$

(9.9.9)

Note for this section $I_x = \pi a^4/4$. The maximum stress occurs at the origin and is given by

$$\tau_{\max} = \tau_{yz}(0,0) = \frac{P}{\pi a^2}\frac{3+2v}{2(1+v)}$$

(9.9.10)

This can be compared to the value developed from strength of materials theory $\tau_{\max} = 4P/3\pi a^2$. Differences in the maximum shear stress between the two theories are small, and for the special case $v = 1/2$, the elasticity solution is the same as the elementary result. Comparison of the shear stress distribution with strength of materials theory for $v = 0.1$ has been given by Sadd (1979), and again differences were found to be small. Displacements for this problem can be determined through the usual integration process (see Exercise 9.32).

Example 9.9 Rectangular section

Our second flexure example involves a beam of rectangular section with end loading ($P_x = 0, P_y = P$) passing through the shear center, as shown in Fig. 9.21. The section dimensions are the same as those given in Fig. 9.12. As in the previous example, there is no torsion ($\alpha = 0$), and so for this case $\phi = 0$ and $F = \psi$. Formulation equations (9.8.11) then give

$$\nabla^2 \psi = -\frac{\nu}{1+\nu} \frac{P}{I_x} x \text{ in } R$$

$$\frac{d\psi}{ds} = -\frac{1}{2} \frac{P}{I_x} y^2 \frac{dx}{ds} \text{ on } S$$

(9.9.11)

For the rectangular section

$$\frac{d\psi}{ds} = \begin{cases} \pm\dfrac{d\psi}{dy} = 0, & x = \pm a \\[2mm] \mp\dfrac{d\psi}{dx} = \pm\dfrac{1}{2}\dfrac{P}{I_x}b^2, & y = \pm b \end{cases}$$

(9.9.12)

Based on these boundary relations we are motivated to select a solution of the form

$$\psi = \frac{P}{I_x}\left[f - \frac{1}{6}\frac{\nu}{1+\nu}(x^3 - a^2 x) - \frac{b^2 x}{2} \right]$$

(9.9.13)

with the harmonic function f satisfying

$$f(x,y) = \begin{cases} 0, & x = \pm a \\[2mm] \dfrac{\nu}{6(1+\nu)}(x^3 - a^2 x), & y = \pm b \end{cases}$$

(9.9.14)

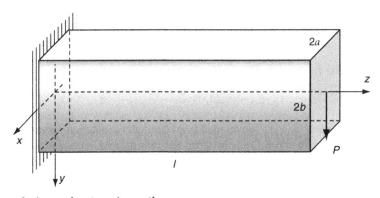

FIG. 9.21 Flexure of a beam of rectangular section.

Because we expect τ_{yz} to be an even function in x and y, and τ_{xz} to be odd in y, we look for a harmonic solution for f in the form

$$f(x,y) = \sum_{n=1}^{\infty} A_n \sin\frac{n\pi x}{a} \cosh\frac{n\pi y}{a} \tag{9.9.15}$$

This form satisfies $(9.9.14)_1$ identically, while $(9.9.14)_2$ implies that

$$\sum_{n=1}^{\infty} b_n \sin\frac{n\pi x}{a} = \frac{\nu}{6(1+\nu)}\left(x^3 - a^2 x\right) \tag{9.9.16}$$

where $b_n = A_n \cosh(n\pi b/a)$. Relation (9.9.16) is recognized as a Fourier sine series, and thus the coefficients follow from standard theory (8.2.28) and are given by $b_n = 2\nu a^3(-1)^n/(1+\nu)n^3\pi^3$. Putting these results back together gives the final form of the stress function

$$\psi = \frac{P}{I_x}\left[-\frac{1}{6}\frac{\nu}{1+\nu}\left(x^3 - a^2 x\right) - \frac{b^2 x}{2} + \frac{2\nu a^3}{(1+\nu)\pi^3}\sum_{n=1}^{\infty}\frac{(-1)^n}{n^3}\frac{\sin\dfrac{n\pi x}{a}\cosh\dfrac{n\pi y}{a}}{\cosh\dfrac{n\pi b}{a}}\right] \tag{9.9.17}$$

The stresses then follow to be

$$\tau_{xz} = \frac{2\nu a^2 P}{(1+\nu)\pi^2 I_x}\sum_{n=1}^{\infty}\frac{(-1)^n}{n^2}\frac{\sin\dfrac{n\pi x}{a}\sinh\dfrac{n\pi y}{a}}{\cosh\dfrac{n\pi b}{a}}$$

$$\tau_{yz} = \frac{P}{2I_x}\left(b^2 - y^2\right) + \frac{\nu P}{6(1+\nu)I_x}\left[3x^2 - a^2 - \frac{12a^2}{\pi^2}\sum_{n=1}^{\infty}\frac{(-1)^n}{n^2}\frac{\cos\dfrac{n\pi x}{a}\cosh\dfrac{n\pi y}{a}}{\cosh\dfrac{n\pi b}{a}}\right] \tag{9.9.18}$$

$$\sigma_z = -\frac{P}{I_x}y(l - z)$$

The corresponding results from strength of materials gives $\tau_{yz} = P(b^2 - y^2)/2I_x$, and thus the second term of $(9.9.18)_2$ represents the correction to the elementary theory. Note that if $\nu = 0$, this correction term vanishes, and the two theories predict identical stresses. For the case of a thin rectangular section with $b \gg a$, $\cosh(n\pi b/a) \to \infty$, and it can be shown that the elasticity solution reduces to the strength of materials prediction. A similar result is also found for the case of a thin section with $a \gg b$. Comparison of the shear stress distribution τ_{yz} with strength of materials theory for $\nu = 1/2$ has been presented by Sadd (1979), and differences between the two theories were found to be sizeable. As in the previous example, the maximum stress occurs at $x = y = 0$

$$\tau_{\max} = \tau_{yz}(0,0)\frac{P}{2I_x}b^2 - \frac{\nu P a^2}{6(1+\nu)I_x}\left[1 + \frac{12}{\pi^2}\sum_{n=1}^{\infty}\frac{(-1)^n}{n^2}\operatorname{sech}\frac{n\pi b}{a}\right] \tag{9.9.19}$$

Again, the strength of materials result is given by the first term in relation (9.9.19).

This concludes our brief presentation of flexure examples. Solutions to additional flexure problems are given by Sokolnikoff (1956) and Timoshenko and Goodier (1970).

References

Higgins TJ: A comprehensive review of Saint-Venant torsion problem, *Am J Phys* 10:248–259, 1942.
Higgins TJ: The approximate mathematical methods of applied physics as exemplified by application to Saint-Venant torsion problem, *J Appl Phys* 14:469–480, 1943.
Higgins TJ: Analogic experimental methods in stress analysis as exemplified by Saint-Venant's torsion problem, *Proc Soc Exp Stress Analysis* 2:17–27, 1945.
Kellogg OD: *Foundations of potential theory*, New York, 1969, Dover.
Sadd MH: A comparison of mechanics of materials and theory of elasticity stress analysis, *Mech Eng News ASEE* 16:34–39, 1979.
Sokolnikoff IS: *Mathematical theory of elasticity*, New York, 1956, McGraw-Hill.
Timoshenko SP, Goodier JN: *Theory of elasticity*, New York, 1970, McGraw-Hill.

Exercises

9.1 Under the assumption that $\sigma_x = \sigma_y = \tau_{xy} = 0$, show that equilibrium and compatibility equations with zero body forces reduce to relations (9.1.2). Next integrate relations

$$\frac{\partial^2 \sigma_z}{\partial x^2} = \frac{\partial^2 \sigma_z}{\partial y^2} = \frac{\partial^2 \sigma_z}{\partial z^2} = \frac{\partial^2 \sigma_z}{\partial x \partial y} = 0$$

to justify that $\sigma_z = C_1 x + C_2 y + C_3 z + C_4 xz + C_5 yz + C_6$, where C_i are arbitrary constants.

9.2 During early development of the torsion formulation, Navier attempted to extend Coulomb's theory for bars of circular section and to assume that there is no warping displacement for general cross-sections. Show that although such an assumed displacement field will satisfy all elasticity field equations, it will not satisfy the boundary conditions and thus is not an acceptable solution.

9.3 Referring to Fig. 9.2, if we choose a different reference origin that is located at point (a,b) with respect to the given axes, the displacement field would now be given by

$$u = -\alpha z(y - b), \quad v = \alpha z(x - a), \quad w = w(x, y)$$

where x and y now represent the new coordinates. Show that this new representation leads to an identical torsion formulation as originally developed.

9.4 In terms of a *conjugate function* $\psi(x,y)$ defined by

$$\frac{\partial \psi}{\partial x} = -\frac{1}{\alpha} \frac{\partial w}{\partial y}, \quad \frac{\partial \psi}{\partial y} = \frac{1}{\alpha} \frac{\partial w}{\partial x}$$

show that the torsion problem may be formulated as

$$\nabla^2 \psi = 0 \text{ in } R$$

$$\psi = \frac{1}{2}\left(x^2 + y^2\right) + \text{constant on } S$$

9.5 A function $f(x,y)$ is defined as *subharmonic* in a region R if $\nabla^2 f \geq 0$ at all points in R. It can be proved that the maximum value of a subharmonic function occurs only on the boundary S of region R. For the torsion problem, show that the square of the resultant shear stress $\tau^2 = \tau_{xz}^2 + \tau_{yz}^2$ is a subharmonic function, and thus the maximum shear stress will always occur on the section boundary.

9.6 We wish to reformulate the torsion problem using cylindrical coordinates. First show that the general form of the displacements can be expressed as $u_r = 0$, $u_\theta = \alpha rz$, $u_z = u_z(r,\theta)$. Next show that this leads to the following strain and stress fields

$$e_r = e_\theta = e_z = e_{r\theta} = 0, \quad e_{rz} = \frac{1}{2}\frac{\partial u_z}{\partial r}, \quad e_{\theta z} = \frac{1}{2}\left(\alpha r + \frac{1}{r}\frac{\partial u_z}{\partial \theta}\right)$$

$$\sigma_r = \sigma_\theta = \sigma_z = \tau_{r\theta} = 0, \quad \tau_{rz} = \mu\frac{\partial u_z}{\partial r}, \quad \tau_{\theta z} = \mu\left(\alpha r + \frac{1}{r}\frac{\partial u_z}{\partial \theta}\right)$$

Verify that with no body forces the equilibrium equations reduce to $\partial \tau_{rz}/\partial r + (1/r)\partial \tau_{\theta z}/\partial \theta + \tau_{rz}/r = 0$. Following the same scheme as used in Section 9.3.1, develop the compatibility relation

$$\frac{1}{r}\frac{\partial \tau_{rz}}{\partial \theta} - \frac{\partial \tau_{\theta z}}{\partial r} - \frac{\tau_{\theta z}}{r} = -2\mu\alpha$$

Finally, choosing a stress–stress function relation of the form $\tau_{rz} = (1/r)\partial\phi/\partial\theta$, $\tau_{\theta z} = -\partial\phi/\partial r$, show that the equilibrium is identically satisfied and the compatibility relation gives the expected result

$$\frac{\partial^2 \phi}{\partial r^2} + \frac{1}{r}\frac{\partial \phi}{\partial r} + \frac{1}{r^2}\frac{\partial^2 \phi}{\partial \theta^2} = -2\mu\alpha$$

9.7 Using polar coordinates and the basic results of Exercise 9.6, formulate the torsion of a cylinder of circular section with radius a, in terms of the usual Prandtl stress function. Note for this case, there will be no warping displacement and $\phi = \phi(r)$. Show that the stress function is given by

$$\phi = -\frac{\mu\alpha}{2}\left(r^2 - a^2\right)$$

and the only non-zero stress simplifies to $\tau_{\theta z} = \mu\alpha r$. Check these results with the solution given for the elliptical section for the case with $a = b$.

9.8 Employing the membrane analogy, develop an approximate solution to the torsion problem of a thin rectangular section as shown. Neglecting end effects at $y = \pm b$, the membrane deflection will then depend only on x, and the governing equation can be integrated to give $z = \phi = \mu\alpha(a^2 - x^2)$, thus verifying that the membrane shape is parabolic. Formally compute the maximum membrane slope and volume enclosed to justify relations (9.5.15).

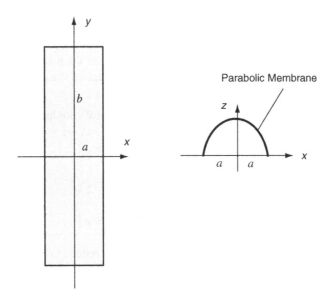

Parabolic Membrane

9.9 Using the stress results for the torsion of the elliptical section, formally integrate the strain–displacement relations and develop the displacement solution (9.4.11).

9.10 For the torsion of an elliptical section, show that the resultant shear stress at any point within the cross-section is tangent to an ellipse that passes through the point and has the same ratio of major to minor axes as that of the boundary ellipse.

9.11 Develop relation (9.4.14) for the load-carrying torque of an equilateral triangular section.

9.12 For the torsion of a bar of elliptical section, express the torque equation (9.4.6) in terms of the polar moment of inertia of the section, and compare this result with the corresponding relation for the equilateral triangular section.

9.13* For the triangular section shown in Fig. 9.9, calculate the resultant shear stress along the line $y = 0$, and plot the result over the range $-2a \leq x \leq a$. Determine and label all maximum and minimum values.

9.14* For the triangular section of Example 9.2, plot contours of constant resultant shear stress

$$\tau = \sqrt{\tau_{xz}^2 + \tau_{yz}^2}$$

Point out how these contours would imply that the maximum shear stress occurs at the midpoints of each boundary side.

9.15 Consider the torsion of a bar of general triangular section as shown in the following figure. Using the boundary equation technique of Section 9.4, attempt a stress function solution of the form

$$\phi = K(x - a)(y - m_1 x)(y + m_2 x)$$

where m_1, m_2, and a are geometric constants defined in the figure and K is a constant to be determined. Show that this form will be a solution to the torsion problem only for the case of an equilateral triangular section.

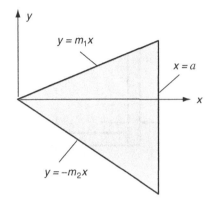

9.16 For the torsion problem in Example 9.3, explicitly justify that the required values for the constants appearing in the stress function are given by $c = 3 - \sqrt{8}$ and $K = -\mu\alpha/[4a^2(1 - \sqrt{2})]$. Also calculate the resulting shear stresses and determine the location and value of the maximum stress.

9.17 Attempt to solve the torsion of a rectangular section shown in Fig. 9.12 by using the boundary equation method. Show that trying a stress function created from the four products of the boundary lines $x = \pm a$ and $y = \pm b$ will not satisfy the governing equation (9.3.9).

9.18 Using the displacement formulation given in Section 9.3.2, use standard separation of variables to solve the torsion problem of an elliptical section shown in Fig. 9.7. For this problem it is only necessary to use results coming from the zero separation constant case. Verify your solution with relation (9.4.11).

9.19 Using the displacement formulation given in Section 9.3.2, use standard separation of variables to solve the torsion problem of a rectangular section shown in Fig. 9.12. Verify your solution with relation (9.5.14).

9.20* Using the torque relation (9.5.12) for the rectangular section, compute the nondimensional load-carrying parameter $T/\mu\alpha b^4$, and plot this as a function of the dimensionless ratio b/a over the range $1 \le b/a \le 10$. For the case where b/a approaches 10, show that the load-carrying behavior can be given by the approximate relation (9.5.15).

9.21 Using the relation (9.5.16), develop an approximate solution for the load-carrying torque of the channel section shown.

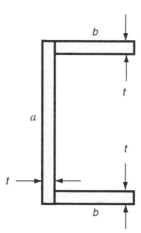

9.22 For the given narrow triangular section, show that the approximate torsional load carrying capacity is given by $T = \frac{2}{3}\mu\alpha c_o^3 h$. Use the parabolic membrane scheme outlined in Exercise 9.8 for any level y in the section.

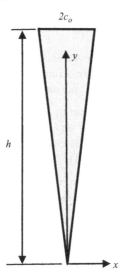

9.23 A circular shaft with a keyway can be approximated by the section shown in the following figure. The keyway is represented by the boundary equation $r = b$, while the shaft has the boundary relation $r = 2a \cos \theta$. Using the technique of Section 9.4, a trial stress function is suggested of the form

$$\phi = K(b^2 - r^2)\left(1 - \frac{2a\cos\theta}{r}\right)$$

where K is a constant to be determined. Show that this form will solve the problem and determine the constant K. Compute the two shear stress components τ_{xz} and τ_{yz}.

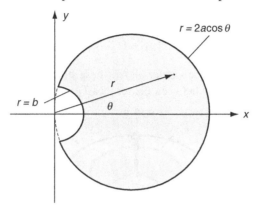

9.24* For the keyway section of Exercise 9.23, show that resultant stresses on the shaft and keyway boundaries are given by

$$\tau_{shaft} = \mu\alpha a\left(\frac{b^2}{4a^2\cos^2\theta} - 1\right), \quad \tau_{keyway} = \mu\alpha(2a\cos\theta - b)$$

Determine the maximum values of these stresses, and show that for $b \ll a$, the magnitude of the maximum keyway stress is approximately twice that of the shaft stress. Finally, make a plot of the stress concentration factor

$$\frac{(\tau_{max})_{keyway}}{(\tau_{max})_{solid\ shaft}}$$

versus the ratio b/a over the range $0 \le b/a \le 1$. Note that $(\tau_{max})_{solid\ shaft}$ is the maximum shear stress for solid shaft of circular section and can be determined from Example 9.1 or strength of materials theory. Show that the stress concentration plot gives

$$\frac{(\tau_{max})_{keyway}}{(\tau_{max})_{solid\ shaft}} \to 2 \quad \text{as } b/a \to 0$$

thus indicating that a small notch will result in a doubling of the stress in a circular section under torsion.

9.25 For the hollow elliptical section in Example 9.5, explicitly show that the given stress function (9.6.1) will yield the constant value specified by (9.6.2) on the inner boundary. Next using the methods from Example 9.1, show that the general torque relation (9.3.31) will also produce the result (9.6.3). Finally verify the maximum shear stress relation (9.6.4).

9.26 The illustrated uniform composite circular rod is composed of a solid inner core with shear modulus μ_1 and a different outer core material of modulus μ_2. The two materials are assumed to be perfectly bonded at interface $r = r_1$, and thus the interfacial tangential displacements must be continuous $u_{\theta_1}(r_1) = u_{\theta_2}(r_1)$. The composite shaft is to be loaded with an overall torque T. First justify that the angle of twists in each material must be the same and that torques in each material must satisfy $T = T_1 + T_2$. Next develop the torque-angle of twist relation

$$T = \frac{\pi}{2}\left[r_1^4\mu_1 + \left(r_2^4 - r_1^4\right)\mu_2\right]\alpha$$

and determine the stress $\tau_{\theta z}$ in each material. Are these stresses continuous across the interface? Finally establish a relationship between the torques T_1 and T_2 carried in each material.

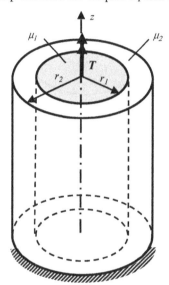

9.27 Example 9.6 provides the torsion solution of a closed thin-walled section shown in Fig. 9.16. Investigate the solution of the identical section for the case where a small cut has been introduced as shown in the following figure. This cut creates an open tube and produces significant changes to the stress function (use membrane analogy), stress field, and load-carrying capacity. The open tube solution can be approximately determined using results (9.5.15) from the thin rectangular solution. For the open tube, develop an equivalent relation as given by (9.6.8) for the closed tube. For identical applied torque, compare the stresses for each case, and justify that the closed tube has much lower stress and is thus much stronger.

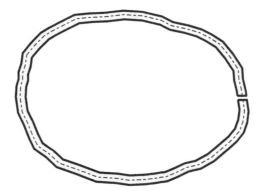

9.28 Consider the torsion of a rod with the half-ring cross-section as shown. Formulating the problem in polar coordinates (see Exercise 9.6), the governing stress function equation becomes

$$\frac{\partial^2 \phi}{\partial r^2} + \frac{1}{r}\frac{\partial \phi}{\partial r} + \frac{1}{r^2}\frac{\partial^2 \phi}{\partial \theta^2} = -2\mu\alpha$$

Using Fourier methods to solve this problem, first show that we can expand the constant right-hand side in a Fourier since series to get $-2\mu\alpha = -\sum_{n=0}^{m} 8\mu\alpha/[(2n+1)\pi]\sin(2n+1)\theta$. Based on this, use the series solution from $\phi(r,\theta) = \sum_{n=0}^{m} F_n(r)\sin(2n+1)\theta$ and show that the governing equation leads to the ordinary differential equation

$$\frac{d^2 F_n}{dr^2} + \frac{1}{r}\frac{dF_n}{dr} - \frac{(2n+1)^2}{r^2}F_n = -\frac{8\mu\alpha}{(2n+1)\pi}$$

Show that the solution to this equation to give by

$$F_n = A_n r^{2n+1} + B_n r^{-2n-1} + K_n r^2, \quad \text{where } K_n = \frac{8\mu\alpha}{(2n+1)(2n-1)(2n+3)\pi}$$

Note that the stress function form already satisfies the zero boundary condition on $\theta = 0$ and π. Finally apply the remaining boundary conditions on r_1 and r_2, to determine the constants A_n and B_n and show that the stresses can be written as

$$\tau_{rz} = \frac{1}{r}\frac{\partial \phi}{\partial \theta} = \sum_{n=0}^{\infty} \left[A_n r^{2n} + B_n r^{-2n-2} + K_n r\right](2n+1)\cos(2n+1)\theta$$

$$\tau_{\theta z} = -\frac{\partial \phi}{\partial r} = -\sum_{n=0}^{\infty} \left[A_n(2n+1)r^{2n} - B_n(2n+1)r^{-2n-2} + 2K_n r\right]\sin(2n+1)\theta$$

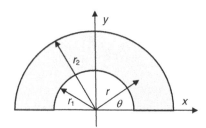

9.29 The potential energy per unit length for the torsion problem was given in Exercise 6.16. Using the principle of minimum potential energy, $\delta\Pi = 0$ and this leads to a minimization of the following integral expression

$$I = \iint_R \left[\left(\frac{\partial \phi}{\partial x} \right)^2 + \left(\frac{\partial \phi}{\partial y} \right)^2 - 4\mu\alpha\phi \right] dxdy$$

From variational calculates this is equivalent to satisfying the governing differential equation $\nabla^2 \phi = -2\mu\alpha$. Using this result we can apply a Rayleigh–Ritz scheme (see Section 6.7) and choose an approximate set of terms for the stress function $\phi = a_1\phi_1 + a_2\phi_2 + \ldots$ and minimize the potential energy by using relations $\partial I/\partial a_i = 0$. Note that these trial ϕ_i functions must satisfy the boundary conditions $\phi_i = 0$. Using this proposed scheme, develop an approximate solution for the torsion of the "L shaped" section, by using only a single term of the form $\phi_1 = \sin 4\pi x \sin 4\pi y$. Show that the Ritz coefficient is given by $a_1 = -\mu\alpha/7\pi^4$, leading to a load carrying torque capacity of $T = \mu\alpha/14\pi^6$. Using just a single term will not yield a very accurate approximate solution, and incorporating additional terms will provide better results.

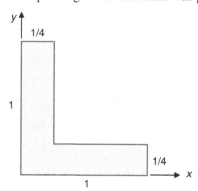

9.30 For the solution of the conical shaft given in Example 9.7, compare the maximum shearing stress $\tau_{\theta z}$ with the corresponding result from strength of materials theory. Specifically, consider the case with a cone angle $\varphi = 20°$ with $z = l$, and compare dimensionless values of $\tau_{\theta z}l^3/T$.

9.31* Make a comparison study of the torsion of a conical shaft given in Example 9.7 with corresponding results from mechanics of materials. First develop the $\tau_{\theta z}$ shear stress relations for each theory, and then make a comparison plot of the maximum section shear stress (normalized by the torque, T) for the case with cone angle $\varphi = 30°$ over the range $4 \le z \le 10$.

9.32 Determine the displacement field for the flexure problem of a beam of circular section given in Example 9.8. Starting with the stress solution (9.9.9), integrate the strain–displacement relations and use boundary conditions that require the displacements and rotations to vanish at $z = 0$. Compare the elasticity results with strength of materials theory. Also investigate whether the elasticity displacements indicate that plane sections remain plane.

9.33* Make a comparison of theory of elasticity and strength of materials shear stresses for the flexure of a beam of rectangular section from Example 9.9. For each theory, calculate and plot the dimensionless shear stress $\tau_{yz}(0, y)b^2/P$ versus y/b for an aspect ratio $b/a = 1$.

9.34 Solve the flexure problem without twist of an elastic beam of elliptical section as shown in Fig. 9.7 with $P_y = P$. Show that the stress results reduce to (9.9.9) for the circular case with $a = b$.

Advanced applications

Complex variable methods

Complex variable theory provides a very powerful tool for the solution of many problems in elasticity. Such applications include solutions of the torsion problem and most importantly the plane problem discussed in Chapters 7 and 8. The technique is also useful for cases involving anisotropic and thermoelastic materials, and these are discussed in subsequent chapters. Employing complex variable methods enables many problems to be solved that would be intractable by other schemes. The method is based on reduction of the elasticity boundary-value problem to a formulation in the complex domain. This formulation then allows many powerful mathematical techniques available from complex variable theory to be applied to the elasticity problem. Such applications were originally formulated by Filon (1903), Kolosov (1909), and additional Russian researchers further expanded the use of this technique. Comprehensive texts on this solution method include Muskhelishvili (1953, 1963), Milne-Thomson (1960), Green and Zerna (1968), and England (1971). Additional briefer sources of information can also be found in Sokolnikoff (1956) and Little (1973). The purpose of this chapter is to introduce the basics of the method and to investigate its application to particular problems of engineering interest. We shall first briefly review complex variable theory to provide a general background needed to develop elasticity solutions. Further and more detailed information on complex variables may be found in the mathematical texts by Churchill (1960) or Kreyszig (2011).

10.1 Review of complex variable theory

A *complex variable* z is defined by two real variables x and y in the form

$$z = x + iy \qquad (10.1.1)$$

where $i = \sqrt{-1}$ is called the *imaginary unit*, x is known as the *real part* of z, that is, $x = \mathrm{Re}(z)$, while y is called the *imaginary part*, $y = \mathrm{Im}(z)$. This definition can also be expressed in *polar form* by

$$z = r(\cos\theta + i\sin\theta) = re^{i\theta} \qquad (10.1.2)$$

where $r = \sqrt{x^2 + y^2}$ is known as the *modulus* of z and $\theta = \tan^{-1}(y/x)$ is the *argument*. These definitions may be visualized in a plot of the *complex plane*, as shown in Fig. 10.1, where the variable z may be thought of as a point in the plane, and definitions of r and θ have obvious graphical meaning. Because a complex variable includes two quantities (real and imaginary parts), it can be used in a similar fashion as a two-dimensional vector with x and y components. This type of representation is used several times in our plane elasticity applications. The *complex conjugate* $\bar{z}$ of the variable z is defined by

$$\bar{z} = x - iy = re^{-i\theta} \qquad (10.1.3)$$

Elasticity. https://doi.org/10.1016/B978-0-12-815987-3.00010-4

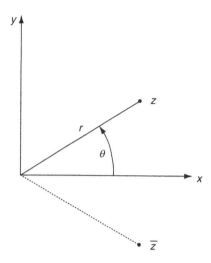

FIG. 10.1 Complex plane.

It should be apparent that this quantity is simply a result of changing the sign of the imaginary part of z, and in the complex plane (see Fig. 10.1) is a reflection of z about the real axis. Note that $r = \sqrt{z\bar{z}}$.

Using the definitions (10.1.1) and (10.1.3), the following differential operators can be developed

$$\frac{\partial}{\partial x} = \frac{\partial}{\partial z} + \frac{\partial}{\partial \bar{z}}, \quad \frac{\partial}{\partial y} = i\left(\frac{\partial}{\partial z} - \frac{\partial}{\partial \bar{z}}\right)$$

$$\frac{\partial}{\partial z} = \frac{1}{2}\left(\frac{\partial}{\partial x} - i\frac{\partial}{\partial y}\right), \quad \frac{\partial}{\partial \bar{z}} = \frac{1}{2}\left(\frac{\partial}{\partial x} + i\frac{\partial}{\partial y}\right)$$

(10.1.4)

Addition, subtraction, multiplication, and division of complex numbers z_1 and z_2 are defined by

$$z_1 + z_2 = (x_1 + x_2) + i(y_1 + y_2)$$

$$z_1 - z_2 = (x_1 - x_2) + i(y_1 - y_2)$$

$$z_1 z_2 = (x_1 x_2 - y_1 y_2) + i(y_1 x_2 + x_1 y_2)$$

$$\frac{z_1}{z_2} = \frac{x_1 + iy_1}{x_2 + iy_2} = \frac{x_1 x_2 + y_1 y_2}{x_2^2 + y_2^2} + i\frac{y_1 x_2 - x_1 y_2}{x_2^2 + y_2^2}$$

(10.1.5)

A *function of a complex variable* z may be written as

$$f(z) = f(x + iy) = u(x,y) + iv(x,y)$$

(10.1.6)

where $u(x,y)$ and $v(x,y)$ are the real and imaginary parts of the complex function $f(z)$. An example of this definition is given by

$$f(z) = az + bz^2 = a(x + iy) + b(x + iy)^2 = (ax + bx^2 - by^2) + i(ay + 2bxy)$$

thus $u(x,y) = ax + bx^2 - by^2$ and $v(x,y) = ay + 2bxy$, where we have assumed that a and b are real constants.

The complex conjugate of the complex function is defined by

$$\overline{f(z)} = \bar{f}(\bar{z}) = u(x,y) - iv(x,y) \tag{10.1.7}$$

and thus for the previous example of $f(z) = az + bz^2$

$$\overline{f(z)} = \left(\overline{az + bz^2} \right) = a\bar{z} + b\bar{z}^2$$

$$= a(x - iy) + b(x - iy)^2$$

$$= \left(ax + bx^2 - by^2 \right) - i(ay + 2bxy)$$

Differentiation of functions of a complex variable follows the usual definitions. Let $f(z)$ be a single-valued continuous function of z in a domain D. The function f is differentiable at point z_o in D if

$$f'(z_o) = \lim_{\Delta z \to 0} \left(\frac{f(z_o + \Delta z) - f(z_o)}{\Delta z} \right) \tag{10.1.8}$$

exists and is independent of how $\Delta z \to 0$. If the function is differentiable at all points in a domain D, then it is said to be *holomorphic*, *analytic*, or *regular* in D. Points where the function is not analytic are called *singular points*.

Using the representation (10.1.6) with differential relations (10.1.4), the derivative of f can be expressed by

$$f'(z) = \frac{\partial}{\partial z}(u + iv) = \frac{1}{2}\left(\frac{\partial u}{\partial x} + \frac{\partial v}{\partial y} \right) + i\frac{1}{2}\left(\frac{\partial v}{\partial x} - \frac{\partial u}{\partial y} \right) \tag{10.1.9}$$

Because the derivative limit must be the same regardless of the path associated with $\Delta z \to 0$, relation (10.1.9) must be valid for the individual cases of $\Delta x = 0$ and $\Delta y = 0$, and thus

$$f'(z) = \frac{1}{2}\left(\frac{\partial u}{\partial x} \right) + i\frac{1}{2}\left(\frac{\partial v}{\partial x} \right) = \frac{1}{2}\left(\frac{\partial v}{\partial y} \right) + i\frac{1}{2}\left(-\frac{\partial u}{\partial y} \right) \tag{10.1.10}$$

Equating real and imaginary parts in relations (10.1.10) gives

$$\frac{\partial u}{\partial x} = \frac{\partial v}{\partial y}, \quad \frac{\partial u}{\partial y} = -\frac{\partial v}{\partial x} \tag{10.1.11}$$

which are called the *Cauchy–Riemann equations*. In polar coordinate form, these relations may be written as

$$\frac{\partial u}{\partial r} = \frac{1}{r}\frac{\partial v}{\partial \theta}, \quad \frac{1}{r}\frac{\partial u}{\partial \theta} = -\frac{\partial v}{\partial r} \tag{10.1.12}$$

Note that by simple differentiation of these relations, it can be shown that

$$\nabla^2 u = 0, \quad \nabla^2 v = 0 \tag{10.1.13}$$

and thus the real and imaginary parts of any analytic function of a complex variable are solutions to Laplace's equation and are *harmonic functions*. It can also be observed that relations (10.1.11) allow the differential of u to be expressed in terms of the variable v, that is

$$du = \frac{\partial u}{\partial x}dx + \frac{\partial u}{\partial y}dy = \frac{\partial v}{\partial y}dx - \frac{\partial v}{\partial x}dy \qquad (10.1.14)$$

and so if we know v, we could calculate u by integrating relation (10.1.14). In this discussion, the roles of u and v could be interchanged, and therefore if we know one of these functions, the other can be determined. This behavior establishes u and v as *conjugate functions*.

Next consider some concepts and results related to integration in the complex plane shown in Fig. 10.2. The line integral over a curve C from z_1 to z_2 is given by

$$\int_C f(z)dz = \int_C (u + iv)(dx + idy) = \int_C ((udx - vdy) + i(udy + vdx)) \qquad (10.1.15)$$

Using the Cauchy–Riemann relations, we can show that if the function f is analytic in a region D that encloses the curve C, then the line integral is independent of the path taken between the end points z_1 and z_2. This fact leads to two useful theorems in complex variable theory.

Cauchy Integral Theorem: If a function f(z) is analytic at all points interior to and on a closed curve C, then

$$\oint_C f(z)dz = 0 \qquad (10.1.16)$$

Cauchy Integral Formula: If f(z) is analytic everywhere within and on a closed curve C, and if z_o is any point interior to C, then

$$f(z_o) = \frac{1}{2\pi i}\oint_C \frac{f(z)}{z - z_o}dz \qquad (10.1.17)$$

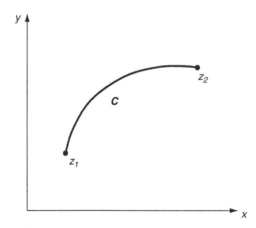

FIG. 10.2 Contour in the complex plane.

The integration over C is to be taken in the positive sense with the enclosed region to the left as the curve is traversed. Notice that the Cauchy integral formula provides a method to express the value of an analytic function at interior points of a domain in terms of values on the boundary.

It is often convenient to express functions of a complex variable in a power series. If $f(z)$ is analytic at all points *within* a circle C with center at $z = a$, then at each point inside C the function admits the *Taylor series expansion*

$$f(z) = f(a) + f'(a)(z-a) + \cdots + \frac{f^{(n)}(a)}{n!}(z-a)^n + \cdots \qquad (10.1.18)$$

about point $z = a$. For the special case where $a = 0$, the representation is referred to as the *Maclaurin series*. These results are useful for expansions in interior regions. A generalization of series representations for an *annular region* also exists. If $f(z)$ is analytic on two concentric circles C_1 and C_2 and throughout the region between these circles ($C_1 > C_2$), then the function may be expressed by the *Laurent series*

$$f(z) = \sum_{n=0}^{\infty} A_n (z-a)^n + \sum_{n=1}^{\infty} \frac{B_n}{(z-a)^n} \qquad (10.1.19)$$

where

$$A_n = \frac{1}{2\pi i} \oint_{C_1} \frac{f(z)}{(z-z_0)^{n+1}} dz, \quad n = 0, 1, 2, \ldots$$

$$B_n = \frac{1}{2\pi i} \oint_{C_2} \frac{f(z)}{(z-z_0)^{-n+1}} dz, \quad n = 1, 2, \ldots \qquad (10.1.20)$$

Recall that points where a complex function is not analytic are called singular points or singularities. We now wish to discuss briefly one particular type of singularity called a *pole*. If $f(z)$ is singular at $z = a$, but the product $(z-a)^n f(z)$ is analytic for some integer value of n, then $f(z)$ is said to have a *pole of order n* at $z = a$. For this case, the analytic product form can be expanded in a Taylor series about $z = a$

$$(z-a)^n f(z) = \sum_{k=0}^{\infty} A_k (z-a)^k$$

$$A_k = \frac{1}{k!} \frac{d^k}{dz^k} \left\{ (z-a)^n f(z) \right\} \Big|_{z=a}$$

Rewriting this series for $f(z)$

$$f(z) = \sum_{k=0}^{\infty} A_k \frac{(z-a)^k}{(z-a)^n}$$

Integrating this expression around a closed contour C that encloses the point a and using the Cauchy integral formula reduces the right-hand side to a single term

$$\oint_C f(z) dz = 2\pi i A_{n-1}$$

The quantity A_{n-1} is called the *residue* of $f(z)$ at the pole $z = a$, and this result would allow the calculation of the integral by knowing the residue of the pole inside the contour C. Thus, if $f(z)$ is analytic except for a pole of order n at $z = a$ in a region enclosed by an arbitrary contour C, then the integral of $f(z)$ around C is given by

$$\oint_C f(z)dz = 2\pi i \left[\frac{1}{(n-1)!} \frac{d^{n-1}}{dz^{n-1}} \{(z-a)^n f(z)\} \Big|_{z=a} \right] \qquad (10.1.21)$$

If more than one pole exists in the domain enclosed by C, then the integral is evaluated using relation (10.1.21) by including the summation of the residues of all poles in the domain. This procedure is called the *calculus of residues* and is useful to evaluate complex integrals. Using this scheme along with the Cauchy integral formula, the following useful integral relation may be developed

$$\frac{1}{2\pi i} \oint_C \frac{1}{\zeta^n(\zeta - z)} d\zeta = \begin{cases} 0, & n > 0 \\ 1, & n = 0 \end{cases} \qquad (10.1.22)$$

where C is the contour around the unit circle and z is inside the circle.

Another type of nonanalytic, singular behavior involves multivalued complex functions. Examples of such behavior are found in the functions $z^{1/2}$ and $\log z$. Consider in more detail the logarithmic function

$$\log z = \log(re^{i\theta}) = \log r + i\theta$$

It is observed that this function is multivalued in θ. This multivaluedness can be eliminated by restricting the range of θ to $-\pi < \theta \leq \pi$, and this results in the *principal value* of $\log z$. For this case, the function is single-valued at all points except for the origin and the negative real axis. The origin is then referred to as a *branch point*, and the negative real axis is a *branch* cut. By restricting the function to the domain $r > 0$ and $-\pi < \theta \leq \pi$, the singular behavior is avoided, and the function is analytic in the restricted range. Because of the common occurrence of functions involving $\sqrt[n]{(\cdot)}$ and $\log$, branch points and branch cuts are present in many applications in elasticity.

Consider next the issue of the *connectivity* of the plane domain. Recall that a *simply connected* region is one where any closed curve can be continuously shrunk to a point without going outside the region, and for two dimensions this simply means a region with no holes in it. Fig. 10.3 illustrates a

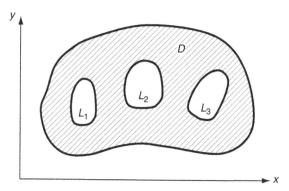

FIG. 10.3 Multiply connected domain.

multiply connected region D with several internal boundaries L_k. It can be shown that analytic functions in such multiply connected regions need not be single-valued. Note, however, that such regions can be made simply connected by making appropriate cuts joining each of the internal boundaries with the outer boundary.

The final topic in our complex variable review involves the powerful method of *conformal transformation* or *mapping*. This transformation concept provides a convenient means to find elasticity solutions to interior and exterior problems of complex shape. The concept starts with a general relationship between two complex variables z and ζ

$$z = w(\zeta), \quad \zeta = f(z) \tag{10.1.23}$$

The transformation w is assumed to be analytic in the domain of interest, and this establishes a one-to-one mapping of points in the ζ-plane to points in the z-plane, as shown in Fig. 10.4. Thus, the region R is mapped onto the region D by the relation $\zeta = f(z)$. The term *conformal* is associated with the property that angles between line elements are preserved under the transformation.

Many plane elasticity problems rely on solutions related to the unit circle, and thus the conformal mapping of a region R in the z-plane into a unit circle in the ζ-plane is commonly used. This case is shown in Fig. 10.5. The particular transformation that accomplishes this mapping is given by the following

$$z = \sum_{k=0}^{\infty} c_k \zeta^k \tag{10.1.24}$$

where the constants c_k would be determined by the specific shape of the domain R. Another useful transformation is that which maps the *exterior* of region R into the unit circle, and this is of the form

$$z = \frac{C}{\zeta} + \sum_{k=0}^{\infty} c_k \zeta^k \tag{10.1.25}$$

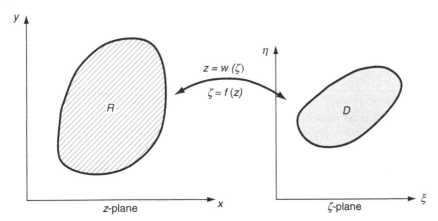

FIG. 10.4 Conformal mapping.

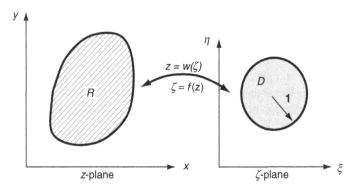

FIG. 10.5 Conformal mapping of a region onto the unit circle.

where as before the constants C and c_k would be determined by the shape of R. Other special mappings are presented as this theory is applied to specific elasticity problems in later sections. A large number of conformal mappings have been developed for various applications in many branches of engineering science; see, for example, Kober (1952).

10.2 Complex formulation of the plane elasticity problem

The general plane problem of elasticity formulated in Chapter 7 establishes the two theories of *plane strain* and *plane stress*. Although each case is related to a completely different two-dimensional model, the basic formulations are quite similar. By simple changes in elastic constants, solutions to each case were shown to be interchangeable (see Table 7.1).

The basic equations for *plane strain* include expressions for the stresses

$$\sigma_x = \lambda\left(\frac{\partial u}{\partial x} + \frac{\partial v}{\partial y}\right) + 2\mu\frac{\partial u}{\partial x}$$

$$\sigma_y = \lambda\left(\frac{\partial u}{\partial x} + \frac{\partial v}{\partial y}\right) + 2\mu\frac{\partial v}{\partial y} \qquad (10.2.1)$$

$$\tau_{xy} = \mu\left(\frac{\partial u}{\partial y} + \frac{\partial v}{\partial x}\right)$$

while the Navier equations reduced to

$$\mu\nabla^2 \boldsymbol{u} + (\lambda + \mu)\boldsymbol{\nabla}(\boldsymbol{\nabla}\cdot\boldsymbol{u}) = 0 \qquad (10.2.2)$$

where the Laplacian is given by $\nabla^2 = (\ )_{xx} + (\ )_{yy}$, with subscripts representing partial differentiation. For both plane strain and plane stress with *zero body forces*, the stresses were expressed in a self-equilibrated form using the Airy stress function ϕ

$$\sigma_x = \frac{\partial^2 \phi}{\partial y^2}, \sigma_y = \frac{\partial^2 \phi}{\partial x^2}, \tau_{xy} = -\frac{\partial^2 \phi}{\partial x\partial y} \qquad (10.2.3)$$

and from the compatibility relations, ϕ satisfied the biharmonic equation

$$\nabla^4\phi = \phi_{xxxx} + 2\phi_{xxyy} + \phi_{yyyy} = 0 \tag{10.2.4}$$

Thus, the stress formulation to the plane problem reduced to solving the biharmonic equation.

We now wish to represent the Airy stress function in terms of functions of a complex variable and transform the plane problem into one involving complex variable theory. Using relations (10.1.1) and (10.1.3), the variables x and y can be expressed in terms of z and $\bar{z}$, and thus functions of x and y can be expressed as functions of z and $\bar{z}$. Applying this concept to the Airy stress function, we can write $\phi = \phi(z, \bar{z})$. Repeated use of the differential operators defined in Eq. (10.1.4) allows the following representation of the harmonic and biharmonic operators

$$\nabla^2() = 4\frac{\partial^2()}{\partial z\partial\bar{z}}, \ \nabla^4() = 16\frac{\partial^4()}{\partial z^2\partial\bar{z}^2} \tag{10.2.5}$$

Therefore, the governing biharmonic elasticity equation (10.2.4) can be expressed as

$$\frac{\partial^4\phi}{\partial z^2\partial\bar{z}^2} = 0 \tag{10.2.6}$$

Integrating this result yields

$$\phi(z,\bar{z}) = \frac{1}{2}\left(\overline{z\gamma(z)} + \bar{z}\gamma(z) + \chi(z) + \overline{\chi(z)}\right) \tag{10.2.7}$$

$$= \mathrm{Re}(\bar{z}\gamma(z) + \chi(z))$$

where γ and χ are arbitrary functions of the indicated variables, and we have invoked the fact that ϕ must be real. This result demonstrates that the Airy stress function can be formulated in terms of *two functions of a complex variable*.

Following along another path, we consider the governing Navier equations (10.2.2) and introduce the complex displacement $U = u + iv$ to get

$$(\lambda + \mu)\frac{\partial}{\partial\bar{z}}\left(\frac{\partial U}{\partial z} + \frac{\overline{\partial U}}{\partial z}\right) + 2\mu\frac{\partial^2 U}{\partial\bar{z}\partial z} = 0 \tag{10.2.8}$$

Integrating this expression yields a solution form for the complex displacement

$$2\mu U = \kappa\gamma(z) - z\overline{\gamma'(z)} - \overline{\psi(z)} \tag{10.2.9}$$

where again $\gamma(z)$ and $\psi(z) = \chi'(z)$ are arbitrary functions of a complex variable and the Kolosov parameter κ depends only on Poisson's ratio ν

$$\kappa = \begin{cases} 3 - 4\nu, & \text{plane strain} \\ \dfrac{3-\nu}{1+\nu}, & \text{plane strain} \end{cases} \tag{10.2.10}$$

Result (10.2.9) is the complex variable formulation for the displacement field and is written in terms of two arbitrary functions of a complex variable.

Using relations (10.2.3) and (10.2.7) yields the *fundamental stress combinations*

$$\sigma_x + \sigma_y = 2\left(\gamma'(z) + \overline{\gamma'(z)}\right)$$
$$\sigma_y - \sigma_x + 2i\tau_{xy} = 2(\bar{z}\gamma''(z) + \psi'(z)) \tag{10.2.11}$$

By adding and subtracting and equating real and imaginary parts, relations (10.2.11) can be easily solved for the individual stresses (see Exercise 10.5). Using standard transformation laws (see Exercise 3.3), the stresses and displacements in polar coordinates can be written as

$$\sigma_r + \sigma_\theta = \sigma_x + \sigma_y$$
$$\sigma_\theta - \sigma_r + 2i\tau_{r\theta} = (\sigma_y - \sigma_x + 2i\tau_{xy})e^{2i\theta} \tag{10.2.12}$$
$$u_r + iu_\theta = (u + iv)e^{-i\theta}$$

From the original definition of the traction vector, we can express these components as

$$T_x^n = \sigma_x n_x + \tau_{xy} n_y = \phi_{yy} n_x - \phi_{xy} n_y = \phi_{yy}\frac{dy}{ds} + \phi_{xy}\frac{dx}{ds} = \frac{d}{ds}\left(\frac{\partial\phi}{\partial y}\right)$$
$$T_y^n = \tau_{xy} n_y + \sigma_y n_y = -\phi_{xy} n_x + \phi_{xx} n_y = -\left(\phi_{xy}\frac{dy}{ds} + \phi_{xx}\frac{dx}{ds}\right) = -\frac{d}{ds}\left(\frac{\partial\phi}{\partial x}\right) \tag{10.2.13}$$

and thus

$$T_x^n + iT_y^n = \frac{d}{ds}\left(\frac{\partial\phi}{\partial y} - i\frac{\partial\phi}{\partial x}\right) = -i\frac{d}{ds}\left(\frac{\partial\phi}{\partial x} + i\frac{\partial\phi}{\partial y}\right)$$
$$= -i\frac{d}{ds}\left(\gamma(z) + z\overline{\gamma'(z)} + \overline{\psi(z)}\right) \tag{10.2.14}$$

Therefore, we have demonstrated that all of the basic variables in plane elasticity are expressible in terms of two arbitrary functions of a complex variable. These two functions $\gamma(z)$ and $\psi(z)$ are commonly referred to as the *Kolosov–Muskhelishvili potentials*. The solution to particular problems is then reduced to finding the appropriate potentials that satisfy the boundary conditions. This solution technique is greatly aided by mathematical methods of complex variable theory.

Example 10.1 Constant stress state example

Consider the complex potentials $\gamma(z) = Az$, $\psi(z) = Bz$, where A and B are complex constants. We wish to determine the stresses and displacements and explicitly show that this example corresponds to a uniform stress field. Using the stress combinations (10.2.11)

$$\sigma_x + \sigma_y = 2\left(\gamma'(z) + \overline{\gamma'(z)}\right) = 2(A + \overline{A}) = 4\text{Re}A = 4A_R$$
$$\sigma_y - \sigma_x + 2i\tau_{xy} = 2(\bar{z}\gamma''(z) + \psi'(z)) = 2B = 2(B_R + iB_I)$$

Equating real and imaginary parts in the second relation gives

$$\sigma_y - \sigma_x = 2B_R$$
$$\tau_{xy} = B_I$$

and this allows the individual stresses to be calculated as.

$$\sigma_x = 2A_R - B_R, \quad \sigma_y = 2A_R + B_R, \quad \tau_{xy} = B_I$$

If these stresses are to be a uniform state $\sigma_x = \sigma_x^o, \sigma_y = \sigma_y^o, \tau_{xy} = \tau_{xy}^o$ then the constants must take the form

$$A_R = \frac{\sigma_x^o + \sigma_y^o}{4}, B_R = \frac{\sigma_y^o + \sigma_x^o}{2}, \quad B_I = \tau_{zy}^o$$

Note that the imaginary part of A is not determined by the stress state.
The polar coordinate stresses can easily be calculated by using relation (10.2.12)

$$\sigma_r + \sigma_\theta = 4A_R$$
$$\sigma_\theta - \sigma_r + 2i\tau_{r\theta} = 2(B_R + iB_I)e^{2i\theta} = 2(B_R + iB_I)(\cos 2\theta + i \sin 2\theta)$$

Again, separating and equating real and imaginary parts gives the individual stresses

$$\sigma_r = 2A_R - B_R \cos 2\theta + B_I \sin 2\theta$$
$$\sigma_\theta = 2A_R + B_R \cos 2\theta - B_I \sin 2\theta$$
$$\tau_{r\theta} = B_R \sin 2\theta + B_I \cos 2\theta$$

Finally, the displacements follow from equation (10.2.9)

$$2\mu(u + iv) = \kappa A z - z\overline{A} - \overline{B}\overline{z} = \kappa(A_R + iA_I)(x + iy)$$
$$- (x + iy)(A_R - iA_I) - (B_R - iB_I)(x - iy)$$

Equating real and imaginary parts, the individual components can be determined as

$$u = \frac{1}{2\mu}[(A_R(\kappa - 1) - B_R)x + (B_I - A_I(\kappa + 1))y]$$

$$v = \frac{1}{2\mu}[(A_I(\kappa + 1) + B_I)x + (A_R(\kappa - 1) + B_R)y]$$

10.3 Resultant boundary conditions

The final formulation step in the complex variable approach is to develop expressions to handle general resultant boundary conditions, and this involves methods to determine the resultant force and moment acting on arbitrary boundary segments. Consider the boundary segment AB for an interior simply

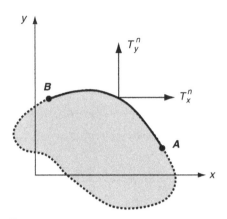

FIG. 10.6 Resultant boundary loading.

connected domain problem shown in Fig. 10.6. The resultant force components produced by tractions acting on this segment may be expressed in complex form as

$$F_x + iF_y = \int_A^B \left(T_x^n + iT_y^n\right)ds$$

$$= -i\int_A^B d\left[\gamma(z) + z\overline{\gamma'(z)} + \overline{\psi(z)}\right] \tag{10.3.1}$$

$$= -i\left[\gamma(z) + z\overline{\gamma'(z)} + \overline{\psi(z)}\right]_A^B$$

Again, the direction of the boundary integration is always taken to keep the region to the left. Similarly, the resultant moment M with respect to the coordinate origin is given by

$$M = \int_A^B \left(xT_y^n - yT_x^n\right)ds$$

$$= -\int_A^B \left[xd\left(\frac{\partial\phi}{\partial x}\right) + yd\left(\frac{\partial\phi}{\partial y}\right)\right] \tag{10.3.2}$$

$$= -\left[x\frac{\partial\phi}{\partial x} + y\frac{\partial\phi}{\partial y}\right]_A^B + \phi|_A^B$$

$$= \mathrm{Re}[\chi(z) - z\psi(z) - z\bar{z}\gamma'(z)]_A^B$$

where $\chi'(z) = \psi(z)$.

10.4 General structure of the complex potentials

It has been shown that the solution to plane elasticity problems involves determination of two complex potential functions $\gamma(z)$ and $\psi(z)$. These potentials have some general properties and structures that we now wish to investigate. First, by examining relations for the stresses and displacements, a particular *indeterminacy* or *arbitrariness* of the potentials can be found. From the first stress relation in set (10.2.11), it is observed that an arbitrary imaginary constant iC may be added to the potential $\gamma'(z)$ without affecting the stresses. From the second stress relation $(10.2.11)_2$, an arbitrary complex constant can be added to the potential $\psi(z)$ without changing the stresses. These two observations indicate that without changing the state of stress, a new set of complex potentials $\gamma^*(z)$ and $\psi^*(z)$ could be written as

$$\gamma^*(z) = \gamma(z) + iCz + A$$
$$\psi^*(z) = \psi(z) + B \tag{10.4.1}$$

Using these new forms in relation (10.2.9) yields a displacement field that differs from the original form by the terms

$$2\mu(U^* - U) = (\kappa + 1)iCz + \kappa A - \overline{B} \tag{10.4.2}$$

These difference terms correspond to *rigid-body motions* [see relations (2.2.9)], and thus as expected the stresses determine the displacements only up to rigid-body motions.

Particular general forms of these potentials exist for regions of different topology. Most problems of interest involve *finite simply connected*, *finite multiply connected*, and *infinite multiply connected* domains as shown in Fig. 10.7. We now present specific forms for each of these cases.

10.4.1 Finite simply connected domains

Consider the finite simply connected region shown in Fig. 10.7A. For this case, the potential functions $\gamma(z)$ and $\psi(z)$ are analytic and single-valued in the domain and this allows the following power series representation

$$\gamma(z) = \sum_{n=0}^{\infty} a_n z^n$$
$$\psi(z) = \sum_{n=0}^{\infty} b_n z^n \tag{10.4.3}$$

where a_n and b_n are constants to be determined by the boundary conditions of the problem under study.

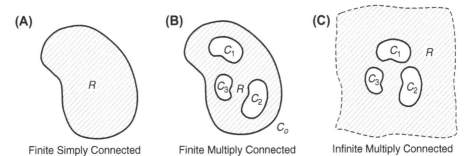

(A) Finite Simply Connected
(B) Finite Multiply Connected
(C) Infinite Multiply Connected

FIG. 10.7 Typical domains of interest.

10.4.2 Finite multiply connected domains

For the general region shown in Fig. 10.7B, it is assumed that the domain has k internal boundaries as shown. For this topology, the potential functions need not be single-valued, as can be demonstrated by considering the behaviors of the stresses and displacements around each of the n contours C_k in region R. For the present case, we shall limit the problem and assume that the displacements and stresses are continuous and single-valued everywhere. Multivalued displacements lead to the theory of *dislocations*, and this is discussed at a later stage in Chapter 15. The resultant force on a typical *internal boundary C_k* may be determined by using relation (10.3.1)

$$F_k = X_k + iY_k = \oint_{C_k} \left(T_x^n + iT_y^n \right) ds = i \left[\gamma(z) + z\overline{\gamma'(z)} + \overline{\psi(z)} \right]_{C_k} \tag{10.4.4}$$

where $[f(z,\bar{z})]_{C_k}$ is referred to as the *cyclic function* of f and represents the change of the function f around closed contour C_k. Note that in relation (10.4.4), the internal boundary circuit C_k is traversed with the region on the left, thus leading to a *clockwise* circuit and a change of sign from relation (10.3.1). Of course, the cyclic function of a single-valued expression is zero; further details on properties of cyclic functions may be found in Milne-Thomson (1960). Because the resultant force on a given internal boundary will not necessarily be zero, the cyclic function on the right-hand side of relation (10.4.4) should properly produce this result. Therefore, the potential functions $\gamma(z)$ and $\psi(z)$ must have appropriate multivalued behavior. It can be shown that the logarithmic function previously discussed in Section 10.1 can provide the necessary multivaluedness, because

$$[\log(z - z_k)]_{C_k} = 2\pi i \tag{10.4.5}$$

where z_k is a point within the contour C_k and the cyclic evaluation is taken in the counterclockwise sense for the usual right-handed coordinate system with θ measured counterclockwise. Including such logarithmic terms for each of the two complex potentials and employing (10.4.4) for all contours within the region R in Fig. 10.7B develops the following general forms

$$\gamma(z) = -\sum_{k=1}^{n} \frac{F_k}{2\pi(1+\kappa)} \log(z - z_k) + \gamma^*(z)$$

$$\psi(z) = \sum_{k=1}^{n} \frac{\kappa \overline{F}_k}{2\pi(1+\kappa)} \log(z - z_k) + \psi^*(z)$$
$$\tag{10.4.6}$$

where F_k is the resultant force on each contour C_k, $\gamma^*(z)$ and $\psi^*(z)$ are arbitrary analytic functions in R, and κ is the material constant defined by Eq. (10.2.10).

10.4.3 Infinite domains

For the region shown in Fig. 10.7C, the general form of the potentials is determined in an analogous manner as in the previous case. The logarithmic terms in Eq. (10.4.6) may be expanded in the region *exterior to a circle enclosing all m contours* C_k to get

$$\log(z - z_k) = \log z + \log\left(1 - \frac{z_k}{z}\right) = \log z - \left(\frac{z_k}{z} + \frac{1}{2}\left(\frac{z_k}{z}\right)^2 + \cdots\right)$$

$$= \log z + (\text{arbitrary analytic function})$$

Combining this result with the requirement that the stresses remain bounded at infinity gives the general form for this case

$$\gamma(z) = -\frac{\sum\limits_{k=1}^{m} F_k}{2\pi(1+\kappa)}\log z + \frac{\sigma_x^\infty + \sigma_y^\infty}{4}z + \gamma^{**}(z)$$

$$\psi(z) = -\frac{\kappa\sum\limits_{k=1}^{m} \overline{F}_k}{2\pi(1+\kappa)}\log z + \frac{\sigma_y^\infty - \sigma_x^\infty + 2i\tau_{xy}^\infty}{2}z + \psi^{**}(z)$$

(10.4.7)

where σ_x^∞, σ_y^∞, and τ_{xy}^∞ are the stresses at infinity, and $\gamma^{**}(z)$ and $\psi^{**}(z)$ are arbitrary analytic functions outside the region enclosing all *m* contours. Using power series theory, these analytic functions can be expressed as

$$\gamma^{**}(z) = \sum_{n=1}^{\infty} a_n z^{-n}$$

$$\psi^{**}(z) = \sum_{n=1}^{\infty} b_n z^{-n}$$

(10.4.8)

An examination of the displacements at infinity would indicate unbounded behavior unless all stresses at infinity vanish and $\Sigma F_k = \Sigma \overline{F}_k = 0$. This fact occurs because even a bounded strain over an infinite length will produce infinite displacements. Note that the case of a simply connected, infinite domain is obtained by dropping the summation terms in Eq. (10.4.7).

10.5 Circular domain examples

We now develop some solutions of particular plane elastic problems involving regions of a circular domain. The process starts by developing a general solution to a circular region with arbitrary edge loading, as shown in Fig. 10.8. The region $0 \leq r \leq R$ is to have arbitrary boundary loadings at $r = R$ specified by $\sigma_r = f_1(\theta)$, $\tau_{r\theta} = -f_2(\theta)$, which can be written in complex form as

$$f = f_1(\theta) + if_2(\theta) = \sigma_r - i\tau_{r\theta}\big|_{r=R}$$

(10.5.1)

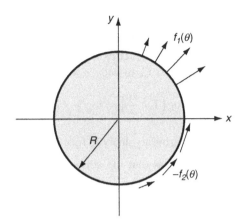

FIG. 10.8 Circular disk problem.

The fundamental stress combinations and displacements in polar coordinates were given in relations (10.2.12). The tractions given by Eq. (10.2.14) may be expressed in polar form as

$$T_x^r + iT_y^r = -i\frac{d}{ds}\left(\gamma(z) + z\overline{\gamma'(z)} + \overline{\psi(z)}\right)\bigg|_{r=R} \tag{10.5.2}$$

Integrating this result around the boundary $r = R$ $(ds = R d\theta)$ gives

$$i\int \left(T_x^r + iT_y^r\right) R d\theta = \left(\gamma(z) + z\overline{\gamma'(z)} + \overline{\psi(z)}\right)\bigg|_{r=R} = g \tag{10.5.3}$$

where the boundary function g depends only on θ. Using general form (10.4.3) for the complex potentials, the stress resultant becomes

$$\sigma_r - i\tau_{r\theta} = \gamma'(z) + \overline{\gamma'(z)} - e^{2i\theta}[\bar{z}\gamma''(z) + \psi'(z)]$$

$$= \sum_{n=1}^{\infty}\left(a_n n z^{n-1} + \bar{a}_n n\bar{z}^{n-1} - e^{2i\theta}\left[\bar{z}a_n n(n-1)z^{n-2} + b_n n z^{n-1}\right]\right)$$

$$= a_1 + \bar{a}_1 + \sum_{k=1}^{\infty}\left(-\left[a_{k+1}(k^2-1)r^k + b_{k-1}(k-1)r^{k-2}\right]e^{ik\theta} + \bar{a}_{k+1}(k+1)r^k e^{-ik\theta}\right)$$

$$\tag{10.5.4}$$

This relation can be recognized as the *complex Fourier series* expansion for $\sigma_r - i\tau_{r\theta}$. On the boundary $r = R$, the complex boundary-loading function f can also be expanded in a similar Fourier series as

$$f(\theta) = \sum_{k=-\infty}^{\infty} C_k e^{ik\theta}$$

$$\tag{10.5.5}$$

$$C_k = \frac{1}{2\pi}\int_0^{2\pi} f(\theta)e^{-ik\theta}d\theta$$

Matching (10.5.4) with (10.5.5) on the boundary and equating like powers of exponentials of θ yields the system

$$a_1 + \bar{a}_1 = C_o = 2\text{Re}(a_1)$$
$$\bar{a}_{k+1}(k+1)R^k = C_{-k}, (k > 0)$$
$$a_{k+1}(k^2 - 1)R^k + b_{k-1}(k-1)R^{k-2} = C_k, (k > 0)$$

$$(10.5.6)$$

Equating real and imaginary parts in relations (10.5.6) generates a system of equations to determine the constants a_k and b_k. This solution is essentially the same as the Michell solution previously discussed in Section 8.3. Note that the annulus ($r_i \le r \le r_o$) and the exterior ($r \ge R$) domain problems may be solved in a similar fashion.

This solution scheme only duplicates previous methods based on Fourier analysis. A more powerful use of complex variable techniques involves the application of Cauchy integral formulae. In order to discuss this method, consider again the circular region with unit boundary radius. Relation (10.5.3) becomes

$$\left(\gamma(z) + z\overline{\gamma'(z)} + \overline{\psi(z)} \right)\Big|_{z=\zeta} = g$$

$$(10.5.7)$$

where $\zeta = z|_{r=1} = e^{i\theta}$ and $\bar{\zeta} = e^{-i\theta} = 1/\zeta$. Multiplying (10.5.7) by $1/2\pi i(\zeta - z)$ and integrating around the boundary contour C ($r = 1$) yields

$$\frac{1}{2\pi i} \oint_C \frac{\gamma(\zeta)}{\zeta - z} d\zeta + \frac{1}{2\pi i} \oint_C \zeta \frac{\overline{\gamma'(\zeta)}}{\zeta - z} d\zeta$$
$$+ \frac{1}{2\pi i} \oint_C \frac{\overline{\psi(\zeta)}}{\zeta - z} d\zeta = \frac{1}{2\pi i} \oint_C \frac{g(\zeta)}{\zeta - z} d\zeta$$

$$(10.5.8)$$

Using the Cauchy integral formula, the first term in Eq. (10.5.8) is simply $\gamma(z)$. Using the general series form (10.4.3) for the potentials and employing the integral formula (10.1.22), the remaining two terms on the left-hand side of (10.5.8) can be evaluated, and the final result reduces to

$$\gamma(z) + \bar{a}_1 z + 2\bar{a}_2 + \overline{\psi(0)} = \frac{1}{2\pi i} \oint_C \frac{g(\zeta)}{\zeta - z} d\zeta$$

$$(10.5.9)$$

We also find that $a_n = 0$ for $n > 2$, and so $\gamma(z) = a_o + a_1 z + a_2 z^2$. These results can be used to solve for the remaining terms in order to determine the final form for the potential $\gamma(z)$. Using a similar scheme but starting with the complex conjugate of (10.5.7), the potential $\psi(z)$ may be found. Dropping the constant terms that do not contribute to the stresses, the final results are summarized as

$$\gamma(z) = \frac{1}{2\pi i} \oint_C \frac{g(\zeta)}{\zeta - z} d\zeta - \bar{a}_1 z, \quad a_1 + \bar{a}_1 = \frac{1}{2\pi i} \oint_C \frac{g(\zeta)}{\zeta^2} d\zeta$$
$$\psi(z) = \frac{1}{2\pi i} \oint_C \frac{\overline{g(\zeta)}}{\zeta - z} d\zeta - \frac{\gamma'(z)}{z} + \frac{a_1}{z}$$

$$(10.5.10)$$

Note that the preceding solution is valid only for the unit disk. For the case of a disk of radius a, the last two terms for $\psi(z)$ should be multiplied by a^2.

We now consider a couple of specific examples using this general solution.

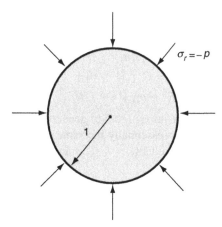

FIG. 10.9 Disk under uniform compression.

Example 10.2 Disk under uniform compression

Consider the case of uniform compression loading of the circular disk, as shown in Fig. 10.9. The boundary tractions for this case become

$$T_x^r + iT_y^r = (\sigma_r + i\tau_{r\theta})e^{i\theta} = -pe^{i\theta}$$

and thus the boundary-loading function defined by Eq. (10.5.3) reduces to

$$g = i\int_0^\theta \left(T_x^r + iT_y^r\right)d\theta = -i\int_0^\theta pe^{i\theta}d\theta = -pe^{i\theta} = -p\zeta$$

Substituting into relation (10.5.10)₁ gives

$$\gamma(z) = -\frac{1}{2\pi i}\oint_C \frac{p\zeta}{\zeta - z}d\zeta - \bar{a}_1 z = -pz - \bar{a}_1 z$$

$$a_1 + \bar{a}_1 = -\frac{1}{2\pi i}\oint_C \frac{p}{\zeta}d\zeta = -p$$

(10.5.11)

Finally, substituting these results into relation (10.5.10)₂ gives the result for the second potential function

$$\psi(z) = -\frac{1}{2\pi i}\oint_C \frac{p}{\zeta(\zeta - z)}d\zeta + \frac{p + \bar{a}_1}{z} + \frac{a_1}{z} = 0$$

(10.5.12)

With the potentials now explicitly determined, the stress combinations can be calculated from (10.2.11) and (10.2.12), giving

$$\sigma_r + \sigma_\theta = 2(-p - \bar{a}_1 - p - a_1) = -2p$$

$$\sigma_\theta - \sigma_r + 2i\tau_{r\theta} = 0$$

Separating the real and imaginary parts gives individual stresses

$$\sigma_r = \sigma_\theta = -p, \quad \tau_{r\theta} = 0 \tag{10.5.13}$$

Of course, this hydrostatic state of stress is the expected result that is easily verified as a special case of Example 8.6.

Example 10.3 Circular plate with concentrated edge loading

Consider next the circular plate of radius a under symmetric concentrated edge loadings F, as shown in Fig. 10.10.

For this case, the boundary condition on $|z| = a(\zeta = ae^{i\theta})$ may be expressed as

$$\sigma_r + i\tau_{r\theta} = \frac{Fe^{-i\alpha}}{a}\delta(\theta - \alpha) + \frac{Fe^{i\alpha}}{a}\delta(\theta - \pi - \alpha) \tag{10.5.14}$$

The expression $\delta()$ is the *Dirac delta function*, which is a special defined function that is zero everywhere except at the origin, where it is singular and has the integral property $\int_{-d}^{d} f(x) \, \delta(x - \xi)dx = f(\xi)$ for any parameter d and continuous function f. Using this representation, the resultant boundary-loading function can be expressed as

$$g = i\int_0^\theta \left(T_x^r + iT_y^r\right)ad\theta = \begin{cases} 0, & 0 \le \theta < \alpha \\ iF, & \alpha \le \theta \le \pi - \alpha \\ 0, & \pi - \alpha \le \theta \le 2\pi \end{cases} \tag{10.5.15}$$

Continued

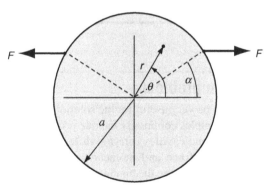

FIG. 10.10 Circular plate with edge loading.

Thus, using the general solution (10.5.10) then gives

$$a_1 + \bar{a}_1 = \frac{F}{2\pi} \int_{ae^{i\alpha}}^{ae^{i(\pi-\alpha)}} \frac{d\zeta}{\zeta^2} = -\frac{F}{2\pi} \frac{1}{\zeta}\Big|_{ae^{i\alpha}}^{ae^{i(\pi-\alpha)}} = \frac{F}{\pi a}\cos\alpha$$

and the expressions for the potential functions become

$$\gamma(z) = \frac{F}{2\pi} \int_{ae^{i\alpha}}^{ae^{i(\pi-\alpha)}} \frac{d\zeta}{\zeta - z} - \bar{a}_1 z = \frac{F}{2\pi}\log(\zeta - z)\Big|_{ae^{i\alpha}}^{ae^{i(\pi-\alpha)}}$$

$$-\bar{a}_1 z = \frac{F}{2\pi}\log\left(\frac{z + ae^{-i\alpha}}{z - ae^{i\alpha}}\right) - \bar{a}_1 z \tag{10.5.16}$$

$$\psi(z) = -\frac{F}{2\pi}\log\left(\frac{z + ae^{-i\alpha}}{z - ae^{i\alpha}}\right) + \frac{Fa^3\cos\alpha}{\pi z(z + ae^{-i\alpha})(z - ae^{i\alpha})} + \frac{a_1 + \bar{a}_1}{z}a^2$$

The stress resultant then becomes

$$\sigma_r + \sigma_\theta = 2\left(\gamma'(z) + \overline{\gamma'(z)}\right)$$

$$= -\frac{2Fa\cos\alpha}{\pi}\left[\frac{1}{(z + ae^{-i\alpha})(z - ae^{i\alpha})} + \frac{1}{(\bar{z} + ae^{i\alpha})(\bar{z} - ae^{-i\alpha})} + \frac{1}{a^2}\right] \tag{10.5.17}$$

Note that for the case with $\alpha = 0$ (diametrical compression), we get

$$\sigma_r + \sigma_\theta = \sigma_x + \sigma_y = -\frac{2Fa}{\pi}\left[\frac{1}{(z^2 - a^2)} + \frac{1}{(\bar{z}^2 - a^2)} + \frac{1}{a^2}\right] \tag{10.5.18}$$

which was the problem previously solved in Example 8.10, giving the stresses specified in relations (8.4.69). Solutions to many other problems of circular domain can be found in Muskhelishvili (1963), Milne-Thomson (1960), and England (1971).

10.6 Plane and half-plane problems

Complex variable methods prove to be very useful for the solution of a large variety of full-space and half-space problems. Full-space examples commonly include problems with various types of internal concentrated force systems and internal cavities carrying different loading conditions. Typical half-space examples include concentrated force and moment systems applied to the free surface and indentation contact mechanics problems where the boundary conditions may be in terms of the stresses or displacements, or of mixed type over a portion of the free surface. This general class of problems involves infinite domains and requires the general solution form given by Eq. (10.4.7).

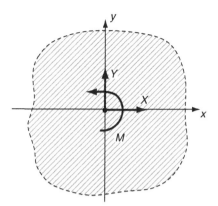

FIG. 10.11 Concentrated force system in an infinite medium.

Example 10.4 Concentrated force—moment system in an infinite plane

We now investigate the elasticity solution to the full plane with a concentrated force and moment acting at the origin, as shown in Fig. 10.11.

Using the general potential solutions (10.4.7) with no stresses at infinity, we choose the particular form

$$\gamma(z) = -\frac{X + iY}{2\pi(1 + \kappa)} \log z$$

$$\psi(z) = \frac{\kappa(X - iY)}{2\pi(1 + \kappa)} \log z + \frac{iM}{2\pi z} \tag{10.6.1}$$

The stress combinations become

$$\sigma_x + \sigma_y = 2\left(\gamma'(z) + \overline{\gamma'(z)}\right) = -\frac{1}{\pi(1 + \kappa)}\left(\frac{X + iY}{z} + \frac{X - iY}{\bar{z}}\right)$$

$$\sigma_y - \sigma_x + 2i\tau_{xy} = 2(\bar{z}\gamma''(z) + \psi'(z)) = \frac{X + iY}{\pi(1 + \kappa)}\frac{\bar{z}}{z^2} + \frac{\kappa(X - iY)}{\pi(1 + \kappa)}\frac{1}{z} - \frac{iM}{\pi z^2} \tag{10.6.2}$$

while the resulting displacements are

$$2\mu U = \kappa\gamma(z) - z\overline{\gamma'(z)} - \overline{\psi(z)}$$

$$= -\frac{\kappa(X + iY)}{2\pi(1 + \kappa)}(\log z + \log\bar{z}) + \frac{X - iY}{2\pi(1 + \kappa)}\frac{z}{\bar{z}} + \frac{iM}{x\pi\bar{z}} \tag{10.6.3}$$

Using relations (10.3.1) and (10.3.2), the resultant force and moment on any internal circle C enclosing the origin is given by

$$\oint_C \left(T_x^n + iT_y^n\right)ds = i\left[\gamma(z) + z\overline{\gamma'(z)} + \overline{\psi(z)}\right]_C = X + iY$$

$$\oint_C \left(xT_y^n - yT_x^n\right)ds = -\text{Re}[\chi(z) - z\psi(z) - z\bar{z}\gamma'(z)]_C = M \tag{10.6.4}$$

Continued

Note that appropriate sign changes have been made as a result of integrating around an internal cavity in the clockwise sense. Thus, the proper resultant match is attained with the applied loading for any circle, and in the limit, as the circle radius goes to zero, the concentrated force system in the problem is realized.

For the special case of $X = P$ and $Y = M = 0$, the stresses reduce to

$$\sigma_x = -\frac{Px}{2\pi(1+\kappa)r^2}\left[4\frac{x^2}{r^2}+\kappa-1\right]$$

$$\sigma_y = \frac{Px}{2\pi(1+\kappa)r^2}\left[4\frac{x^2}{r^2}+\kappa-5\right] \tag{10.6.5}$$

$$\tau_{xy} = \frac{Py}{2\pi(1+\kappa)r^2}\left[4\frac{y^2}{r^2}-3-\kappa\right], \quad r^2 = x^2 + y^2$$

Example 10.5 Concentrated force system on the surface of a half plane

Consider now the half plane carrying a general concentrated force system on the free surface, as shown in Fig. 10.12. Recall this Flamant problem was previously solved using Fourier methods in Example 8.8 (Section 8.4.7).

Following the solution pattern from Example 10.4, the complex potentials can be written as

$$\gamma(z) = -\frac{X+iY}{2\pi}\log z$$

$$\psi(z) = \frac{(X-iY)}{2\pi}\log z \tag{10.6.6}$$

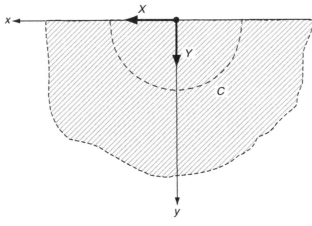

FIG. 10.12 Concentrated force system on a half-space.

The stress combinations then become

$$\sigma_r + \sigma_\theta = 2\left[\gamma'(z) + \overline{\gamma'(z)}\right] = -\frac{1}{\pi}\left(\frac{X+iY}{z} + \frac{X-iY}{\bar{z}}\right)$$

$$\sigma_\theta - \sigma_r + 2i\tau_{r\theta} = 2e^{2i\theta}[\bar{z}\gamma''(z) + \psi'(z)] = 2e^{2i\theta}\left(\frac{X+iY}{2\pi}\frac{\bar{z}}{z^2} + \frac{X-iY}{2\pi}\frac{1}{z}\right)$$

which can be reduced to

$$\sigma_r + \sigma_\theta = -\frac{2}{\pi r}(X\cos\theta + Y\sin\theta)$$

$$\sigma_\theta - \sigma_r + 2i\tau_{r\theta} = \frac{2}{\pi r}(X\cos\theta + Y\sin\theta)$$

(10.6.7)

Solving for the individual stresses gives

$$\sigma_r = \frac{2}{\pi r}(X\cos\theta + Y\sin\theta)$$

$$\sigma_\theta = \tau_{r\theta} = 0$$

(10.6.8)

This result matches with our previous solution to this problem in Example 8.8; see relations (8.4.34). Again, it is somewhat surprising that both σ_θ and $\tau_{r\theta}$ vanish even with the tangential surface loading X.

The boundary condition related to the concentrated force involves integrating the tractions around the contour C (a semicircle of arbitrary radius centered at the origin), as shown in Fig. 10.12. Thus, using (10.4.4)

$$\oint_C \left(T_x^n + iT_y^n\right)ds = i\left[\gamma(z) + z\overline{\gamma'(z)} + \overline{\psi(z)}\right]_C = X + iY$$

which verifies the appropriate boundary condition. By using the moment relation (10.3.2), it can also be shown that the resultant tractions on the contour C give zero moment.

For the special case $X = 0$ and $Y = P$, the individual stresses can be extracted from result (10.6.8) to give

$$\sigma_r = -\frac{2P}{\pi r}\sin\theta, \quad \sigma_\theta = \tau_{r\theta} = 0$$

(10.6.9)

Again, this case was previously presented in Example 8.8 by relation (8.4.35).

By employing analytic continuation theory and Cauchy integral representations, other more complicated surface boundary conditions can be handled. Such cases typically arise from contact mechanics problems involving the indentation of an elastic half space by another body. Some of these problems were previously discussed in Section 8.5. Such a problem is illustrated in Fig. 10.13, and the boundary conditions under the indenter could involve stresses and/or displacements depending on the contact conditions specified. These problems are discussed in Muskhelishvili (1963), Milne-Thomson (1960), and England (1971).

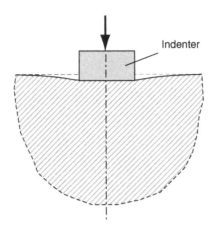

FIG. 10.13 Typical indentation problem.

Example 10.6 Stressed infinite plane with a circular hole

The final example in this section is a full plane containing a stress-free circular hole, and the problem is loaded with a general system of uniform stresses at infinity, as shown in Fig. 10.14. A special case of this problem was originally investigated in Example 8.7.

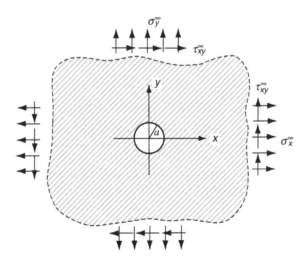

FIG. 10.14 Stress-free hole under general far-field loading.

The general solution form (10.4.7) is again used; however, for this problem the terms with stresses at infinity are retained while the logarithmic terms are dropped because the hole is stress free. The complex potentials may then be written as

$$\gamma(z) = \frac{\sigma_x^\infty + \sigma_y^\infty}{4} z + \sum_{n=1}^{\infty} a_n z^{-n}$$

$$\psi(z) = \frac{\sigma_y^\infty - \sigma_x^\infty + 2i\tau_{zy}^\infty}{2} z + \sum_{n=1}^{\infty} b_n z^{-n}$$

(10.6.10)

Using relation (10.5.4), the stress-free condition on the interior of the hole may be written as

$$(\sigma_r - i\tau_{r\theta})_{r=a} = \left(\gamma'(z) + \overline{\gamma'(z)} - e^{2i\theta} [\bar{z}\gamma''(z) + \psi'(z)] \right)_{r=a} = 0$$

(10.6.11)

Substituting the general form (10.6.10) in this condition gives

$$\frac{\sigma_x^\infty + \sigma_y^\infty}{2} - \frac{\sigma_y^\infty - \sigma_x^\infty + 2i\tau_{zy}^\infty}{2} e^{2i\theta}$$

$$= \sum_{n=1}^{\infty} \left(\frac{1}{a^{n+1}} \left[na_n \left(e^{(n+1)i\theta} + e^{-(n+1)i\theta} + (n+1)e^{-(n+1)i\theta} \right) - nb_n e^{-(n-1)i\theta} \right] \right)$$

Equating like powers of $e^{in\theta}$ gives relations for the coefficients a_n and b_n

$$a_1 = -\frac{\sigma_y^\infty - \sigma_x^\infty + 2i\tau_{xy}^\infty}{2} a^2, \quad a_n = 0(n \geq 2)$$

$$b_1 = -\frac{\sigma_x^\infty + \sigma_y^\infty}{2} a^2, \quad b_2 = 0, \quad b_3 = a^2 a_1, \quad b_n = 0(n \geq 4)$$

(10.6.12)

The potential functions are now determined and the stresses and displacements can easily be found using the standard relations in Section 10.2. Exercise 10.18 further explores a specific loading case. Recall that our previous work using Fourier methods in Example 8.7 investigated several special cases of this problem with uniaxial $\sigma_x^\infty = S, \sigma_y^\infty = \tau_{xy}^\infty = 0$ and biaxial $\sigma_x^\infty = \sigma_y^\infty = S$, $\tau_{xy}^\infty = 0$ loadings.

10.7 Applications using the method of conformal mapping

The method of conformal mapping discussed in Section 10.1 provides a very powerful tool to solve plane problems with complex geometry. The general concept is to establish a mapping function, which will transform a complex region in the z-plane (actual domain) into a simple region in the ζ-plane. If the elasticity solution is known for the geometry in the ζ-plane, then through appropriate transformation formulae the solution for the actual problem can be easily determined. Because we have established the general solution for the interior unit disk problem in Section 10.5, mapping functions that transform regions onto the unit disk (see Fig. 10.5) will be most useful. Specific mapping examples are discussed later.

To establish the appropriate transformation relations, we start with the general mapping function

$$z = w(\zeta)$$

(10.7.1)

where w is an analytic single-valued function. Using this result, the derivatives are related by

$$dz = \frac{dw}{d\zeta} d\zeta \tag{10.7.2}$$

Now the complex potentials are to be transformed into functions of ζ through the relations

$$\gamma(z) = \gamma(w(\zeta)) = \gamma_1(\zeta), \psi(z) = \psi(w(\zeta)) = \psi_1(\zeta) \tag{10.7.3}$$

and thus

$$\frac{d\gamma}{dz} = \frac{d\gamma_1}{d\zeta} \frac{d\zeta}{dz} = \frac{\gamma_1'(\zeta)}{w'(\zeta)} \tag{10.7.4}$$

These relations allow the stress combinations to be expressed in the ζ-plane as

$$\sigma_\rho + \sigma_\varphi = \sigma_x + \sigma_y = 2\left(\frac{\gamma_1'(\zeta)}{w'(\zeta)} + \overline{\frac{\gamma_1'(\zeta)}{w'(\zeta)}} \right)$$

$$\sigma_\varphi - \sigma_\rho + 2i\tau_{\rho\varphi} = \frac{2\zeta^2}{\rho^2 \overline{w'(\zeta)}} \left(\overline{w(\zeta)} \left[\frac{\gamma_1''(\zeta)}{w'(\zeta)} - \frac{\gamma_1'(\zeta) w''(\zeta)}{[w'(\zeta)]^2} \right] + \psi_1'(\zeta) \right) \tag{10.7.5}$$

where in the transformed plane $\zeta = \rho e^{i\varphi}$ and $e^{2i\varphi} = \dfrac{\zeta^2 \overline{w'(\zeta)}}{\rho^2 w'(\zeta)}$. The boundary tractions become

$$i \int \left(T_x^n + i T_y^n \right) ds = \gamma_1(\zeta) + \frac{w(\zeta)}{\overline{w'(\zeta)}} \overline{\gamma_1'(\zeta)} + \overline{\psi_1(\zeta)} \tag{10.7.6}$$

The complex displacement transforms to

$$2\mu \left(u_\rho + i u_\varphi \right) = \kappa \gamma_1(\zeta) - \frac{w(\zeta)}{\overline{w'(\zeta)}} \overline{\gamma_1'(\zeta)} + \overline{\psi_1(\zeta)} \tag{10.7.7}$$

To proceed further we must establish the form of the complex potentials, and this requires information on the problem geometry in order to determine an appropriate mapping function. Although many types of problems can be handled by this scheme, we specialize to the particular case of an *infinite domain bounded internally by an arbitrary closed curve C*, as shown in Fig. 10.15. This case has important applications to problems of stress concentration around holes and cracks in extended planes. We

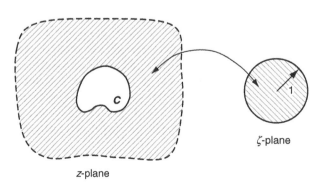

FIG. 10.15 General mapping for an infinite plane with an interior hole.

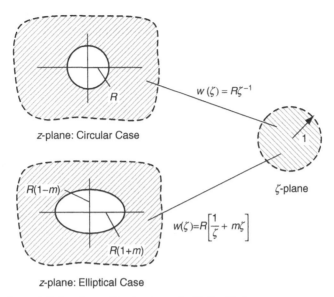

FIG. 10.16 Mappings for an infinite plane with circular and elliptical holes.

choose the particular transformation that maps the region exterior to boundary C onto the interior of the unit disk. Some authors use a scheme that maps the region onto the exterior of the unit disk in the ζ-plane. Either mapping scheme can be used for problem solution by incorporating the appropriate interior or exterior solution for the unit disk problem. Mappings for the special cases of circular and elliptical holes are shown in Fig. 10.16, and additional examples can be found in Milne-Thomson (1960) and Little (1973).

For the exterior problem, the potential functions are given by relations (10.4.7) and (10.4.8), and when applied to the case under study give

$$\gamma(z) = -\frac{F}{2\pi(1+\kappa)}\log[w(\zeta)] + \frac{\sigma_x^\infty + \sigma_y^\infty}{4}w(\zeta) + \gamma^*(\zeta)$$

$$\psi(z) = -\frac{\kappa\overline{F}}{2\pi(1+\kappa)}\log[w(\zeta)] + \frac{\sigma_y^\infty - \sigma_x^\infty + 2i\tau_{xy}^\infty}{2}w(\zeta) + \psi^*(\zeta)$$

(10.7.8)

where F is the resultant force on the internal boundary C, and the functions $\gamma^*(\zeta)$ and $\psi^*(\zeta)$ are analytic in the interior of the unit circle. For the geometry under investigation, the mapping function will always have the general form $w(\zeta) = C\zeta^{-1} + $ (analytic function), and thus the logarithmic term in Eq. (10.7.8) can be written as $\log w = -\log \zeta + $ (analytic function). This allows the potentials to be expressed as

$$\gamma_1(\zeta) = \frac{F}{2\pi(1+\kappa)}\log \zeta + \frac{\sigma_x^\infty + \sigma_y^\infty}{4}\frac{C}{\zeta} + \gamma^*(\zeta)$$

$$\psi_1(\zeta) = -\frac{\kappa\overline{F}}{2\pi(1+\kappa)}\log \zeta + \frac{\sigma_y^\infty - \sigma_x^\infty + 2i\tau_{xy}^\infty}{2}\frac{C}{\zeta} + \psi^*(\zeta)$$

(10.7.9)

We now investigate a specific case of an elliptical hole in a stressed plane.

Example 10.7 Stressed infinite plane with an elliptical hole

Consider the problem of a stress-free elliptical hole in an infinite plane subjected to uniform stress $\sigma_x^\infty = S$, $\sigma_y^\infty = \tau_{xy}^\infty = 0$, as shown in Fig. 10.17. The mapping function is given in Fig. 10.16 as

$$w(\zeta) = R\left(\frac{1}{\zeta} + m\zeta\right) \tag{10.7.10}$$

where the major and minor axes are related to the parameters R and m by

$$R = \frac{a+b}{2}, \quad m = \frac{a-b}{a+b} \Rightarrow a = R(1+m), \quad b = R(1-m)$$

For this case, relations (10.7.9) give the potentials

$$\gamma_1(\zeta) = \frac{S}{4}\frac{R}{\zeta} + \gamma^*(\zeta)$$

$$\psi_1(\zeta) = -\frac{S}{2}\frac{R}{\zeta} + \psi^*(\zeta) \tag{10.7.11}$$

where $\gamma^*(\zeta)$ and $\psi^*(\zeta)$ are analytic in the unit circle. These functions may be determined by using either Fourier or Cauchy integral methods as outlined in Section 10.5. Details on this procedure may be found in Little (1973), Muskhelishvilli (1963), or Milne-Thomson (1960). The result is

$$\gamma^*(\zeta) = \frac{SR}{4}(2-m)\zeta$$

$$\psi^*(\zeta) = \frac{SR}{2}\frac{\zeta}{(m\zeta^2-1)}\left(m^2 - 1 - \zeta^2 - m\right) \tag{10.7.12}$$

The stress combination in the ζ-plane is then given by

$$\sigma_\rho + \sigma_\varphi = S\,\mathrm{Re}\left(\frac{\left(2\zeta^2 - m\zeta^2 - 1\right)\left(m\bar{\zeta}^2 - 1\right)}{m^2\zeta^2\bar{\zeta}^2 - m\left(\zeta^2 + \bar{\zeta}^2\right) + 1}\right) \tag{10.7.13}$$

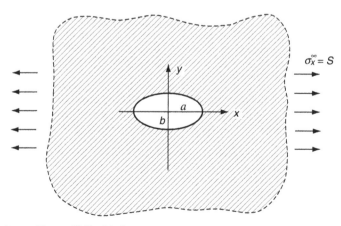

FIG. 10.17 Infinite plane with an elliptical hole.

On the boundary $\zeta = e^{i\varphi}$, $\sigma\rho = 0$ and the circumferential stress is given by

$$\sigma_\varphi(\varphi) = S\left(\frac{2m + 1 - 2\cos 2\varphi - m^2}{m^2 - 2m\cos 2\varphi + 1}\right) \tag{10.7.14}$$

The maximum value of this stress is found at $\varphi = \pm\pi/2$ with a value

$$(\sigma_\varphi)_{max} = -S\left(\frac{m - 3}{m + 1}\right) = S\left(1 + 2\frac{b}{a}\right) \tag{10.7.15}$$

Note the case $m = 0$ corresponds to the circular hole ($a = b = R$) and gives a stress concentration factor of 3, as found previously in Example 8.7. The case $m = 1$ gives $b = 0$, and thus the hole reduces to a *line crack* of length $2a$ parallel to the applied loading. This gives $(\sigma_\varphi)_{max} = S$ with no stress concentration effect. The most interesting case occurs when $m = -1$ because this gives $a = 0$ and reduces the elliptical hole to a line crack of length $2b$ *perpendicular* to the direction of applied stress. As expected for this case, the maximum value of σ_φ at the tip of a crack becomes *unbounded*. Because of the importance of this topic, we further investigate the nature of the stress distribution around cracks in the next section. A plot of the stress concentration factor $(\sigma_\varphi)_{max}/S$ versus the aspect ratio b/a is shown in Fig. 10.18. It is interesting to observe that this relationship is actually linear. For aspect ratios less than 1, the concentration is smaller than that of the circular case, while very high concentrations exist for $b/a > 1$.

A more general and complete solution to the elliptical hole problem has been given by Bonfoh et al. (2007) using *Mathematica* to handle the complex algebraic manipulations required to determine full closed-form analytical expressions for the stress and displacement fields. Gao (1996) also provides a similar complete analytical solution. Further details on such stress concentration problems for holes of different shape can be found in Savin (1961). Numerical techniques employing the finite element method are applied to this stress concentration problem in Chapter 16 (see Example 16.2 and Fig. 16.5).

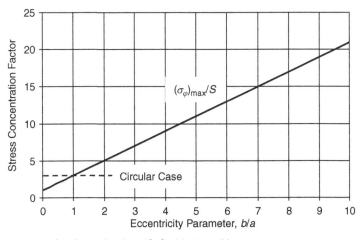

FIG. 10.18 Stress concentration factor for the elliptical hole problem.

10.8 Applications to fracture mechanics

As shown in the previous example and in Section 8.4.10, the elastic stress field around crack tips can become unbounded. For brittle solids, this behavior can lead to rapid crack propagation resulting in catastrophic failure of the structure. Therefore, the nature of such elevated stress distributions around cracks is important in engineering applications, and the general study of such problems forms the basis of *linear elastic fracture mechanics*. Complex variable methods provide a convenient and powerful means to determine stress and displacement fields for a large variety of crack problems. We therefore wish to investigate some of the basic procedures for such applications.

Several decades ago, Westergaard (1937) presented a specialized complex variable method to determine the stresses around cracks. The method used a single complex potential now respectfully called the *Westergaard stress function*. Although this scheme is not a complete representation for all plane elasticity problems, it was widely used to solve many practical problems of engineering interest. More recently, Sih (1966) and Sanford (1979) have re-examined the Westergaard method and established appropriate modifications to extend the validity of this specialized technique. More detailed information on the general method can be found in Sneddon and Lowengrub (1969) and Sih (1973), and an extensive collection of solutions to crack problems has been given by Tada et al. (2000).

Crack problems in elasticity introduce singularities and discontinuities with two important and distinguishing characteristics. The first is involved with the unbounded nature of the stresses at the crack tip, especially in the *type of singularity* of the field. The second feature is that the displacements along the crack surface are commonly multivalued. For open cracks, the crack surface will be stress free. However, some problems may have loadings that can produce crack closure, leading to complicated interface conditions. In order to demonstrate the basic complex variable application for such problems, we now consider a simple example of a crack in an infinite plane under uniform tension loading.

Example 10.8 Infinite plane with a central crack

Consider the problem of an infinite plane containing a stress-free crack of length $2a$ lying along the x-axis, as shown in Fig. 10.19. The plane is subjected to uniform tension S in the y direction, and thus the problem has symmetries about the coordinate axes.

The solution to this problem follows the general procedures of the previous section using the mapping function

$$z = \frac{a}{2}\left(\zeta^{-1} + \zeta\right) \tag{10.8.1}$$

Note this relation is somewhat different than our previous work in that it maps the exterior problem in the z-plane onto the *exterior of the unit circle* in the ζ-plane. Inverting this relation gives

$$\zeta = \frac{1}{a}\left(z + \sqrt{z^2 - a^2}\right) \tag{10.8.2}$$

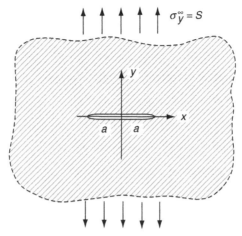

FIG. 10.19 Central crack in an infinite plane.

where the positive sign for the radical term has been chosen because we are interested in the exterior mapping. Using this result, we can eliminate ζ from expressions in our previous work and express the potentials in terms of z

$$\gamma(z) = \frac{S}{4}\left(2\sqrt{z^2 - a^2} - z\right)$$

$$\psi(z) = \frac{S}{2}\left(z - \frac{a^2}{\sqrt{z^2 - a^2}}\right)$$

(10.8.3)

For this case the stress combinations become

$$\sigma_x + \sigma_y = S\ \mathrm{Re}\left(\frac{z}{\sqrt{z^2 - a^2}} + \frac{\bar{z}}{\sqrt{\bar{z}^2 - a^2}} - 1\right)$$

$$\sigma_y - \sigma_x + 2i\tau_{xy} = S\left(\frac{\bar{z}}{\sqrt{z^2 - a^2}} - \frac{\bar{z}z^2}{\left(z^2 - a^2\right)^{3/2}} + \frac{za^2}{\left(z^2 - a^2\right)^{3/2}} + 1\right)$$

$$= Sa^2\left(\frac{z - \bar{z}}{\left(z^2 - a^2\right)^{3/2}} + \frac{1}{a^2}\right)$$

(10.8.4)

Fracture mechanics applications are normally interested in the solution near the crack tip. In order to extract this information, consider the geometry of the crack neighborhood, as shown in Fig. 10.20. For this case we define a polar coordinate system centered on the crack tip at $z = a$ and

Continued

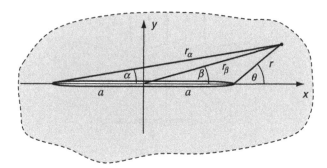

FIG. 10.20 Crack geometry.

wish to develop the crack-tip solution for small values of r. In terms of the given geometry, we note that $z = r_\beta\, e^{i\beta}$, $z - a = re^{i\theta}$, $z + a = r_\alpha\, e^{i\alpha}$, and $r \sin \theta = r_\alpha \sin \alpha = r_\beta \sin \beta$.

Using these new geometric variables, the stress combinations and displacements can be written as

$$\sigma_x + \sigma_y = S\left(\frac{2r_\beta}{\sqrt{rr_\alpha}}\cos\left[\beta - \frac{\theta+\alpha}{2}\right] - 1\right)$$

$$\sigma_y - \sigma_x + 2i\tau_{xy} = \frac{2Sa^2 ir_\beta \sin\beta}{(rr_\alpha)^{3/2}}\left(\cos\left[\frac{3(\theta+\alpha)}{2}\right] - i\sin\left[\frac{3(\theta+\alpha)}{2}\right]\right) + S$$

(10.8.5)

Evaluating these relations for small r gives

$$\sigma_x + \sigma_y = \frac{2Sa}{\sqrt{2ar}}\cos\frac{\theta}{2}$$

$$\sigma_y - \sigma_x + 2i\tau_{xy} = \frac{2Sa}{\sqrt{2ar}}\sin\frac{\theta}{2}\cos\frac{\theta}{2}\left(\sin\frac{3\theta}{2} + i\cos\frac{3\theta}{2}\right)$$

(10.8.6)

and solving for the individual stresses produces the following

$$\sigma_x = \frac{K_I}{\sqrt{2\pi r}}\cos\frac{\theta}{2}\left(1 - \sin\frac{\theta}{2}\sin\frac{3\theta}{2}\right)$$

$$\sigma_y = \frac{K_I}{\sqrt{2\pi r}}\cos\frac{\theta}{2}\left(1 + \sin\frac{\theta}{2}\sin\frac{3\theta}{2}\right)$$

(10.8.7)

$$\tau_{xy} = \frac{K_I}{\sqrt{2\pi r}}\sin\frac{\theta}{2}\cos\frac{\theta}{2}\cos\frac{3\theta}{2}$$

where the parameter $K_I = S\sqrt{\pi a}$ and is referred to as the *stress intensity factor*. Using relation (10.2.9), the corresponding crack-tip displacements can be expressed by

$$u = \frac{K_I}{\mu} \frac{\sqrt{r}}{2\pi} \cos\frac{\theta}{2} \left(\frac{\kappa - 1}{2} + \sin^2\frac{\theta}{2} \right)$$

$$v = \frac{K_I}{\mu} \frac{\sqrt{r}}{2\pi} \sin\frac{\theta}{2} \left(\frac{\kappa + 1}{2} - \cos^2\frac{\theta}{2} \right)$$

$$(10.8.8)$$

As observed in Section 8.4.10, these results indicate that the crack-tip stresses have an $r^{-1/2}$ singularity, while the displacements behave as $r^{1/2}$. The *stress intensity factor* K_I is a measure of the intensity of the stress field near the crack tip under the *opening mode (mode I) deformation*. Two additional *shearing modes* also exist for such crack problems, and the crack-tip stress and displacement fields for these cases have the same r dependence but different angular distributions (see Eq. (8.4.57) and Exercise 8.44). For the central crack problem considered in this example, the stress intensity factor was proportional to $\sqrt{a}$; however, for other crack geometries, this factor will be related to problem geometry in a more complex fashion. Comparing the vertical displacements on the top and bottom crack surfaces indicates that $v(r,\pi) = -v(r,-\pi)$. This result illustrates the expected multivalued discontinuous behavior on each side of the crack surface under opening mode deformation.

10.9 Westergaard method for crack analysis

As mentioned, Westergaard (1937) developed a specialized complex variable technique to handle a restricted class of plane problems. The method uses a single complex potential, and thus the scheme is not a complete representation for all plane elasticity problems. Nevertheless, the technique has been extensively applied to many practical problems in fracture mechanics dealing with the determination of stress fields around cracks. As previously mentioned, Shi (1966) and Sanford (1979) have re-examined the Westergaard method and established appropriate modifications to extend its validity.

In order to develop the procedure, consider again the central crack problem shown in Fig. 10.19. Because this is a symmetric problem, the shear stresses must vanish on $y = 0$, and thus from relation (10.2.11)

$$\text{Im}[\bar{z}\gamma''(z) + \psi'(z)] = 0 \text{ on } y = 0 \tag{10.9.1}$$

This result can be satisfied by taking

$$z\gamma''(z) + \psi'(z) = A \tag{10.9.2}$$

where we have used $z = \bar{z}$ on $y = 0$, and A is a real constant determined by the boundary conditions. Eq. (10.9.2) can be integrated to give the result

$$\psi(z) = \gamma(z) - z\gamma'(z) + Az \tag{10.9.3}$$

where the constant of integration has been dropped. This provides a relation to express one potential function in terms of the other, and thus we can eliminate one function for this class of problem.

Using the stress combination definitions (10.2.11), we eliminate the ψ potential and find

$$
\begin{aligned}
\sigma_x &= 2\,\mathrm{Re}[\gamma'(z)] - 2y\mathrm{Im}[\gamma''(z)] - A \\
\sigma_y &= 2\,\mathrm{Re}[\gamma'(z)] + 2y\mathrm{Im}[\gamma''(z)] + A \\
\tau_{xy} &= -2y\,\mathrm{Re}[\gamma''(z)]
\end{aligned}
\tag{10.9.4}
$$

Defining the *Westergaard stress function* $Z(z) = 2\gamma'(z)$, the stresses can now be written as

$$
\begin{aligned}
\sigma_x &= \mathrm{Re}Z(z) - y\mathrm{Im}Z'(z) - A \\
\sigma_y &= \mathrm{Re}Z(z) + y\mathrm{Im}Z'(z) + A \\
\tau_{xy} &= -y\mathrm{Re}Z'(z)
\end{aligned}
\tag{10.9.5}
$$

Note that this scheme is sometimes referred to as the *modified Westergaard stress function formulation*.

This method can be applied to the central crack problem of Example 10.8. For this case, the Westergaard function is given by

$$
Z(z) = \frac{Sz}{\sqrt{z^2 - a^2}} - \frac{S}{2}
\tag{10.9.6}
$$

with $A = S/2$. The stresses follow from equation (10.9.5) and would give identical values as previously developed.

The Westergaard method can also be developed for skew-symmetric crack problems in which the normal stress σ_y vanishes along $y = 0$. Exercise 10.27 explores this formulation.

References

Bonfoh N, Tiem S, Carmasol A: Analytical solutions of plane elasticity problems, Part I: elastic regions weakened by elliptic holes, *J Sci Pour l'Ingenieur* 8:62–73, 2007.

Churchill RV: *Complex variables and applications*, New York, 1960, McGraw-Hill.

England AH: *Complex variable methods in elasticity*, New York, 1971, John Wiley.

Filon LNG: On the approximate solution for the bending of a beam of rectangular cross-section, *Phil Trans* 201: 63–155, 1903.

Gao XL: A general solution of an infinite elastic plate with an elliptical hole under biaxial loading, *Int J Press Invest Piping* 67:95–104, 1996.

Green AE, Zerna W: *Theoretical elasticity*, ed 3, London, 1968, Oxford University Press.

Kober H: *Dictionary of conformal representations*, New York, 1952, Dover.

Kolosov GV: *On the application of complex function theory to a plane problem of the mathematical theory of elasticity* (Dissertation). Dorpat University.

Kreyszig E: *Advanced engineering mathematics*, New York, 2011, John Wiley.

Little RW: *Elasticity*, Englewood Cliffs, NJ, 1973, Prentice Hall.

Milne-Thomson LM: *Plane elastic systems*, Berlin, 1960, Springer.

Muskhelishvili NJ: *Singular integral equations*, trans. JRM Radok and P Noordhoff The Netherlands, 1953. Groningen.

Muskhelishvili NJ: *Some basic problems of the theory of elasticity*, The Netherlands, 1963. Groningen, trans. JRM Radok and P Noordhoff.

Sanford RJ: A critical re-examination of the Westergaard method for solving opening-mode crack problems, *Mech Res Commun* 6:289–294, 1979.

Savin GN: *Stress concentration around holes*, New York, 1961, Pergamon.

Sih GC: On the Westergaard method of crack analysis, *Int J Fract Mech* 2:628–630, 1966.

Sih GC, editor: *Mechanics of fracture I methods of analysis and solutions of crack problems*, Leyden, The Netherlands, 1973, Noordhoff International.

Sneddon IN, Lowengrub M: *Crack problems in the classical theory of elasticity*, New York, 1969, John Wiley.

Sokolnikoff IS: *Mathematical theory of elasticity*, New York, 1956, McGraw-Hill.

Tada H, Paris PC, Irwin GR: *The stress analysis of cracks handbook*, ed 3, New York, 2000, American Society of Mechanical Engineers.

Westergaard HM: Bearing pressures and cracks, *J Appl Mech* 6, 1937. A49–53.

Exercises

10.1 Derive the relations (10.1.4) and (10.2.5)

$$\frac{\partial}{\partial x} = \frac{\partial}{\partial z} + \frac{\partial}{\partial \bar{z}}, \quad \frac{\partial}{\partial y} = i\left(\frac{\partial}{\partial z} - \frac{\partial}{\partial \bar{z}}\right)$$

$$\frac{\partial}{\partial z} = \frac{1}{2}\left(\frac{\partial}{\partial x} - i\frac{\partial}{\partial y}\right), \quad \frac{\partial}{\partial \bar{z}} = \frac{1}{2}\left(\frac{\partial}{\partial x} + i\frac{\partial}{\partial y}\right)$$

$$\nabla^2() = 4\frac{\partial^2()}{\partial z \partial \bar{z}}, \quad \nabla^4() = 16\frac{\partial^4()}{\partial z^2 \partial \bar{z}^2}$$

10.2 Formally integrate relation (10.2.6) and establish the result

$$\phi(z, \bar{z}) = \frac{1}{2}\left(\overline{z\gamma(z)} + \bar{z}\gamma(z) + \psi(z) + \overline{\psi(z)}\right)$$

$$= \mathrm{Re}(\bar{z}\gamma(z) + \psi(z))$$

10.3 Starting with the Navier equations (10.2.2) for plane strain, introduce the complex displacement $U = u + iv$, and show that

$$(\lambda + \mu)\frac{\partial}{\partial \bar{z}}\left(\frac{\partial U}{\partial z} + \frac{\overline{\partial U}}{\partial z}\right) + 2\mu\frac{\partial^2 U}{\partial \bar{z} \partial z} = 0$$

Integrate this result with respect to $\bar{z}$ to get

$$(\lambda + \mu)\left(\frac{\partial U}{\partial z} + \frac{\overline{\partial U}}{\partial z}\right) + 2\mu\frac{\partial U}{\partial z} = f'(z)$$

where $f'(z)$ is an arbitrary analytic function of z. Next, combining both the preceding equation and its conjugate, solve for $\partial U/\partial z$, and simplify to get form (10.2.9).

10.4 Establish the relations

$$\frac{\partial}{\partial \bar{z}}[2\mu(u + iv)] = -\left[\overline{z\gamma''(z)} + \overline{\psi'(z)}\right] = \mu(e_x - e_y + 2ie_{xy})$$

where e_x, e_y, and e_{xy} are the strain components.

10.5 Explicitly derive the fundamental stress combinations

$$\sigma_x + \sigma_y = 2\left[\gamma'(z) + \overline{\gamma'(z)}\right]$$

$$\sigma_y - \sigma_x + 2i\tau_{xy} = 2[\bar{z}\gamma''(z) + \psi'(z)]$$

Next solve these relations for the individual stresses

$$\sigma_x = 2\,\mathrm{Re}\left[\gamma'(z) - \frac{1}{2}\bar{z}\gamma''(z) - \frac{1}{2}\psi'(z)\right]$$

$$\sigma_y = 2\,\mathrm{Re}\left[\gamma'(z) + \frac{1}{2}\bar{z}\gamma''(z) + \frac{1}{2}\psi'(z)\right]$$

$$\tau_{xy} = \mathrm{Im}[\bar{z}\gamma''(z) + \psi'(z)]$$

10.6 Develop the polar coordinate transformation relations for the stress combinations and complex displacements given in equations (10.2.12).

10.7 Determine the Cartesian stresses and displacements in a rectangular domain ($-a \leq x \leq a$; $-b \leq y \leq b$) from the potentials $\gamma(z) = Aiz^2$, $\psi(z) = -Aiz^2$, where A is an arbitrary constant. Discuss the boundary values of these stresses, and show that this particular case could be used to solve a pure bending problem.

10.8 Determine the polar coordinate stresses corresponding to the complex potentials $\gamma(z) = Az$ and $\psi(z) = Bz^{-1}$, where A and B are arbitrary constants. Show that these potentials could solve the plane problem of a cylinder with both internal and external pressure loadings.

10.9 Show that the potentials $\gamma(z) = 0$, $\psi(z) = A/z$ will solve the problem of a circular hole of radius a with uniform pressure loading p in an infinite elastic plane. Determine the constant A and the stress and displacement fields for $r \geq a$.

10.10 Consider the problem geometry described in Exercise 10.9. Show that the suggested potentials (with different A) can also be used to solve the problem of a rigid inclusion of radius $a + \delta$, which is forced into the hole of radius a. Determine the new constant A and the stress and displacement fields for $r \geq a$.

10.11 From Section 10.4, the complex potentials

$$\gamma(z) = -\frac{X + iY}{2\pi(1 + \kappa)}\log z, \quad \psi(z) = \frac{\kappa(X - iY)}{2\pi(1 + \kappa)}\log z$$

would be the appropriate forms for a problem in which the body contains a hole surrounding the origin (i.e., multiply connected). Show for this case that the complex displacement U is unbounded as $|z| \to 0$ and $|z| \to \infty$. Also, explicitly verify that the resultant force across any contour surrounding the origin is $X + iY$. Finally, determine the stress distribution on the circle $r = a$.

10.12 Show that the resultant moment caused by tractions on a boundary contour AB is given by relation (10.3.2)

$$M = \mathrm{Re}[\chi(z) - z\psi(z) - z\bar{z}\gamma'(z)]_A^B, \quad \text{where } \chi'(z) = \psi(z)$$

10.13 An infinite elastic medium $|z| \geq a$ is bonded over its internal boundary $|z| = a$ to a *rigid inclusion of radius a*. The inclusion is acted upon by a *force* $X + iY$ and a *moment M* about its center. Show that the problem is solved by the potentials

$$\gamma(z) = -\frac{X + iY}{2\pi(1 + \kappa)} \log z$$

$$\psi(z) = \frac{X - iY}{2\pi(1 + \kappa)} \kappa \log z + \frac{X + iY}{2\pi(1 + \kappa)} \frac{a^2}{z^2} + \frac{iM}{2\pi z}$$

Finally, show that the rigid-body motion of the inclusion is given by

$$u_o = \frac{-\kappa X \log a}{2\pi\mu(1 + \kappa)}, \quad v_o = \frac{-\kappa Y \log a}{2\pi\mu(1 + \kappa)}, \quad \theta_o = \frac{M}{4\pi\mu a^2}$$

10.14 An infinite isotropic sheet contains a perfectly bonded, rigid inclusion of radius a and is under uniform tension T in the x-direction as shown. Show that this problem is solved by the following potentials

$$\gamma(z) = \frac{T}{4}\left(z - \frac{2a^2}{\kappa z}\right), \quad \psi(z) = -\frac{T}{2}\left(z - \frac{\kappa - 1}{2z}a^2 + \frac{a^4}{\kappa z^3}\right)$$

Note that because of the problem symmetry, the inclusion will not move. In solving this problem, you should verify boundary conditions on $r = a$ and at $r \to \infty$, and the appropriateness of the potential forms as per Section 10.4.

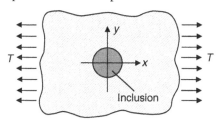

Inclusion

10.15 Consider the unit disk problem with *displacement boundary conditions* $u_r + iu_\theta = h(\zeta)$ on $C: \zeta = e^{i\theta}$. Using Cauchy integral methods described in Section 10.5, determine the form of the potentials $\gamma(z)$ and $\psi(z)$.

10.16 For Example 10.3 with $\alpha = 0$, verify that the stresses from equation (10.5.18) reduce to those previously given in Eq. (8.4.69).

10.17 Consider the concentrated force system problem shown in Fig. 10.11. Verify for the special case of $X = P$ and $Y = M = 0$ that the stress field reduces to relations (10.6.5). Also determine the corresponding stresses in polar coordinates.

10.18 For the stress concentration problem shown in Fig. 10.14, solve the problem with the far-field loadings $\sigma_x^\infty = \sigma_y^\infty = S$, $\tau_{xy}^\infty = 0$, and compute the stress concentration factor. Verify your solution with that given in Eqs. (8.4.9) and (8.4.10).

10.19 Verify the mappings shown in Fig. 10.16 by explicitly investigating points on the boundaries and the point at infinity in the z-plane.

10.20* Consider relation (10.7.14) for the circumferential stress σ_φ on the boundary of the elliptical hole shown in Fig. 10.17. Explicitly verify that the maximum stress occurs at $\varphi = \pi/2$. Next plot the distribution of σ_φ versus φ for the cases of $m = 0, \pm0.5, \pm1$.

10.21 Using the solution from Example 10.7, apply the principle of superposition and solve the problem of a stress-free elliptical hole under uniform biaxial tension as shown. In particular show that the circumferential stress is given by

$$\sigma_\varphi(\varphi) = S_x\left(\frac{2m + 1 - 2\cos 2\varphi - m^2}{m^2 - 2m\cos 2\varphi + 1}\right) + S_y\left(\frac{-2m + 1 + 2\cos 2\varphi - m^2}{m^2 - 2m\cos 2\varphi + 1}\right)$$

Finally for the special case of $S_x = S_y = S$, show that

$$\sigma_\varphi(\varphi) = S\left(\frac{2(1 - m^2)}{m^2 - 2m\cos 2\varphi + 1}\right)$$

and justify that for $a > b$, $\sigma_{max} = \sigma_\varphi(0) = 2S(a/b)$, while for $b > a$, $\sigma_{max} = \sigma_\varphi(\pi/2) = 2S(b/a)$.

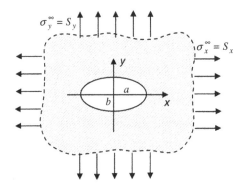

10.22* Consider the problem of an infinite plate containing a stress-free elliptical hole with $\sigma_x^\infty = \sigma_y^\infty = 0$, $\tau_{xy}^\infty = S$. For this problem, the derivative of the complex potential has been developed by Milne-Thomson (1960)

$$\frac{d\gamma(\zeta)}{dz} = \frac{iS}{m - \zeta^2}$$

Show that the stress on the boundary of the hole is given by

$$\sigma_\varphi = -\frac{4S \sin 2\varphi}{m^2 - 2m \cos 2\varphi + 1}$$

Determine and plot σ_φ vs. φ for the cases $m = 0, 0.5,$ and 1. Identify maximum stress values and locations.

10.23 Verify the crack-tip stress distributions given by Eqs. (10.8.6) and (10.8.7).

10.24 Construct a contour plot of the crack tip stress component σ_y from Eq. (10.8.7). Normalize values with respect to K_I.

10.25 Verify that the crack-tip displacements are given by Eq. (10.8.8).

10.26 Show that the Westergaard stress function

$$Z(z) = \frac{Sz}{\sqrt{(z^2 - a^2)}} - \frac{S}{2}$$

with $A = S/2$ solves the central crack problem shown in Fig. 10.19.

10.27 Following similar procedures as in Section 10.9, establish a Westergaard stress function method for *skew-symmetric* crack problems. For this case, assume that loadings are applied skew-symmetrically with respect to the crack, thereby establishing that the normal stress σ_y vanishes along $y = 0$. Show that this leads to the following relations

$$\sigma_x = 2\text{Re}Z(z) - y\text{Im}Z'(z)$$
$$\sigma_y = y\text{Im}Z'(z)$$
$$\tau_{xy} = -\text{Im}Z(z) - y\text{Re}Z'(z) - B$$

where $Z(z) = 2\gamma'(z)$ and B is a real constant.

Anisotropic elasticity

11

It has long been recognized that deformation behavior of many materials depends upon orientation; that is, the stress–strain response of a sample taken from the material in one direction will be different than if the sample were taken in a different direction. The term *anisotropic* is generally used to describe such behaviors. Early investigators of these phenomena were motivated by the response of naturally occurring anisotropic materials such as wood and crystalline solids. Today, extensive use of engineered composites (see Jones, 1998; Swanson, 1997) has brought forward many new types of fiber- and particle-reinforced materials with anisotropic response. Thus, knowledge of stress distributions in anisotropic materials is very important for proper use of these new high-performance materials in structural applications. Our previous development of the linear elastic stress–strain relations in Section 4.2 began with the general case of inhomogeneous and anisotropic behavior. However, this generality was quickly eliminated, and only the homogeneous isotropic case was subsequently developed in detail. We now wish to go back and further investigate the anisotropic homogeneous case and develop applications for particular elasticity problems including torsion and plane problems. Much of the original work in this field was done by Lekhnitskii (1968, 1981), while Love (1934) and Hearmon (1961) also made early contributions. More recently, texts by Ting (1996a) and Rand and Rovenski (2005) provide modern and comprehensive accounts on this subject.

11.1 Basic concepts

The directional-dependent behaviors found in anisotropic solids normally result from particular microstructural features within the material. Our previous isotropic model neglected these effects, thus resulting in a material that behaved the same in all directions. Micro features commonly arise in natural and synthetic materials in such a way as to produce a stress–strain response with particular symmetries. This concept is based on the *Neumann principle* (Love, 1934) that *symmetry in material microgeometry corresponds to identical symmetry in the constitutive response.* We can qualitatively gain an understanding of this concept from some simple two-dimensional cases shown in Fig. 11.1. The figure illustrates idealized internal microstructures of two crystalline solids and one fiber composite. The two crystalline materials correspond to special atomic packing arrangements that lead to identical behaviors in the indicated directions of the arrows. The fiber-reinforced composite has a 90° fiber layout, which again produces identical behaviors in the layout directions. Many other material symmetries exist for more complicated microstructures, and some follow a curvilinear reference system such as that found in wood. These symmetries generally lead to a reduction in the complexity of the stress–strain constitutive relation, and examples of this are shown in the next section. On a related topic, for multiphase or porous materials, many researchers (e.g., Cowin, 1985) have been

Elasticity. https://doi.org/10.1016/B978-0-12-815987-3.00011-6

331

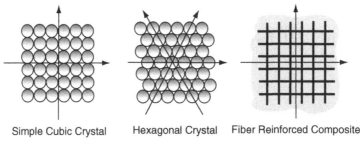

Simple Cubic Crystal   Hexagonal Crystal   Fiber Reinforced Composite
(Arrows indicate material symmetry directions)

FIG. 11.1 Material microstructures.

trying to establish relationships between the elasticity tensor C_{ijkl} and a *fabric tensor* that characterizes the arrangement of microstructural material components (also see Sadd, 2019).

From Section 4.2, the general form of Hooke's law was given by

$$\sigma_{ij} = C_{ijkl}e_{kl} \tag{11.1.1}$$

The fourth-order elasticity tensor C_{ijkl} contains all of the elastic stiffness moduli, and we have previously established the following symmetry properties

$$C_{ijkl} = C_{jikl} = C_{ijlk} = C_{klij} \tag{11.1.2}$$

The first two symmetries in relation (11.1.2) come from the symmetry of the stress and strain tensors, while the final relation comes from arguments based on the existence of the strain energy function (see Section 6.1). Relations (11.1.2) reduce the original 81 independent elastic constants within C_{ijkl} to a set of 21 *elastic moduli* for the general case. We shall assume that the material is homogeneous and thus the moduli are independent of spatial position. On occasion we may wish to invert (11.1.1) and write strain in terms of stress

$$e_{ij} = S_{ijkl}\sigma_{kl} \tag{11.1.3}$$

where S_{ijkl} is the *elastic compliance tensor*, which has identical symmetry properties as those in relations (11.1.2).

Because of the various pre-existing symmetries, stress–strain relations (11.1.1) and (11.1.3) contain many superfluous terms and equations. To avoid these, a convenient contracted notation has been developed, sometimes referred to as *Voigt matrix notation*

$$\begin{bmatrix} \sigma_x \\ \sigma_y \\ \sigma_z \\ \tau_{yz} \\ \tau_{zx} \\ \tau_{xy} \end{bmatrix} = \begin{bmatrix} C_{11} & C_{12} & \cdots & C_{16} \\ C_{21} & . & \cdots & . \\ . & . & \cdots & . \\ . & . & \cdots & . \\ . & . & \cdots & . \\ C_{61} & . & \cdots & C_{66} \end{bmatrix} \begin{bmatrix} e_x \\ e_y \\ e_z \\ 2e_{yz} \\ 2e_{zx} \\ 2e_{xy} \end{bmatrix} \tag{11.1.4}$$

or in compact notation

$$\sigma_i = C_{ij} e_j \tag{11.1.5}$$

where σ_i and e_i are defined by comparing relations (11.1.4) and (11.1.5). Note that the symmetry imposed by strain energy implies that the 6×6 **C** matrix is symmetric; that is, $C_{ij} = C_{ji}$, and thus only 21 independent elastic constants exist. The two elasticity stiffness tensors are related by the expression

$$C_{ij} = \begin{bmatrix} C_{1111} & C_{1122} & C_{1133} & C_{1123} & C_{1131} & C_{1112} \\ \cdot & C_{2222} & C_{2233} & C_{2223} & C_{2231} & C_{2212} \\ \cdot & \cdot & C_{3333} & C_{3323} & C_{3331} & C_{3312} \\ \cdot & \cdot & \cdot & C_{2323} & C_{2331} & C_{2312} \\ \cdot & \cdot & \cdot & \cdot & C_{3131} & C_{3112} \\ \cdot & \cdot & \cdot & \cdot & \cdot & C_{1212} \end{bmatrix} \tag{11.1.6}$$

A similar scheme can be established for relation (11.1.3), and a compliance matrix S_{ij} can be defined by

$$e_i = S_{ij} \sigma_j \tag{11.1.7}$$

11.2 Material symmetry

From the previous section, we determined that for the general anisotropic case (sometimes referred to as *triclinic material*), 21 independent elastic constants are needed to characterize the material response. However, as per our discussion related to Fig. 11.1, most real materials have some types of symmetry, which further reduces the required number of independent elastic moduli. Orientations for which an anisotropic material has the same stress–strain response can be determined by coordinate transformation (rotation) theory previously developed in Sections 1.4 and 1.5. Such particular transformations are sometimes called the *material symmetry group*. Further details on this topic have been presented by Zheng and Spencer (1993) and Cowin and Mehrabadi (1995). In order to determine various material symmetries, it is more convenient to work in the noncontracted form. Thus, applying this theory, Hooke's law (11.1.1) can be expressed in a new coordinate system as

$$\sigma'_{ij} = C'_{ijkl} e'_{kl} \tag{11.2.1}$$

Now because the stress and strain must transform as second-order tensors

$$\begin{aligned} \sigma'_{ij} &= Q_{ik} Q_{jl} \sigma_{kl}, \quad \sigma_{ij} = Q_{ki} Q_{lj} \sigma'_{kl} \\ e'_{ij} &= Q_{ik} Q_{jl} e_{kl}, \quad e_{ij} = Q_{ki} Q_{lj} e'_{kl} \end{aligned} \tag{11.2.2}$$

Combining Eqs. (11.2.1) and (11.2.2) and using the orthogonality conditions (1.4.9) and (1.4.10) yields the transformation law for the elasticity tensor

$$C'_{ijkl} = Q_{im} Q_{jn} Q_{kp} Q_{lq} C_{mnpq} \tag{11.2.3}$$

If under a specific transformation **Q** the material response is to be the same, relation (11.2.3) reduces to

$$C_{ijkl} = Q_{im} Q_{jn} Q_{kp} Q_{lq} C_{mnpq} \tag{11.2.4}$$

This material symmetry relation will provide a system of equations that allows reduction in the number of independent elastic moduli. It has been shown that the number of symmetry planes required to determine the structure of particular elasticity matrices is usually less than the number of symmetries that exist within the material. We now consider some specific cases of practical interest.

11.2.1 Plane of symmetry (monoclinic material)

We first investigate the case of a material with a *plane of symmetry*. Such a medium is commonly referred to as a *monoclinic material*. We consider the case of symmetry with respect to the x,y-plane as shown in Fig. 11.2.

For this particular symmetry, the required transformation is simply a mirror reflection about the x,y-plane and is given by

$$Q_{ij} = \begin{bmatrix} 1 & 0 & 0 \\ 0 & 1 & 0 \\ 0 & 0 & -1 \end{bmatrix} \qquad (11.2.5)$$

Note that this transformation is not a simple rotation that preserves the right-handedness of the coordinate system; that is, it is not a proper orthogonal transformation. Nevertheless, it can be used for our symmetry investigations. Using this specific transformation in relation (11.2.4) gives $C_{ijkl} = -C_{ijkl}$ if the index 3 appears an *odd* number of times, and thus these particular moduli would have to vanish. In terms of the contracted notation, this gives

$$C_{i4} = C_{i5} = C_{46} = C_{56} = 0 \ (i = 1,2,3) \qquad (11.2.6)$$

Thus, the elasticity matrix takes the form

$$C_{ij} = \begin{bmatrix} C_{11} & C_{12} & C_{13} & 0 & 0 & C_{16} \\ . & C_{22} & C_{23} & 0 & 0 & C_{26} \\ . & . & C_{33} & 0 & 0 & C_{36} \\ . & . & . & C_{44} & C_{45} & 0 \\ . & . & . & . & C_{55} & 0 \\ . & . & . & . & . & C_{66} \end{bmatrix} \qquad (11.2.7)$$

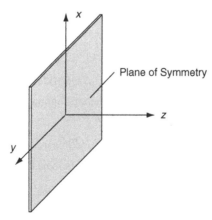

FIG. 11.2 Plane of symmetry for a monoclinic material.

It is therefore observed that *13 independent elastic moduli* are needed to characterize monoclinic materials.

11.2.2 Three perpendicular planes of symmetry (orthotropic material)

A material with three mutually perpendicular planes of symmetry is called *orthotropic*. Common examples of such materials include wood and fiber-reinforced composites. To investigate the material symmetries for this case, it is convenient to let the symmetry planes correspond to coordinate planes as shown in Fig. 11.3.

The symmetry relations can be determined by using 180° rotations about each of the coordinate axes. Another convenient scheme is to start with the reduced form from the previous monoclinic case, and reapply the same transformation with respect to, say, the *y,z*-plane. This results in the additional elastic moduli being reduced to zero

$$C_{16} = C_{26} = C_{36} = C_{45} = 0 \tag{11.2.8}$$

Thus, the elasticity matrix for the orthotropic case reduces to having only *nine independent* stiffnesses given by

$$C_{ij} = \begin{bmatrix} C_{11} & C_{12} & C_{13} & 0 & 0 & 0 \\ . & C_{22} & C_{23} & 0 & 0 & 0 \\ . & . & C_{33} & 0 & 0 & 0 \\ . & . & . & C_{44} & 0 & 0 \\ . & . & . & . & C_{55} & 0 \\ . & . & . & . & . & C_{66} \end{bmatrix} \tag{11.2.9}$$

It should be noted that only two transformations were needed to develop the final reduced constitutive form (11.2.9). The material also must satisfy a third required transformation that the properties would be the same under a reflection of the *x,z*-plane. However, attempting this transformation would only give relations that are identically satisfied. Thus, for some materials the reduced constitutive form may be developed by using only a portion of the total material symmetries [see Ting (1996a) for more on

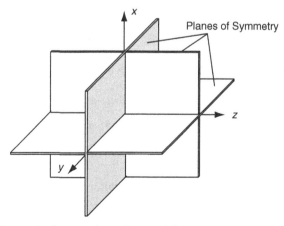

FIG. 11.3 Three planes of symmetry for an orthotropic material.

this topic]. On another issue for orthotropic materials, vanishing shear strains imply vanishing shear stresses, and thus the principal axes of stress coincide with the principal axes of strain. This result is of course not true for general anisotropic materials; see, for example, the monoclinic constitutive form (11.2.7).

For orthotropic materials, the compliance matrix has similar form but is commonly written using notation related to isotropic theory

$$S_{ij} = \begin{bmatrix} \dfrac{1}{E_1} & -\dfrac{\nu_{21}}{E_2} & -\dfrac{\nu_{31}}{E_3} & 0 & 0 & 0 \\[2mm] -\dfrac{\nu_{12}}{E_1} & \dfrac{1}{E_2} & -\dfrac{\nu_{32}}{E_3} & 0 & 0 & 0 \\[2mm] -\dfrac{\nu_{13}}{E_1} & -\dfrac{\nu_{23}}{E_2} & \dfrac{1}{E_3} & 0 & 0 & 0 \\[2mm] \cdot & \cdot & \cdot & \dfrac{1}{\mu_{23}} & 0 & 0 \\[2mm] \cdot & \cdot & \cdot & \cdot & \dfrac{1}{\mu_{31}} & 0 \\[2mm] \cdot & \cdot & \cdot & \cdot & \cdot & \dfrac{1}{\mu_{12}} \end{bmatrix} \tag{11.2.10}$$

where E_i are Young's moduli in the three directions of material symmetry, ν_{ij} are Poisson's ratios defined by $-e_j/e_i$ for a stress in the i direction, and μ_{ij} are the shear moduli in the i, j-planes. Symmetry of this matrix requires that $\nu_{ij}/E_i = \nu_{ji}/E_j$.

11.2.3 Axis of symmetry (transversely isotropic material)

Another common form of material symmetry is with respect to rotations about an axis. This concept can be specified by stating that a material possesses an *axis of symmetry of order n* when the elastic moduli remain unchanged for rotations of $2\pi/n$ radians about the axis. This situation is shown schematically in Fig. 11.4. The hexagonal packing crystalline case shown in Fig. 11.1 has such a symmetry about the axis perpendicular to the page for $n = 6$ (60° increments). It can be shown (Lekhnitskii, 1981) that the only possible symmetries for this case are for orders 2, 3, 4, 6, and infinity. The order 2 case is equivalent to a plane of symmetry previously discussed, and order 6 is equivalent to the infinite case.

The transformation for arbitrary rotations θ about the z-axis is given by

$$Q_{ij} = \begin{bmatrix} \cos\theta & \sin\theta & 0 \\ -\sin\theta & \cos\theta & 0 \\ 0 & 0 & 1 \end{bmatrix} \tag{11.2.11}$$

Using this transformation and invoking symmetry for arbitrary rotations corresponds to the case of $n \to \infty$, and such materials are called *transversely isotropic*. The elasticity stiffness matrix for this case reduces to

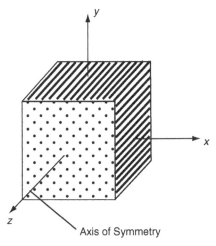

FIG. 11.4 Axis of symmetry for a transversely isotropic material.

$$C_{ij} = \begin{bmatrix} C_{11} & C_{12} & C_{13} & 0 & 0 & 0 \\ \cdot & C_{11} & C_{13} & 0 & 0 & 0 \\ \cdot & \cdot & C_{33} & 0 & 0 & 0 \\ \cdot & \cdot & \cdot & C_{44} & 0 & 0 \\ \cdot & \cdot & \cdot & \cdot & C_{44} & 0 \\ \cdot & \cdot & \cdot & \cdot & \cdot & (C_{11} - C_{12})/2 \end{bmatrix} \tag{11.2.12}$$

Thus, for transversely isotropic materials, only *five* independent elastic constants exist.

11.2.4 Cubic symmetry

Another common anisotropic symmetry applies to many face centered cubic (FCC) and body centered cubic (BCC) metal crystals. Such materials have a four-fold symmetry that combines orthotropic results with an additional symmetry around [1,1,1]. This case is referred to as *cubic symmetry* and modifies the orthotropic results to yield

$$C_{11} = C_{22} = C_{33} \ , \ \ C_{12} = C_{23} = C_{31} \ , \ \ C_{44} = C_{55} = C_{66} \tag{11.2.13}$$

Hence the elasticity matrix for the cubic case reduces to having only *three* independent stiffnesses given by

$$C_{ij} = \begin{bmatrix} C_{11} & C_{12} & C_{12} & 0 & 0 & 0 \\ \cdot & C_{11} & C_{12} & 0 & 0 & 0 \\ \cdot & \cdot & C_{11} & 0 & 0 & 0 \\ \cdot & \cdot & \cdot & C_{44} & 0 & 0 \\ \cdot & \cdot & \cdot & \cdot & C_{44} & 0 \\ \cdot & \cdot & \cdot & \cdot & \cdot & C_{44} \end{bmatrix} \tag{11.2.14}$$

Notice that this result is similar to the isotropic case (discussed next) but without the relationship between C_{11}, C_{12} and C_{44}.

11.2.5 Complete symmetry (isotropic material)

For the case of complete symmetry, the material is referred to as *isotropic*, and the fourth-order elasticity tensor has been previously given by

$$C_{ijkl} = \lambda \delta_{ij}\delta_{kl} + \mu(\delta_{ik}\delta_{jl} + \delta_{il}\delta_{jk}) \qquad (11.2.15)$$

This form can be determined by invoking symmetry with respect to two orthogonal axes, which implies symmetry about the remaining axis. In contracted matrix form, this result would be expressed as

$$C_{ij} = \begin{bmatrix} \lambda + 2\mu & \lambda & \lambda & 0 & 0 & 0 \\ \cdot & \lambda + 2\mu & \lambda & 0 & 0 & 0 \\ \cdot & \cdot & \lambda + 2\mu & 0 & 0 & 0 \\ \cdot & \cdot & \cdot & \mu & 0 & 0 \\ \cdot & \cdot & \cdot & \cdot & \mu & 0 \\ \cdot & \cdot & \cdot & \cdot & \cdot & \mu \end{bmatrix} \qquad (11.2.16)$$

Thus, as shown previously, only *two* independent elastic constants exist for isotropic materials. For each case presented, a similar compliance elasticity matrix could be developed. Our brief presentation does not include all cases of material symmetry, and in fact we have only investigated what is generally referred to as *rectilinear anisotropy*. Based on symmetry planes, Ting (2003) has proven that there are only eight symmetries for linear anisotropic elastic materials for the general case. Of course, a large variety of curvilinear material symmetries also exist in various biological and synthetic materials and these will be discussed further in Section 11.7.

Example 11.1 Hydrostatic compression of a monoclinic cube

In order to demonstrate the difference in behavior between isotropic and anisotropic materials, consider a simple example of a cube of monoclinic material under hydrostatic compression. For this case, the state of stress is given by $\sigma_{ij} = -p\delta_{ij}$, and the monoclinic Hooke's law in compliance form would read as follows

$$\begin{bmatrix} e_x \\ e_y \\ e_z \\ 2e_{yz} \\ 2e_{zx} \\ 2e_{xy} \end{bmatrix} = \begin{bmatrix} S_{11} & S_{12} & S_{13} & 0 & 0 & S_{16} \\ \cdot & S_{22} & S_{23} & 0 & 0 & S_{26} \\ \cdot & \cdot & S_{33} & 0 & 0 & S_{36} \\ \cdot & \cdot & \cdot & S_{44} & S_{45} & 0 \\ \cdot & \cdot & \cdot & \cdot & S_{55} & 0 \\ \cdot & \cdot & \cdot & \cdot & \cdot & S_{66} \end{bmatrix} \begin{bmatrix} -p \\ -p \\ -p \\ 0 \\ 0 \\ 0 \end{bmatrix} \qquad (11.2.17)$$

Expanding this matrix relation gives the following deformation field components

$$e_x = -(S_{11} + S_{12} + S_{13})p$$

$$e_y = -(S_{12} + S_{22} + S_{23})p$$

$$e_z = -(S_{13} + S_{23} + S_{33})p$$

$$e_{yz} = 0 \qquad\qquad (11.2.18)$$

$$e_{zx} = 0$$

$$e_{xy} = -\frac{1}{2}(S_{16} + S_{26} + S_{36})p$$

The corresponding strains for the isotropic case would be given by $e_x = e_y = e_z = -[(1 - 2\nu)/E]p$, $e_{yz} = e_{zx} = e_{xy} = 0$. Thus, the response of the monoclinic material is considerably different from isotropic behavior and yields a nonzero shear strain even under uniform hydrostatic stress. Additional examples using simple shear and/or bending deformations can also be used to demonstrate the complexity of anisotropic stress–strain behavior (see Sendeckyj, 1975). It should be apparent that laboratory testing methods attempting to characterize anisotropic materials would have to be more involved than those used for isotropic solids.

11.3 Restrictions on elastic moduli

Several general restrictions exist on particular combinations of elastic moduli for the anisotropic material classes discussed previously. These restrictions follow from arguments based on rotational invariance and the positive definiteness of the strain energy function.

Consider first the idea of rotational invariance. This concept has already been discussed in Section 1.6, where it was shown that for all 3×3 symmetric matrices or tensors there exist three invariants given by relations (1.6.3). This general concept may be applied to symmetric square matrices of any order including the general elasticity matrix C_{ij}. One scheme to generate such invariant relationships is to employ the rotational transformation (11.2.11) about the z-axis. Using this transformation, we can show that

$$C_{44}' = C_{44} \cos^2\theta - 2C_{45} \sin\theta \cos\theta + C_{55} \sin^2\theta$$
$$C_{55}' = C_{44} \sin^2\theta + 2C_{45} \sin\theta \cos\theta + C_{55} \cos^2\theta$$

Adding these individual equations together gives the simple result

$$C_{44}' + C_{55}' = C_{44} + C_{55}$$

and thus this sum must be an *invariant* with respect to such rotations. Other invariants can also be found using this type of rotational transformation scheme, and the results include the following invariant forms

$$C_{11} + C_{22} + 2C_{12}$$

$$C_{66} - C_{12}$$

$$C_{44} + C_{55}$$

$$C_{13} + C_{23} \tag{11.3.1}$$

$$C_{34}^2 + C_{35}^2$$

$$C_{11} + C_{22} + C_{33} + 2(C_{12} + C_{23} + C_{13})$$

Next consider modulus restrictions based on strain energy concepts. In terms of the contracted notation, the strain energy function can be written as

$$U = \frac{1}{2} \sigma_{ij} e_{ij} = \frac{1}{2} C_{ij} e_i e_j = \frac{1}{2} S_{ij} \sigma_i \sigma_j \tag{11.3.2}$$

Now the strain energy is always *positive definite*, $U \geq 0$ with equality only if the stresses or strains vanish. This result implies that both the C_{ij} and S_{ij} must be *positive definite symmetric matrices*. From matrix theory (see, for example, Ting, 1996a,b), it can be shown that for such a case all *eigenvalues* and *principal minors* of C_{ij} and S_{ij} must be positive and nonzero. The leading principal minors p_i of C_{ij} are defined by the collection of six determinants

$$p_1 = C_{11}, p_2 = \begin{vmatrix} C_{11} & C_{12} \\ C_{12} & C_{22} \end{vmatrix}, p_3 = \begin{vmatrix} C_{11} & C_{12} & C_{13} \\ C_{12} & C_{22} & C_{23} \\ C_{13} & C_{23} & C_{33} \end{vmatrix}, ..., p_6 = \det[C] \tag{11.3.3}$$

and thus $p_i > 0$. For the triclinic or monoclinic cases, (11.3.3) will yield six very lengthy and complicated relations. However, for orthotropic, transversely isotropic, cubic and isotropic cases,

Table 11.1 Positive definite restrictions on elastic moduli.

Material	Restriction inequalities
Orthotropic	$C_{11} > 0$, $C_{44} > 0$, $C_{55} > 0$, $C_{66} > 0$, $C_{11}C_{22} - C_{12}^2 > 0$
	$C_{11}C_{22}C_{33} - C_{11}C_{23}^2 - C_{22}C_{13}^2 - C_{33}C_{12}^2 + 2C_{12}C_{23}C_{31} > 0$
Transversely Isotropic	$C_{11} > 0$, $C_{44} > 0$, $C_{11}^2 - C_{12}^2 > 0$, $C_{11} - C_{12} > 0$
	$C_{11}C_{22}C_{33} - C_{11}C_{23}^2 - C_{22}C_{13}^2 - C_{33}C_{12}^2 + 2C_{12}C_{23}C_{31} > 0$
Cubic	$C_{11} > 0$, $C_{44} > 0$, $C_{11}^2 - C_{12}^2 > 0$,
	$C_{11}^3 - 3C_{11}C_{12}^2 + 2C_{12}^3 > 0$
	or $(C_{11} - C_{12})^2 (C_{11} + 2C_{12}) > 0 \Rightarrow C_{11} + 2C_{12} > 0$
Isotropic	$\lambda + 2\mu > 0$, $\lambda + \mu > 0$, $3\lambda + 2\mu > 0$, $\mu > 0$

Table 11.2 Typical Elastic Moduli for Some Planar Orthotopic Composite Materials

Material	E_1 (GPa)	E_2 (GPa)	ν_{12}	μ_{12} (GPa)
S-Glass/Epoxy	50	17	0.27	7
Boron/Epoxy	205	20	0.23	6.5
Carbon/Epoxy	205	10	0.26	6
Kevlar49/Epoxy	76	5.5	0.34	2.2

Note: Direction 1 corresponds to fiber layout axis.

specific compact results can be obtained. These restriction relations are shown in Table 11.1. It should be mentioned that similar bounds can also be developed for the compliance matrix S_{ij}.

Note that for the isotropic case, relations (6.3.11) from Section 6.3 give the same results as those shown in Table 11.1. Typical values of elastic moduli for some planar orthotropic composite materials are given in Table 11.2. These values represent average properties, and in some cases considerable variation may occur depending on the type and percentage of fibers and/or resin used in the composite mix.

11.4 Torsion of a solid possessing a plane of material symmetry

As our first example, consider the torsion of a prismatic bar of arbitrary cross-section, as shown in Fig. 11.5. The isotropic problem was investigated in Chapter 9, and we now wish to formulate and develop solutions to the problem where the bar material is anisotropic with a plane of material symmetry normal to the bar axis (z-axis). For this case, the x,y-plane is the symmetry plane (similar to Fig. 11.2), and the elasticity matrix takes the reduced form for a monoclinic material as given by Eq. (11.2.7).

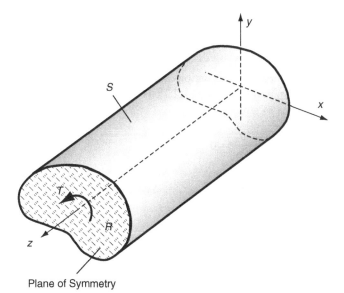

FIG. 11.5 Torsion of an anisotropic bar.

Following the usual procedure for torsion problems, boundary conditions are formulated on the lateral surface S and on the end sections R. Conditions on the lateral surfaces of the bar are to be stress free, and these traction conditions are expressed as

$$\sigma_x n_x + \tau_{xy} n_y = 0$$
$$\tau_{xy} n_x + \sigma_y n_y = 0 \qquad (11.4.1)$$
$$\tau_{xz} n_x + \tau_{yz} n_y = 0$$

where n_x and n_y are the components of the unit normal vector to surface S. The loadings on the end sections (or any bar cross-section R) reduce to a single resultant moment T about the z-axis, and this is formulated as

$$\int_R \sigma_z dA = \int_R \tau_{xz} dA = \int_R \tau_{yz} dA = 0$$

$$\int_R x\sigma_z dA = \int_R y\sigma_z dA = 0 \qquad (11.4.2)$$

$$\int_R (x\tau_{yz} - y\tau_{xz}) dA = T$$

11.4.1 Stress formulation

Following our previous approach in Chapter 9, we seek the torsion solution using the *Saint–Venant semi-inverse method*. Based on the boundary conditions (11.4.1), we assume as before that $\sigma_x = \sigma_y = \tau_{xy} = 0$. Using the equilibrium equations, we find that τ_{xz} and τ_{yz} are independent of z, and the remaining z equation reduces to

$$\frac{\partial \tau_{xz}}{\partial x} + \frac{\partial \tau_{yz}}{\partial y} = 0 \qquad (11.4.3)$$

Next we employ the strain compatibility equations and substitute for the strains using the appropriate form of Hooke's law $e_i = S_{ij}\sigma_j$ coming from (11.2.7). For the anisotropic problem, this yields a new form of compatibility in terms of stress given by

$$\frac{\partial^2 \sigma_z}{\partial x^2} = \frac{\partial^2 \sigma_z}{\partial y^2} = \frac{\partial^2 \sigma_z}{\partial z^2} = \frac{\partial^2 \sigma_z}{\partial x \partial y} = 0$$

$$\frac{\partial}{\partial x}\left[-\frac{\partial}{\partial x}(S_{44}\tau_{yz} + S_{45}\tau_{xz}) + \frac{\partial}{\partial y}(S_{54}\tau_{yz} + S_{55}\tau_{xz}) + S_{63}\sigma_z \right] = 2S_{13}\frac{\partial^2 \sigma_z}{\partial y \partial z} \qquad (11.4.4)$$

$$\frac{\partial}{\partial y}\left[-\frac{\partial}{\partial y}(S_{54}\tau_{yz} + S_{55}\tau_{xz}) + \frac{\partial}{\partial x}(S_{44}\tau_{yz} + S_{45}\tau_{xz}) + S_{63}\sigma_z \right] = 2S_{23}\frac{\partial^2 \sigma_z}{\partial x \partial z}$$

As found in Chapter 9, the first line of (11.4.4) can be integrated, giving the result

$$\sigma_z = C_1 x + C_2 y + C_3 z + C_4 xz + C_5 yz + C_6 \tag{11.4.5}$$

However, using this result in boundary conditions (11.4.2) gives $C_i = 0$, and thus σ_z must vanish. This result simplifies the remaining compatibility relations in (11.4.4), and these may be integrated to give the single equation

$$-\frac{\partial}{\partial x}(S_{44}\tau_{yz} + S_{45}\tau_{xz}) + \frac{\partial}{\partial y}(S_{54}\tau_{yz} + S_{55}\tau_{xz}) = C \tag{11.4.6}$$

where C is an arbitrary constant of integration. Using Hooke's law and strain–displacement and rotation relations, it can be shown that the constant is given by the simple result

$$C = -2\alpha \tag{11.4.7}$$

where α is the *angle of twist* per *unit length* of the bar.

The stress formulation to this problem is then given by Eqs. (11.4.3) and (11.4.6), and the solution to this system is conveniently found by employing a stress function approach similar to the Prandtl stress function formulation given in Section 9.3 for the isotropic case. Following the usual approach, we seek a solution form that identically satisfies equilibrium

$$\tau_{xz} = \alpha\frac{\partial\psi}{\partial y}, \quad \tau_{yz} = -\alpha\frac{\partial\psi}{\partial x} \tag{11.4.8}$$

where ψ is the *stress function* for our anisotropic problem. With equilibrium satisfied identically, the compatibility relation (11.4.6) yields the governing equation for the stress function

$$S_{44}\frac{\partial^2\psi}{\partial x^2} - 2S_{45}\frac{\partial^2\psi}{\partial x\partial y} + S_{55}\frac{\partial^2\psi}{\partial y^2} = -2 \tag{11.4.9}$$

The remaining boundary condition on the lateral surface $(11.4.1)_3$ becomes

$$\frac{\partial\psi}{\partial y}n_x - \frac{\partial\psi}{\partial x}n_y = 0 \tag{11.4.10}$$

From our previous investigation in Section 9.3, the components of the normal vector can be expressed in terms of derivatives of the boundary arc length measure [see Eq. (9.3.11)], and this allows (11.4.10) to be written as

$$\frac{\partial\psi}{\partial x}\frac{dx}{ds} + \frac{\partial\psi}{\partial y}\frac{dy}{ds} = \frac{d\psi}{ds} = 0 \tag{11.4.11}$$

Thus, it follows as before that the stress function $\psi(x,y)$ is a *constant on the boundary*. For solid cross-sections, this constant can be set to zero without loss of generality. However, for hollow bars with sections containing internal cavities (see Fig. 9.4), the constant can still be selected as zero on the outer boundary, but it will take on different constant values on each of the internal boundaries. More details on this are given later in the discussion.

Following similar steps as in Section 9.3, the resultant moment boundary condition on the ends can be expressed as

$$T = \int_R (x\tau_{yz} - y\tau_{xz})dA = -\alpha \int_R \left(x\frac{\partial\psi}{\partial x} + y\frac{\partial\psi}{\partial y} \right)dA$$

$$= 2\alpha \int_R \psi dA - \alpha \int_R \left(\frac{\partial(x\psi)}{\partial x} + \frac{\partial(y\psi)}{\partial y} \right)dA \qquad (11.4.12)$$

$$= 2\alpha \int_R \psi dA + 2\alpha \sum_{k=1}^{N} \psi_k A_k$$

where ψ_k is the constant value of the stress function on internal contour C_k enclosing area A_k. If the section is simply connected (no holes), then the summation term in relation (11.4.12) is dropped. One can show that all other boundary conditions are now satisfied using the assumed stress field, and thus the problem formulation in terms of the stress function is now complete.

11.4.2 Displacement formulation

Next consider the displacement formulation of the anisotropic torsion problem. Again, following similar arguments as given in Section 9.3, we assume a displacement field with one unknown component of the form

$$u = -\alpha yz, \quad v = \alpha xz, \quad w = w(x, y) \qquad (11.4.13)$$

where α is the angle of twist per unit length and w is the *warping displacement*.

This displacement field gives the following strain components

$$e_x = e_y = e_z = e_{xy} = 0$$

$$e_{xz} = \frac{1}{2}\left(\frac{\partial w}{\partial x} - \alpha y \right), \quad e_{yz} = \frac{1}{2}\left(\frac{\partial w}{\partial y} + \alpha x \right) \qquad (11.4.14)$$

and using Hooke's law, the stresses become

$$\sigma_x = \sigma_y = \sigma_z = \tau_{xy} = 0$$

$$\tau_{xz} = C_{55}\left(\frac{\partial w}{\partial x} - \alpha y \right) + C_{45}\left(\frac{\partial w}{\partial y} + \alpha x \right) \qquad (11.4.15)$$

$$\tau_{yz} = C_{45}\left(\frac{\partial w}{\partial x} - \alpha y \right) + C_{44}\left(\frac{\partial w}{\partial y} + \alpha x \right)$$

Substituting these stresses into the equilibrium equations yields the following governing equation for the warping displacement

$$C_{55}\frac{\partial^2 w}{\partial x^2} + 2C_{45}\frac{\partial^2 w}{\partial x \partial y} + C_{44}\frac{\partial^2 w}{\partial y^2} = 0 \qquad (11.4.16)$$

For this formulation, the boundary conditions on the lateral surface give the result

$$\left[C_{55}\left(\frac{\partial w}{\partial x} - \alpha y \right) + C_{45}\left(\frac{\partial w}{\partial y} + \alpha x \right) \right] n_x$$

$$+ \left[C_{45}\left(\frac{\partial w}{\partial x} - \alpha y \right) + C_{44}\left(\frac{\partial w}{\partial y} + \alpha x \right) \right] n_y = 0 \tag{11.4.17}$$

and the moment condition on the ends is given by

$$T = \alpha \int_R \left[C_{44}x^2 + C_{55}y^2 - 2C_{45}xy + C_{44}x\frac{\partial w}{\partial y} \right.$$

$$\left. - C_{55}y\frac{\partial w}{\partial x} + C_{45}\left(x\frac{\partial w}{\partial x} - y\frac{\partial w}{\partial y} \right) dA \right] \tag{11.4.18}$$

Note that all other boundary conditions in set (11.4.1) and (11.4.2) are satisfied.

Comparison of the stress and displacement formulations for the anisotropic torsion problem results in similar conclusions found for the isotropic case in Chapter 9. The stress function is governed by a slightly more complicated nonhomogeneous differential equation but with a simpler boundary condition. This fact commonly favors using the stress function approach for problem solution.

11.4.3 General solution to the governing equation

The governing equation for both the stress and displacement formulations of the torsion problem can be written as

$$au_{xx} + 2bu_{xy} + cu_{yy} = d \tag{11.4.19}$$

where the constants a, b, and c are related to appropriate elastic moduli, and d is either zero or -2, depending on the formulation. Of course, for the nonhomogeneous case, the general solution is the sum of the particular plus homogeneous solutions.

To investigate the general solution to (11.4.19) for the homogeneous case, consider solutions of the form $u(x,y) = f(x + \lambda y)$, where λ is a parameter. Using this form in (11.4.19) gives

$$\left(a + 2b\lambda + c\lambda^2 \right)f'' = 0$$

Since f'' cannot be zero, the term in parentheses must vanish, giving the *characteristic equation*

$$c\lambda^2 + 2b\lambda + a = 0 \tag{11.4.20}$$

Solving the quadratic characteristic equation gives roots

$$\lambda_{1,2} = \frac{-b \pm \sqrt{b^2 - ac}}{c} \tag{11.4.21}$$

Using these roots, the original differential equation (11.4.19) can be written in operator form as

$$D_1 D_2 u(x, y) = 0$$

$$\text{where } D_k = \frac{\partial}{\partial y} - \lambda_k \frac{\partial}{\partial x} \tag{11.4.22}$$

It is apparent that the characteristic equation (11.4.20) has *complex conjugate roots* whenever $b^2 < ac$. As per our discussion in Section 11.3, elastic moduli for materials possessing a strain energy function must satisfy relations $C_{44}C_{55} > C_{45}^2$ and $S_{44}S_{55} > S_{45}^2$, and this implies that all roots to (11.4.20) will be complex conjugate pairs of the form $\lambda_1 = \lambda$ and $\lambda_2 = \bar{\lambda}$. The general solution to (11.4.22) then becomes

$$u(x,y) = f_1(x + \lambda y) + f_2(x + \bar{\lambda}y) \tag{11.4.23}$$

where f_1 and f_2 are arbitrary functions to be determined.

Because $u(x,y)$ must be real, f_1 and f_2 must be complex conjugates of each other, and so (11.4.23) can be written in the simplified form

$$u(x,y) = 2\text{Re}[f_1(x + \lambda y)] \tag{11.4.24}$$

Because λ is a complex number, we can introduce the complex variable $z^* = x + \lambda y$, and the previous solution form can be written as

$$u(x,y) = 2\text{Re}[f(z^*)] \tag{11.4.25}$$

This formulation then allows the method of complex variables to be applied to the solution of the torsion problem. As discussed in the previous chapter, this method is very powerful and can solve many problems, some of which are intractable by other schemes. We will not, however, pursue the formal use of this method for our limited discussion of the anisotropic torsion problem.

These results then provide the general solution for the homogeneous case. To complete our discussion we need the particular solution to (11.4.19). Using the structure of the equation, a simple particular solution is given by

$$u_p(x,y) = \frac{d(x^2 + y^2)}{2(a + c)} \tag{11.4.26}$$

Example 11.2 Torsion of an elliptical orthotropic bar

Consider the torsion of a bar with elliptical cross-section as shown in Fig. 9.7. Recall that this problem was previously solved for the isotropic case in Example 9.1. Here, we wish to solve the problem for the case of an orthotropic material. For convenience, the coordinate system is taken to coincide with the material symmetry axes, and this will yield the reduced stiffness matrix given in relation (11.2.9). Note that for this case $C_{45} = S_{45} = 0$. Because of the simple section geometry and the expected correspondence with the isotropic case, the solution method will not employ the complex variable scheme discussed previously. Rather, we will use the boundary equation method presented in Section 9.4.

Consider first the solution using the stress function formulation. Using the scheme for the isotropic case, we choose a stress function form that will vanish on the boundary of the elliptical cross-section

$$\psi = K\left(\frac{x^2}{a^2} + \frac{y^2}{b^2} - 1\right)$$

where the constant K is to be determined. Substituting this form into the governing equation (11.4.9) determines the value of K and gives the final solution

$$\psi = \frac{a^2 b^2 - b^2 x^2 - a^2 y^2}{S_{55}a^2 + S_{44}b^2} \tag{11.4.27}$$

The stresses then follow from relations (11.4.8)

$$\tau_{xz} = -\frac{2\alpha a^2 y}{S_{55}a^2 + S_{44}b^2}$$
$$\tau_{yz} = \frac{2\alpha b^2 x}{S_{55}a^2 + S_{44}b^2} \tag{11.4.28}$$

which reduce to Eqs. (9.4.8) for the isotropic case with $S_{44} = S_{55} = 1/\mu$. The load-carrying torque may be determined from result (11.4.12)

$$T = \frac{\alpha \pi a^3 b^3}{S_{55}a^2 + S_{44}b^2} \tag{11.4.29}$$

The warping displacement again follows from integrating relations (11.4.15)$_{2,3}$, giving the result

$$w(x,y) = \frac{b^2 C_{55} - a^2 C_{44}}{a^2 C_{44} + b^2 C_{55}} xy \tag{11.4.30}$$

which again reduces appropriately to the isotropic case given by (9.4.11). With the warping displacement determined, the twisting moment can also be calculated from relation (11.4.18).

11.5 Plane deformation problems

We now wish to investigate the solution of two-dimensional problems of an anisotropic elastic solid. The material is chosen to have a plane of material symmetry that coincides with the plane of reference for the deformation field. Plane problems were first discussed in Chapter 7, and this leads to the formulation of two theories: *plane strain* and *plane stress*. The assumed displacement field for plane strain was given in Section 7.1, and the corresponding assumptions on the stress field for plane stress were specified in Section 7.2. These general assumptions still apply for this case with a plane of material symmetry, and each theory produces similar governing equations for anisotropic materials. Ultimately, a complex variable formulation similar to that of Chapter 10 will be established. Further details on this formulation can be found in Milne-Thomson (1960), Lekhnitskii (1981), and Sendeckyj (1975). We begin with the case of plane stress in the x,y-plane. For this case, the elasticity stiffness matrix is given by relation (11.2.7) and a similar form would exist for the compliance matrix. Under the usual *plane stress* (or *generalized plane stress*) assumptions $\sigma_z = \tau_{xz} = \tau_{yz} = 0$, and Hooke's law would then read

$$e_x = S_{11}\sigma_x + S_{12}\sigma_y + S_{16}\tau_{xy}$$
$$e_y = S_{12}\sigma_x + S_{22}\sigma_y + S_{26}\tau_{xy} \tag{11.5.1}$$
$$2e_{xy} = S_{16}\sigma_x + S_{26}\sigma_y + S_{66}\tau_{xy}$$

For *plane strain*, the usual assumptions give $e_z = e_{xz} = e_{yz} = 0$, and Hooke's law in terms of the stiffness matrix would read

$$e_x = B_{11}\sigma_x + B_{12}\sigma_y + B_{16}\tau_{xy}$$
$$e_y = B_{12}\sigma_x + B_{22}\sigma_y + B_{16}\tau_{xy} \tag{11.5.2}$$
$$2e_{xy} = B_{16}\sigma_x + B_{26}\sigma_y + B_{66}\tau_{xy}$$

where the constants B_{ij} may be expressed in terms of the compliances S_{ij} by the relations

$$B_{11} = \frac{S_{11}S_{33} - S_{13}^2}{S_{33}}, \; B_{12} = \frac{S_{12}S_{33} - S_{13}S_{23}}{S_{33}}$$

$$B_{22} = \frac{S_{22}S_{33} - S_{23}^2}{S_{33}}, \; B_{16} = \frac{S_{16}S_{33} - S_{13}S_{36}}{S_{33}} \tag{11.5.3}$$

$$B_{66} = \frac{S_{66}S_{33} - S_{36}^2}{S_{33}}, \; B_{26} = \frac{S_{26}S_{33} - S_{23}S_{36}}{S_{33}}$$

Comparing stress–strain relations (11.5.1) and (11.5.2), it is observed that they are of the same form, and a simple interchange of the elastic moduli S_{ij} with the corresponding B_{ij} will transform the plane stress relations into those of plane strain. This is a similar result as found earlier for the isotropic case. Because of this transformation property, we proceed only with the plane stress case, realizing that any of the subsequent developments can be easily converted to plane strain results.

The *Airy stress function* $\phi(x,y)$ can again be introduced, and for the case with zero body forces, we have the usual relations

$$\sigma_x = \frac{\partial^2\phi}{\partial y^2}, \; \sigma_y = \frac{\partial^2\phi}{\partial x^2}, \; \tau_{xy} = -\frac{\partial^2\phi}{\partial x\partial y} \tag{11.5.4}$$

This stress field automatically satisfies the equilibrium equations, and using this form in (11.5.1) yields the corresponding strain field in terms of the stress function. As before, the only remaining nonzero compatibility relation is

$$\frac{\partial^2 e_x}{\partial y^2} + \frac{\partial^2 e_y}{\partial x^2} = 2\frac{\partial^2 e_{xy}}{\partial x\partial y} \tag{11.5.5}$$

and substituting the strain field into this relation gives the governing equation for the stress function

$$S_{22}\frac{\partial^4\phi}{\partial x^4} - 2S_{26}\frac{\partial^4\phi}{\partial x^3\partial y} + (2S_{12} + S_{66})\frac{\partial^4\phi}{\partial x^2\partial y^2} - 2S_{16}\frac{\partial^4\phi}{\partial x\partial y^3} + S_{11}\frac{\partial^4\phi}{\partial y^4} = 0 \tag{11.5.6}$$

The case with nonzero body forces has been given by Sendeckyj (1975).

The general solution to Eq. (11.5.6) can be found using methods of characteristics as discussed previously in the torsion problem formulation; see result (11.4.23). The process starts by looking for solutions of the form $\phi = \phi(x + \mu y)$, where μ is a parameter. Using this in (11.5.6) gives the characteristic equation

$$S_{11}\mu^4 - 2S_{16}\mu^3 + (2S_{12} + S_{66})\mu^2 - 2S_{26}\mu + S_{22} = 0 \qquad (11.5.7)$$

The four roots of this equation are related to the elastic compliances by the relations

$$\mu_1\mu_2\mu_3\mu_4 = S_{22}/S_{11}$$

$$\mu_1\mu_2\mu_3 + \mu_2\mu_3\mu_4 + \mu_3\mu_4\mu_1 + \mu_4\mu_1\mu_2 = 2S_{26}/S_{11}$$

$$\mu_1\mu_2 + \mu_2\mu_3 + \mu_3\mu_4 + \mu_4\mu_1 + \mu_1\mu_3 + \mu_2\mu_4 = (2S_{12} + S_{66})/S_{11} \qquad (11.5.8)$$

$$\mu_1 + \mu_2 + \mu_3 + \mu_4 = 2S_{16}/S_{11}$$

Using this formulation, the governing equation (11.5.6) can be written in operator form

$$D_1 D_2 D_3 D_4 \phi = 0$$

$$\text{where } D_k = \frac{\partial}{\partial y} - \mu_k \frac{\partial}{\partial x} \qquad (11.5.9)$$

It can be shown (Lekhnitskii, 1981) that the roots of the characteristic Eq. (11.5.7) must be complex. Because complex roots always occur in conjugate pairs, this leads to two particular cases

$$\text{Case 1: } \mu_1 = \alpha_1 + i\beta_1, \ \mu_2 = \alpha_2 + i\beta_2, \ \mu_3 = \bar{\mu}_1, \ \mu_4 = \bar{\mu}_2$$
$$\text{Case 2: } \mu_1 = \mu_2 = \alpha + i\beta, \ \mu_3 = \mu_4 = \bar{\mu}_1 \qquad (11.5.10)$$

With the equality condition, the second case rarely occurs, and it can be shown that it will reduce to an isotropic formulation. We therefore do not consider this case further. Note for the *orthotropic case*, $S_{16} = S_{26} = 0$, and the roots of the characteristic equation become *purely complex*, that is, $\alpha_i = 0$ (see Exercise 11.15).

For the unequal complex conjugate root case, (11.5.9) can be separated into four equations and integrated in a similar fashion as done to get result (11.4.23). This then leads to the general solution

$$\phi(x,y) = F_1(x + \mu_1 y) + F_2(x + \mu_2 y) + F_3(x + \mu_3 y) + F_4(x + \mu_4 y)$$
$$= F_1(x + \mu_1 y) + F_2(x + \mu_2 y) + F_3(x + \bar{\mu}_1 y) + F_4(x + \bar{\mu}_2 y)$$
$$= 2\text{Re}[F_1(x + \mu_1 y) + F_2(x + \mu_2 y)] \qquad (11.5.11)$$
$$= 2\text{Re}[F_1(z_1) + F_2(z_2)], \quad z_1 = x + \mu_1 y, \quad z_2 = x + \mu_2 y$$

where we have used similar arguments as in the development of the torsion solution (11.4.24) and (11.4.25). Thus, we have now established that the general solution to the anisotropic plane problem is given in terms of two arbitrary functions of the complex variables z_1 and z_2.

We now wish to express the remaining elasticity equations in terms of these two complex potential functions. It is generally more convenient to introduce two new complex potentials that are simply the derivatives of the original pair

$$\Phi_1(z_1) = \frac{dF_1}{dz_1}, \quad \Phi_2(z_2) = \frac{dF_2}{dz_2} \tag{11.5.12}$$

In terms of these potentials, the in-plane stresses can be written as

$$\sigma_x = 2\text{Re}\left[\mu_1^2\Phi_1'(z_1) + \mu_2^2\Phi_2'(z_2)\right]$$
$$\sigma_y = 2\text{Re}\left[\Phi_1'(z_1) + \Phi_2'(z_2)\right] \tag{11.5.13}$$
$$\tau_{xy} = -2\text{Re}\left[\mu_1\Phi_1'(z_1) + \mu_2\Phi_2'(z_2)\right]$$

where primes indicate derivatives with respect to argument. Using Hooke's law, the strains may be determined, and the displacements follow from integration of the strain–displacement relations, giving the result

$$u(x,y) = 2\text{Re}[p_1\Phi_1(z_1) + p_2\Phi_2(z_2)]$$
$$v(x,y) = 2\text{Re}[q_1\Phi_1(z_1) + q_2\Phi_2(z_2)] \tag{11.5.14}$$

where we have dropped the rigid-body motion terms and

$$p_i = S_{11}\mu_i^2 - S_{16}\mu_i + S_{12}$$
$$q_i = S_{12}\mu_i - S_{26} + S_{22}/\mu_i) \tag{11.5.15}$$

In *polar coordinates*, the stresses and displacements take the form

$$\sigma_r = 2\text{Re}\left[(\sin\theta - \mu_1\cos\theta)^2\Phi_1'(z_1) + (\sin\theta - \mu_2\cos\theta)^2\Phi_2'(z_2)\right]$$
$$\sigma_\theta = 2\text{Re}\left[(\cos\theta + \mu_1\sin\theta)^2\Phi_1'(z_1) + (\cos\theta + \mu_2\sin\theta)^2\Phi_2'(z_2)\right]$$
$$\tau_{r\theta} = 2\text{Re}\left[(\sin\theta - \mu_1\cos\theta)(\cos\theta + \mu_1\sin\theta)\Phi_1'(z_1) \right.$$
$$\left. + (\sin\theta - \mu_2\cos\theta)(\cos\theta + \mu_2\sin\theta)\Phi_2'(z_2)\right] \tag{11.5.16}$$
$$u_r = 2\text{Re}[(p_1\cos\theta + q_1\sin\theta)\Phi_1(z_1) + (p_2\cos\theta + q_2\sin\theta)\Phi_2(z_2)]$$
$$u_\theta = 2\text{Re}[(q_1\cos\theta - p_1\sin\theta)\Phi_1(z_1) + (q_2\cos\theta - p_2\sin\theta)\Phi_2(z_2)] \tag{11.5.17}$$

Next we wish to establish the usual boundary conditions in terms of the complex potentials. Results developed in the previous chapter, Eq. (10.2.13), are also valid here, and thus the traction vector can be written as

$$T_x^n = \sigma_x n_x + \tau_{xy}n_y = \frac{d}{ds}\left(\frac{\partial\phi}{\partial y}\right)$$
$$\tag{11.5.18}$$
$$T_y^n = \tau_{xy}n_x + \sigma_y n_y = -\frac{d}{ds}\left(\frac{\partial\phi}{\partial x}\right)$$

Integrating this result over the boundary gives the boundary forces

$$\int_S T_x^n ds + C_1 = \frac{\partial \phi}{\partial y} = 2\mathrm{Re}[\mu_1 \Phi_1(z_1) + \mu_2 \Phi_2(z_2)] = p_x(s)$$

$$\int_S T_y^n ds + C_2 = -\frac{\partial \phi}{\partial x} = -2\mathrm{Re}[\Phi_1(z_1) + \Phi_2(z_2)] = p_y(s)$$

(11.5.19)

where $p_x(s)$ and $p_y(s)$ are the prescribed boundary tractions and C_1 and C_2 are arbitrary constants of integration that do not affect the stresses, and thus can be chosen as any convenient value. The displacement boundary conditions follow directly from Eq. (11.5.14)

$$u(s) = 2\mathrm{Re}[p_1 \Phi_1(z_1) + p_2 \Phi_2(z_2)]$$
$$v(s) = 2\mathrm{Re}[q_1 \Phi_1(z_1) + q_2 \Phi_2(z_2)]$$

(11.5.20)

where $u(s)$ and $v(s)$ are the prescribed boundary displacements.

Therefore, we have now formulated the plane anisotropic problem in terms of two arbitrary functions of the complex variables z_1 and z_2. In regard to the general structure of these complex potentials, many of the conclusions from the isotropic case covered previously in Section 10.4 would still hold for the anisotropic formulation. We now investigate the use of this formulation for the solution to several problems of engineering interest.

Example 11.3 Uniform tension of an anisotropic sheet

Consider first the simple problem shown in Fig. 11.6 of an anisotropic plane under uniform tension T acting at an angle α measured from the horizontal. For this problem, we already know the solution, namely a uniform stress field given by

$$\sigma_x = T\cos^2\alpha$$
$$\sigma_y = T\sin^2\alpha$$
$$\tau_{xy} = T\sin\alpha\cos\alpha$$

(11.5.21)

Complex potential functions corresponding to such a constant stress field would take the form $\Phi_1(z_1) = A_1 z_1$, $\Phi_2(z_2) = A_2 z_2$, where A_1 and A_2 are constants that may be complex. Using this form in relations (11.5.13) gives

$$\sigma_x = 2\mathrm{Re}[\mu_1^2 A_1 + \mu_2^2 A_2]$$
$$\sigma_y = 2\mathrm{Re}[A_1 + A_2]$$
$$\tau_{xy} = -2\mathrm{Re}[\mu_1 A_1 + \mu_2 A_2]$$

(11.5.22)

Equating (11.5.21) with (11.5.22) gives

$$T\cos^2\alpha = 2\mathrm{Re}[\mu_1^2 A_1 + \mu_2^2 A_2]$$
$$T\sin^2\alpha = 2\mathrm{Re}[A_1 + A_2]$$
$$T\sin\alpha\cos\alpha = -2\mathrm{Re}[\mu_1 A_1 + \mu_2 A_2]$$

(11.5.23)

Continued

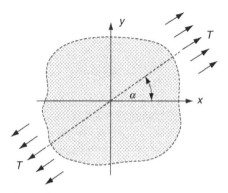

FIG. 11.6 Uniform tension of an anisotropic plane.

Because the complex constants A_1 and A_2 each have real and imaginary parts, the previous three relations cannot completely determine these four values. Another condition is needed, and it is commonly chosen as $A_1 = \bar{A}_1$. Using this constraint, (11.5.23) can now be solved to yield

$$A_1 = \frac{T(\cos \alpha + \mu_2 \sin \alpha)(\cos \alpha + \bar{\mu}_2 \sin \alpha)}{(\mu_1 - \bar{\mu}_2)(\mu_1 - \mu_2) + (\bar{\mu}_1 - \bar{\mu}_2)(\mu_1 - \mu_2)}$$

$$A_2 = \frac{T(\cos \alpha + \bar{\mu}_1 \sin \alpha)(\cos \alpha + \bar{\mu}_2 \sin \alpha) - (\mu_1 - \bar{\mu}_1)(\mu_1 - \bar{\mu}_2)A_1}{(\mu_2 - \bar{\mu}_1)(\mu_2 - \bar{\mu}_2)}$$

(11.5.24)

Example 11.4 Concentrated force system in an infinite plane

Consider next the problem of an infinite anisotropic plane containing a concentrated force system at the origin, as shown in Fig. 11.7. The problem is similar to Example 10.4, which investigated the isotropic case.

Guided by our previous isotropic analysis, we choose the logarithmic form for the complex potentials

$$\Phi_1(z_1) = A_1 \log z_1$$
$$\Phi_2(z_2) = A_2 \log z_2$$

(11.5.25)

The stresses from these potentials are

$$\sigma_x = 2\text{Re}\left[\mu_1^2 \frac{A_1}{z_1} + \mu_2^2 \frac{A_2}{z_2}\right]$$

$$\sigma_y = 2\text{Re}\left[\frac{A_1}{z_1} + \frac{A_2}{z_2}\right]$$

(11.5.26)

$$\tau_{xy} = -2\text{Re}\left[\mu_1 \frac{A_1}{z_1} + \mu_2 \frac{A_2}{z_2}\right]$$

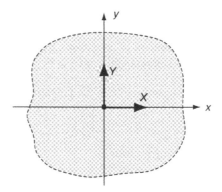

FIG. 11.7 Concentrated force system in an infinite plane.

Consider the boundary loading on a circle C enclosing the origin. Using the general result (11.5.19), the resultant loadings are given by

$$-X = \oint_C T_x^n ds = 2\text{Re}[\mu_1 \Phi_1(z_1) + \mu_2 \Phi_2(z_2)]_C$$

$$-Y = \oint_C T_y^n ds = -2\text{Re}[\Phi_1(z_1) + \Phi_2(z_2)]_C$$

(11.5.27)

where we have dropped the arbitrary constants. Substituting in the complex potentials, and using the cyclic properties of logarithmic functions [see (10.4.5)], the preceding relations become

$$-X = 4\pi\text{Re}[\mu_1 A_1 i + \mu_2 A_2 i]$$

$$-Y = -4\pi\text{Re}[A_1 i + A_2 i]$$

(11.5.28)

This system is not sufficient to determine completely the complex constants A_1 and A_2, and additional relations can be found by invoking the condition of single-valued displacements. If the displacements are to be single-valued, then the cyclic function (defined in Section 10.4) of relations (11.5.14) must be zero

$$\text{Re}[p_1 \Phi_1(z_1) + p_2 \Phi_2(z_2)]_C = 0$$

$$\text{Re}[q_1 \Phi_1(z_1) + q_2 \Phi_2(z_2)]_C = 0$$

(11.5.29)

and for this case gives the result

$$\text{Re}[p_1 A_1 i + p_2 A_2 i] = 0$$

$$\text{Re}[q_1 A_1 i + q_2 A_2 i] = 0$$

(11.5.30)

Relations (11.5.28) and (11.5.30) now provide sufficient relations to complete the problem.

Example 11.5 Concentrated force system on the surface of a half-plane

We now develop the solution to the problem of an anisotropic half-plane carrying a general force system at a point on the free surface. The problem shown in Fig. 11.8 was originally solved for the isotropic case in Example 10.5.

Again guided by our previous isotropic solution, the potential functions are chosen as

$$\Phi_1(z_1) = A_1 \log z_1$$
$$\Phi_2(z_2) = A_2 \log z_2$$

(11.5.31)

The stresses from these potentials are then given by

$$\sigma_x = 2\mathrm{Re}\left[\mu_1^2\frac{A_1}{z_1} + \mu_2^2\frac{A_2}{z_2}\right]$$

$$\sigma_y = 2\mathrm{Re}\left[\frac{A_1}{z_1} + \frac{A_2}{z_2}\right]$$

(11.5.32)

$$\tau_{xy} = -2\mathrm{Re}\left[\mu_1\frac{A_1}{z_1} + \mu_2\frac{A_2}{z_2}\right]$$

Following the procedures from Example 10.5, we consider the boundary loading on a semicircle C lying in the half-space domain and enclosing the origin. Using the general result (11.5.19), the resultant loadings are given by

$$-X = \oint_C T_x^n ds = 2\mathrm{Re}[\mu_1\Phi_1(z_1) + \mu_2\Phi_2(z_2)]_C$$

$$-Y = \oint_C T_y^n ds = -2\mathrm{Re}[\Phi_1(z_1) + \Phi_2(z_2)]_C$$

(11.5.33)

Substituting in the complex potentials, and again using the cyclic properties of the logarithmic function, we find

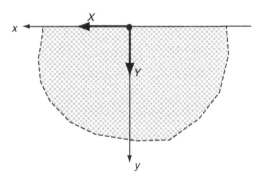

FIG. 11.8 Concentrated force system on a half-plane.

$$-X = 2\pi \text{Re}[\mu_1 A_1 i + \mu_2 A_2 i]$$
$$-Y = -2\pi \text{Re}[A_1 i + A_2 i] \qquad (11.5.34)$$

As before, this system is not sufficient to determine completely the complex constants A_1 and A_2. Additional relations can be found by invoking the stress-free boundary condition on surface $y = 0$, giving the result

$$\sigma_y(x,0) = 2\text{Re}\left[\frac{A_1}{z_1} + \frac{A_2}{z_2}\right]\bigg|_{y=0} = 0$$

$$\tau_{xy}(x,0) = -2\text{Re}\left[\mu_1 \frac{A_1}{z_1} + \mu_2 \frac{A_2}{z_2}\right]\bigg|_{y=0} = 0 \qquad (11.5.35)$$

Solving relations (11.5.34) and (11.5.35), the constants are found to be

$$A_1 = \frac{(X + \mu_2 Y)}{2i\pi(\mu_2 - \mu_1)}$$

$$A_2 = \frac{(X + \mu_1 Y)}{2i\pi(\mu_1 - \mu_2)} \qquad (11.5.36)$$

With the constants determined, the stresses can easily be calculated using (11.5.32). Using polar coordinates, we can again show the surprising result that $\sigma_\theta = \tau_{r\theta} = 0$, and thus the stress state will be only *radial*. This result matches our findings for the corresponding isotropic case given by relations (8.4.34) and/or (10.6.8). Exercise 11.17 computes and compares σ_r stress components for orthotropic and isotropic cases, and significant differences between the two cases are found.

Example 11.6 Infinite plate with an elliptical hole

Let us now investigate the solution to a class of problems involving an elliptical hole in an infinite anisotropic plate, as shown in Fig. 11.9. Although we develop solutions only to a couple of cases in this example, Savin (1961) provides many additional solutions to problems of this type. We first construct the general solution for arbitrary loading on the hole surface for the case where the loading produces no net force or moment. Finally, a specific case of a pressure loading is investigated in detail.

Employing the usual conformal mapping concept, consider the mapping function that transforms the *exterior of the ellipse to the exterior of a unit circle*

$$z = w(\zeta) = \frac{a+b}{2}\zeta + \frac{a-b}{2\zeta} \qquad (11.5.37)$$

The complex variables z_1 and z_2 can be expressed in terms of z and $\bar{z}$ as

Continued

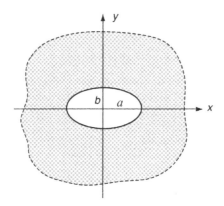

FIG. 11.9 Elliptical hole in an infinite anisotropic plane.

$$z_1 = x + \mu_1 y = \frac{1}{2}(1 - i\mu_1)(x + iy) + \frac{1}{2}(1 + i\mu_1)(x - iy) = \gamma_1 z + \delta_1 \bar{z}$$

$$z_2 = x + \mu_2 y = \frac{1}{2}(1 - i\mu_2)(x + iy) + \frac{1}{2}(1 + i\mu_2)(x - iy) = \gamma_2 z + \delta_2 \bar{z}$$

(11.5.38)

where $\gamma_i = (1 - i\mu_i)/2$, $\delta_i = (i + \mu_i)/2$. Relations (11.5.38) lead to the concept of *induced mappings* whereby the transformation (11.5.37) induces mappings in the variables z_1 and z_2

$$z_1 = \gamma_1 w(\zeta) + \delta_1 \overline{w(\zeta)} = w_1(\zeta_1)$$

$$z_2 = \gamma_2 w(\zeta) + \delta_2 \overline{w(\zeta)} = w_2(\zeta_2)$$

(11.5.39)

Using the specified transformation (11.5.37) in (11.5.38), the mapped variables ζ_1, ζ_2 can be determined as

$$\zeta_1 = \frac{z_1 + \sqrt{z_1^2 - a^2 - \mu_1^2 b^2}}{a - i\mu_1 b}$$

$$\zeta_2 = \frac{z_2 + \sqrt{z_2^2 - a^2 - \mu_2^2 b^2}}{a - i\mu_2 b}$$

(11.5.40)

Note that on the boundary of the hole, $\zeta_1 = \zeta_2 = e^{i\theta}$.

Using the general results (11.5.19), we assume these boundary loadings can be expanded in a complex Fourier series on the elliptic boundary

$$p_x(s) = \sum_{m=1}^{\infty} \left(A_m e^{im\theta} + \bar{A}_m e^{-im\theta} \right)$$

$$p_y(s) = \sum_{m=1}^{\infty} \left(B_m e^{im\theta} + \bar{B}_m e^{-im\theta} \right)$$

(11.5.41)

where A_m and B_m are complex constants to be determined by the specific boundary loading. Following our experience from the previous chapter for the isotropic case, we expect our solution to be given by potential functions of the series form

$$\Phi_1(z_1) = \sum_{m=1}^{\infty} a_m \zeta_1^{-m}$$

$$\Phi_2(z_2) = \sum_{m=1}^{\infty} b_m \zeta_2^{-m}$$

(11.5.42)

where a_m and b_m are complex constants and ζ_1 and ζ_2 are given by relations (11.5.40). Substituting these potential forms into the boundary loading relations (11.5.19) and combining with (11.5.41) allows the determination of the constants a_m and b_m in terms of boundary loading. This then provides the final general solution form

$$\Phi_1(z_1) = \sum_{m=1}^{\infty} \frac{\overline{B}_m + \mu_2 \overline{A}_m}{\mu_1 - \mu_2} \zeta_1^{-m}$$

$$\Phi_2(z_2) = -\sum_{m=1}^{\infty} \frac{\overline{B}_m + \mu_1 \overline{A}_m}{\mu_1 - \mu_2} \zeta_2^{-m}$$

(11.5.43)

11.5.1 Uniform pressure loading case

Consider now the specific case of a pressure p acting uniformly on the entire elliptical cavity. For this case, the boundary tractions are given by

$$T_x^n = -pn_x, \quad T_y^n = -pn_y$$

where n_x and n_y are the usual normal vector components. The boundary loading functions are then determined from relations (11.5.19), giving the result

$$p_x(s) = -\int_0^s p\,dy + C_1 = -pb\sin\theta + C_1$$

$$p_y(s) = \int_0^s p\,dx + C_2 = pa\cos\theta - pa + C_2$$

(11.5.44)

The arbitrary constants can now be chosen for convenience as $C_1 = 0$ and $C_2 = pa$. Using these results in boundary relation (11.5.41) determines the Fourier coefficients as

$$\overline{A}_1 = ipb/2, \quad \overline{B}_1 = pa/2$$

$$A_m = B_m = 0, \quad m = 2, 3, 4, \ldots$$

(11.5.45)

This then determines the complex potentials, and the stresses and displacements can be calculated from previous relations (11.5.13) and (11.5.14).

The maximum stresses are most important for applications, and these occur as tangential stresses on the boundary of the elliptical hole. It can be shown that this tangential stress on the elliptical cavity is given by

$$\sigma_\theta = \frac{p}{a^2 \sin^2\theta + b^2 \cos^2\theta} \text{Re}\left\{\frac{ie^{-i\theta}}{(a\sin^2\theta + \mu_1 b\cos\theta)(a\sin\theta - \mu_2 b\cos\theta)}\right.$$

$$\cdot \left[(\mu_1\mu_2 a - i\mu_1 b - i\mu_2 b)a^3\sin^3\theta + i(\mu_1\mu_2 - 2)a^2 b^2\sin^2\theta\cos\theta \right. \tag{11.5.46}$$

$$\left.\left. + (2\mu_1\mu_2 - 1)a^2 b^2 \sin\theta\cos^2\theta + (\mu_1 a + \mu_2 a - ib)b^3\cos^3\theta\right]\right\}$$

For the circular case ($a = b$), this result becomes

$$\sigma_\theta = p\text{Re}\left\{\frac{ie^{-i\theta}}{(\sin\theta + \mu_1\cos\theta)(\sin\theta - \mu_2\cos\theta)}\right.$$

$$\cdot \left[(\mu_1\mu_2 - i\mu_1 - i\mu_2)\sin^3\theta + i(\mu_1\mu_2 - 2)\sin^2\theta\cos\theta \right. \tag{11.5.47}$$

$$\left.\left. + (2\mu_1\mu_2 - 1)\sin\theta\cos^2\theta + (\mu_1 + \mu_2 - i)\cos^3\theta\right]\right\}$$

We can extract the *isotropic limit* by choosing the case $\mu_1 = \mu_2 = i$, and result (11.5.47) becomes simply $\sigma_\theta = p$, which is the correct value for a pressurized circular hole in an isotropic sheet (see Section 8.4.1). It should be noted that this scheme of developing the isotropic limit must be done on the *final relations* for the stresses and displacements. For example, if the expression $\mu_1 = \mu_2 = i$ had been substituted into, say, relation (11.5.43) for the potential functions, a meaningless result would occur. Exercise 11.18 explores σ_θ numerical results for the orthotropic case and demonstrates that at particular field points, anisotropy will increase this hoop stress component compared to the isotropic value.

Example 11.7 Stressed infinite plate with an elliptical hole

Consider next an infinite anisotropic plate with a stress-free elliptical hole. The plate is loaded in the x direction, as shown in Fig. 11.10. Recall that the isotropic case was previously solved in Example 10.7.

The potentials for this problem can be determined by our previously developed conformal mapping procedures from Example 11.6. The details for this and other cases are given in Savin (1961), and the final result may be written as

$$\Phi_1(z_1) = A_1 z_1 - \frac{iSb}{2(\mu_1 - \mu_2)}\frac{a - i\mu_1 b}{z_1 + \sqrt{z_1^2 - (a^2 + \mu_1^2 b^2)}}$$

$$\tag{11.5.48}$$

$$\Phi_2(z_2) = A_2 z_2 + \frac{iSb}{2(\mu_1 - \mu_2)}\frac{a - i\mu_2 b}{z_2 + \sqrt{z_2^2 - (a^2 + \mu_2^2 b^2)}}$$

The first term in each expression corresponds to the uniform tension case discussed in Example 11.3. For tension in the x direction, the constants become

$$A_1 = \frac{S}{2\left[(\alpha_2 - \alpha_1)^2 + (\beta_2^2 - \beta_1^2)\right]}$$

$$A_2 = \frac{-S}{2\left[(\alpha_2 - \alpha_1)^2 + (\beta_2^2 - \beta_1^2)\right]} + i\frac{(\alpha_1 - \alpha_2)S}{2\beta_2\left[(\alpha_2 - \alpha_1)^2 + (\beta_2^2 - \beta_1^2)\right]}$$

(11.5.49)

with parameters α_i and β_i defined by Eq. (11.5.10)$_1$.
The stresses for this case follow from (11.5.13)

$$\sigma_x = S + \mathrm{Re}\left[-\frac{iSb\mu_1^2}{(\mu_1 - \mu_2)(a + i\mu_1 b)}\left[\frac{z_1}{\sqrt{z_1^2 - (a^2 + \mu_1^2 b^2)}} - 1\right]\right.$$

$$\left. + \frac{iSb\mu_2^2}{(\mu_1 - \mu_2)(a + i\mu_2 b)}\left[\frac{z_2}{\sqrt{z_2^2 - (a^2 + \mu_2^2 b^2)}} - 1\right]\right]$$

$$\sigma_y = \mathrm{Re}\left[-\frac{iSb}{(\mu_1 - \mu_2)(a + i\mu_1 b)}\left[\frac{z_1}{\sqrt{z_1^2 - (a^2 + \mu_1^2 b^2)}} - 1\right]\right.$$

(11.5.50)

$$\left. + \frac{iSb}{(\mu_1 - \mu_2)(a + i\mu_2 b)}\left[\frac{z_2}{\sqrt{z_2^2 - (a^2 + \mu_2^2 b^2)}} - 1\right]\right]$$

$$\tau_{xy} = -\mathrm{Re}\left[-\frac{iSb\mu_1}{(\mu_1 - \mu_2)(a + i\mu_1 b)}\left[\frac{z_1}{\sqrt{z_1^2 - (a^2 + \mu_1^2 b^2)}} - 1\right]\right.$$

$$\left. + \frac{iSb\mu_2}{(\mu_1 - \mu_2)(a + i\mu_2 b)}\left[\frac{z_2}{\sqrt{z_2^2 - (a^2 + \mu_2^2 b^2)}} - 1\right]\right]$$

Consider now the special case of an orthotropic material with $\mu_i = i\beta_i$. For this case, the stress σ_x along the y-axis ($x = 0$) is given by

$$\sigma_x(0, y) = S + \frac{Sb}{(\beta_1 - \beta_2)}\left[-\frac{\beta_1^2}{(a - \beta_1 b)}\left[\frac{\beta_1 y}{\sqrt{(a^2 + \beta_1^2(y^2 - b^2)}} - 1\right]\right.$$

(11.5.51)

$$\left. + \frac{\beta_2^2}{(a - \beta_2 b)}\left[\frac{\beta_2 y}{\sqrt{(a^2 + \beta_2^2(y^2 - b^2)}} - 1\right]\right]$$

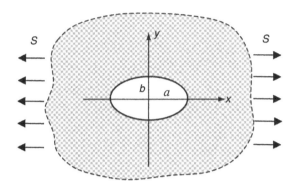

FIG. 11.10 Infinite anisotropic plate with an elliptical hole.

Investigating the value of this stress at the edge of the ellipse ($y = b$), we get

$$\sigma_x(0,b) = S\left[1 + (\beta_1 + \beta_2)\frac{b}{a}\right] \qquad (11.5.52)$$

The isotropic limit of this result is found by setting $\beta_1 = \beta_2 = 1$, which gives

$$\sigma_x(0,b) = S\left[1 + 2\frac{b}{a}\right] \qquad (11.5.53)$$

and this matches with the isotropic case given previously in Eq. (10.7.15). For many materials, $\beta_1 + \beta_2 > 2$ (see Exercise 11.15), and thus the stress concentration for the anisotropic case is commonly greater than the corresponding isotropic material. Exercise 11.19 explores σ_x numerical results and demonstrates that this stress component for the orthotropic case will be larger than the corresponding isotropic value.

11.6 Applications to fracture mechanics

The elastic stress and displacement distribution around cracks in anisotropic media has important applications in the fracture behavior of composite materials. Similar to our previous study in Sections 10.8 and 10.9, we now wish to develop solutions to some basic plane problems of anisotropic materials containing cracks. As discussed before, a crack can be regarded as the limiting case of an elliptical cavity as one axis is reduced to zero. Thus, in some cases the solution to the crack problem can be determined from a corresponding elliptical cavity problem. There exists, however, more direct methods for solving crack problems in anisotropic materials. Original work on this topic was developed by Sih et al. (1965), and further information may be found in Sih and Liebowitz (1968).

The first problem we wish to investigate is that of a pressurized crack in an infinite medium. The solution to this problem can be conveniently determined from our solution of the pressurized elliptical cavity problem in Example 11.6. The crack case follows by simply letting the semiminor axis $b \to 0$. From (11.5.45) we find $\bar{A}_1 = 0$, and relations (11.5.43) then give the potential functions

$$\Phi_1(z_1) = \frac{pa^2\mu_2}{2(\mu_1 - \mu_2)|}\left[z_1 + \sqrt{z_1^2 - a^2}\right]^{-1} = \frac{-p\mu_2}{2(\mu_1 - \mu_2)}\left[\sqrt{z_1^2 - a^2} - z_1\right]$$

$$\Phi_2(z_2) = -\frac{pa^2\mu_1}{2(\mu_1 - \mu_2)}\left[z_2 + \sqrt{z_2^2 - a^2}\right]^{-1} = \frac{p\mu_1}{2(\mu_1 - \mu_2)}\left[\sqrt{z_2^2 - a^2} - z_2\right]$$

(11.6.1)

The stresses follow from relations (11.5.13)

$$\sigma_x = -p\mathrm{Re}\left[\frac{\mu_1^2\mu_2}{\mu_1 - \mu_2}\left[\frac{z_1}{\sqrt{z_1^2 - a^2}} - 1\right] - \frac{\mu_1\mu_2^2}{\mu_1 - \mu_2}\left[\frac{z_2}{\sqrt{z_2^2 - a^2}} - 1\right]\right]$$

$$\sigma_y = -p\mathrm{Re}\left[\frac{\mu_2}{\mu_1 - \mu_2}\left[\frac{z_1}{\sqrt{z_1^2 - a^2}} - 1\right] - \frac{\mu_1}{\mu_1 - \mu_2}\left[\frac{z_2}{\sqrt{z_2^2 - a^2}} - 1\right]\right]$$

(11.6.2)

$$\tau_{xy} = p\mathrm{Re}\left[\frac{\mu_1\mu_2}{\mu_1 - \mu_2}\left[\frac{z_1}{\sqrt{z_1^2 - a^2}} - \frac{z_2}{\sqrt{z_2^2 - a^2}}\right]\right]$$

Evaluating these stresses on the x-axis ($z_1 = z_2 = x$) gives

$$\sigma_x = -p\mathrm{Re}\left(\mu_1\mu_2\left[\frac{x}{\sqrt{x^2 - a^2}} - 1\right]\right)$$

$$\sigma_y = p\mathrm{Re}\left(\frac{x}{\sqrt{x^2 - a^2}} - 1\right)$$

(11.6.3)

$$\tau_{xy} = 0$$

For the case $|x_1| > a$, the stresses can be written as

$$\sigma_x = -p\left[\frac{x}{\sqrt{x^2 - a^2}} - 1\right]\mathrm{Re}\{\mu_1\mu_2\}$$

$$\sigma_y = p\left[\frac{x}{\sqrt{x^2 - a^2}} - 1\right]$$

(11.6.4)

$$\tau_{xy} = 0$$

The stresses depend on the material properties only through the term $\mathrm{Re}\{\mu_1\mu_2\}$. Note that for the isotropic case $\mu_1 = \mu_2 = i$, and thus

$$\sigma_x = \sigma_y = p\left[\frac{x}{\sqrt{x^2 - a^2}} - 1\right], \quad \tau_{xy} = 0 \tag{11.6.5}$$

Notice that both the anisotropic and isotropic stresses are singular at $x = \pm a$, which corresponds to each crack tip. In the neighborhood of the crack tip $x = a$, we can use the usual approximations $x + a \approx a$, $x - a \approx r$ (see Fig. 10.20), and for this case equations (11.6.4) and (11.6.5) indicate that the crack-tip stress field has the $1/\sqrt{r}$ singularity for both the anisotropic and isotropic cases.

Next let us investigate the restricted problem of determining the stress and displacement solution in the vicinity of a crack tip in an infinite medium, as shown in Fig. 11.11. We assume that the problem has uniform far-field loading in the y direction normal to the crack.

Considering only the solution in the neighborhood of the crack tip (i.e., small $|z|$), it can be shown that the potential functions can be reduced to the following form

$$\Phi_1'(z_1) = A_1 z_1^{-1/2}, \quad \Phi_2'(z_2) = A_2 z_2^{-1/2} \tag{11.6.6}$$

where A_1 and A_2 are arbitrary constants. Using this result in Eqs. (11.5.13) and (11.5.14) gives the following stress and displacement fields

$$\sigma_x = \frac{K_1}{\sqrt{2r}}\mathrm{Re}\left[\frac{\mu_1\mu_2}{\mu_1 - \mu_2}\left(\frac{\mu_2}{\sqrt{\cos\theta - \mu_2\sin\theta}} - \frac{\mu_1}{\sqrt{\cos\theta - \mu_1\sin\theta}}\right)\right]$$

$$\sigma_y = \frac{K_1}{\sqrt{2r}}\mathrm{Re}\left[\frac{1}{\mu_1 - \mu_2}\left(\frac{\mu_1}{\sqrt{\cos\theta - \mu_2\sin\theta}} - \frac{\mu_2}{\sqrt{\cos\theta - \mu_1\sin\theta}}\right)\right]$$

$$\tau_{xy} = \frac{K_1}{\sqrt{2r}}\mathrm{Re}\left[\frac{\mu_1\mu_2}{\mu_1 - \mu_2}\left(\frac{1}{\sqrt{\cos\theta - \mu_1\sin\theta}} - \frac{1}{\sqrt{\cos\theta - \mu_2\sin\theta}}\right)\right] \tag{11.6.7}$$

$$u = K_1\sqrt{2r}\,\mathrm{Re}\left[\frac{1}{\mu_1 - \mu_2}\left(\mu_1 p_2\sqrt{\cos\theta - \mu_2\sin\theta} - \mu_2 p_1\sqrt{\cos\theta - \mu_1\sin\theta}\right)\right]$$

$$v = K_1\sqrt{2r}\,\mathrm{Re}\left[\frac{1}{\mu_1 - \mu_2}\left(\mu_1 q_2\sqrt{\cos\theta - \mu_2\sin\theta} - \mu_2 q_1\sqrt{\cos\theta - \mu_1\sin\theta}\right)\right]$$

where for convenience we have chosen

$$A_1 = \frac{\mu_2}{2\sqrt{2}(\mu_2 - \mu_1)}K_1$$

$$A_2 = \frac{\mu_1}{2\sqrt{2}(\mu_1 - \mu_2)}K_1 \tag{11.6.8}$$

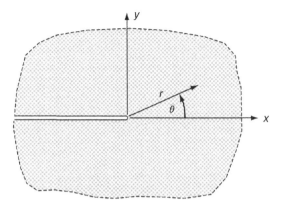

FIG. 11.11 Crack in an infinite anisotropic plane.

Similar to the isotropic case, the parameter K_1 is referred to as the stress intensity factor. It is important to note from stress relations in (11.6.7) that the crack-tip stress singularity is of order $1/\sqrt{r}$, which is *identical to the isotropic case*. This result holds for all plane problems with a plane of material symmetry (sometimes referred to as *rectilinear anisotropy*). However, it has been shown that the nature of this singularity does change for materials with more complex anisotropy. It can also be observed from (11.6.7) that, unlike the isotropic case, variation of the local stress and displacement field depends upon material properties through the roots μ_i. Finally, similar to the isotropic case, the stress and displacement field near the crack tip depends on remote boundary conditions only through the stress intensity factor.

Next let us consider a more specific fracture mechanics problem of a crack of length $2a$ lying along the x-axis in an infinite medium with far-field stress $\sigma_y^\infty = S$, as illustrated in Fig. 11.12.

For this problem, the complex potentials are given by Sih et al. (1965) as

$$\Phi_1(z_1) = A_1 z_1 + \frac{S a^2 \mu_2}{2(\mu_1 - \mu_2)}\left[z_1 + \sqrt{z_1^2 - a^2}\right]^{-1}$$

$$\Phi_2(z_2) = A_2 z_2 - \frac{S a^2 \mu_1}{2(\mu_1 - \mu_2)}\left[z_2 + \sqrt{z_2^2 - a^2}\right]^{-1}$$

(11.6.9)

where A_1 and A_2 are again constants. Substituting this form into relations (11.5.13) gives the following stress field in the vicinity of the crack tip

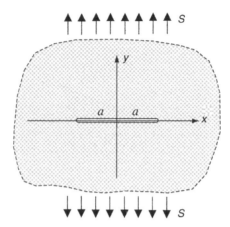

FIG. 11.12 Central crack in an infinite anisotropic plane.

$$\sigma_x = \frac{S\sqrt{a}}{\sqrt{2r}}\text{Re}\left[\frac{\mu_1\mu_2}{\mu_1 - \mu_2}\left(\frac{\mu_2}{\sqrt{\cos\theta - \mu_2\sin\theta}} - \frac{\mu_1}{\sqrt{\cos\theta - \mu_1\sin\theta}}\right)\right]$$

$$\sigma_y = \frac{S\sqrt{a}}{\sqrt{2r}}\text{Re}\left[\frac{1}{\mu_1 - \mu_2}\left(\frac{\mu_1}{\sqrt{\cos\theta - \mu_2\sin\theta}} - \frac{\mu_2}{\sqrt{\cos\theta - \mu_1\sin\theta}}\right)\right] \qquad (11.6.10)$$

$$\tau_{xy} = \frac{S\sqrt{a}}{\sqrt{2r}}\text{Re}\left[\frac{\mu_1\mu_2}{\mu_1 - \mu_2}\left(\frac{1}{\sqrt{\cos\theta - \mu_1\sin\theta}} - \frac{1}{\sqrt{\cos\theta - \mu_2\sin\theta}}\right)\right]$$

Note the similarity of this result with the relations developed in (11.6.7). For this case, the stress intensity factor is then given by $K_1 = S\sqrt{a}$.

The previous two examples include only *opening mode deformation* of the crack tip. Other loading cases can produce a *shearing deformation mode*, and these cases introduce a new stress field with a different stress intensity factor, commonly denoted by K_2. Sih et al. (1965) provide additional information on these examples. The analytically simpler crack problem for anisotropic antiplane strain deformation is given in Exercise 11.23 and the results are comparable to the isotropic problem developed in Exercise 8.44.

11.7 Curvilinear anisotropic problems

As mentioned earlier, many materials have an anisotropic microstructure that would require a *curvilinear anisotropic model*. Biological examples of such cases would include wood coming from trees that grow in approximately cylindrical fashion, and various tissue and bone material. There are also many cases of synthetic composite materials with such curvilinear microstructure. As done previously for rectilinear anisotropy cases in Section 11.2, we would expect that curvilinear anisotropy would also occur with some symmetries in material structure. This would lead to a convenient modeling scheme of incorporating Hooke's law within an orthogonal curvilinear co-ordinate system using, for example, cylindrical or spherical coordinates. Lekhnitskii (1981), Gal-mudi and Dvorkin (1995), Horgan and Baxter (1996), and others have developed solutions to these types of problems, and we will now explore such solutions.

11.7.1 Two-dimensional polar-orthotropic problem

Following the work of Galmudi and Dvorkin (1995) and Horgan and Baxter (1996), we first limit the discussion to the two-dimensional case using a polar coordinate system model. Therefore, consider the curvilinear microstucture shown in Fig. 11.13. We assume that the material has a uniform micro-stucture such that properties have orthogonal symmetry with respect to the r and θ directions as shown. Under this assumption, the material would be classified as being *polar-orthotropic,* and following the basic form of relation (11.2.10), we could write Hooke's law for the plane stress case as

$$\sigma_r = \frac{E_r}{1 - \nu_{\theta r}\nu_{r\theta}}(e_r + \nu_{\theta r}e_\theta), \sigma_\theta = \frac{E_\theta}{1 - \nu_{\theta r}\nu_{r\theta}}(e_\theta + \nu_{r\theta}e_r) \qquad (11.7.1)$$

We also assume axisymmetry so that stresses will only depend on the radial coordinate and $\tau_{r\theta} = 0$. Note that from the discussion in Section 11.2.2

$$\frac{\nu_{\theta r}}{E_\theta} = \frac{\nu_{r\theta}}{E_r} \qquad (11.7.2)$$

Using the strain—displacement relations, the stresses can be expressed as

$$\sigma_r = \frac{E_r}{1 - \nu_{\theta r}\nu_{r\theta}}\left(\frac{du}{dr} + \nu_{\theta r}\frac{u}{r}\right), \quad \sigma_\theta = \frac{E_\theta}{1 - \nu_{\theta r}\nu_{r\theta}}\left(\frac{u}{r} + \nu_{r\theta}\frac{du}{dr}\right) \qquad (11.7.3)$$

In polar coordinates with no body forces, the equilibrium equation is

$$\frac{d\sigma_r}{dr} + \frac{\sigma_r - \sigma_\theta}{r} = 0 \qquad (11.7.4)$$

Using relations (11.7.3) and (11.7.2), this equation can be written as

$$\frac{d^2u}{dr^2} + \frac{1}{r}\frac{du}{dr} - n^2\frac{u}{r^2} = 0 \qquad (11.7.5)$$

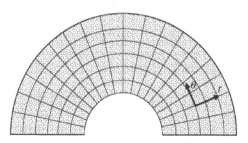

FIG. 11.13 Material with an idealized orthogonal curvilinear microstructure leading to polar orthotropy.

where $n^2 = E_\theta/E_r = \nu_{\theta r}/\nu_{r\theta}$. The parameter n provides a measure of the amount of material anisotropy, and with $n > 1 \Rightarrow E_\theta > E_r$ and the material may be classified as *circumferentially orthotropic*, while with $n < 1 \Rightarrow E_r > E_\theta$ and the material is classified as *radially orthotropic* (Horgan and Baxter, 1996). The isotropic case is found by setting $n = 1$. Eq. (11.7.5) is a Cauchy–Euler differential equation and is similar to the isotropic result previously given in (8.3.10). The equation can be easily solved giving the result for the radial displacement

$$u = Ar^n + Br^{-n} \qquad (11.7.6)$$

where A and B are arbitrary constants. This solution allows the stresses to be expressed by the general form

$$\sigma_r = C_1 r^{n-1} + C_2 r^{-n-1}, \quad \sigma_\theta = C_1 n r^{n-1} - C_2 n r^{-n-1} \qquad (11.7.7)$$

where $C_1 = A \dfrac{E_r}{1 - \nu_{\theta r}\nu_{r\theta}}(n - \nu_{\theta r})$ and $C_2 = -B \dfrac{E_r}{1 - \nu_{\theta r}\nu_{r\theta}}(n + \nu_{\theta r})$ are new, appropriately defined arbitrary constants.

Consider now the specific problem of the thick-walled cylindrical domain problem with internal and external pressure loadings previously shown in Fig. 8.8. For this case we redefine the annular domain with $a \le r \le b$, and consider pressure loading equal to p only on the outer boundary $r = b$. Under these boundary conditions, the arbitrary constants C_1 and C_2 can be easily determined and the stresses become

$$\sigma_r = -\frac{pb^{n+1}}{b^{2n} - a^{2n}}\left(r^{n-1} - a^{2n}r^{-n-1}\right)$$
$$\sigma_\theta = -\frac{pb^{n+1}n}{b^{2n} - a^{2n}}\left(r^{n-1} + a^{2n}r^{-n-1}\right) \qquad (11.7.8)$$

It can be shown that both of these normal stresses will be compressive in the region $a \le r \le b$, and that for $n > 1$, $|\sigma_\theta| > |\sigma_r|$.

Dimensionless distribution plots of the radial and hoop stresses are shown in Fig. 11.14 for cases of $n = 0.5$, 1.0, and 1.5 with $b/a = 5$. It is seen that for $n > 1$, the magnitude of the radial stress will be less than the isotropic value; Galmudi and Dvorkin (1995) refer to this as *stress shielding*. The opposite behavior occurs for the case $n < 1$, where the radial stress magnitude is greater than the isotropic value, thereby leading to *stress amplification*. Both of these effects can be viewed as being related to the decay of boundary conditions, and thus could have importance to the applicability of Saint–Venant's principle for anisotropic problems (see comments at the end of Section 5.6). Notice also that the hoop stress magnitude at the inner boundary ($r = a$) decreases with increasing values of the anisotropic parameter, n. In general, the stresses are significantly affected by the curvilinear anisotropy.

Going back to the isotropic case ($n = 1$), relations (11.7.8) reduce to

$$\sigma_r = -\frac{pb^2}{b^2 - a^2}\left(1 - \frac{a^2}{r^2}\right)$$
$$\sigma_\theta = -\frac{pb^2}{b^2 - a^2}\left(1 + \frac{a^2}{r^2}\right) \qquad (11.7.9)$$

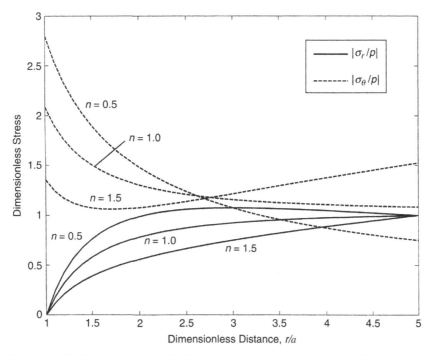

FIG. 11.14 Stress distributions in a polar orthotropic annular domain ($a \leq r \leq b$) with external pressure with $b/a = 5$.

For this case, taking the limit as $b \to \infty$ and letting $p = -T$, the result is

$$\sigma_r = T\left(1 - \frac{a^2}{r^2}\right), \quad \sigma_\theta = T\left(1 + \frac{a^2}{r^2}\right) \tag{11.7.10}$$

which matches with our previous result (8.4.9) and corresponds to a stress-free hole in an infinite medium under equal far-field biaxial tensile loading T. Note the isotropic problem then generates a stress concentration factor of 2.

As was first pointed out by Galmudi and Dvorkin (1995), attempting to do the same limiting analysis for the anisotropic case ($n \neq 1$) will not produce a converged solution. This surprising result is related to the fact that for $n < 1$, the stresses will become unbounded as $b \to \infty$. A similar result for isotropic inhomogeneous materials will be shown in Chapter 14. Although we cannot analytically evaluate the limiting case $b \to \infty$, we can still explore this situation by evaluating relations (11.7.8) for the case with large but finite b/a ratios. Fig. 11.15 illustrates the dimensionless hoop stress for the case with $b/a = 50$ for several values of $n \leq 1$. Values shown at $r = a$ actually illustrate the stress concentration factors for a small stress-free hole in a large sheet under equal far-field biaxial loading. It can be seen that the stress concentration significantly increases as the anisotropy parameter n is reduced from the isotropic value of unity. Note for this case $E_\theta < E_r$, and n decreasing from 1.0 would correspond to a material where the hoop modulus becomes increasingly smaller than the radial

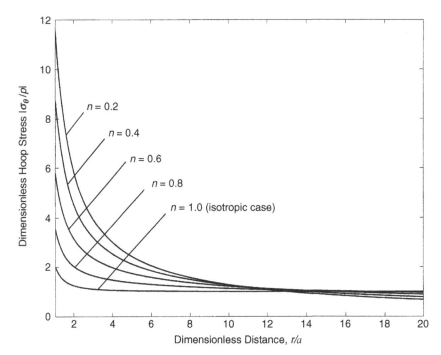

FIG. 11.15 Dimensionless hoop stresses in annular region $b/a = 50$ **simulating a small hole in a large sheet with** $n \leq 1$.

modulus. A similar plot of the hoop stress, for cases with $n > 1$, would show a further decrease in the local stress concentration but would now predict an increasing stress field with radial distance from the hole (see Fig. 11.14 with $n = 1.5$).

11.7.2 Three-dimensional spherical-orthotropic problem

Following the work of Lekhnitskii (1981) and Horgan and Baxter (1996), we next explore a three-dimensional case using a spherical coordinate system as shown in Fig. B.1. Similar to our previous example, we assume that the material has a uniform microstucture such that properties have orthogonal symmetry with respect to the R, ϕ, θ directions. Under these assumptions, the material would be classified as being *spherical-orthotropic*. We make one further simplification, and assume that properties in the ϕ and θ directions are the same. Under these conditions, relation (11.2.10) would allow us to express Hooke's law in the form

$$e_R = \frac{1}{E_R}\sigma_R - \frac{\nu_{R\theta}}{E_R}\left(\sigma_\phi + \sigma_\theta\right)$$

$$e_\phi = \frac{1}{E_\theta}\sigma_\phi - \frac{\nu_{\theta R}}{E_\theta}\sigma_\theta - \frac{\nu_{R\theta}}{E_R}\sigma_R$$

$$e_\theta = \frac{1}{E_\theta}\sigma_\theta - \frac{\nu_{R\theta}}{E_R}\sigma_R - \frac{\nu_{\theta R}}{E_\theta}\sigma_\phi \tag{11.7.11}$$

$$e_{\phi\theta} = \tau_{\phi\theta}/2\mu_\theta = \frac{1+\nu_{\theta R}}{E_\theta}\tau_{\phi\theta}$$

$$e_{R\phi} = \tau_{R\phi}/2\mu_R, \quad e_{\theta R} = \tau_{\theta R}/2\mu_R$$

Note that because of common properties in the ϕ and θ directions, we have only introduced four elastic constants and similar to the previous two-dimensional case, these must satisfy the usual relation

$$\frac{\nu_{\theta R}}{E_\theta} = \frac{\nu_{R\theta}}{E_R} \tag{11.7.12}$$

Equation form (11.7.11) can be inverted to express the stresses in terms of the strains as

$$\sigma_R = C_{11}e_R + C_{12}\left(e_\phi + e_\theta\right)$$

$$\sigma_\phi = C_{12}e_R + C_{22}e_\phi + C_{23}e_\theta$$

$$\sigma_\theta = C_{12}e_R + C_{23}e_\phi + C_{22}e_\theta \tag{11.7.13}$$

$$\tau_{R\phi} = C_{44}e_{R\phi}, \quad \tau_{\phi\theta} = (C_{22} - C_{23})e_{\phi\theta}, \quad \tau_{\theta R} = C_{44}e_{\theta R}$$

where the elastic moduli C_{ij} are related to E, ν, and μ forms by the relations

$$C_{11} = \frac{E_R(1 - \nu_{\theta R})}{m}, \quad C_{12} = \frac{E_\theta\nu_{R\theta}}{m}$$

$$C_{22} = \frac{E_\theta}{(1 + \nu_{\theta R})m}\left(1 - \nu_{R\theta}^2\frac{E_\theta}{E_R}\right)$$

$$C_{23} = \frac{E_\theta}{(1 + \nu_{\theta R})m}\left(\nu_{\theta R} + \nu_{R\theta}^2\frac{E_\theta}{E_R}\right) \tag{11.7.14}$$

$$C_{44} = 2\mu_R, \quad C_{22} - C_{23} = 2\mu_\theta = \frac{E_\theta}{(1 + \nu_{\theta R})}$$

$$m = 1 - \nu_{\theta R} - 2\nu_{R\theta}^2\frac{E_\theta}{E_R}$$

With the given symmetry in the material response, we can now focus on problems that will only produce spherically symmetric deformations yielding a single radial displacement $u_R = u(R)$. Using relations (A.3), the strains thus become

$$e_R = \frac{du}{dR}, \quad e_\phi = e_\theta = \frac{u}{R}, \quad e_{R\phi} = e_{\phi\theta} = e_{\theta R} = 0 \tag{11.7.15}$$

and from Hooke's law (11.7.13) the stresses are

$$\sigma_R = C_{11}\frac{du}{dR} + 2C_{12}\frac{u}{R}$$

$$\sigma_\phi = \sigma_\theta = C_{12}\frac{du}{dR} + (C_{22} + C_{23})\frac{u}{R} \tag{11.7.16}$$

$$\tau_{R\phi} = \tau_{\phi\theta} = \tau_{\theta R} = 0$$

With no body forces, the equilibrium Eq. (A.6) then reduce to a single relation

$$\frac{d\sigma_R}{dR} + \frac{2}{R}(\sigma_R - \sigma_\phi) = 0 \tag{11.7.17}$$

Unfortunately this result contains two unknown stresses, and thus it is more convenient to move to a displacement formulation in terms of the single unknown $u_R = u(R)$. Substituting (11.7.16) into (11.7.17) then gives

$$\frac{d^2u}{dR^2} + \frac{2}{R}\frac{du}{dR} - 2\frac{C_{22} + C_{23} - C_{12}}{C_{11}}\frac{u}{R^2} = 0 \tag{11.7.18}$$

As in the previous polar-orthotropic case, Eq. (11.7.18) is a Cauchy–Euler differential equation and is similar to the isotropic result given in Exercise 13.22. The equation can be easily solved giving the result

$$u = AR^{n-(1/2)} + BR^{-n-(1/2)} \tag{11.7.19}$$

where A and B are arbitrary constants to be determined from the boundary conditions, and the anisotropy parameter n is given by

$$n^2 = \frac{1}{4} + 2\frac{C_{22} + C_{23} - C_{12}}{C_{11}} = \frac{1}{4} + \frac{2E_\theta(1 - \nu_{R\theta})}{E_R(1 - \nu_{\theta R})} = \frac{1}{4} + \frac{2\nu_{\theta R}(1 - \nu_{R\theta})}{\nu_{R\theta}(1 - \nu_{\theta R})} \tag{11.7.20}$$

Note the isotropic case ($E_R = E_\theta$, $\nu_{R\theta} = \nu_{\theta R}$) is retained with $n = 3/2$.

The problem of a spherical shell $a \leq R \leq b$ with both internal an external pressures can now be solved and the results are given in Lekhnitskii (1981). The case to be investigated here will include only the external loading case as presented by Horgan and Baxter (1996), with boundary conditions $\sigma_R(a) = 0$, $\sigma_R(b) = -p$. Applying these two conditions determines the constants A and B and gives the following solution

$$\sigma_R = -\frac{pb^{n+(3/2)}}{b^{2n} - a^{2n}} \left[R^{n-(3/2)} - a^{2n} R^{-(n+(3/2))} \right]$$

$$\sigma_\phi = \sigma_\theta = -\frac{pb^{n+(3/2)}}{b^{2n} - a^{2n}} \left[C_n R^{n-(3/2)} - C_{-n} a^{2n} R^{-(n+(3/2))} \right]$$

(11.7.21)

with

$$C_n = \frac{1 + [n - (1/2)]\nu_{R\theta}}{\nu_{R\theta}[2 + (1 - \nu_{\theta R})(n - (1/2))]/\nu_{\theta R}]}$$

$$C_{-n} = \frac{1 - [n + (1/2)]\nu_{R\theta}}{\nu_{R\theta}[2 - (1 - \nu_{\theta R})(n + (1/2))]/\nu_{\theta R}]}$$

(11.7.22)

Dimensionless distribution plots of the radial and hoop stresses are shown in Fig. 11.16 for cases of $n = 1.0$, 1.5 (isotropic value), and 2.0 with $b/a = 5$. Results are quite similar to the two-dimensional case shown previously in Fig. 11.14. It is seen that with $n > 1.5$, locally the magnitude of the stresses will be less than the isotropic value (stress shielding), while for $n < 1.5$ the stresses will be greater than the isotropic value (stress amplification). Again, these effects can be viewed as being related to the decay of boundary conditions, and thus could have applicability to Saint–Venant's principle for anisotropic problems. Notice also that the hoop stress magnitude at the inner boundary ($R = a$)

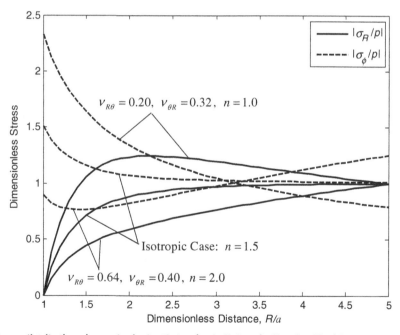

FIG. 11.16 Stress distributions in a spherical orthotropic shell domain ($1 \leq R \leq 5$) with external pressure.

decreases with increasing values of the anisotropic parameter, n. In general, these three-dimensional stresses are significantly affected by the spherical curvilinear anisotropy.

Additional features of this and other similar problems with polar and spherical curvilinear anisotropy are discussed by Galmudi and Dvorkin (1995) and Horgan and Baxter (1996).

References

Cowin SC: The relationship between the elasticity tensor and the fabric tensor, *Mech Materials* 4:137−147, 1985.

Cowin SC, Mehrabadi MM: Anisotropic symmetries of linear elasticity, *Appl Mech Rev ASME* 48:247−285, 1995.

Galmudi D, Dvorkin J: Stresses in anisotropic cylinders, *Mech Res Commun* 22:109−113, 1995.

Hearmon RFS: *An introduction to applied anisotropic elasticity*, London, 1961, Oxford University Press.

Horgan CO, Baxter SC: Effects of curvilinear anisotropy on radially symmetric stresses in anisotropic linearly elastic solids, *J Elast* 42:31−48, 1996.

Jones RM: *Mechanics of composite materials*, New York, 1998, Taylor & Francis.

Lekhnitskii SG: *Anisotropic plates*, New York, 1968, Gordon & Breach.

Lekhnitskii SG: *Theory of elasticity of an anisotropic elastic body*, Moscow, 1981, Mir Publishers.

Love AEH: *A treatise of the mathematical theory of elasticity*, ed 4, London, 1934, Cambridge University Press.

Milne-Thomson LM: *Plane elastic systems*, Berlin, 1960, Springer.

Rand O, Rovenski V: *Analytical methods in anisotropic elasticity*, Boston, 2005, Birkhäuser.

Sadd MH: *Continuum mechanics modeling of material behavior*, 2019, Elsevier.

Savin GN: *Stress concentration around holes*, New York, 1961, Pergamon Press.

Sendeckyj GP: Some topics of anisotropic elasticity. In Chamis CC, editor: Composite Materials, vol. 7. *Structural Design and Analysis Part I*, New York, 1975, Academic Press.

Sih GC, Liebowitz H: Mathematical theories of brittle fracture. In Liebowitz H, editor: *Fracture*, vol. 2, chap 2, New York, 1968, Academic Press.

Sih GC, Paris PC, Irwin GR: On cracks in rectilinearly anisotropic bodies, *Int J Fract Mech* 1:189−203, 1965.

Swanson SR: *Introduction to design and analysis with advanced composite materials*, Upper Saddle River, NJ, 1997, Prentice Hall.

Ting TCT: *Anisotropic elasticity, theory and applications*, New York, 1996a, Oxford University Press.

Ting TCT: Positive definiteness of anisotropic elastic constants, *Math Mech Solids* 1:301−314, 1996b.

Ting TCT: Generalized Cowin−Mehrabadi theorems and a direct proof that the number of linear elastic symmetries is eight, *Int J Solids Structures* 40:7129−7142, 2003.

Zheng QS, Spencer AJM: Tensors which characterize anisotropes, *Int J Eng Sci* 31:679−693, 1993.

Exercises

11.1 From strain energy arguments in Section 6.1, it was found that $\partial\sigma_{ij}/\partial e_{kl} = \partial\sigma_{kl}/\partial e_{ij}$. Show that these results imply that $C_{ij} = C_{ji}$, therefore justifying that only 21 independent elastic moduli are needed to characterize the most general anisotropic material.

11.2 Using material symmetry through 180° rotations about each of the three coordinate axes, explicitly show the reduction of the elastic stiffness matrix to nine independent components for *orthotropic materials*. Also demonstrate that after two rotations, the third transformation is actually already satisfied.

11.3 A *transversely isotropic material* with an x_3-*axis of symmetry* was specified by the elasticity matrix given in Eq. (11.2.12). Under an arbitrary θ rotation about the x_3-axis given by relation (11.2.11), all components of this elasticity matrix should remain the same. Explicitly show this property for the 55 and 22 components of the C matrix. Use the translation relation (11.1.6) and the given structure of the C matrix such as $C_{16} = C_{1112} = 0$, $C_{44} = C_{55} \dots$

11.4 Verify the inequality restrictions on the elastic moduli for orthotropic, transversely isotropic, cubic and isotropic materials given by Table 11.1.

11.5 For the orthotropic case, show that by using arguments of a positive definite strain energy function, $v_{ij}^2 < (E_i/E_j)$. Next, using typical values for E_1 and E_2 from Table 11.2, justify that this theory could allow the unexpected result that $v_{12} > 1$.

11.6 Similar to Example 11.1, consider the hydrostatic compression of a material with cubic symmetry. Determine the general expressions for the strains. We now specify that the material is aluminum with material properties $C_{11} = 108$, $C_{12} = 61$, $C_{44} = 28$ *GPa*. If the dilatation was found to be -2×10^{-5}, determine the hydrostatic stress.

11.7 For the torsion of cylinders discussed in Section 11.4, show that with $\sigma_x = \sigma_y = \sigma_z = \tau_{xy} = 0$, the compatibility equations yield

$$-\frac{\partial}{\partial x}(S_{44}\tau_{yz} + S_{45}\tau_{xz}) + \frac{\partial}{\partial y}(S_{54}\tau_{yz} + S_{55}\tau_{xz}) = C$$

where C is a constant.

11.8 In terms of the stress function ψ, the torsion problem was governed by Eq. (11.4.9)

$$S_{44}\psi_{xx} - 2S_{45}\psi_{xy} + S_{55}\psi_{yy} = -2$$

Show that the homogeneous counterpart of this equation may be written as

$$\left(\frac{\partial}{\partial y} - \mu_1\frac{\partial}{\partial x}\right)\left(\frac{\partial}{\partial y} - \mu_2\frac{\partial}{\partial x}\right)\psi = 0$$

where $\mu_{1,2}$ are the roots of the *characteristic equation*

$$S_{55}\mu^2 - 2S_{45}\mu + S_{44} = 0$$

11.9 Explicitly justify relationships (11.5.3) between the compliances of the plane stress and plane strain theories.

11.10 Investigate case 2 ($\mu_1 = \mu_2$) in Eq. (11.5.10), and determine the general form of the Airy stress function. Show that this case is actually an isotropic formulation.

11.11 Determine the roots of the characteristic equation (11.5.7) for S-Glass/Epoxy material with properties given in Table 11.2. Justify that they are purely imaginary.

11.12 Recall that for the plane anisotropic problem, the Airy stress function was found to be

$$\phi = F_1(z_1) + \overline{F_1(z_1)} + F_2(z_2) + \overline{F_2(z_2)}$$

where $z_1 = x + \mu_1 y$ and $z_2 = x + \mu_2 y$. Explicitly show that the in-plane stresses are given by

$$\sigma_x = 2\text{Re}\left[\mu_1^2 F_1''(z_1) + \mu_2^2 F_2''(z_2)\right]$$

$$\sigma_y = 2\text{Re}\left[F_1''(z_1) + F_2''(z_2)\right]$$

$$\tau_{xy} = -2\text{Re}\left[\mu_1 F_1''(z_1) + \mu_2 F_2''(z_2)\right]$$

11.13 For the plane stress case, in terms of the two complex potentials Φ_1 and Φ_2, compute the two in-plane displacements u and v and thus justify relations (11.5.14).

11.14 Determine the polar coordinate stresses and displacements in terms of the complex potentials Φ_1 and Φ_2, as given by Eqs. (11.5.16) and (11.5.17).

11.15* For the plane problem with an orthotropic material, show that the characteristic Eq. (11.5.7) reduces to the quadratic equation in μ^2

$$S_{11}\mu^4 + (2S_{12} + S_{66})\mu^2 + S_{22} = 0$$

Explicitly solve this equation for the roots μ_i, and show that they are purely complex and thus can be written as $\mu_{1,2} = i\beta_{1,2}$, where

$$\beta_{1,2}^2 = -\frac{1}{2S_{11}}\left\{-(2S_{12} + S_{66}) \pm \sqrt{(2S_{12} + S_{66})^2 - 4S_{11}S_{22}}\right\}$$

Justify the isotropic case where $\beta_{1,2} = 1$. Finally, determine $\beta_{1,2}$ for each of the four composite materials given in Table 11.2. See MATLAB code C.10 for numerical methods to calculate β-parameters.

11.16 Consider an anisotropic *monoclinic* material symmetric about the x,y-plane (see Fig. 11.2) and subject to an *antiplane deformation* specified by $u = v = 0$, $w = w(x,y)$. Show that in the absence of body forces, the out-of-plane displacement must satisfy the Navier equation

$$C_{55}\frac{\partial^2 w}{\partial x^2} + 2C_{45}\frac{\partial^2 w}{\partial x \partial y} + C_{44}\frac{\partial^2 w}{\partial y^2} = 0$$

Next looking for solutions that are of the form $w = F(x + \mu y)$, show that this problem is solved by

$$w = 2\text{Re}\{F(z^*)\}$$

$$\tau_{xz} = 2\text{Re}\{(\mu C_{45} + C_{55})F'(z^*)\}$$

$$\tau_{yz} = 2\text{Re}\{(\mu C_{44} + C_{45})F'(z^*)\}$$

where $z^* = x + \mu y$ and μ are the roots of the equation $C_{44}\mu^2 + 2C_{45}\mu + C_{55} = 0$. Note that for this case, positive definite strain energy implies that $C_{44}C_{55} > C_{45}^2$; therefore the roots will occur in complex conjugate pairs.

11.17* For Example 11.5, consider the case of only a normal boundary load ($X = 0$), and assume that the material is orthotropic with $\mu_i = i\beta_i$ (see Exercise 11.15). Show that the resulting stress field is given by

$$\sigma_r = -\frac{Y\beta_1\beta_2(\beta_1 + \beta_2)\sin\theta}{\pi r\left(\cos^2\theta + \beta_1^2\sin\theta\right)\left(\cos^2\theta + \beta_2^2\sin^2\theta\right)}, \qquad \sigma_\theta = \tau_{r\theta} = 0$$

Next compare the stress component σ_r with the corresponding isotropic value by plotting the stress contours $\sigma_r/Y = $ constant for each case. Use orthotropic material values for the Carbon/Epoxy composite given in Table 11.2, and compare with the corresponding isotropic case.

11.18* Consider the case of the pressurized circular hole in an anisotropic sheet. Using orthotropic material properties given in Table 11.2 for Carbon/Epoxy, compute and plot the boundary hoop stress σ_θ as a function of θ. Compare with the isotropic case.

11.19* Investigate the case of a circular hole of radius a in Example 11.7. Use orthotropic material properties given in Table 11.2 for Carbon/Epoxy with the 1-axis along the direction of loading. Compute and plot the stress $\sigma_x(0, y)$ for $y > a$. Also compare with the corresponding isotropic case.

11.20 Consider the elliptical hole problem in Example 11.7. By letting $a \to 0$, determine the stress field for the case where the hole reduces to a line crack of length $2b$. Demonstrate the nature of the singularity for this case.

11.21 The potentials

$$\Phi_1(z_1) = A_1 z_1 + \frac{Sa^2\mu_2}{2(\mu_1 - \mu_2)}\left[z_1 + \sqrt{z_1^2 - a^2}\right]^{-1}$$

$$\Phi_2(z_2) = A_2 z_2 - \frac{Sa^2\mu_1}{2(\mu_1 - \mu_2)}\left[z_2 + \sqrt{z_2^2 - a^2}\right]^{-1}$$

were proposed to solve the plane extension of an anisotropic panel containing a crack of length $2a$ (see Fig. 11.12). Recall that the constants A_1 and A_2 correspond to the uniform tension case, and for stress S in the y direction

$$A_1 = \frac{(\alpha_2^2 + \beta_2^2)S}{2\left[(\alpha_2 - \alpha_1)^2 + (\beta_2^2 - \beta_2^2)\right]}$$

$$A_2 = \frac{(\alpha_1^2 + \beta_1^2 - 2\alpha_1\alpha_2)S}{2\left[(\alpha_2 - \alpha_1)^2 + (\beta_2^2 - \beta_1^2)\right]} + i\,\frac{\left[\alpha_2(\alpha_1^2 - \beta_1^2) - \alpha_1(\alpha_2^2 - \beta_2^2)\right]S}{2\beta_2\left[(\alpha_2 - \alpha_1)^2 + (\beta_2^2 - \beta_1^2)\right]}$$

(a) Determine the general stress field and verify the far-field behavior.

(b) Show that the stress field is singular at each crack tip.

(c) Using the limiting procedures as related to Fig. 10.20, verify that the crack-tip stress field is given by (11.6.10).

11.22* Construct a contour plot of the crack tip stress component σ_y from solution (11.6.7). This result could be compared with the equivalent isotropic problem from Exercise 10.24.

11.23 Consider the case of a crack problem in an anisotropic monoclinic material under antiplane deformation as described in Exercise 11.16. Following relation (11.6.6), choose the complex potential form as $F(z^*) = A\sqrt{z^*}$, where $A = -\sqrt{2}\,K_3\mu/(C_{55} + \mu C_{45})$ and K_3 is a real constant. Using this form, show that the nonzero displacement and stresses in the vicinity of the crack tip (see Fig. 11.11) are given by

$$w = K_3\sqrt{2r}\mathrm{Re}\left\{\frac{\sqrt{\cos\theta + \mu\sin\theta}}{C_{45} + \mu C_{44}}\right\}$$

$$\tau_{xz} = -\frac{K_3}{\sqrt{2r}}\mathrm{Re}\left\{\frac{\mu}{\sqrt{\cos\theta + \mu\sin\theta}}\right\}$$

$$\tau_{yz} = \frac{K_3}{\sqrt{2r}}\mathrm{Re}\left\{\frac{1}{\sqrt{\cos\theta + \mu\sin\theta}}\right\}$$

Note that the parameter K_3 will be related to the stress intensity factor for this case. Verify that shear stress τ_{yz} vanishes on each side of the crack face, $\theta = \pm\pi$. These results can be compared to the corresponding solution for the isotropic case given in Exercise 8.44.

11.24 Explicitly develop the governing Navier equation (11.7.5) for the polar orthotropic problem. Verify that its solution is given by (11.7.6) and show how this leads to the stress solution (11.7.7). Finally, confirm that the problem with only external pressure loading is given by (11.7.8).

11.25 For the spherically orthotropic problem, justify that Hooke's law (11.7.11) can be inverted into form (11.7.13) under the relations (11.7.14).

11.26 Under the stated symmetry conditions in Section 11.7.2, explicitly show that in the absence of body forces the general equilibrium equations reduce to forms (11.7.17) and (11.7.18). Verify the general displacement solution given by (11.7.19) and (11.7.20), and the particular stress solution (11.7.21) and (11.7.22) for the external loading case.

11.27 For the rotating disk problem given previously in Example 8.11, the governing equilibrium equation was given by (8.4.74). Since this equation is also valid for anisotropic materials, consider the polar-orthotropic case and use equations (11.7.3) to express the equilibrium equation in terms of the displacement as

$$r^2\frac{d^2u}{dr^2} + r\frac{du}{dr} + \frac{E_\theta}{E_r}u = -\frac{(1 - \nu_{r\theta}\nu_{\theta r})}{E_r}\rho\omega^2 r^3$$

Next show that the general solution to this equation is given by

$$u = C_1 r^n + C_2 r^{-n} - \frac{(1 - \nu_{r\theta}\nu_{\theta r})}{E_r[9 - (E_\theta/E_r)]}\rho\omega^2 r^3, \quad n = \sqrt{E_\theta/E_r}$$

where C_1 and C_2 are arbitrary constants and $n \neq 3$. Note for the case of a solid disk, $C_2 = 0$.

11.28 Using the results from Exercise 11.27, show that the stresses in a rotating solid circular polar-orthotropic disk of radius a with boundary condition $\sigma_r(a) = 0$ are given by

$$\sigma_r = \frac{3 + \nu_{\theta r}}{9 - n^2} \rho \omega^2 \left(a^{3-n} r^{n-1} - r^2\right)$$

$$\sigma_\theta = \frac{\rho \omega^2}{9 - n^2} \left[n(3 + \nu_{\theta r})a^{3-n} r^{n-1} - \left(n^2 + 3\nu_{\theta r}\right)r^2\right]$$

Discuss issues with the case $n < 1$

11.29 Consider the cantilever beam problem shown previously in Exercise 8.2 with no axial force ($N = 0$). Assume a plane stress anisotropic model given by Hooke's law (11.5.1) and governed by the Airy stress function equation (11.5.6). Show that the stress function

$$\phi = \frac{3P}{4c^3} \left[c^2 xy - \frac{xy^3}{3} + \frac{S_{16}}{6S_{11}} \left(2c^2 y^2 - y^4\right)\right]$$

satisfies the governing equation and gives the following stress field

$$\sigma_x = -\frac{3P}{2c^3} xy + \frac{3P}{2c^3} \frac{S_{16}}{S_{11}} \left(\frac{c^2}{3} - y^2\right), \quad \sigma_y = 0, \quad \tau_{xy} = -\frac{3P}{4c} \left(1 - \frac{y^2}{c^2}\right)$$

Next show that these stresses satisfy the problem boundary conditions in the usual sense with exact pointwise specification on $y = \pm c$, and only resultant force conditions on the ends $x = 0$ and $x = L$. What happens to this solution if we let the material become orthotropic?

Thermoelasticity

Many important stress analysis problems involve structures that are subjected to both mechanical and thermal loadings. Thermal effects within an elastic solid produce heat transfer by conduction, and this flow of thermal energy establishes a temperature field within the material. Most solids exhibit a volumetric change with temperature variation, and thus the presence of a temperature distribution generally induces stresses created from boundary or internal constraints. If the temperature variation is sufficiently high, these stresses can reach levels that may lead to structural failure, especially for brittle materials. Thus, for many problems involving high temperature variation, the knowledge of thermal stress analysis can be very important.

The purpose of this chapter is to provide an introduction to thermoelasticity; that is, elasticity with thermal effects. We develop the basic governing equations for isotropic materials and investigate several solutions to problems of engineering interest. We have already briefly discussed the form of Hooke's law for this case in Section 4.4. More detailed information may be found in several texts devoted entirely to the subject such as Boley and Weiner (1960), Nowacki (1962), Parkus (1976), Kovalenko (1969), Nowinski (1978), and Burgreen (1971). We start our study with some developments of heat conduction in solids and the energy equation.

12.1 Heat conduction and the energy equation

As mentioned, the flow of heat in solids is associated with temperature differences within the material. This process is governed by the *Fourier law of heat conduction*, which is the constitutive relation between the heat flux vector q and the temperature gradient ∇T. This theory formulates a linear relationship that is given by

$$q_i = -k_{ij}T_{,j} \tag{12.1.1}$$

where k_{ij} is the *thermal conductivity tensor*. It can be shown that this tensor is symmetric, that is, $k_{ij} = k_{ji}$. For the isotropic case $k_{ij} = k\delta_{ij}$, and thus

$$q_i = -kT_{,i} \tag{12.1.2}$$

where k is a material constant called the *thermal conductivity*. Note the flow of heat moves against the temperature gradient, that is, flows from hot to cold regions.

In order to properly establish thermoelasticity theory, particular thermal variables such as temperature and heat flux must be included, and this requires incorporation of the *energy equation*. Previous to this point, our purely mechanical theory did not require this field relation. The energy equation represents the *principle of conservation of energy*, and this concept is to be applied for the

Elasticity. https://doi.org/10.1016/B978-0-12-815987-3.00012-8

special case of an elastic solid continuum. Details of the equation derivation will not be presented here, and the interested reader is referred to Boley and Weiner (1960), Fung (1965) or Sadd (2019) for a more complete discussion on the thermodynamic development. We consider an elastic solid that is *stress free* at a uniform temperature T_o when all external forces are zero. This stress-free state is referred to as the *reference state*, and T_o is called the *reference temperature*. For this case, the energy equation can be written as

$$\rho\dot{\varepsilon} = \sigma_{ij}v_{i,j} - q_{i,i} + \rho h \tag{12.1.3}$$

where ρ is the mass density, ε is the *internal energy*, v_i is the velocity field, and h is any *prescribed energy source term*. From thermodynamic theory, the internal energy rate may be simplified to

$$\dot{\varepsilon} = c\dot{T} \tag{12.1.4}$$

where c is the *specific heat capacity at constant volume*.

Recall that the stress follows from the *Duhamel–Neumann constitutive relation* given previously in (4.4.5) as

$$\sigma_{ij} = C_{ijkl}e_{kl} + \beta_{ij}(T - T_o) \tag{12.1.5}$$

and for the isotropic case this reduces to

$$\sigma_{ij} = \lambda e_{kk}\delta_{ij} + 2\mu e_{ij} - (3\lambda + 2\mu)\alpha(T - T_o)\delta_{ij}$$
$$e_{ij} = \frac{1+\nu}{E}\sigma_{ij} - \frac{\nu}{E}\sigma_{kk}\delta_{ij} + \alpha(T - T_o)\delta_{ij} \tag{12.1.6}$$

where α is the *coefficient of thermal expansion*.

Using results (12.1.4) and (12.1.6) in the energy equation and linearizing yields

$$kT_{,ii} = \rho c\dot{T} + (3\lambda + 2\mu)\alpha T_o\dot{e}_{ii} - \rho h \tag{12.1.7}$$

Note that the expression $(3\lambda + 2\mu)\alpha T_o\dot{e}_{ii}$ involves both thermal and mechanical variables, and consequently is referred to as the *coupling term* in the energy equation. It has been shown (see, for example, Boley and Weiner, 1960) that for most materials under static or quasi-static loading conditions, this coupling term is small and can be neglected. For this case, we establish the so-called *uncoupled conduction equation*

$$kT_{,ii} = \rho c\dot{T} - \rho h \tag{12.1.8}$$

For our applications, we consider only uncoupled theory and normally with no sources ($h = 0$). Another simplification is to consider only *steady-state conditions*, and for this case the conduction equation reduces to the Laplace equation

$$T_{,ii} = \nabla^2 T = \frac{\partial^2 T}{\partial x^2} + \frac{\partial^2 T}{\partial y^2} + \frac{\partial^2 T}{\partial z^2} = 0 \tag{12.1.9}$$

It should be noted that for the uncoupled, no-source case the energy equation (12.1.8) reduces to a single *parabolic* partial differential equation, while for the steady state case the reduction leads to an *elliptic* equation (12.1.9) for the temperature distribution. For either case, with appropriate thermal boundary conditions, the temperature field can be determined *independent of the stress-field calculations*. Once the temperature is obtained, elastic stress analysis procedures can then be employed to complete the problem solution.

12.2 General uncoupled formulation

Let us now formulate the general uncoupled thermoelastic problem. Many of our previous equations are still valid and remain unchanged, including the strain-displacement relations

$$e_{ij} = \frac{1}{2}\left(u_{i,j} + u_{j,i}\right) \tag{12.2.1}$$

the strain—compatibility equations

$$e_{ij,kl} + e_{kl,ij} - e_{ik,jl} - e_{jl,ik} = 0 \tag{12.2.2}$$

and the equilibrium equations

$$\sigma_{ij,j} + F_i = 0 \tag{12.2.3}$$

These are to be used with the new form of Hooke's law

$$\sigma_{ij} = \lambda e_{kk}\delta_{ij} + 2\mu e_{ij} - (3\lambda + 2\mu)\alpha(T - T_o)\delta_{ij} \tag{12.2.4}$$

and the energy equation

$$\rho c \dot{T} = k T_{,ii} \tag{12.2.5}$$

The 16 equations (12.2.1) and (12.2.3)−(12.2.5) constitute the fundamental set of field equations for uncoupled thermoelasticity for the 16 unknowns u_i, e_{ij}, σ_{ij}, and T. As before, it proves to be very helpful for problem solution to further reduce this set to a *displacement* and/or *stress formulation* as previously done for the isothermal case. Recall that the compatibility equations are used only for the stress formulation. These further reductions are not carried out at this point, but will be developed in the next section for the two-dimensional formulation. Boundary conditions for the mechanical problem are identical as before, while thermal boundary conditions normally take the form of specifying the temperatures or heat fluxes on boundary surfaces.

12.3 Two-dimensional formulation

The basic two-dimensional thermoelasticity formulation follows in similar fashion as done previously for the isothermal case in Chapter 7, leading to the usual *plane strain* and *plane stress* problems. Each of these formulations is now briefly developed. Some parts of the ensuing presentation are identical to the isothermal formulation, while other results create new terms or equations. It is important to pay special attention to these new contributions and to be able to recognize them in the field equations and boundary conditions.

12.3.1 Plane strain

The basic assumption for plane strain in the x,y-plane was given by the displacement field

$$u = u(x, y), \quad v = v(x, y), \quad w = 0 \tag{12.3.1}$$

Recall that this field is a reasonable approximation for cylindrical bodies with a large z dimension, as shown previously in Fig. 7.1. This leads to the following strain and stress fields

$$e_x = \frac{\partial u}{\partial x}, \quad e_y = \frac{\partial v}{\partial y}, \quad e_{xy} = \frac{1}{2}\left(\frac{\partial u}{\partial y} + \frac{\partial v}{\partial x}\right)$$

$$e_z = e_{xz} = e_{yz} = 0$$

(12.3.2)

$$\sigma_x = \lambda\left(\frac{\partial u}{\partial x} + \frac{\partial v}{\partial y}\right) + 2\mu\frac{\partial u}{\partial x} - \alpha(3\lambda + 2\mu)(T - T_o)$$

$$\sigma_y = \lambda\left(\frac{\partial u}{\partial x} + \frac{\partial v}{\partial y}\right) + 2\mu\frac{\partial v}{\partial y} - \alpha(3\lambda + 2\mu)(T - T_o)$$

(12.3.3)

$$\tau_{xy} = \mu\left(\frac{\partial u}{\partial y} + \frac{\partial v}{\partial x}\right)$$

$$\sigma_z = \nu(\sigma_x + \sigma_y) - E\alpha(T - T_o)$$

$$\tau_{xz} = \tau_{yz} = 0$$

In the absence of body forces, the equilibrium equations become

$$\frac{\partial \sigma_x}{\partial x} + \frac{\partial \tau_{xy}}{\partial y} = 0$$

(12.3.4)

$$\frac{\partial \tau_{xy}}{\partial x} + \frac{\partial \sigma_y}{\partial y} = 0$$

and in terms of displacements these equations reduce to

$$\mu\nabla^2 u + (\lambda + \mu)\frac{\partial}{\partial x}\left(\frac{\partial u}{\partial x} + \frac{\partial v}{\partial y}\right) - (3\lambda + 2\mu)\alpha\frac{\partial T}{\partial x} = 0$$

(12.3.5)

$$\mu\nabla^2 v + (\lambda + \mu)\frac{\partial}{\partial y}\left(\frac{\partial u}{\partial x} + \frac{\partial v}{\partial y}\right) - (3\lambda + 2\mu)\alpha\frac{\partial T}{\partial y} = 0$$

where $\nabla^2 = \frac{\partial^2}{\partial x^2} + \frac{\partial^2}{\partial y^2}$. Comparing this result with the equivalent isothermal equations (7.1.5), it is noted that thermoelasticity theory creates additional thermal terms in Navier's relations (12.3.5).

The only nonzero compatibility equation for plane strain is given by

$$\frac{\partial^2 e_x}{\partial y^2} + \frac{\partial^2 e_y}{\partial x^2} = 2\frac{\partial^2 e_{xy}}{\partial x \partial y}$$

(12.3.6)

Using Hooke's law in this result gives

$$\nabla^2(\sigma_x + \sigma_y) + \frac{E\alpha}{1 - \nu}\nabla^2 T = 0$$

(12.3.7)

Again, note the additional thermal term in this relation when compared to the isothermal result given by (7.1.7). The additional terms in both (12.3.5) and (12.3.7) can be thought of as thermal body forces that contribute to the generation of the stress, strain, and displacement fields. Relations (12.3.5) would be used for the displacement formulation, while (12.3.4) and (12.3.7) would be incorporated in the stress formulation.

The boundary conditions for the plane strain problem are normally specified for either the stresses

$$T_x^n = \sigma_x n_x + \tau_{xy} n_y = \left(T_x^n\right)_s$$
$$T_y^n = \tau_{xy} n_x + \sigma_y n_y = \left(T_y^n\right)_s \tag{12.3.8}$$

or the displacements

$$u = u_s(x, y)$$
$$v = v_s(x, y) \tag{12.3.9}$$

where $\left(T_x^n\right)_s$, $\left(T_y^n\right)_s$, u_s, and v_s are the specified boundary tractions and displacements on the lateral surfaces. Note that these specified values must be independent of z and the temperature field must also depend only on the in-plane coordinates; that is, $T = T(x, y)$. It should be recognized that using Hooke's law (12.3.3) in the traction boundary conditions (12.3.8) will develop relations that include the temperature field.

12.3.2 Plane stress

The fundamental starting point for plane stress (and/or generalized plane stress) in the x, y-plane is an assumed stress field of the form

$$\sigma_x = \sigma_x(x, y), \quad \sigma_y = \sigma_y(x, y), \quad \tau_{xy} = \tau_{xy}(x, y)$$
$$\sigma_z = \tau_{xz} = \tau_{yz} = 0 \tag{12.3.10}$$

As per our previous discussion in Section 7.2 this field is an appropriate approximation for bodies thin in the z direction (see Fig. 7.3). The thermoelastic strains corresponding to this stress field come from Hooke's law

$$e_x = \frac{1}{E}(\sigma_x - \nu\sigma_y) + \alpha(T - T_o)$$

$$e_y = \frac{1}{E}(\sigma_y - \nu\sigma_x) + \alpha(T - T_o)$$

$$e_{xy} = \frac{1+\nu}{E}\tau_{xy} \tag{12.3.11}$$

$$e_z = -\frac{\nu}{E}(\sigma_x + \sigma_y) + \alpha(T - T_o)$$

$$e_{xz} = e_{yz} = 0$$

The equilibrium and strain compatibility equations for this case are identical to the plane strain model; that is, Eqs. (12.3.4) and (12.3.6). However, because of the differences in the form of Hooke's law, plane stress theory gives slightly different forms for the displacement equilibrium equations and stress compatibility relations. However, as we discovered previously for the isothermal case, differences between plane stress and plane strain occur only in particular coefficients involving the elastic

Table 12.1 Elastic moduli transformation relations for conversion between plane stress and plane strain for thermoelastic problems.

	E	ν	α
Plane stress to plane strain	$\dfrac{E}{1-\nu^2}$	$\dfrac{\nu}{1-\nu}$	$(1+\nu)\alpha$
Plane strain to plane stress	$\dfrac{E(1+2\nu)}{(1+\nu)^2}$	$\dfrac{\nu}{1+\nu}$	$\dfrac{1+\nu}{1+2\nu}\alpha$

constants, and by simple interchange of elastic moduli one theory can be transformed into the other (see Table 7.1). This result also holds for the thermoelastic case, and the specific transformation rules are given in Table 12.1.

Using these transformation results, the displacement equilibrium equations for plane stress follow from (12.3.5)

$$\mu\nabla^2 u + \frac{E}{2(1-\nu)}\frac{\partial}{\partial x}\left(\frac{\partial u}{\partial x}+\frac{\partial v}{\partial y}\right) - \frac{E}{1-\nu}\alpha\frac{\partial T}{\partial x} = 0$$

$$\mu\nabla^2 v + \frac{E}{2(1-\nu)}\frac{\partial}{\partial y}\left(\frac{\partial u}{\partial x}+\frac{\partial v}{\partial y}\right) - \frac{E}{1-\nu}\alpha\frac{\partial T}{\partial y} = 0$$

(12.3.12)

and the plane stress compatibility relation becomes

$$\nabla^2(\sigma_x+\sigma_y) + E\alpha\nabla^2 T = 0 \tag{12.3.13}$$

The boundary conditions for plane stress are similar in form to those of plane strain specified by relations (12.3.8) and (12.3.9), and these would apply on the lateral edges of the domain.

Reviewing plane strain theory, it is observed that the temperature effect is equivalent to adding an additional body force $-(3\lambda+2\mu)\alpha\dfrac{\partial T}{\partial x}$ to Navier's equations of equilibrium and adding a traction term $(3\lambda+2\mu)\alpha(T-T_o)n_i$ to the applied boundary tractions. A similar statement could be made about the plane stress theory, and in fact this concept can be generalized to three-dimensional theory.

12.4 Displacement potential solution

We now present a general scheme for the solution to the thermoelastic displacement problem. Although this scheme can be employed for the three-dimensional case (see Timoshenko and Goodier, 1970), only the plane problem will be considered here. We introduce a *displacement potential* Ψ, such that the displacement vector is given by

$$u = \nabla\Psi \tag{12.4.1}$$

Further details on potential methods are discussed in Chapter 13. Using this representation in Navier's equations for plane stress (12.3.12) with no body forces gives the result

$$\frac{\partial}{\partial x}\left(\frac{\partial^2 \Psi}{\partial x^2} + \frac{\partial^2 \Psi}{\partial y^2}\right) = (1+\nu)\alpha\frac{\partial T}{\partial x}$$

$$\frac{\partial}{\partial y}\left(\frac{\partial^2 \Psi}{\partial x^2} + \frac{\partial^2 \Psi}{\partial y^2}\right) = (1+\nu)\alpha\frac{\partial T}{\partial y}$$

(12.4.2)

These equations can be integrated to give

$$\frac{\partial^2 \Psi}{\partial x^2} + \frac{\partial^2 \Psi}{\partial y^2} = (1+\nu)\alpha T$$

(12.4.3)

where the constant of integration has been dropped and T denotes the temperature change from the stress-free reference value. Note for the plane strain case, the coefficient on the temperature term would become $\alpha(1+\nu)/(1-\nu)$.

The general solution to (12.4.3) can be written as the sum of a particular integral plus the solution to the homogeneous equation

$$\Psi = \Psi^{(p)} + \Psi^{(h)}$$

(12.4.4)

with

$$\nabla^2 \Psi^{(h)} = 0$$

(12.4.5)

The particular integral of the Poisson equation (12.4.3) is given by standard methods of potential theory (see, for example, Kellogg, 1953)

$$\Psi^{(p)} = \frac{1}{2\pi}(1+\nu)\alpha\iint_R T(\xi,\eta)\log r \, d\xi d\eta$$

(12.4.6)

where $r = [(x-\xi)^2 + (y-\eta)^2]^{1/2}$ and R is the two-dimensional domain of interest.

The displacement field corresponding to these two solutions may be expressed as

$$u_i = u_i^{(p)} + u_i^{(h)}$$

(12.4.7)

It is noted that the homogeneous solution field satisfies the *Navier equation*

$$\mu u_{i,kk}^{(h)} + \frac{E}{2(1-\nu)}u_{k,ki}^{(h)} = 0$$

(12.4.8)

which corresponds to an *isothermal problem*. The boundary conditions for the solution $u_i^{(h)}$ are determined from the original conditions by subtracting the contributions of the particular integral solution $u_i^{(p)}$. Thus, with the particular integral known, the general problem is then reduced to solving an isothermal case.

12.5 Stress function formulation

Let us now continue the plane problem formulation and pursue the usual stress function method of solution. As before, we can introduce the Airy stress function defined by

$$\sigma_x = \frac{\partial^2 \phi}{\partial y^2}, \quad \sigma_y = \frac{\partial^2 \phi}{\partial x^2}, \quad \tau_{xy} = -\frac{\partial^2 \phi}{\partial x \partial y} \tag{12.5.1}$$

Recall that this representation satisfies the equilibrium equations identically. Using this form in the compatibility equation (12.3.13) for the plane stress case gives

$$\nabla^4 \phi + E\alpha \nabla^2 T = 0$$

or

$$\frac{\partial^4 \phi}{\partial x^4} + 2\frac{\partial^4 \phi}{\partial x^2 \partial y^2} + \frac{\partial^4 \phi}{\partial y^4} + E\alpha \left(\frac{\partial^2 T}{\partial x^2} + \frac{\partial^2 T}{\partial y^2} \right) = 0 \tag{12.5.2}$$

The corresponding equation for plane strain follows by using the transformation relations in Table 12.1.

The general solution to (12.5.2) can be written in the form $\phi = \phi^{(p)} + \phi^{(h)}$, where $\phi^{(h)}$ satisfies the homogeneous equation

$$\nabla^4 \phi^{(h)} = 0 \tag{12.5.3}$$

and for plane stress $\phi^{(p)}$ is a particular solution of the equation

$$\nabla^2 \phi^{(p)} + E\alpha T = 0 \tag{12.5.4}$$

A similar result can be obtained for the plane strain case. Note that for the steady-state problem, the temperature field is harmonic, and thus (12.5.2) reduces to the homogeneous equation.

The general traction boundary conditions are expressible as

$$T_x^n = \sigma_x n_x + \tau_{xy} n_y = \frac{\partial^2 \phi}{\partial y^2} \frac{dy}{ds} + \frac{\partial^2 \phi}{\partial x \partial y} \frac{dx}{ds} = \frac{d}{ds} \left(\frac{\partial \phi}{\partial y} \right)$$

$$T_y^n = \tau_{xy} n_x + \sigma_y = -\frac{\partial^2 \phi}{\partial x \partial y} \frac{dy}{ds} - \frac{\partial^2 \phi}{\partial x^2} \frac{dx}{ds} = -\frac{d}{ds} \left(\frac{\partial \phi}{\partial x} \right) \tag{12.5.5}$$

which are identical to the isothermal relations (10.2.13). Integrating these results over a particular portion of the boundary C gives

$$\int_C T_x^n ds + C_1 = \frac{\partial \phi}{\partial y}$$

$$\int_C T_y^n ds + C_2 = -\frac{\partial \phi}{\partial x} \tag{12.5.6}$$

where C_1 and C_2 are arbitrary constants of integration.

Combining this result with the total differential definition $d\phi = \frac{\partial \phi}{\partial x} dx + \frac{\partial \phi}{\partial y} dy$ and integrating over C from 0 to s gives

$$\phi(s) = -x \int_0^s T_y^n ds + y \int_0^s T_x^n ds + \int_0^s \left(xT_y^n - yT_x^n \right) ds \tag{12.5.7}$$

where we have dropped constants of integration because they will not contribute to the stress field. Likewise, using the directional derivative definition $d\phi/dn = \nabla\phi \cdot \boldsymbol{n}$ gives the result

$$\frac{d\phi}{dn} = -\frac{dx}{ds}\int_0^s T_x^n ds - \frac{dy}{ds}\int_0^s T_y^n ds = -\boldsymbol{t} \cdot \boldsymbol{F} \tag{12.5.8}$$

where $\boldsymbol{t}$ is the unit tangent vector to the boundary curve and $\boldsymbol{F}$ is the resultant boundary force. For many applications, the boundary conditions are simply expressed in terms of specific stress components, and for the Cartesian case we can use the defining relations (12.5.1) to develop appropriate conditions necessary to solve the problem.

Note that for the case of zero surface tractions $T_x^n = T_y^n = 0$, these boundary conditions imply that

$$\phi = \frac{d\phi}{dn} = 0 \text{ on the boundary} \tag{12.5.9}$$

For this case under steady-state conditions, the solution to the homogeneous form of (12.5.2) is the trivial solution $\phi \equiv 0$. Thus, we can conclude the rather surprising result: *For simply connected regions, a steady temperature distribution with zero boundary tractions will not affect the in-plane stress field.* Note, however, for multiply connected bodies, we must add additional equations ensuring the single-valuedness of the displacement field. When including these additional relations, a steady temperature field normally gives rise to in-plane stresses. Additional information on analysis of multiply connected regions can be found in Kovalenko (1969).

Example 12.1 Thermal stresses in an elastic strip

Consider the thermoelastic problem in a rectangular domain as shown in Fig. 12.1. We assume that the vertical dimension of the domain is much larger than the horizontal width $(2a)$, and thus the region may be described as an infinite strip of material. For this problem, assume that the temperature is independent of x and given by $T = T_o \sin \beta y$, where T_o and β are constants. Note that by using Fourier methods and superposition we could generate a more general temperature field.

Considering the plane stress case, the governing stress function equation becomes

$$\nabla^4 \phi = E\alpha T_o \beta^2 \sin \beta y \tag{12.5.10}$$

The particular solution to this equation is easily found to be

$$\phi^{(p)} = \frac{E\alpha T_o}{\beta^2} \sin\beta y \tag{12.5.11}$$

For the homogeneous solution we try the separation of variables approach and choose $\phi^{(h)} = f(x) \sin \beta y$. Using this form in the homogeneous biharmonic equation gives an auxiliary equation for the function $f(x)$

$$f'''' - 2\beta^2 f'' + \beta^4 f = 0$$

The general solution to this differential equation is

$$f = C_1 \sinh \beta x + C_2 \cosh \beta x + C_3 x \sinh \beta x + C_4 x \cosh \beta x \tag{12.5.12}$$

Continued

Now since the temperature field was symmetric in x, we expect the stresses to also exhibit the same symmetry. Thus, the stress function must also be symmetric in x and so $C_1 = C_4 = 0$. Combining the particular and homogeneous solutions, the resulting stresses become

$$\sigma_x = -\beta^2 [C_2 \cosh \beta x + C_3 x \sinh \beta x] \sin \beta y - E\alpha T_o \sin \beta y$$

$$\sigma_y = \beta^2 \left[C_2 \cosh \beta x + C_3 \left(x \sinh \beta x + \frac{2}{\beta} \cosh \beta x \right) \right] \sin \beta y$$

$$\tau_{xy} = -\beta^2 \left[C_2 \sinh \beta x + C_3 \left(x \cosh \beta x + \frac{1}{\beta} \sinh \beta x \right) \right] \cos \beta y$$
(12.5.13)

These stress results can then be further specified by employing boundary conditions on the lateral sides of the strip at $x = \pm a$. For example, we could specify stress-free conditions $\sigma_x(\pm a, y) = \tau_{xy}(\pm a, y) = 0$, and this would determine the constants C_2 and C_3 (see Exercise 12.6).

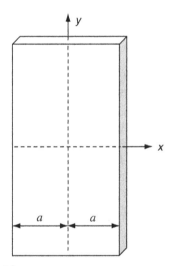

FIG. 12.1 Thermoelastic rectangular strip.

12.6 Polar coordinate formulation

We now wish to list the basic plane thermoelastic equations in polar coordinates. Recall that the isothermal results were previously given in Section 7.6. Following the same notational scheme as before, the strain–displacement relations are given by

$$e_r = \frac{\partial u_r}{\partial r}, \quad e_\theta = \frac{u_r}{r} + \frac{1}{r}\frac{\partial u_\theta}{\partial \theta}$$

$$e_{r\theta} = \frac{1}{2}\left(\frac{1}{r}\frac{\partial u_r}{\partial \theta} + \frac{\partial u_\theta}{\partial r} - \frac{u_\theta}{r}\right)$$

(12.6.1)

For the case of plane stress, Hooke's law becomes

$$\sigma_r = \frac{E}{1-v^2}[e_r + v e_\theta - (1+v)\alpha(T-T_o)]$$

$$\sigma_\theta = \frac{E}{1-v^2}[e_\theta + v e_r - (1+v)\alpha(T-T_o)]$$

(12.6.2)

$$\tau_{r\theta} = \frac{E}{1+v}\,e_{r\theta}$$

In the absence of body forces, the equilibrium equations reduce to

$$\frac{\partial \sigma_r}{\partial r} + \frac{1}{r}\frac{\partial \tau_{r\theta}}{\partial \theta} + \frac{\sigma_r - \sigma_\theta}{r} = 0$$

$$\frac{\partial \tau_{r\theta}}{\partial r} + \frac{1}{r}\frac{\partial \sigma_\theta}{\partial \theta} + \frac{2\tau_{r\theta}}{r} = 0$$

(12.6.3)

The Airy stress function definition now becomes

$$\sigma_r = \frac{1}{r}\frac{\partial \phi}{\partial r} + \frac{1}{r^2}\frac{\partial^2 \phi}{\partial \theta^2}$$

$$\sigma_\theta = \frac{\partial^2 \phi}{\partial r^2}$$

(12.6.4)

$$\tau_{r\theta} = -\frac{\partial}{\partial r}\left(\frac{1}{r}\frac{\partial \phi}{\partial \theta}\right)$$

which again satisfies (12.6.3) identically. The governing stress function equation given previously by (12.5.2)₁ still holds with the Laplacian and biharmonic operators specified by

$$\nabla^2 = \frac{\partial^2}{\partial r^2} + \frac{1}{r}\frac{\partial}{\partial r} + \frac{1}{r^2}\frac{\partial^2}{\partial \theta^2}$$

$$\nabla^4 = \nabla^2\nabla^2 = \left(\frac{\partial^2}{\partial r^2} + \frac{1}{r}\frac{\partial}{\partial r} + \frac{1}{r^2}\frac{\partial^2}{\partial \theta^2}\right)\left(\frac{\partial^2}{\partial r^2} + \frac{1}{r}\frac{\partial}{\partial r} + \frac{1}{r^2}\frac{\partial^2}{\partial \theta^2}\right)$$

(12.6.5)

12.7 Radially symmetric problems

We now investigate some particular thermoelastic solutions to plane stress problems with radially symmetric fields. For this case we assume that all field quantities depend only on the radial coordinate;

that is, $\sigma_r = \sigma_r(r)$, $\sigma_\theta = \sigma_\theta(r)$, $\tau_{r\theta} = \tau_{r\theta}(r)$, $T = T(r)$. Similarly, the stress function also has this reduced dependency, and thus the stresses are specified by

$$\sigma_r = \frac{1}{r}\frac{d\phi}{dr}$$

$$\sigma_\theta = \frac{d^2\phi}{dr^2} = \frac{d}{dr}(r\sigma_r) \qquad (12.7.1)$$

$$\tau_{r\theta} = 0$$

The governing equation in terms of the stress function simplifies to

$$\frac{1}{r}\frac{d}{dr}\left\{r\frac{d}{dr}\left[\frac{1}{r}\frac{d}{dr}\left(r\frac{d\phi}{dr}\right)\right]\right\} + E\alpha\frac{1}{r}\frac{d}{dr}\left(r\frac{dT}{dr}\right) = 0 \qquad (12.7.2)$$

This relation can be recast in terms of the radial stress by using $(12.7.1)_1$, giving the result

$$\frac{1}{r}\frac{d}{dr}\left\{r\frac{d}{dr}\left[\frac{1}{r}\frac{d}{dr}(r^2\sigma_r)\right]\right\} = -E\alpha\frac{1}{r}\frac{d}{dr}\left(r\frac{dT}{dr}\right) \qquad (12.7.3)$$

which can be directly integrated to give

$$\sigma_r = \frac{C_3}{r^2} + C_2 + \frac{C_1}{4}(2\log r - 1) - \frac{E\alpha}{r^2}\int Trdr \qquad (12.7.4)$$

The constants of integration C_i are normally determined from the boundary conditions, and the temperature appearing in the integral is again the temperature difference from the reference state. Note that C_1 and C_3 must be set to zero for domains that include the origin. Combining this result with $(12.7.1)_2$ gives the hoop stress, and thus the two nonzero stress components are determined.

Considering the displacement formulation for the radially symmetric case, $u_r = u(r)$ and $u_\theta = 0$. Going back to the equilibrium equations (12.6.3), it is observed that the second equation vanishes identically. Using Hooke's law and strain-displacement relations in the first equilibrium equation gives

$$\frac{d}{dr}\left[\frac{1}{r}\frac{d}{dr}(ru)\right] = (1+\nu)\alpha\frac{dT}{dr} \qquad (12.7.5)$$

This equation can be directly integrated, giving the displacement solution

$$u = A_1 r + \frac{A_2}{r} + \frac{(1+\nu)\alpha}{r}\int Trdr \qquad (12.7.6)$$

where A_i are constants of integration determined from the boundary conditions, and as before T is the temperature difference from the reference state. The general displacement solution (12.7.6) can then be used to determine the strains from relations (12.6.1) and stresses from Hooke's law (12.6.2). As found in Section 8.3 for the isothermal case, the stresses developed from the displacement solution do not contain the logarithmic term found in relation (12.7.4). Thus, the logarithmic term is inconsistent with single-valued displacements, and further discussion on this point is given in Section 8.3. We commonly drop this term for most problem solutions, but an exception to this is given in Exercise 12.11.

Example 12.2 Circular plate problems

Let us investigate the thermal stress problem in an annular circular plate shown in Fig. 12.2. The solid plate solution is determined as a special case as $r_i \to 0$. The problem is to be radially symmetric, and we choose stress-free inner and outer boundaries.

After dropping the log term, the general stress solution (12.7.4) gives

$$\sigma_r = \frac{C_3}{r^2} + C_2 - \frac{E\alpha}{r^2} \int Tr\,dr \tag{12.7.7}$$

Using the boundary conditions $\sigma_r(r_i) = \sigma_r(r_o) = 0$ determines the two constants C_2 and C_3. Incorporating these results, the stresses become

$$\sigma_r = \frac{E\alpha}{r^2} \left\{ \frac{r^2 - r_i^2}{r_o^2 - r_i^2} \int_{r_i}^{r_o} T(\xi)\xi d\xi - \int_{r_i}^{r} T(\xi)\xi d\xi \right\}$$

$$\tag{12.7.8}$$

$$\sigma_\theta = \frac{E\alpha}{r^2} \left\{ \frac{r^2 + r_i^2}{r_o^2 - r_i^2} \int_{ri}^{r_o} T(\xi)\xi d\xi + \int_{r_i}^{r} T(\xi)\xi d\xi - Tr^2 \right\}$$

and the corresponding displacement solution is given by

$$u = \frac{\alpha}{r} \left\{ (1+\nu) \int_{r_i}^{r} T(\xi)\xi d\xi + \frac{(1-\nu)r^2 + (1+\nu)r_i^2}{r_o^2 - r_i^2} \int_{r_i}^{r_o} T(\xi)\xi d\xi \right\} \tag{12.7.9}$$

In order to explicitly determine the stress and displacement fields, the temperature distribution must be determined. As mentioned, this is calculated from the energy or conduction equation. Assuming steady-state conditions, the conduction equation was given by (12.1.9), and for the radially symmetric case this reduces to

$$\frac{1}{r} \frac{d}{dr} \left(r \frac{dT}{dr} \right) = 0 \tag{12.7.10}$$

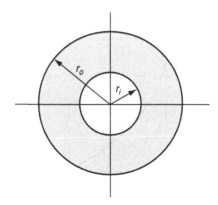

FIG. 12.2 Annular plate geometry.

This equation is easily integrated directly, giving the solution

$$T = A_1 \log r + A_2 \qquad (12.7.11)$$

Choosing thermal boundary conditions $T(r_i) = T_i$, $T(r_o) = 0$, the constants A_1 and A_2 can be determined, and the temperature solution is obtained as

$$T = \frac{T_i}{\log\left(\frac{r_i}{r_o}\right)} \log\left(\frac{r}{r_o}\right) = \frac{T_i}{\log\left(\frac{r_o}{r_i}\right)} \log\left(\frac{r_o}{r}\right) \qquad (12.7.12)$$

For the case $T_i > 0$, this distribution is shown schematically in Fig. 12.3. Substituting this temperature distribution into the stress solution (12.7.8) gives

$$\sigma_r = \frac{E\alpha T_i}{2\log(r_o/r_i)}\left\{ -\log\left(\frac{r_o}{r}\right) - \frac{r_i^2}{r_o^2 - r_i^2}\left(1 - \frac{r_o^2}{r^2}\right)\log\left(\frac{r_o}{r_i}\right) \right\}$$

$$\sigma_\theta = \frac{E\alpha T_i}{2\log(r_o/r_i)}\left\{ 1 - \log\left(\frac{r_o}{r}\right) - \frac{r_i^2}{r_o^2 - r_i^2}\left(1 + \frac{r_o^2}{r^2}\right)\log\left(\frac{r_o}{r_i}\right) \right\} \qquad (12.7.13)$$

Note for this solution when $T_i > 0$, $\sigma_r < 0$, and the hoop stress σ_θ takes on maximum values at the inner and outer boundaries of the plate. For the specific case $r_o/r_i = 3$, the stress distribution through the plate is illustrated in Fig. 12.4. For steel material ($E = 200$ GPa, $\alpha = 13 \times 10^{-6}/°\mathrm{C}$) with $T_i = 100\,°\mathrm{C}$, the maximum hoop stress on the inner boundary is about -174 MPa.

For the case of a thin ring plate where $r_o \approx r_i$, we can write $r_o/r_i \approx 1 + \varepsilon$, where ε is a small parameter. The logarithmic term can be simplified using

$$\log\left(\frac{r_o}{r_i}\right) \approx \log(1 + \varepsilon) \approx \varepsilon - \frac{\varepsilon^2}{2} + \frac{\varepsilon^3}{3} - \cdots$$

and this yields the following approximation

$$\sigma_\theta(r_i) \approx -\frac{E\alpha T_i}{2}\left(1 + \frac{\varepsilon}{3}\right) \approx -\frac{E\alpha T_i}{2}$$

$$\sigma_\theta(r_o) \approx \frac{E\alpha T_i}{2}\left(1 - \frac{\varepsilon}{3}\right) \approx \frac{E\alpha T_i}{2} \qquad (12.7.14)$$

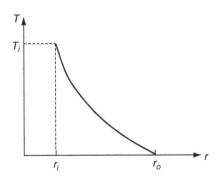

FIG. 12.3 Temperature distribution in an annular plate.

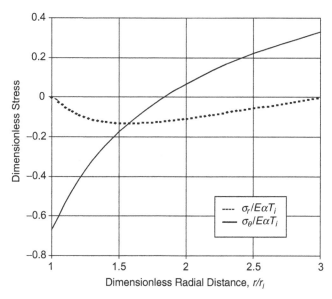

FIG. 12.4 Stress distribution in an annular plate ($r_o/r_i = 3$).

Finally, by allowing the inner radius r_i to reduce to zero, we obtain the solution for a solid circular plate. For this case, the constant C_3 in solution (12.7.7) must be set to zero for finite stresses at the origin. The resulting stress field for zero boundary loading becomes

$$\sigma_r = E\alpha \left\{ \frac{1}{r_o^2} \int_0^{r_o} Tr\,dr - \frac{1}{r^2} \int_0^r Tr\,dr \right\}$$

$$\sigma_\theta = E\alpha \left\{ \frac{1}{r_o^2} \int_0^{r_o} Tr\,dr + \frac{1}{r^2} \int_0^r Tr\,dr - T \right\}$$

(12.7.15)

Casual inspection of the integral term $\frac{1}{r^2} \int_0^r Tr\,dr$ indicates the possibility of unbounded behavior at the origin. This term can be investigated using l'Hospital's rule, and it can be shown that

$$\lim_{r \to 0} \left(\frac{1}{r^2} \int_0^r Tr\,dr \right) = \frac{1}{2} T(0)$$

Because we expect the temperature at the origin to be finite, this limit then implies that the stresses will also be finite at $r = 0$. Using a temperature boundary condition $T(r_o) = T_o$, the general solution (12.7.11) predicts a uniform temperature $T = T_o$ throughout the entire plate. For this case, relations (12.7.15) give $\sigma_r = \sigma_\theta = 0$, and thus the plate is stress free. This particular result verifies the general discussion in Section 12.5 that a steady temperature distribution in a simply connected region with zero boundary tractions gives rise to zero stress. The previous results for plane stress can be easily converted to plane strain by using the appropriate conversion of elastic constants.

The general thermoelastic plane problem (without axial symmetry) can be developed using methods of Fourier analysis; see, for example, Boley and Weiner (1960). The results lead to a similar solution pattern as developed in Section 8.3. Instead of pursuing this development, we look at the use of complex variable methods for the general plane problem.

12.8 Complex variable methods for plane problems

We now wish to develop a complex variable technique for the solution to plane problems in thermoelasticity. As demonstrated in Chapter 10, the complex variable method is a very powerful tool for solution of two-dimensional problems. This method may be extended to handle problems involving thermal stress; see Bogdanoff (1954) and Timoshenko and Goodier (1970).

For the steady-state case, the scheme starts by defining a *complex temperature*

$$T^*(z) = T + iT_I \tag{12.8.1}$$

where the actual temperature T is the real part of T^* and T_I is the conjugate of T. As before, these temperatures actually represent the change with respect to the stress-free reference state. Further define the integrated temperature function

$$t^*(z) = \int T^*(z)dz = t_R + it_I \tag{12.8.2}$$

Using the Cauchy–Riemann equations

$$\frac{\partial t_R}{\partial x} = \frac{\partial t_I}{\partial y}, \quad \frac{\partial t_R}{\partial y} = -\frac{\partial t_I}{\partial x} \tag{12.8.3}$$

Note that these results imply that the temperature can be expressed as

$$T = \frac{\partial t_R}{\partial x} = \frac{\partial t_I}{\partial y} \tag{12.8.4}$$

Next decompose the two-dimensional displacement field as

$$u = u' + \beta t_R$$
$$v = iv' + \beta t_I \tag{12.8.5}$$

where β is a constant to be determined. Substituting these displacements into Hooke's law (12.3.3) for plane strain yields the following stress field

$$\sigma_x = \lambda\left(\frac{\partial u'}{\partial x} + \frac{\partial v'}{\partial y}\right) + 2\mu\frac{\partial u'}{\partial x} + [2\beta(\lambda+\mu) - \alpha(3\lambda+2\mu)]T$$

$$\sigma_y = \lambda\left(\frac{\partial u'}{\partial x} + \frac{\partial v'}{\partial y}\right) + 2\mu\frac{\partial v'}{\partial y} + [2\beta(\lambda+\mu) - \alpha(3\lambda+2\mu)]T \tag{12.8.6}$$

$$\tau_{xy} = \mu\left(\frac{\partial u'}{\partial y} + \frac{\partial v'}{\partial x}\right)$$

By choosing

$$\beta = \begin{cases} (1+v)\alpha, & \text{plane strain} \\ \alpha, & \text{plane strain} \end{cases} \tag{12.8.7}$$

the temperature terms in (12.8.6) are eliminated and thus the problem reduces to the *isothermal case* in terms of the displacements u', v'. This reduction indicates that the general thermoelastic plane problem can be formulated in terms of complex variable theory by the relations

$$\sigma_x + \sigma_y = 2\left(\gamma'(z) + \overline{\gamma'(z)}\right)$$

$$\sigma_y - \sigma_x + 2i\tau_{xy} = 2(\bar{z}\gamma''(z) + \psi'(z))$$

$$2\mu(u + iv) = \kappa\gamma(z) - z\overline{\gamma'(z)} - \overline{\psi(z)} + 2\mu\beta t^*(z) \tag{12.8.8}$$

$$T_x^n + iT_y^n = -i\frac{d}{ds}\left(\gamma(z) + z\overline{\gamma'(z)} + \overline{\psi(z)}\right)$$

where we have used many of the relations originally developed in Section 10.2. The material parameter κ was given by (10.2.10) and β is specified in (12.8.7). Thus, the problem is solved by superposition of an isothermal state with appropriate boundary conditions and a displacement field given by $u + iv = \beta t^*(z)$. For the nonsteady case, the temperature is no longer harmonic, and we would have to represent the complex temperature in the more general scheme $T^* = T^*(z, \bar{z})$.

Example 12.3 Annular plate problem

Consider again the annular plate problem shown in Fig. 12.2. Assume a complex temperature potential of the form

$$T^*(z) = -C\frac{1}{z} \tag{12.8.9}$$

where C is a real constant. The actual temperature field follows as

$$T = -C\,Re\left(\frac{1}{z}\right) = -\frac{C}{r}\cos\theta \tag{12.8.10}$$

and it is easily verified that this temperature is a harmonic function, thus indicating a steady-state field. Note that the boundary temperatures on the inner and outer surfaces for this case become

$$T(r_i) = -\frac{C\cos\theta}{r_i}, \quad T(r_o) = -\frac{C\cos\theta}{r_o}$$

and this would have to match with the assumed temperature boundary conditions. Of course, we could use Fourier superposition methods to handle a more general boundary distribution. Using relation (12.8.5), it is found that this temperature field produces a logarithmic term in the displacement distribution, and this leads to a discontinuity when evaluating the cyclic behavior. This displacement discontinuity must be removed by adding an additional field with the opposite cyclic behavior. Based on our previous experience from Chapter 10, we therefore choose an additional field with the following potentials

$$\gamma(z) = A \log z + \gamma_o(z)$$
$$\psi(z) = B \log z + \psi_o(z)$$

(12.8.11)

where $\gamma_o(z)$ and $\psi_o(z)$ are single-valued and analytic in the domain $(r_i \leq r \leq r_o)$. For single-valued displacements in the region, we can use equations $(12.8.8)_3$ to evaluate and set the cyclic displacement to zero, thus giving

$$\kappa A + B = 2\mu \beta C$$

(12.8.12)

where we have taken A and B to be real.

Again, choosing stress-free boundaries at r_i and r_o and using results from (10.2.11) and (10.2.12), we can write

$$(\sigma_r - i\tau_{r\theta})_{r=r_i,r_o} = \left(\gamma'(z) + \overline{\gamma'(z)} - e^{2i\theta} [\bar{z}\gamma''(z) + \psi'(z)] \right)_{r=r_i,r_o} = 0$$

(12.8.13)

and this is satisfied by potentials with the following properties

$$A = B$$

$$\gamma_o(z) = -A \frac{z^2}{r_i^2 + r_o^2}$$

(12.8.14)

$$\psi_o(z) = -A \frac{r_i^2 r_o^2}{z^2 (r_i^2 + r_o^2)}$$

Thus, the final form of the potentials becomes

$$\gamma(z) = \frac{2\mu\beta C}{1+\kappa} \left(\log z - \frac{z^2}{r_i^2 + r_o^2} \right)$$

$$\psi(z) = \frac{2\mu\beta C}{1+\kappa} \left(\log z - \frac{r_i^2 r_o^2}{z^2 (r_i^2 + r_o^2)} \right)$$

(12.8.15)

The stresses follow from (12.8.8), and the radial stress at $\theta = 0$ is given by

$$\sigma_r|_{\theta=0} = \sigma_x(x,0) = \frac{4\mu\beta C}{(1+\kappa)(r_i^2 + r_o^2)} \left(1 - \frac{r_i^2}{x^2} \right) \left(1 - \frac{r_o^2}{x^2} \right) x$$

(12.8.16)

Example 12.4 Circular hole in an infinite plane under uniform heat flow

We now investigate the localized thermal stresses around a traction-free circular cavity in a plane of infinite extent. The thermal loading is taken to be a uniform heat flow q in the vertical direction, and the circular hole is to be insulated from heat transfer. The plane stress problem shown in Fig. 12.5 was originally solved by Florence and Goodier (1959). Such problems have applications to stress concentration and thermal fracture in structures carrying high thermal gradients.

If the plane had no hole, the temperature distribution for uniform heat flow in the negative y direction would be $T = qy/k$. The presence of the insulated hole locally disturbs this linear distribution. This arises from the thermal boundary condition on $r = a$ given by the Fourier conduction law (12.1.2)

$$q_n(a, \theta) = -k\frac{\partial T}{\partial r}(a, \theta) = 0 \tag{12.8.17}$$

where we have introduced the usual polar coordinates. The form of the complex temperature follows from theory discussed in Chapter 10. A far-field behavior term is added to a series form, which is analytic in the region exterior to the circular hole to form the expression

$$T^*(z) = -\frac{iqz}{k} + \sum_{n=1}^{\infty} a_n z^{-n} \tag{12.8.18}$$

Applying boundary condition (12.8.17) determines the coefficients a_n and gives the final form

$$T^*(z) = -\frac{iq}{k}\left(z - \frac{a^2}{z}\right) \tag{12.8.19}$$

which yields the actual temperature field

$$T(r, \theta) = \frac{q}{k}\left(r + \frac{a^2}{r}\right)\sin\theta \tag{12.8.20}$$

This solution can also be determined using separation of variables and Fourier methods on the heat conduction equation (12.1.9) in polar coordinates (see Exercise 12.16).

Using (12.8.8)$_3$, the displacements resulting from this temperature distribution are

$$(u + iv) = \beta\int T^*(z)dz = -\frac{iq\alpha}{k}\left(\frac{z^2}{2} - a^2\log z\right) \tag{12.8.21}$$

Evaluating the cyclic function of this complex displacement around a contour C enclosing the hole, we find

$$[(u + iv)]_C = -\frac{iq\alpha}{k}\left[\frac{z^2}{2} - a^2\log z\right]_C = -\frac{2q\alpha\pi a^2}{k} \tag{12.8.22}$$

Thus, this temperature field creates a displacement discontinuity, and this must be annulled by superimposing an isothermal dislocation solution that satisfies zero tractions on $r = a$, with stresses that vanish at infinity. It can be shown that these conditions are satisfied by potentials of the following form

$$\gamma(z) = A\log z$$

$$\psi(z) = -A\left(\frac{a^2}{z^2} + \log z + 1\right) \tag{12.8.23}$$

Continued

with

$$A = -\frac{2i\mu qa^2\alpha}{(1+\kappa)k}$$

Using our previous polar coordinate stress combinations (10.2.12), we find

$$\sigma_r + \sigma_\theta = -\frac{E\alpha qa^2}{kr}\sin\theta$$

$$\sigma_\theta - \sigma_r + 2i\tau_{r\theta} = -\frac{E\alpha qa^4}{kr^3}\sin\theta + i\frac{E\alpha qa}{k}\left(\frac{a}{r} - \frac{a^3}{r^3}\right)\cos\theta$$

(12.8.24)

and the individual stresses then become

$$\sigma_r = -\frac{1}{2}\frac{E\alpha qa}{k}\left(\frac{a}{r} - \frac{a^3}{r^3}\right)\sin\theta$$

$$\sigma_\theta = -\frac{1}{2}\frac{E\alpha qa}{k}\left(\frac{a}{r} + \frac{a^3}{r^3}\right)\sin\theta$$

(12.8.25)

$$\tau_{r\theta} = \frac{1}{2}\frac{E\alpha qa}{k}\left(\frac{a}{r} - \frac{a^3}{r^3}\right)\cos\theta$$

The largest stress is given by the hoop stress on the boundary of the hole

$$\sigma_{max} = \sigma_\theta(a, \theta) = -\frac{E\alpha qa}{k}\sin\theta$$

(12.8.26)

Notice that this expression takes on maximum values of $\mp E\alpha qa/k$ at $\theta = \pm\pi/2$ and predicts a maximum compressive stress on the hot side of the hole $\theta = \pi/2$ and maximum tensile stress on the cold side $\theta = -\pi/2$. For the case of a *steel plate* with properties $E = 200$ GPa and $\alpha = 13 \times 10^{-6}/°C$, and with $qa/k = 100\,°C$, the maximum stress is 260 MPa.

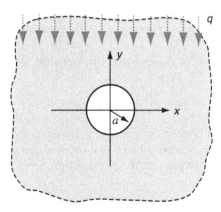

FIG. 12.5 Flow of heat around a circular hole in an infinite plane.

Example 12.5 Elliptical hole in an infinite plane under uniform heat flow

Similar to the previous example, we now investigate the localized thermal stresses around a traction-free elliptical hole (with semiaxes a and b) in a plane of infinite extent, as shown in Fig. 12.6. The thermal loading is again taken to be a uniform heat flow q in the vertical direction, and the hole is to be insulated from heat transfer. The plane stress solution to this problem again comes from the work of Florence and Goodier (1960), who solved the more general case of an ovaloid hole with heat flow at an arbitrary angle. This problem is solved by complex variable methods employing conformal transformation (see Section 10.7).

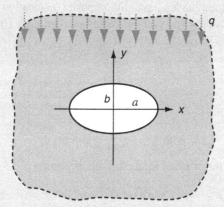

FIG. 12.6 Flow of heat around an elliptical hole in an infinite plane.

As discussed in Chapter 10, conformal mapping provides a very useful tool for this type of problem, and the appropriate mapping function

$$z = w(\zeta) = R\left(\zeta + \frac{m}{\zeta}\right) \tag{12.8.27}$$

transforms the region exterior to the unit circle in the ζ-plane onto the region exterior to the ellipse in the z-plane. The ellipse major and minor axes are related to the mapping parameters by $2Rm = a - b$ and $2R = a + b$. As before, in the transformed plane, $\zeta = \rho e^{i\theta}$.

From our previous example, the temperature distribution for heat flow around an insulated circular hole of unit radius in the ζ-plane may be written as

$$T = \frac{q}{k}R\left(\rho + \frac{1}{\rho}\right)\sin\theta \tag{12.8.28}$$

The complex temperature corresponding to this result is

$$T^*(\zeta) = -\frac{q}{k}iR\left(\zeta - \frac{1}{\zeta}\right) \tag{12.8.29}$$

Again, this temperature field creates a dislocation in the displacement. Following similar steps as in Eqs. (12.8.21) and (12.8.22), the cyclic function of the complex temperature displacement is given by

$$[(u + iv)]_C = \left[\alpha \int T^*(z)dz\right]_C = -\frac{2q\alpha\pi R^2}{k}(1 + m) \tag{12.8.30}$$

where C is the counterclockwise contour around the unit circle enclosing the origin.

Employing conformal transformation, relations (10.7.5)–(10.7.7) can be used to determine the stresses, displacements, and tractions in the ζ-plane. As in the previous example, we now wish to superimpose an isothermal state having equal but opposite dislocation behavior as (12.8.30), with zero tractions on the hole boundary and vanishing stresses at infinity. The appropriate potentials that satisfy these conditions are given by

$$\gamma(\zeta) = A \log \zeta$$

$$\psi(z) = \overline{A} \log \zeta - A \frac{1 + m\zeta^2}{\zeta^2 - m} \tag{12.8.31}$$

$$\text{with } A = -\frac{E\alpha q R^2 i}{4k}(1 + m)$$

Using relations (10.7.5), the stresses in the ζ-plane become

$$\sigma_\rho = -\frac{E\alpha q a}{2kh(\theta)}\rho(\rho^2 + m)\left[\rho^4 - \rho^2(1 + m^2) + m^2\right]\sin\theta$$

$$\sigma_\theta = -\frac{E\alpha q a}{2kh(\theta)}\rho(\rho^2 + m)\left\{\left[\rho^4 + \rho^2(1 + m)^2 + m^2\right]\sin\theta - 2\rho^2 m \sin3\theta\right\} \tag{12.8.32}$$

$$\tau_{\rho\theta} = \frac{E\alpha q a}{2kh(\theta)}\rho(\rho^2 - m)\left[\rho^4 - \rho^2(1 + m^2) + m^2\right]\cos\theta$$

where $h(\theta) = [\rho^4 - 2\rho^2 m \cos2\theta + m^2]^2$. It can be shown that the circular hole case is found by setting $m = 0$, and the stresses will reduce to those given in the previous example in Eq. (12.8.25). Another interesting special case is given by $m = 1$, which corresponds to the elliptical hole reducing to a *line crack* of length $2a$ along the x-axis. For this case the heat flow is perpendicular to the crack, and the stresses become

$$\sigma_\rho = -\frac{E\alpha q a}{2kh(\theta)}\rho(\rho^2 + 1)\left[\rho^4 - 2\rho^2 + 1\right]\sin\theta$$

$$\sigma_\theta = -\frac{E\alpha q a}{2kh(\theta)}\rho(\rho^2 + 1)\left\{\left[\rho^4 + 4\rho^2 + 1\right]\sin\theta - 2\rho^2 \sin3\theta\right\} \tag{12.8.33}$$

$$\tau_{\rho\theta} = \frac{E\alpha q a}{2kh(\theta)}\rho(\rho^2 - 1)\left[\rho^4 - 2\rho^2 + 1\right]\cos\theta$$

with $h(\theta) = [\rho^4 - 2\rho^2 \cos2\theta + 1]^2$. On the surface of the crack ($\rho = 1$), the hoop stress becomes

$$\sigma_\theta(1,\theta) = -\frac{E\alpha qa}{2k}\frac{3\sin\theta - \sin3\theta}{(1 - \cos2\theta)^2} = -\frac{E\alpha qa}{2k\sin\theta} \qquad (12.8.34)$$

and as expected this stress becomes unbounded at the ends of the crack at $\theta = 0, \pi$.

Another interesting result for this case occurs with the shear stress behavior along the positive x-axis ($\theta = 0$)

$$\tau_{\rho\theta}(\rho,0) = \frac{E\alpha qa}{2k}\frac{\rho}{\rho^2 - 1} \qquad (12.8.35)$$

We again observe that this stress component becomes infinite at the crack tip when $\rho = 1$.

As mentioned, Florence and Goodier (1960) solved the more general problem of an ovaloid hole for which the elliptical cavity is a special case. Deresiewicz (1961) solved the general thermal stress problem of a plate with an insulated hole of arbitrary shape and worked out solution details for a triangular hole under uniform heat flow. For the anisotropic case, Sadd and Miskioglu (1978) and Miskioglu (1978) have investigated the problem of an insulated elliptical hole in an anisotropic plane under unidirectional heat flow. Sih (1962) has investigated the singular nature of the thermal stresses at crack tips. He showed that the usual $1/\sqrt{r}$ singularity also exists for this case and that the stress intensity factors [see Eq. (10.8.7)] are proportional to the temperature gradient.

References

Bogdanoff JL: Note on thermal stresses, *J Appl Mech* 76:88, 1954.

Boley BA, Weiner JH: *Theory of thermal stresses*, New York, 1960, John Wiley.

Burgreen D: *Elements of thermal stress analysis*, Jamaica, NY, 1971, C. P. Press.

Deresiewicz H: Thermal stress in a plate due to disturbance of uniform heat flow by a hole of general shape, *J Appl Mech* 28:147−149, 1961.

Florence AL, Goodier JN: Thermal stress at spherical cavities and circular holes in uniform heat flow, *J Appl Mech* 26:293−294, 1959.

Florence AL, Goodier JN: Thermal stresses due to disturbance of uniform heat flow by an insulated ovaloid hole, *J Appl Mech* 27:635−639, 1960.

Fung YC: *Foundations of solid mechanics*, Englewood Cliffs, NJ, 1965, Prentice Hall.

Kellogg OD: *Foundations of potential theory*, New York, 1953, Dover.

Kovalenko AD: *Thermoelasticity*, Groningen, The Netherlands, 1969, Noordhoff.

Miskioglu I: *Thermal stresses around an insulated elliptical hole in an anisotropic medium* (Master's thesis). Mississippi State University.

Nowacki W: *Thermoelasticity*, Reading, MA, 1962, Addison Wesley.

Nowinski JL: *Theory of thermoelasticity with applications*, Groningen, The Netherlands, 1978, Sijthoff-Noordhoff.

Parkus H: *Thermoelasticity*, New York, 1976, Springer.

Sadd MH, Miskioglu I: Temperatures in an anisotropic sheet containing an insulated elliptical hole, *J Heat Transf* 100:553−555, 1978.

Sadd MH: *Continuum Mechanics Modeling of Material Behavior*, 2019, Elsevier.

Sih GC: On the singular character of thermal stresses near a crack tip, *J Appl Mech* 29:587−589, 1962.

Timoshenko SP, Goodier JN: *Theory of elasticity*, New York, 1970, McGraw-Hill.

Exercises

12.1 Using the assumption for isotropic materials that a temperature change produces isotropic thermal strains of the form $\alpha(T - T_o)\delta_{ij}$, develop relations (12.1.6).

12.2 For the general three-dimensional thermoelastic problem with no body forces, explicitly develop the Beltrami–Michell compatibility equations

$$\sigma_{ij,kk} + \frac{1}{(1+\nu)}\sigma_{kk,ij} + \frac{E\alpha}{1+\nu}\left(T_{,ij} + \frac{1+\nu}{1-\nu}\delta_{ij}T_{,kk}\right) = 0$$

12.3 If an isotropic solid is heated nonuniformly to a temperature distribution $T(x,y,z)$ and the material has unrestricted thermal expansion, the resulting strains will be $e_{ij} = \alpha T\delta_{ij}$. Show that this case can only occur if the temperature is a linear function of the coordinates

$$T = ax + by + cz + d$$

12.4 Express the traction boundary condition (12.3.8) in terms of displacement and temperature for the plane stress problem.

12.5 Develop the compatibility equations for plane strain (12.3.7) and plane stress (12.3.13).

12.6* Explicitly develop the stress field Eq. (12.5.13) in Example 12.1 and determine the constants C_2 and C_3 for the case of stress-free edge conditions. Plot the value of σ_y through the thickness (vs. coordinate x) for both high-temperature ($\sin \beta y = 1$) and low-temperature ($\sin \beta y = -1$) cases.

12.7 For the radially symmetric case, verify that the governing stress function equation can be expressed as (12.7.2). Integrate this equation and verify the general solution (12.7.4).

12.8 Verify the equilibrium equation in terms of displacement (12.7.5) for the radially symmetric case and then develop its general solution (12.7.6).

12.9 Consider the axisymmetric plane strain problem of a solid circular bar of radius a with a *constant internal heat generation* specified by h_o. The steady-state conduction equation thus becomes

$$\frac{\partial^2 T}{\partial r^2} + \frac{1}{r}\frac{\partial T}{\partial r} + h_o = 0$$

Using boundary condition $T(a) = T_o$, determine the temperature distribution, and then calculate the resulting thermal stresses for the case with zero boundary stress. Such solutions are useful to determine the thermal stresses in rods made of radioactive materials.

12.10 Using the general displacement solution, solve the thermoelastic problem of a solid circular elastic plate with a restrained boundary edge at $r = a$. For the case of a uniform temperature distribution, show that the displacement and stress fields are zero.

12.11 Consider the thermal stress problem in a *circular ring* as shown in the figure. Assuming the temperature and stress fields depend only on the radial coordinate r, the general solution is given by (12.7.4). If the surfaces $r = a$ and $r = b$ are to be stress free, show that the solution can be written as

$$\sigma_r = \frac{A_1}{r^2} + \frac{A_2}{a^2}\left(2\log\frac{r}{a} + 1\right) + \frac{2A_3}{a^2} - \frac{E\alpha}{r^2}\int_a^r Tr\,dr$$

for appropriate constants A_i. Note for this type of problem the logarithmic term is retained as long as the ring is only a segment and not a full ring. For this case the displacements at each end section need not be continuous.

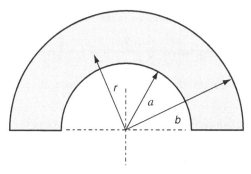

12.12 Consider the thermoelastic problem in spherical coordinates (R, ϕ, θ); see Fig. 1.6. For the case of *spherical symmetry* where all field quantities depend only on the radial coordinate R, develop the general solution

$$u_R = \frac{1+\nu}{1-\nu}\alpha\frac{1}{R^2}\int^R T\xi^2 d\xi + C_1 R + \frac{C_2}{R^2}$$

$$\sigma_R = -\frac{2\alpha E}{1-\nu}\frac{1}{R^3}\int^R T\xi^2 d\xi + \frac{EC_1}{1-2\nu} - \frac{2EC_2}{1+\nu}\frac{1}{R^3}$$

$$\sigma_\phi = \sigma_\theta = \frac{\alpha E}{1-\nu}\frac{1}{R^3}\int^R T\xi^2 d\xi + \frac{EC_1}{1-2\nu} + \frac{EC_2}{1+\nu}\frac{1}{R^3} - \frac{\alpha ET}{1-\nu}$$

Note that any convenient lower limit may be placed on the integral terms to aid in problem solution.

12.13 Use the general solution of Exercise 12.12 to solve the thermal stress problem of a hollow thick-walled spherical shell $(a \leq R \leq b)$ with stress-free boundary conditions. Assuming that the problem is steady state with temperature conditions $T(a) = T_i$, $T(b) = 0$, show that the solution becomes

$$T = \frac{T_i a}{b-a}\left(\frac{b}{R} - 1\right)$$

$$\sigma_R = \frac{\alpha E T_i}{1-\nu}\frac{ab}{b^3-a^3}\left[a+b-\frac{1}{R}\left(b^2+ab+a^2\right)+\frac{a^2 b^2}{R^3}\right]$$

$$\sigma_\phi = \sigma_\theta = \frac{\alpha E T_i}{1-\nu}\frac{ab}{b^3-a^3}\left[a+b-\frac{1}{2R}\left(b^2+ab+a^2\right)-\frac{a^2 b^2}{2R^3}\right]$$

For the case of a thin spherical shell, let $b = a(1+\varepsilon)$, where ε is a small parameter. Show that using this formulation, the hoop stresses at the inner and outer surfaces become

$$\sigma_\phi = \sigma_\theta = \frac{\alpha E T_i}{2(1-\nu)}\left(\mp 1 - \frac{2}{3}\varepsilon\right)$$

and if we neglect the ε term, these values match those of the cylindrical shell given by relations (12.7.14).

12.14 Explicitly develop relations (12.8.6) and verify that by using the value of β given in (12.8.7) the temperature terms will drop out of these relations.

12.15 For Example 12.3, verify that the potentials $\gamma_o(z)$, $\psi_o(z)$ given by relations (12.8.14) satisfy the stress-free boundary conditions on the problem.

12.16 Using separation of variables and Fourier methods, solve the conduction equation and verify that the temperature distribution (12.8.20) in Example 12.4 does indeed satisfy insulated conditions on the circular hole and properly matches conditions at infinity.

12.17 For Example 12.4, explicitly develop the stresses (12.8.25) from the complex potentials given by Eq. (12.8.23).

12.18* Plot the isotherms (contours of constant temperature) for Examples 12.4 and 12.5.

12.19 For the elliptical hole problem in Example 12.5, show that by letting $m = 0$, the stress results will reduce to those of the circular hole problem given in Example 12.4.

12.20* In Example 12.5, show that the dimensionless hoop stress around the boundary of the hole is given by

$$\overline{\sigma}_\theta = \frac{\sigma_\theta}{E\alpha q a/k} = -\frac{(1+m)\left[(1+m+m^2)\sin\theta - m\sin3\theta\right]}{(1-2m\cos2\theta+m^2)^2}$$

For the cases $m = 0,\ \pm1/2,\ \pm1$, plot and compare the behaviour of $\overline{\sigma}_\theta$ versus $\theta(0 \leq \theta \leq 2\pi)$.

12.21   Show that for the *plane anisotropic problem*, the heat-conduction equation for uncoupled steady-state conditions is given by

$$k_{xx}\frac{\partial^2 T}{\partial x^2} + 2k_{xy}\frac{\partial^2 T}{\partial x \partial y} + k_{yy}\frac{\partial^2 T}{\partial y^2} = 0$$

Looking for solutions that are of the form $T = F(x + \lambda y)$, show that this leads to the quadratic characteristic equation

$$k_{yy}\lambda^2 + 2k_{xy}\lambda + k_{xx} = 0$$

Using the fact that $k_{xx}k_{yy} > k_{xy}^2$, demonstrate that the two roots to this equation will be complex conjugate pairs; therefore, since the temperature must be real, the final form of the solution will be

$$T = 2Re\{F(z^*)\}$$

where $z^* = x + \lambda y$. This problem is mathematically similar to Exercise 11.16.

Displacement potentials and stress functions: applications to three-dimensional problems

We now wish to investigate the method of potentials to generate solutions to elasticity problems. Several different potential techniques have been developed in order to solve problems within both displacement and stress formulations. Methods related to the displacement formulation include scalar and vector potentials from the *Helmholtz decomposition, Galerkin vector, Papkovich-Neuber functions* and *Naghdi-Hsu solution*. These schemes provide general solution forms for Navier's equations. Potentials used in the stress formulation are those related to *Maxwell and Morera stress functions*, and these lead to Airy and other common stress functions that we have already used for solution of particular elasticity problems. As previously observed, these stress functions normally satisfy the equilibrium equations identically and when combined with the compatibility relations they yield a simpler and more tractable system of equations.

For either displacement or stress formulations, these solution schemes bring up the question − *are all solutions of elasticity expressible by the particular potential representation*? This issue is normally referred to as the *completeness* of the representations, and over the past several decades these theoretical questions have generally been answered in the affirmative. Wang et al. (2008) provide a comprehensive review of this past work. For many cases these approaches are useful to solve particular three-dimensional elasticity problems, and we will investigate several such solutions. Some potential methods are also particularly useful in formulating and solving dynamic elasticity problems involving wave propagation (Fung, 1965 or Graff, 1991).

13.1 Helmholtz displacement vector representation

A useful relation called the *Helmholtz theorem* states that any sufficiently continuous vector field can be represented as the sum of the gradient of a scalar potential plus the curl of a vector potential. Using this representation for the displacement field, we can write

$$u = \nabla\phi + \nabla \times \boldsymbol{\varphi} \tag{13.1.1}$$

where ϕ is the scalar potential and $\boldsymbol{\varphi}$ is the vector potential. The gradient term in the decomposition has a zero curl and is referred to as the *lamellar* or *irrotational* part, while the curl term in (13.1.1) has no divergence and is called *solenoidal*. Note that this representation specifies three displacement components in terms of four potential components, and furthermore the divergence of $\boldsymbol{\varphi}$ is arbitrary. In order to address these problems, it is common to choose $\boldsymbol{\varphi}$ with zero divergence

$$\nabla \cdot \boldsymbol{\varphi} = 0 \tag{13.1.2}$$

Elasticity. https://doi.org/10.1016/B978-0-12-815987-3.00013-X

It can be easily shown that the volume dilatation ϑ and the rotation vector ω are related to these potentials by

$$\vartheta = e_{kk} = \phi_{i,kk}, \quad \omega_i = -\frac{1}{2}\varphi_{i,kk} \tag{13.1.3}$$

General solutions of these relations can be determined (see Fung, 1965), and thus the scalar and vector potentials can be expressed in terms of the displacement field.

Using representation (13.1.1) in the general three-dimensional Navier equations (5.4.4), we find

$$(\lambda + 2\mu)\boldsymbol{\nabla}\left(\nabla^2\phi\right) + \mu\boldsymbol{\nabla} \times \left(\nabla^2\boldsymbol{\varphi}\right) + \boldsymbol{F} = 0 \tag{13.1.4}$$

Notice that if the divergence and curl are taken of the previous equation with zero body forces, the following relations are generated

$$\nabla^2\nabla^2\phi = \nabla^4\phi = 0, \quad \nabla^2\nabla^2\boldsymbol{\varphi} = \nabla^4\boldsymbol{\varphi} = 0 \tag{13.1.5}$$

and thus we find that both potential functions are biharmonic functions. Further reduction of (13.1.4) will now be made for specific applications.

13.2 Lamé's strain potential

It is noted that for the case of zero body forces, special solutions of (13.1.4) occur with $\nabla^2\phi = $ constant and $\nabla^2\boldsymbol{\varphi} = $ constant. We consider the special case with

$$\nabla^2\phi = \text{constant}, \quad \boldsymbol{\varphi} = 0 \tag{13.2.1}$$

Because our goal is to determine simply a particular solution, we can choose the constant to be zero, and thus the potential ϕ will be a harmonic function. For this case, the displacement representation is commonly written as

$$2\mu u_i = \phi_{,i} \tag{13.2.2}$$

and the function ϕ is called *Lamé's strain potential*. Using this form, the strains and stresses are given by the simple relations

$$e_{ij} = \frac{1}{2\mu}\phi_{,ij}$$
$$\tag{13.2.3}$$
$$\sigma_{ij} = \phi_{,ij}$$

In Cartesian coordinates, these expressions would give

$$u = \frac{1}{2\mu}\frac{\partial\phi}{\partial x}, \quad v = \frac{1}{2\mu}\frac{\partial\phi}{\partial y}, \quad w = \frac{1}{2\mu}\frac{\partial\phi}{\partial z}$$

$$e_x = \frac{1}{2\mu}\frac{\partial^2\phi}{\partial x^2}, \quad e_y = \frac{1}{2\mu}\frac{\partial^2\phi}{\partial y^2}, \dots \tag{13.2.4}$$

$$\sigma_x = \frac{\partial^2\phi}{\partial x^2}, \quad \sigma_y = \frac{\partial^2\phi}{\partial y^2}, \quad \tau_{xy} = \frac{\partial^2\phi}{\partial x\partial y}, \dots$$

Thus, for this case any harmonic function can be used for Lamé's potential. Typical forms of harmonic functions are easily determined, and some examples include

$$x^2 - y^2, \quad xy, \quad r^n \cos n\theta, \quad \log r, \quad \frac{1}{R}, \quad \log(R+z)$$

$$\text{with} \quad r = \sqrt{x^2 + y^2}, \quad \theta = \tan^{-1}\frac{y}{x}, \quad R = \sqrt{x^2 + y^2 + z^2} \tag{13.2.5}$$

13.3 Galerkin vector representation

In the previous sections, the displacement vector was represented in terms of first derivatives of the potential functions ϕ and φ. Galerkin (1930) showed that it is also useful to represent the displacement in terms of second derivatives of a *single vector function*. The proposed representation is given by

$$2\mu\boldsymbol{u} = 2(1 - \nu)\nabla^2\boldsymbol{V} - \nabla(\nabla \cdot \boldsymbol{V}) \tag{13.3.1}$$

where the potential function $\boldsymbol{V}$ is called the *Galerkin vector*. Substituting this form into Navier's equation gives the result

$$\nabla^4\boldsymbol{V} = -\frac{\boldsymbol{F}}{1 - \nu} \tag{13.3.2}$$

Note that for the case of zero body forces, the Galerkin vector is biharmonic. Thus, we have reduced Navier's equation to a simpler fourth-order vector equation.

By comparing the representations given by (13.1.1) with that of (13.3.1), the Helmholtz potentials can be related to the Galerkin vector by

$$\phi = \frac{1 - 2\nu}{2\mu}(\nabla \cdot \boldsymbol{V}), \quad \boldsymbol{\varphi} = -\frac{1 - \nu}{\mu}(\nabla \times \boldsymbol{V}) \tag{13.3.3}$$

Notice that if $\boldsymbol{V}$ is taken to be harmonic, then the curl of $\boldsymbol{\varphi}$ will vanish and the scalar potential ϕ will also be harmonic. This case then reduces to Lamé's strain potential presented in the previous section. With zero body forces, the stresses corresponding to the Galerkin representation are given by

$$\sigma_x = 2(1-\nu)\frac{\partial}{\partial x}\nabla^2 V_x + \left(\nu\nabla^2 - \frac{\partial^2}{\partial x^2}\right)\nabla \cdot \boldsymbol{V}$$

$$\sigma_y = 2(1-\nu)\frac{\partial}{\partial y}\nabla^2 V_y + \left(\nu\nabla^2 - \frac{\partial^2}{\partial y^2}\right)\nabla \cdot \boldsymbol{V}$$

$$\sigma_z = 2(1-\nu)\frac{\partial}{\partial z}\nabla^2 V_z + \left(\nu\nabla^2 - \frac{\partial^2}{\partial z^2}\right)\nabla \cdot \boldsymbol{V}$$

$$\tau_{xy} = (1-\nu)\left(\frac{\partial}{\partial y}\nabla^2 V_x + \frac{\partial}{\partial x}\nabla^2 V_y\right) - \frac{\partial^2}{\partial x\partial y}\nabla \cdot \boldsymbol{V} \tag{13.3.4}$$

$$\tau_{yz} = (1-\nu)\left(\frac{\partial}{\partial z}\nabla^2 V_y + \frac{\partial}{\partial y}\nabla^2 V_z\right) - \frac{\partial^2}{\partial y\partial z}\nabla \cdot \boldsymbol{V}$$

$$\tau_{zx} = (1-\nu)\left(\frac{\partial}{\partial x}\nabla^2 V_z + \frac{\partial}{\partial z}\nabla^2 V_x\right) - \frac{\partial^2}{\partial z\partial x}\nabla \cdot \boldsymbol{V}$$

As previously mentioned, for no body forces the Galerkin vector must be biharmonic. In Cartesian coordinates, the general biharmonic vector equation would decouple, and thus each component of the Galerkin vector would satisfy the scalar biharmonic equation. However, in curvilinear coordinate systems (such as cylindrical or spherical), the unit vectors are functions of the coordinates, and this will not in general allow such a simple decoupling. Eq. (1.9.18) provides the general form for the Laplacian of a vector, and the expression for polar coordinates is given in Example 1.5 by relation (1.9.21)$_7$. Therefore, in curvilinear coordinates the individual components of the Galerkin vector do not necessarily satisfy the biharmonic equation. For cylindrical coordinates, only the z component of the Galerkin vector satisfies the biharmonic equation, while the other components satisfy a more complicated fourth-order partial differential equation (see Exercise 13.8 for details).

Before moving on to specific applications, we investigate a few useful relationships dealing with harmonic and biharmonic functions. Consider the following identity

$$\nabla^2(xf) = x\nabla^2 f + 2\frac{\partial f}{\partial x}$$

Taking the Laplacian of this expression gives

$$\nabla^4(xf) = \nabla^2(x\nabla^2 f) + 2\frac{\partial}{\partial x}(\nabla^2 f)$$

and thus if f is harmonic, the product xf is biharmonic. Obviously, for this result the coordinate x could be replaced by y or z. Likewise we can also show by standard differentiation that the product $R^2 f$ will be biharmonic if f is harmonic, where $R^2 = x^2 + y^2 + z^2$. Using these results, we can write the following generalized representation for a biharmonic function g as

$$g = f_o + xf_1 + yf_2 + zf_3 + \frac{1}{2}R^2 f_4 \tag{13.3.5}$$

where f_i are arbitrary harmonic functions. It should be pointed out that not all of the last four terms of (13.3.5) are independent.

Consider now the special Galerkin vector representation where only the z component of V is nonvanishing; that is, $V = V_z e_z$. For this case, the displacements are given by

$$2\mu u = 2(1-v)\nabla^2 V_z e_z - \nabla\left(\frac{\partial V_z}{\partial z}\right) \tag{13.3.6}$$

With zero body forces, V_z will be biharmonic, and this case is commonly referred to as *Love's strain potential*. A special case of this form was introduced by Love (1944) in studying solids of revolution under axisymmetric loading.

For this case the displacements and stresses in Cartesian coordinates become

$$2\mu u = -\frac{\partial^2 V_z}{\partial x \partial z}, \quad 2\mu v = -\frac{\partial^2 V_z}{\partial y \partial z}, \quad 2\mu w = 2(1-\nu)\nabla^2 V_z - \frac{\partial^2 V_z}{\partial z^2}$$

$$\sigma_x = \frac{\partial}{\partial z}\left(\nu\nabla^2 - \frac{\partial^2}{\partial x^2}\right)V_z, \quad \tau_{xy} = -\frac{\partial^3 V_z}{\partial x \partial y \partial z}$$

$$\sigma_y = \frac{\partial}{\partial z}\left(\nu\nabla^2 - \frac{\partial^2}{\partial y^2}\right)V_z, \quad \tau_{yz} = \frac{\partial}{\partial y}\left((1-\nu)\nabla^2 - \frac{\partial^2}{\partial z^2}\right)V_z \qquad (13.3.7)$$

$$\sigma_z = \frac{\partial}{\partial z}\left((2-\nu)\nabla^2 - \frac{\partial^2}{\partial z^2}\right)V_z, \quad \tau_{zx} = \frac{\partial}{\partial x}\left((1-\nu)\nabla^2 - \frac{\partial^2}{\partial z^2}\right)V_z$$

The corresponding relations in cylindrical coordinates are given by

$$2\mu u_r = -\frac{\partial^2 V_z}{\partial r \partial z}, \quad 2\mu u_\theta = -\frac{1}{r}\frac{\partial^2 V_z}{\partial \theta \partial z}, \quad 2\mu u_z = 2(1-\nu)\nabla^2 V_z - \frac{\partial^2 V_z}{\partial z^2}$$

$$\sigma_r = \frac{\partial}{\partial z}\left(\nu\nabla^2 - \frac{\partial^2}{\partial r^2}\right)V_z, \quad \tau_{r\theta} = -\frac{\partial^3}{\partial r \partial \theta \partial z}\left(\frac{V_z}{r}\right)$$

$$\sigma_\theta = \frac{\partial}{\partial z}\left(\nu\nabla^2 - \frac{1}{r}\frac{\partial}{\partial r} - \frac{1}{r^2}\frac{\partial^2}{\partial \theta^2}\right)V_z, \quad \tau_{\theta z} = \frac{1}{r}\frac{\partial}{\partial \theta}\left((1-\nu)\nabla^2 - \frac{\partial^2}{\partial z^2}\right)V_z \qquad (13.3.8)$$

$$\sigma_z = \frac{\partial}{\partial z}\left((2-\nu)\nabla^2 - \frac{\partial^2}{\partial z^2}\right)V_z, \quad \tau_{zr} = \frac{\partial}{\partial r}\left((1-\nu)\nabla^2 - \frac{\partial^2}{\partial z^2}\right)V_z$$

We now consider some example applications for axisymmetric problems where the field variables are independent of θ.

Example 13.1 Kelvin's problem—concentrated force acting in the interior of an infinite solid

Consider the problem (commonly referred to as *Kelvin's problem*) of a single concentrated force acting at a point in the interior of an unbounded elastic solid. For convenience we choose a co-ordinate system such that the force is applied at the origin and acts in the z direction (see Fig. 13.1). The general boundary conditions on this problem would require that the stress field vanish at infinity, be singular at the origin, and give the resultant force system Pe_z on any surface enclosing the origin.

    The symmetry of the problem suggests that we can choose the Love/Galerkin potential as an axisymmetric form $V_z(r, z)$. In the absence of body forces, this function is biharmonic, and using the last term in representation (13.3.5) with $f_4 = 1/R$ gives the trial potential

$$V_z = AR = A\sqrt{r^2 + z^2} \qquad (13.3.9)$$

Continued

where A is an arbitrary constant to be determined. We shall now show that this potential produces the correct stress field for the concentrated force problem under study.

The displacement and stress fields corresponding to the proposed potential follow from relations (13.3.8)

$$2\mu u_r = \frac{Arz}{R^3}, \quad 2\mu u_\theta = 0, \quad 2\mu u_z = A\left(\frac{2(1-2\nu)}{R} + \frac{1}{R} + \frac{z^2}{R^3}\right)$$

$$\sigma_r = A\left(\frac{(1-2\nu)z}{R^3} - \frac{3r^2z}{R^5}\right), \quad \tau_{r\theta} = 0$$

$$\sigma_\theta = A\frac{(1-2\nu)z}{R^3}, \quad \tau_{\theta z} = 0 \tag{13.3.10}$$

$$\sigma_z = -A\left(\frac{(1-2\nu)z}{R^3} + \frac{3z^3}{R^5}\right), \quad \tau_{zr} = -A\left(\frac{(1-2\nu)r}{R^3} + \frac{3rz^2}{R^5}\right)$$

Clearly, these stresses (and displacements) are singular at the origin and vanish at infinity. To analyze the resultant force condition, consider an arbitrary cylindrical surface enclosing the origin as shown in Fig. 13.1. For convenience, we choose the cylinder to be bounded at $z = \pm a$ and will let the radius tend to infinity. Invoking vertical equilibrium, we can write

$$\int_0^\infty 2\pi r\sigma_z(r,a)dr - \int_0^\infty 2\pi r\sigma_z(r,-a)dr + \int_{-a}^a 2\pi r\tau_{rz}(r,z)dz + P = 0 \tag{13.3.11}$$

The first two terms in (13.3.11) can be combined, and in the limit as $r \to \infty$ the third integral is found to vanish, thus giving

$$P = -2\int_a^\infty 2\pi R\sigma_z(r,a)dR$$

$$= 4\pi A\left[(1-2\nu)a\int_a^\infty \frac{RdR}{R^3} + 3a^3\int_a^\infty \frac{RdR}{R^5}\right] \tag{13.3.12}$$

$$= 8\pi(1-\nu)A$$

The constant is now determined and the problem is solved. Of course, the stress field is linearly related to the applied loading, and typically for such three-dimensional problems the field also depends on Poisson's ratio.

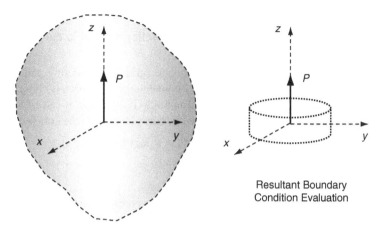

FIG. 13.1 Kelvin's problem: Concentrated force in an infinite medium.

Example 13.2 Boussinesq's problem—concentrated force acting normal to the free surface of a semi-infinite solid

Several other related concentrated force problems can be solved by displacement potential methods. For example, consider *Boussinesq's problem* of a concentrated force acting normal to the free surface of a semi-infinite solid, as shown in Fig. 13.2. Recall that the corresponding two-dimensional problem was solved in Section 8.4.7 (Flamant's problem) and later using complex variables in Example 10.5.

This problem can be solved by combining a Galerkin vector and Lamé's strain potential of the forms

$$V_x = V_y = 0, \quad V_z = AR$$
$$\phi = B \, \log(R + z) \tag{13.3.13}$$

Using similar methods as in the previous example, it is found that the arbitrary constants become

$$A = \frac{P}{2\pi}, \quad B = -\frac{(1 - 2\nu)P}{2\pi} \tag{13.3.14}$$

The displacements and stresses are easily calculated using (13.2.4) and (13.3.7); see Exercise 13.9.

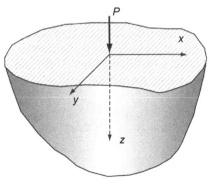

FIG. 13.2 Boussinesq's problem: Normal force on the surface of a half-space.

Example 13.3 Cerruti's problem—concentrated force acting parallel to the free surface of a semi-infinite solid

Another related example is *Cerruti's problem* of a concentrated force acting parallel to the free surface of an elastic half space (see Fig. 13.3). For convenience, the force is chosen to be directed along the x-axis as shown. Although this problem is not axisymmetric, it can be solved by combining a particular Galerkin vector and Lamé's strain potential of the following forms

$$V_x = AR, \quad V_y = 0, \quad V_z = Bx\log(R+z)$$

$$\phi = \frac{Cx}{R+z}$$

(13.3.15)

Again, using methods from the previous examples, the constants are found to be

$$A = \frac{P}{4\pi(1-\nu)}, \quad B = \frac{(1-2\nu)P}{4\pi(1-\nu)}, \quad C = \frac{(1-2\nu)P}{2\pi}$$

(13.3.16)

The displacements and stresses follow from relations (13.2.4) and (13.3.7); see Exercise 13.10.

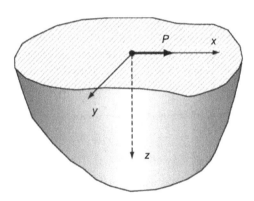

FIG. 13.3 Cerruti's problem: Tangential force on the surface of a half-space.

13.4 Papkovich—Neuber representation

Using scalar and vector potential functions, another general solution to Navier's equations was developed by Papkovich (1932) and later independently by Neuber (1934). The completeness of this representation was shown by Eubanks and Sternberg (1956), and thus all elasticity solutions are representable by this scheme. We outline the development of this solution by first writing Navier's equation in the form

$$\nabla^2 u + \frac{1}{1-2\nu}\nabla(\nabla \cdot u) = -\frac{F}{\mu}$$

(13.4.1)

Using the Helmholtz representation (13.1.1) and relation (13.1.2), this previous equation can be written as

$$\nabla^2 \left[u + \frac{1}{(1 - 2\nu)} \nabla\phi \right] = -\frac{F}{\mu} \tag{13.4.2}$$

Define the vector term in the brackets as

$$h = u + \frac{1}{(1 - 2\nu)} \nabla\phi \tag{13.4.3}$$

We note that

$$\nabla^2 h = -F/\mu, \quad \nabla \cdot h = \frac{2(1 - \nu)}{1 - 2\nu} \nabla^2\phi \tag{13.4.4}$$

Using the identity $\nabla^2(R \cdot h) = R \cdot \nabla^2 h + 2(\nabla \cdot h)$, it can be shown that

$$\nabla \cdot h = \frac{1}{2} \left(\nabla^2(R \cdot h) + \frac{R \cdot F}{\mu} \right) \tag{13.4.5}$$

Combining results (13.4.5) with (13.4.4) gives

$$\nabla^2 \left[\frac{2(1 - \nu)}{1 - 2\nu} \phi - \frac{1}{2} R \cdot h \right] = \frac{R \cdot F}{2\mu} \tag{13.4.6}$$

Defining the term in brackets by scalar h, we get

$$\nabla^2 h = \frac{R \cdot F}{2\mu} \tag{13.4.7}$$

Finally using the definition of h, we can eliminate ϕ from relation (13.4.3) and obtain an expression for the displacement vector.

Redefining new scalar and vector potentials in terms of h and h, we can write

$$2\mu u = A - \nabla \left[B + \frac{A \cdot R}{4(1 - \nu)} \right] \tag{13.4.8}$$

where

$$\nabla^2 A = -2F, \quad \nabla^2 B = \frac{R \cdot F}{2(1 - \nu)} \tag{13.4.9}$$

This general displacement representation is the *Papkovich–Neuber solution* of Navier's equations. For the case with zero body forces, the two potential functions A and B are harmonic. The four individual functions A_x, A_y, A_z, and B, however, are not all independent, and it can be shown that for arbitrary three-dimensional *convex regions*, only three of these functions are independent. Note that a convex region is one in which any two points in the domain may be connected by a line that remains totally within the region.

Comparing the Galerkin vector representation (13.3.1) with the Papkovich solution (13.4.8), it is expected that a relationship between the two solution types should exist, and it can be easily shown that

$$A = 2(1 - \nu)\nabla^2 V, \quad B = \nabla \cdot V - \frac{A \cdot R}{4(1 - \nu)} \tag{13.4.10}$$

As with the Galerkin vector solution, it is convenient to consider the special case of axisymmetry where

$$A_r = A_\theta = 0, \quad A_z = A_z(r, z), \quad B = B(r, z)$$
$$\text{with } \nabla^2 B = 0 \quad \text{and} \quad \nabla^2 A_z = 0 \tag{13.4.11}$$

For this axisymmetric case, B and A_z are commonly called the *Boussinesq potentials*, and as before with zero body forces they are harmonic functions.

As previously pointed out, the three-dimensional Papkovich solution (13.4.8) and (13.4.9) contains four functions to determine three displacement components. In an effort to eliminate this troubling disparity, Naghdi and Hsu (1961) established a general displacement solution representation in terms of just *three potential functions*. This representation can be easily generated from the Papkovich form (zero body force case) by first noting the property

$$\nabla^2 (A \cdot R) = 2\nabla \cdot A \tag{13.4.12}$$

Next employing some basic ideas from potential theory (Kellogg, 1969), a fundamental solution to Poisson's equation $\nabla^2 f(x, y, z) = g(x, y, z)$ is given by

$$f = -\frac{1}{4\pi} \iiint_V \frac{g(\xi, \eta, \zeta)}{R} d\xi d\eta d\zeta, \quad \text{where } R = \sqrt{(x - \xi)^2 + (y - \eta)^2 + (z - \zeta)^2} \tag{13.4.13}$$

Thus we can write the solution to (13.4.12) as

$$A \cdot R = -\frac{1}{2\pi} \iiint_V \frac{\nabla \cdot A}{R} d\xi d\eta d\zeta \tag{13.4.14}$$

If we define a new potential vector as $E = A - \nabla B \Rightarrow \nabla \cdot E = \nabla \cdot A$, $\nabla^2 E = 0$. Finally combining these relations with the Papkovich form (13.4.8) gives

$$2\mu u = E + \frac{1}{8\pi(1 - v)} \nabla \iiint_V \frac{\nabla \cdot E}{R} d\xi d\eta d\zeta \tag{13.4.15}$$

This general form is called the *Naghdi-Hsu solution*, and it is noted that the displacement vector is now written in terms of three components of the potential E. Additional concepts of relationships between Galerkin, Papkovich and Naghdi-Hsu solution forms have been given by Pecknold (1971), Tran-Cong (1981) and Wang (1985).

Example 13.4 Boussinesq's problem revisited

We consider again the problem shown previously in Fig. 13.2 of a concentrated force acting normal to the stress-free surface of a semi-infinite solid. Because the problem is axisymmetric, we use the Boussinesq potentials defined by (13.4.11). These potentials must be harmonic functions of r and z, and using (13.2.5), we try the forms

$$A_z = \frac{C_1}{R}, \quad B = C_2 \log(R + z) \tag{13.4.16}$$

where C_1 and C_2 are constants to be determined.

The boundary conditions on the free surface require that $\sigma_z = \tau_{rz} = 0$ everywhere except at the origin, and that the summation of the total vertical force be equal to P. Calculation of these stresses follows using the displacements from (13.4.8) in Hooke's law, and the result is

$$\sigma_z = -\frac{3C_1 z^3}{4(1-\nu)R^5}$$

$$\tau_{rz} = \frac{r}{R^3}\left(C_2 - \frac{(1-2\nu)}{4(1-\nu)}\,C_1 - \frac{3C_1 z^2}{4(1-\nu)R^2}\right)$$

(13.4.17)

Note that the expression for σ_z vanishes on $z = 0$, but is indeterminate at the origin, and thus this relation will not directly provide a means to determine the constant C_1. Rather than trying to evaluate this singularity at the origin, we pursue the integrated condition on any typical plane $z = $ constant

$$P = -\int_0^\infty \sigma_z(r,z)2\pi r\,dr$$

(13.4.18)

Invoking these boundary conditions determines the two constants

$$C_1 = \frac{2(1-\nu)}{\pi}P, \quad C_2 = \frac{(1-2\nu)}{2\pi}P$$

(13.4.19)

The results for the displacements and stresses are given by

$$u_r = \frac{P}{4\pi\mu R}\left[\frac{rz}{R^2} - \frac{(1-2\nu)r}{R+z}\right]$$

$$u_z = \frac{P}{4\pi\mu R}\left[2(1-\nu) + \frac{z^2}{R^2}\right]$$

(13.4.20)

$$u_\theta = 0$$

$$\sigma_r = \frac{P}{2\pi R^2}\left[-\frac{3r^2 z}{R^3} + \frac{(1-2\nu)R}{R+z}\right]$$

$$\sigma_\theta = \frac{(1-2\nu)P}{2\pi R^2}\left[\frac{z}{R} - \frac{R}{R+z}\right]$$

(13.4.21)

$$\sigma_z = -\frac{3Pz^3}{2\pi R^5}, \quad \tau_{rz} = -\frac{3Prz^2}{2\pi R^5}$$

Notice that the stresses σ_z and τ_{rz} are independent of Poisson's ratio. It has been shown (Abdulaliyev and Ataoglu, 2009) that in general, stress components are independent of ν in sections of simply-connected bodies where the components arising are in equilibrium only with surface tractions.

Many additional problems can be solved using the Papkovich method, and some of these are given in the exercises at the end of this chapter. This technique is also used in Chapter 15 to generate solutions for many singular stress states employed in micromechanics modeling.

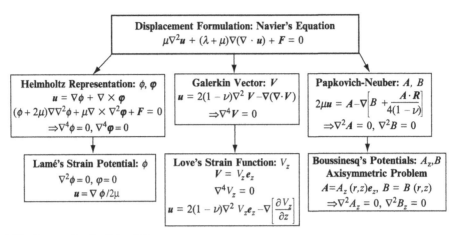

FIG. 13.4 Displacement potential solutions.

An interesting connection can be made for the two-dimensional case between the Papkovich–Neuber scheme and the complex variable method discussed in Chapter 10. For the case of plane deformation in the x,y-plane, we choose

$$A_x = A_x(x, y), \quad A_y = A_y(x, y), \quad A_z = 0, \quad B = B(x, y) \tag{13.4.22}$$

Using the general representation (13.4.8), it can be shown (see Exercise 13.18) that for the plane strain case

$$2\mu(u + iv) = (3 - 4v)\gamma(z) - z\overline{\gamma'(z)} - \overline{\psi(z)} \tag{13.4.23}$$

with appropriate selection of $\gamma(z)$ and $\psi(z)$ in terms of A_x, A_y, and B. It is noted that this form is identical to (10.2.9) found using the complex variable formulation.

A convenient summary flow chart of the various displacement functions discussed in this chapter is shown in Fig. 13.4. The governing equations in terms of the particular potential functions are for the zero body force case. Chou and Pagano (1967) provide additional tables for displacement potentials and stress functions.

13.5 Spherical coordinate formulations

The previous solution examples employing displacement potentials simply used preselected forms of harmonic and biharmonic potentials. We now investigate a more general scheme to determine appropriate potentials for axisymmetric problems described in *spherical coordinates*. Referring to Figs. 1.5 and 1.6, cylindrical coordinates (r,θ,z) are related to spherical coordinates (R,ϕ,θ) through relations

$$R = \sqrt{r^2 + z^2}, \quad \sin\phi = \frac{r}{R}, \quad \cos\phi = \frac{z}{R} \tag{13.5.1}$$

Restricting attention to axisymmetric problems, all quantities are independent of θ, and thus we choose the axisymmetric Galerkin vector representation. Recall that this leads to Love's strain potential V_z, and the displacements and stresses were given by relations (13.3.6) to (13.3.8). Because this potential function was biharmonic, consider first solutions to Laplace's equation. In spherical coordinates the Laplacian operator becomes

$$\nabla^2 = \frac{\partial^2}{\partial R^2} + \frac{2}{R}\frac{\partial}{\partial R} + \frac{1}{R^2}\cot\phi\frac{\partial}{\partial\phi} + \frac{1}{R^2}\frac{\partial^2}{\partial\phi^2} \tag{13.5.2}$$

We first look for separable solutions of the form $R^n\Phi_n(\phi)$, and substituting this into Laplace's equation gives

$$\frac{1}{\sin\phi}\frac{d}{d\phi}\left(\sin\phi\frac{d\Phi_n}{d\phi}\right) + n(n+1)\Phi_n = 0 \tag{13.5.3}$$

Next, making the change of variable $x = \cos\phi$, relation (13.5.3) becomes

$$\left(1-x^2\right)\frac{d^2\Phi_n}{dx^2} - 2x\frac{d\Phi_n}{dx} + n(n+1)\Phi_n = 0 \tag{13.5.4}$$

and this is the well-known *Legendre differential equation*. The two fundamental solutions are the *Legendre functions* $P_n(x)$ and $Q_n(x)$ of the first and second kinds. However, only $P_n(x)$ is continuous for $|x| \le 1$, $(0 \le \phi \le \pi)$, and so we drop the solution $Q_n(x)$. Considering only the case of integer values of parameter n, the solution reduces to the *Legendre polynomials* given by

$$P_n(x) = \frac{1}{2^n n!}\frac{d^n\left(x^2-1\right)^n}{dx^n} \tag{13.5.5}$$

where $P_0 = 1$, $P_1 = x$, $P_2 = \frac{1}{2}(3x^2 - 1)$, $\cdot$ Putting these results together gives the following harmonic solution set

$$\{R^n\Phi_n\} = \left\{1, z, z^2 - \frac{1}{3}\left(r^2 + z^2\right), z^3 - \frac{3}{5}z\left(r^2 + z^2\right), \ldots\right\} \tag{13.5.6}$$

These terms are commonly referred to as *spherical harmonics*.

Our goal, however, is to determine the elasticity solution that requires biharmonic functions for the Love/Galerkin potential. In order to construct a set of biharmonic functions, we employ the last term in relation (13.3.5) and thus argue that if $R^n\Phi_n$ are harmonic, $R^{n+2}\Phi_n$ will be biharmonic. Thus, a representation for the Love strain potential may be written as the linear combination

$$V_z = B_0\left(r^2 + z^2\right) + B_1 z\left(r^2 + z^2\right) + B_2\left(2z^2 - r^2\right)\left(r^2 + z^2\right)$$
$$+ A_0 + A_1 z + A_2\left[z^2 - \frac{1}{3}\left(r^2 + z^2\right)\right] + \cdots \tag{13.5.7}$$

It can be shown that this solution form is useful for general problems with *finite domains*. However, for the case involving *infinite regions*, this form will result in unbounded displacements and stresses at infinity. Therefore, (13.5.7) must be modified for use in regions that extend to infinity. This modification is easily developed by noting that the coefficient $n(n+1)$ in governing equation (13.5.3) will be the same if we were to replace n by $(-n-1)$. This then implies that solution forms

$R^{-n-1}\Phi_{-n-1} = R^{-n-1}\Phi_n$ will also be harmonic functions. Following our previous construction scheme, another set of biharmonic functions for the potential function can then be expressed as

$$V_z = B_0(r^2 + z^2)^{1/2} + B_1 z(r^2 + z^2)^{-1/2} + \cdots$$
$$+ A_0(r^2 + z^2)^{-1/2} + A_1 z(r^2 + z^2)^{-3/2} + \cdots \tag{13.5.8}$$

and this form will be useful for infinite domain problems. For example, the solution to the Kelvin problem in Example 13.1 can be found by choosing only the first term in relation (13.5.8). This scheme can also be employed to construct a set of harmonic functions for the Papkovich potentials; see Little (1973).

Example 13.5 Spherical cavity in an infinite medium subjected to uniform far-field tension

Consider the problem of a stress-free spherical cavity in an infinite elastic solid that is subjected to a uniform tensile stress at infinity. The problem is shown in Fig. 13.5, and for convenience we have oriented the z-axes along the direction of the uniform far-field stress S.

We first investigate the nature of the stress distribution on the spherical cavity caused solely by the far-field stress. For the axisymmetric problem, the spherical stresses are related to the cylindrical components (see Appendix B) by the equations

$$\sigma_R = \sigma_r \sin^2\phi + \sigma_z \cos^2\phi + 2\tau_{rz}\sin\phi\cos\phi$$
$$\sigma_\varphi = \sigma_z \sin^2\phi + \sigma_r \cos^2\phi - 2\tau_{rz}\sin\phi\cos\phi \tag{13.5.9}$$
$$\tau_{R\varphi} = (\sigma_r - \sigma_z)\sin\phi\cos\phi - \tau_{rz}(\sin^2\phi - \cos^2\phi)$$

Therefore, the far-field stress $\sigma_z^\infty = S$ produces normal and shearing stresses on the spherical cavity of the form

$$\sigma_R = S\cos^2\phi, \quad \tau_{R\phi} = -S\sin\phi\cos\phi \tag{13.5.10}$$

Using particular forms from our general solution (13.5.8), we wish to superimpose additional stress fields that will eliminate these stresses and vanish at infinity.

It is found that the superposition of the following three fields satisfies the problem requirements:

1. *Force doublet in z direction.* This state corresponds to a pair of equal and opposite forces in the z direction acting at the origin. The solution is formally determined from the combination of two equal but opposite Kelvin solutions from Example 13.1. The two forces are separated by a distance d, and the limit is taken as $d \to 0$. This summation and limiting process yields a state that is actually the derivative $(\partial/\partial z)$ of the original Kelvin field with a new coefficient of $-Ad$ (see Exercise 13.20). This field's coefficient is denoted as K_1.

2. *Center of dilatation.* This field is the result of three mutually orthogonal double-force pairs from the previous state (1) (see Exercise 13.21). The coefficient of this state is denoted by K_2.

3. *Particular biharmonic term.* A state corresponding to the A_1 term from Eq. (13.5.8).

Combining these three terms with the uniform far-field stress and using the condition of zero stress on the spherical cavity provide sufficient equations to determine the three unknown constants. Details of this process can be found in Timoshenko and Goodier (1970), and the results determine the coefficients of the three superimposed fields

$$K_1 = -\frac{5Sa^3}{2(7-5\nu)}$$

$$K_2 = \frac{S(1-5\nu)a^3}{(7-5\nu)} \qquad (13.5.11)$$

$$A_1 = \frac{Sa^5}{2(7-5\nu)}$$

Using these constants, the stress and displacement fields can be determined. The normal stress on the x,y-plane ($z = 0$) is given by

$$\sigma_z(r,0) = S\left(1 + \frac{4-5\nu}{2(7-5\nu)}\frac{a^3}{r^3} + \frac{9}{2(7-5\nu)}\frac{a^5}{r^5}\right) \qquad (13.5.12)$$

At $r = a$, this result produces the maximum stress

$$\sigma_z(a,0) = (\sigma_z)_{\max} = \frac{27-15\nu}{2(7-5\nu)}S \qquad (13.5.13)$$

Typically, for many metals, $\nu = 0.3$, and this would give a stress concentration factor of

$$\frac{(\sigma_z)_{\max}}{S} = \frac{45}{22} = 2.04 \qquad (13.5.14)$$

It should be noted that in three dimensions the stress concentration factor is generally a function of Poisson's ratio. A plot of this general behavior given by Eq. (13.5.13) is shown in Fig. 13.6. It can be observed that the value of Poisson's ratio produces only small variation on the stress concentration. It is also interesting to note that if the plot were continued for negative values of Poisson's ratio, further decrease in the stress concentration would be found. Exercise 13.28 explores this behavior in more detail.

Note that the corresponding two-dimensional case was previously developed in Example 8.7 and produced a stress concentration factor of 3. Plots of the corresponding two- and three-dimensional stress distributions are shown in Fig. 13.7. For each case the normal stress component in the direction of loading is plotted versus radial distance away from the hole. It is seen that the three-dimensional stresses are always less than two-dimensional predictions. This is to be expected because the three-dimensional field has an additional dimension to decrease the concentration caused by the cavity. Both stress concentrations rapidly decay away from the hole and essentially vanish at $r > 5a$. Additional information on this problem is given by Timoshenko and Goodier (1970).

13.6 Stress functions

In the absence of body forces, the stress formulation of elasticity theory includes the *equilibrium* and *Beltrami–Michell equations*

$$\sigma_{ij,j} = 0 \qquad (13.6.1)$$

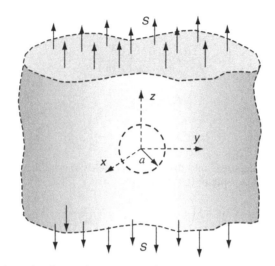

FIG. 13.5 Spherical cavity in an infinite medium under tension.

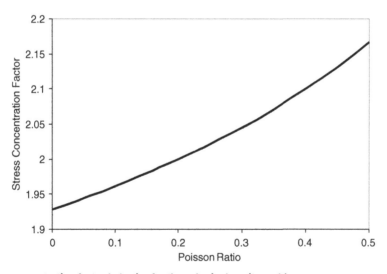

FIG. 13.6 Stress concentration factor behavior for the spherical cavity problem.

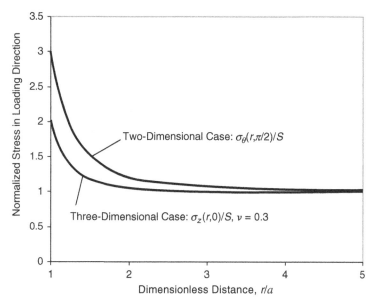

FIG. 13.7 Comparison of two- and three-dimensional stress concentrations around cavities.

$$\sigma_{ij,kk} + \frac{1}{1+\nu}\sigma_{kk,ij} = 0 \tag{13.6.2}$$

In order to develop a general solution to this system, *stress functions* are commonly used. Of course, we have already seen the use of several special stress functions earlier in the text, including Airy's form for the plane problem and Prandtl's function for the torsion example. Here, we investigate the general three-dimensional case and later specialize these results to some of the particular cases just mentioned. The concept of developing a stress function involves the search for a representation of the form

$$\sigma_{ij} = F_{ij}\{\boldsymbol{\Phi}\} \tag{13.6.3}$$

where F_{ij} is some differential operator and $\boldsymbol{\Phi}$ is a tensor-valued variable. Normally, the search looks for forms that automatically satisfy the equilibrium equations (13.6.1), and these are called *self-equilibrated forms*.

It is apparent that the equilibrium equations will be satisfied if σ_{ij} is expressed as the curl of some vector function, because the divergence of a curl vanishes identically. It can be shown that under certain conditions one such equilibrated form that provides a *complete solution* to the elasticity problem is given by

$$\sigma_{ij} = \varepsilon_{imp}\varepsilon_{jkl}\Phi_{mk,pl} \tag{13.6.4}$$

where $\boldsymbol{\Phi}$ is a symmetric second-order tensor. Relation (13.6.4) is sometimes referred to as the *Beltrami representation*, and $\boldsymbol{\Phi}$ is called the *Beltrami stress function*. It has been shown that all elasticity solutions admit this representation; see, for example, Carlson (1966). It is easily demonstrated that (13.6.4) is an equilibrated form, since

$$\sigma_{ij,j} = \left(\varepsilon_{imp}\varepsilon_{jkl}\Phi_{mk,pl}\right)_{,j} = \varepsilon_{imp}\varepsilon_{jkl}\Phi_{mk,plj} = 0$$

because of the product of symmetric and antisymmetric forms in indices jl.

Property (1.3.5) allows expansion of the alternating symbol product, and thus relation (13.6.4) can be expressed as

$$\sigma_{ij} = \delta_{ij}\Phi_{kk,ll} - \delta_{ij}\Phi_{kl,kl} - \Phi_{ij,kk} + \Phi_{li,lj} + \Phi_{lj,li} - \Phi_{kk,ij} \tag{13.6.5}$$

or

$$\begin{aligned}
\sigma_{11} &= \Phi_{33,22} + \Phi_{22,33} - 2\Phi_{23,23} \\
\sigma_{22} &= \Phi_{11,33} + \Phi_{33,11} - 2\Phi_{31,31} \\
\sigma_{33} &= \Phi_{22,11} + \Phi_{11,22} - 2\Phi_{12,12} \\
\sigma_{12} &= -\Phi_{12,33} - \Phi_{33,12} + \Phi_{23,13} + \Phi_{31,23} \\
\sigma_{23} &= -\Phi_{23,11} - \Phi_{11,23} + \Phi_{31,21} + \Phi_{12,31} \\
\sigma_{31} &= -\Phi_{31,22} - \Phi_{22,31} + \Phi_{12,32} + \Phi_{23,12}
\end{aligned} \tag{13.6.6}$$

The first invariant of the stress tensor then becomes

$$\begin{aligned}
\sigma_{nn} &= \varepsilon_{nmp}\varepsilon_{nkl}\Phi_{mk,pl} \\
&= (\delta_{mk}\delta_{pl} - \delta_{ml}\delta_{pk})\Phi_{mk,pl} \\
&= \Phi_{kk,ll} - \Phi_{lk,lk}
\end{aligned} \tag{13.6.7}$$

and thus the compatibility equations (13.6.2) can be expressed in terms of the general stress function as

$$\left(\varepsilon_{imp}\varepsilon_{jkl}\Phi_{mk,pl}\right)_{,nn} + \frac{1}{1+\nu}\left(\Phi_{kk,ll} - \Phi_{lk,lk}\right)_{,ij} = 0 \tag{13.6.8}$$

Not all of the six components of Φ_{ij} are independent. Two alternate ways of generating complete solutions to the stress formulation problem are developed through the use of *reduced forms* that include the *Maxwell* and *Morera* stress function formulations.

13.6.1 Maxwell stress function representation

The *Maxwell stress function representation* considers the reduced form whereby all off-diagonal elements of Φ_{ij} are set to zero

$$\Phi_{ij} = \begin{bmatrix} \Phi_{11} & 0 & 0 \\ 0 & \Phi_{22} & 0 \\ 0 & 0 & \Phi_{33} \end{bmatrix} \tag{13.6.9}$$

which yields a representation

$$\begin{aligned}
\sigma_{11} &= \Phi_{33,22} + \Phi_{22,33} \\
\sigma_{22} &= \Phi_{11,33} + \Phi_{33,11} \\
\sigma_{33} &= \Phi_{22,11} + \Phi_{11,22} \\
\sigma_{12} &= -\Phi_{33,12} \\
\sigma_{23} &= -\Phi_{11,23} \\
\sigma_{31} &= -\Phi_{22,31}
\end{aligned} \tag{13.6.10}$$

Notice that the *Airy stress function* that is used for two-dimensional problems is a special case of this scheme with $\Phi_{11} = \Phi_{22} = 0$ and $\Phi_{33} = \phi(x_1,x_2)$.

13.6.2 Morera stress function representation

The *Morera stress function* method uses the general form with diagonal terms set to zero

$$\Phi_{ij} = \begin{bmatrix} 0 & \Phi_{12} & \Phi_{13} \\ \Phi_{12} & 0 & \Phi_{23} \\ \Phi_{13} & \Phi_{23} & 0 \end{bmatrix} \tag{13.6.11}$$

This approach yields the representation

$$\sigma_{11} = -2\Phi_{23,23}$$
$$\sigma_{22} = -2\Phi_{31,31}$$
$$\sigma_{33} = -2\Phi_{12,12}$$
$$\sigma_{12} = -\Phi_{12,33} + \Phi_{23,13} + \Phi_{13,23} \tag{13.6.12}$$
$$\sigma_{23} = -\Phi_{23,11} + \Phi_{13,21} + \Phi_{12,31}$$
$$\sigma_{31} = -\Phi_{31,22} + \Phi_{12,32} + \Phi_{23,12}$$

It can be observed that for the torsion problem, the *Prandtl stress function* (here denoted by φ) is a special case of this representation with $\Phi_{12} = \Phi_{13} = 0$ and $\Phi_{23,1} = \varphi(x_1, x_2)$.

References

Abdulaliyev Z, Ataoglu S: Effect of Poisson's ratio on three-dimensional stress distributions, *J Appl Mech* 76, 2009.

Carlson DE: On the completeness of the Beltrami stress functions in continuum mechanics, *J Math Anal Appl* 15: 311–315, 1966.

Chou PC, Pagano NJ: *Elasticity tensor, dyadic and engineering approaches*, Princeton, NJ, 1967, D. Van Nostrand.

Eubanks RA, Sternberg E: On the completeness of the Boussinesq–Papkovich stress functions, *J Rational Mech Anal* 5:735–746, 1956.

Fung YC: *Foundations of solid mechanics*, Englewood Cliffs, NJ, 1965, Prentice Hall.

Galerkin B: Contribution à la solution générale du problème de la théorie de l'élasticité dans le cas de trios dimensions, *Comptes Rendus* 190:1047, 1930.

Graff KF: *Wave motion in elastic solids*, New York, 1991, Dover.

Kellogg OD: *Foundations of potential theory*, New York, 1969, Dover.

Lakes RS: Foam structures with a negative Poisson's ratio, *Science* 235:1038–1040, 1987.

Little RW: *Elasticity*, Englewood Cliffs, NJ, 1973, Prentice Hall.

Love AEH: *A treatise on the mathematical theory of elasticity*, ed 4, New York, 1944, Dover.

Naghdi PM, Hsu CS: On a representation of displacements in linear elasticity in terms of three stress functions, *J Math Mech* 10:233–245, 1961.

Neuber H: Ein neurer Censatz zur Losing Raumlicher Probleme der Elastez-etatstheorie, *Z Angew Math Mech* 14: 203, 1934.

Papkovich PF: An expression for a general integral of the equations of the theory of elasticity in terms of harmonic functions, *Izvest Akad Nauk SSSR Ser Matem K estestv neuk* (10).

Peckhold DAW: On the role of the Stokes-Helmholtz decomposition in the derivation of displacement potentials in classical elasticity, *J Elast* 1:171–174, 1971.

Timoshenko SP, Goodier JN: *Theory of elasticity*, New York, 1970, McGraw-Hill.

Tran-Cong T: Notes on some relationships between the Galerkin, Papkovich-Neuber and Naghdi-Hsu solutions in linear elasticity, *Mech Res Commun* 8:207–211, 1981.

Wang MZ: The Naghdi_Hsu solution and the Naghdi-Hsu transformation, *J Elast* 15:103–108, 1985.

Wang MZ, Xu BX, Gao CF: Recent general solutions in linear elasticity and their applications, *Appl Mech Rev ASME* 61, 2008.

Exercises

13.1 Using the Helmholtz representation, determine the displacement field that corresponds to the potentials $\phi = x^2 + 4y^2$, $\boldsymbol{\varphi} = R^2\boldsymbol{e}_3$. Next show that this displacement field satisfies Navier's equation with no body forces.

13.2 Explicitly show that the dilatation and rotation are related to the Helmholtz potentials through relations (13.1.3).

13.3 For the case of zero body forces, show that by using the vector identity $(1.8.5)_9$ Navier's equation can be written as

$$(\lambda + 2\mu)\nabla^2\boldsymbol{u} + (\lambda + \mu)\boldsymbol{\nabla} \times \boldsymbol{\nabla} \times \boldsymbol{u} = 0$$

Using repeated differential operations on this result, show that the displacement vector is biharmonic. Furthermore, because the stress and strain are linear combinations of first derivatives of the displacement, they too will be biharmonic.

13.4 For the case of Lamé's potential, show that strains and stresses are given by (13.2.3).

13.5 Justify that the Galerkin vector satisfies the governing equation (13.3.2).

13.6 Show that the Helmholtz potentials are related to the Galerkin vector by relations (13.3.3).

13.7 Justify relations (13.3.4) for the stress components in terms of the Galerkin vector.

13.8 For the case of zero body forces, the Galerkin vector is biharmonic. However, it was pointed out that in curvilinear coordinate systems, the individual Galerkin vector components might not necessarily be biharmonic. Consider the cylindrical coordinate case where $V = V_r\boldsymbol{e}_r + V_\theta\boldsymbol{e}_\theta + V_z\boldsymbol{e}_z$. Using the results of Section 1.9, first show that the Laplacian operator on each term will give rise to the following relations

$$\nabla^2(V_r\boldsymbol{e}_r) = \left(\nabla^2 V_r - \frac{V_r}{r^2}\right)\boldsymbol{e}_r + \frac{2}{r^2}\frac{\partial V_r}{\partial\theta}\boldsymbol{e}_\theta$$

$$\nabla^2(V_\theta\boldsymbol{e}_\theta) = \left(\nabla^2 V_\theta - \frac{V_\theta}{r^2}\right)\boldsymbol{e}_\theta - \frac{2}{r^2}\frac{\partial V_\theta}{\partial\theta}\boldsymbol{e}_r$$

$$\nabla^2(V_z\boldsymbol{e}_z) = \nabla^2 V_z\boldsymbol{e}_z$$

Using these results, show that the biharmonic components are given by

$$\nabla^2\nabla^2(V_r\boldsymbol{e}_r) = \left[\left(\nabla^2 - \frac{1}{r^2}\right)^2 V_r - \frac{4}{r^4}\frac{\partial^2 V_r}{\partial\theta^2}\right]\boldsymbol{e}_r + \left[\frac{4}{r^2}\left(\nabla^2 - \frac{1}{r^2}\right)\frac{\partial V_r}{\partial\theta}\right]\boldsymbol{e}_\theta$$

$$\nabla^2\nabla^2(V_\theta\boldsymbol{e}_\theta) = \left[-\frac{4}{r^2}\left(\nabla^2 - \frac{1}{r^2}\right)\frac{\partial V_\theta}{\partial\theta}\right]\boldsymbol{e}_r + \left[\left(\nabla^2 - \frac{1}{r^2}\right)^2 V_\theta - \frac{4}{r^4}\frac{\partial^2 V_\theta}{\partial\theta^2}\right]\boldsymbol{e}_\theta$$

$$\nabla^2\nabla^2(V_z\boldsymbol{e}_z) = \nabla^2\nabla^2 V_z\boldsymbol{e}_z$$

and thus only the component V_z will satisfy the scalar biharmonic equation.

13.9 Explicitly show that Boussinesq's problem as illustrated in Fig. 13.2 is solved by the superposition of a Galerkin vector and Lamé's potential given by relation (13.3.13). Verify that the Cartesian displacements and stresses are given by

$$u = \frac{Px}{4\pi\mu R}\left(\frac{z}{R^2} - \frac{1-2\nu}{R+z}\right), \quad v = \frac{Py}{4\pi\mu R}\left(\frac{z}{R^2} - \frac{1-2\nu}{R+z}\right), \quad w = \frac{P}{4\pi\mu R}\left(2(1-\nu) + \frac{z^2}{R^2}\right)$$

$$\sigma_x = -\frac{P}{2\pi R^2}\left[\frac{3x^2 z}{R^3} - (1-2\nu)\left(\frac{z}{R} - \frac{R}{R+z} + \frac{x^2(2R+z)}{R(R+z)^2}\right)\right]$$

$$\sigma_y = -\frac{P}{2\pi R^2}\left[\frac{3y^2 z}{R^3} - (1-2\nu)\left(\frac{z}{R} - \frac{R}{R+z} + \frac{y^2(2R+z)}{R(R+z)^2}\right)\right]$$

$$\sigma_z = -\frac{3Pz^3}{2\pi R^5}, \quad \tau_{xy} = -\frac{P}{2\pi R^2}\left[\frac{3xyz}{R^3} - \frac{(1-2\nu)(2R+z)xy}{R(R+z)^2}\right]$$

$$\tau_{yz} = -\frac{3Pyz^2}{2\pi R^5}, \quad \tau_{xz} = -\frac{3Pxz^2}{2\pi R^5}$$

13.10 Show that Cerruti's problem of Fig. 13.3 is solved by the Galerkin vector and Lamé's potential specified in relations (13.3.15). Develop the expressions for the Cartesian displacements and stresses

$$u = \frac{P}{4\pi\mu R}\left[1 + \frac{x^2}{R^2} + (1-2\nu)\left(\frac{R}{R+z} - \frac{x^2}{(R+z)^2}\right)\right]$$

$$v = \frac{Pxy}{4\pi\mu R}\left(\frac{1}{R^2} - \frac{1-2\nu}{(R+z)^2}\right), \quad w = \frac{Px}{4\pi\mu R}\left(\frac{z}{R^2} + \frac{1-2\nu}{R+z}\right)$$

$$\sigma_x = \frac{Px}{2\pi R^3}\left[-\frac{3x^2}{R^2} + \frac{(1-2\nu)}{(R+z)^2}\left(R^2 - y^2 - \frac{2Ry^2}{R+z}\right)\right]$$

$$\sigma_y = \frac{Px}{2\pi R^3}\left[\frac{-3y^2}{R^2} + \frac{(1-2\nu)}{(R+z)^2}\left(3R^2 - x^2 - \frac{2Rx^2}{R+z}\right)\right]$$

$$\sigma_z = -\frac{3Pxz^2}{2\pi R^5}, \quad \tau_{yz} = -\frac{3Pxyz}{2\pi R^5}, \quad \tau_{xz} = -\frac{3Px^2 z}{2\pi R^5}$$

$$\tau_{xy} = \frac{Py}{2\pi R^3}\left[-\frac{3x^2}{R^2} - \frac{(1-2\nu)}{(R+z)^2}\left(R^2 - x^2 + \frac{2Rx^2}{R+z}\right)\right]$$

13.11 Both the Galerkin vector and Love's strain function were to satisfy the biharmonic equation. If however, these representations were also harmonic, justify that the resulting stresses will all be independent Poisson's ratio for the case of traction-only boundary conditions.

13.12 In three dimensions, often a special case of Poisson's ratio will greatly simplify the problem solution. Show for the special case of $\nu = 1/2$, that the displacement and stress solutions for the Kelvin and Boussinesq problems simplify and that their respective results are very similar.

13.13 Explicitly justify that the Papkovich functions A and B satisfy relations (13.4.9).

13.14 For the axisymmetric case, the Papkovich functions reduced to the Boussinesq potentials B and A_z defined by relations (13.4.11). Show that the general forms for the displacements and stresses in cylindrical coordinates are given by

$$u_r = -\frac{1}{2\mu}\frac{\partial}{\partial r}\left(B + \frac{A_z z}{4(1-\nu)}\right), \quad u_\theta = 0, \quad u_z = \frac{1}{2\mu}\left[A_z - \frac{\partial}{\partial z}\left(B + \frac{A_z z}{4(1-\nu)}\right)\right]$$

$$\sigma_r = -\frac{\nu}{1-2\nu}\nabla^2\left(\frac{A_z z}{4(1-\nu)}\right) + \frac{\nu}{1-2\nu}\frac{\partial A_z}{\partial z} - \frac{\partial^2}{\partial r^2}\left(B + \frac{A_z z}{4(1-\nu)}\right)$$

$$\sigma_\theta = -\frac{\nu}{1-2\nu}\nabla^2\left(\frac{A_z z}{4(1-\nu)}\right) + \frac{\nu}{1-2\nu}\frac{\partial A_z}{\partial z} - \frac{1}{r}\frac{\partial}{\partial r}\left(B + \frac{A_z z}{4(1-\nu)}\right)$$

$$\sigma_z = -\frac{\nu}{1-2\nu}\nabla^2\left(\frac{A_z z}{4(1-\nu)}\right) + \frac{\nu}{1-2\nu}\frac{\partial A_z}{\partial z} + \frac{\partial A_z}{\partial z} - \frac{\partial^2}{\partial z^2}\left(B + \frac{A_z z}{4(1-\nu)}\right)$$

$$\tau_{rz} = 2\mu e_{rz} = \frac{1}{2}\frac{\partial A_z}{\partial r} - \frac{\partial^2}{\partial r\partial z}\left(B + \frac{A_z z}{4(1-\nu)}\right)$$

13.15 Using the results of Exercise 13.14, verify that the displacement and stress fields for the Boussinesq problem of Example 13.4 are given by (13.4.20) and (13.4.21). Note the interesting behavior of the radial displacement, that $u_r > 0$ only for points where $z/R > (1-2\nu)R/(R+z)$. Show that points satisfying this inequality lie inside a cone $\phi \le \phi_o$, with ϕ_o determined by the relation $\cos^2\phi_o + \cos\phi_o - (1-2\nu) = 0$.

13.16* The displacement field for the Boussinesq problem was given by (13.4.20). For this case, construct a displacement vector distribution plot, similar to the two-dimensional case shown in Fig. 8.22. For convenience, choose the coefficient $P/4\pi\mu = 1$ and take $\nu = 0.3$. Compare the two- and three-dimensional results.

13.17 Consider an elastic half-space with $\sigma_z = 0$ on the surface $z = 0$. For the axisymmetric problem, show that the Boussinesq potentials must satisfy the relation $A_z = 2\partial B/\partial z$ within the half-space.

13.18 Consider the Papkovich representation for the two-dimensional plane strain case where $A = A_1(x,y)e_1 + A_2(x,y) e_2$ and $B = B(x,y)$. Show that this representation will lead to the complex variable formulation

$$2\mu(u + iv) = \kappa\gamma(z) - z\overline{\gamma'(z)} - \overline{\psi(z)}$$

with appropriate definitions of $\gamma(z)$ and $\psi(z)$.

13.19 Show that Kelvin's problem of Fig. 13.1 may be solved using the axisymmetric Papkovich functions (Boussinesq potentials)

$$B = 0, \quad A_z = \frac{P}{2\pi R}$$

Verify that the displacements match those given in equations (13.3.10).

13.20 A *force doublet* is commonly defined as two equal but opposite forces acting in an infinite medium as shown in the following figure. Develop the stress field for this problem by superimposing the solution from Example 13.1 onto that of another single force of $-P$ acting at the point $z = -d$. In particular, consider the case as $d \to 0$ such that the product $Pd \to D$, where D is a constant. This summation and limiting process yield a solution that is simply the derivative of the original Kelvin state. For example, the superposition of the radial stress component gives

$$\lim_{d \to 0} [\sigma_r(r,z) - \sigma_r(r, z+d)] = -d \frac{\partial \sigma_r}{\partial z}$$

$$= -\frac{D}{8\pi(1-v)} \frac{\partial}{\partial z} \left[(1-2v)z(r^2+z^2)^{-3/2} - 3r^2z(r^2+z^2)^{-5/2} \right]$$

The other stress components follow in an analogous manner. Using relations (13.5.9), show that the stress components in spherical coordinates can be expressed as

$$\sigma_R = -\frac{(1+v)D}{4\pi(1-v)R^3} \left[-\sin^2\phi + \frac{2(2-v)}{1+v}\cos^2\phi \right]$$

$$\tau_{R\phi} = -\frac{(1+v)D}{4\pi(1-v)R^3} \sin\phi\cos\phi$$

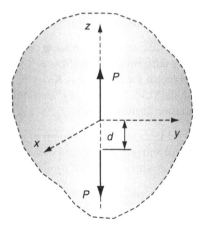

13.21 Using the results of Exercise 13.20, continue the superposition process by combining three force doublets in each of the coordinate directions. This results in a *center of dilatation* at the origin as shown in the figure. Using spherical coordinate components, show that the stress field for this problem is given by

$$\sigma_R = -\frac{(1-2v)D}{2\pi(1-v)R^3} = \frac{C}{R^3}, \quad \tau_{R\phi} = 0$$

where C is another arbitrary constant, and thus the stresses will be symmetrical with respect to the origin.

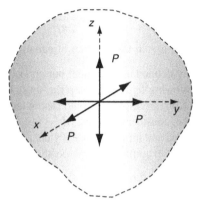

13.22 Using the basic field equations for spherical coordinates given in Appendix A, formulate the elasticity problem for the *spherically symmetric* case, where $u_R = u(R)$, $u_\phi = u_\theta = 0$. In particular, show that the governing equilibrium equation with zero body forces becomes

$$\frac{d^2 u}{dR^2} + \frac{2}{R}\frac{du}{dR} - \frac{2}{R^2}u = 0$$

Next solve this equation and show that the general solution can be expressed as

$$u = C_1 R + \frac{C_2}{R^2}, \quad \sigma_R = K_1 - \frac{2K_2}{R^3}, \quad \sigma_\phi = \sigma_\theta = K_1 + \frac{K_2}{R^3}$$

where C_1, C_2, K_1, and K_2 are arbitrary constants.

13.23 Using the results of Exercise 13.22, solve the problem of a stress-free spherical cavity in an infinite elastic medium under uniform far-field stress $\sigma_x^\infty = \sigma_y^\infty = \sigma_z^\infty = S$. Explicitly show that the stress concentration factor for this case is $K = 1.5$, and compare this value with the corresponding two-dimensional case. Explain why we would expect such a difference between these two concentration factors.

13.24 Using the results of Exercise 13.22, solve the problem of a thick-walled spherical shell with inner radius R_1 loaded with uniform pressure p_1, and with outer radius R_2 loaded with uniform pressure p_2. For the special case with $p_1 = p$ and $p_2 = 0$, show that the stresses are given by

$$\sigma_R = \frac{pR_1^3}{R_2^3 - R_1^3}\left(1 - \frac{R_2^3}{R^3}\right)$$

$$\sigma_\phi = \sigma_\theta = \frac{pR_1^3}{R_2^3 - R_1^3}\left(1 + \frac{R_2^3}{2R^3}\right)$$

13.25* Using the general solution forms of Exercise 13.22, solve the problem of a rigid spherical inclusion of radius a perfectly bonded to the interior of an infinite body subjected to uniform stress at infinity of $\sigma_R^\infty = S$. Explicitly show that the stress field is given by

$$\sigma_R = S\left[1 + 2\frac{1-2v}{1+v}\left(\frac{a}{R}\right)^3\right], \quad \sigma_\phi = \sigma_\theta = S\left[1 - \frac{1-2v}{1+v}\left(\frac{a}{R}\right)^3\right]$$

Determine the nature of the stress field for the incompressible case with $v = 1/2$. Finally, explore the stresses on the boundary of the inclusion ($R = a$), and plot them as a function of Poisson's ratio.

13.26 Consider the three-dimensional stress concentration problem given in Example 13.5. Recall that the maximum stresses occur on the boundary of the spherical cavity ($r = a$). With respect to the problem geometry shown in Fig. 13.5, the maximum stress component was found to be

$$\sigma_z(a, z = 0) = \frac{27 - 15v}{2(7 - 5v)}S$$

Other stress components can also be determined from the solution method outlined in the problem, and two particular components on the cavity boundary are

$$\sigma_\phi(a, \phi = 0) = -\frac{3 + 15v}{2(7 - 5v)}S, \quad \sigma_\theta(a, \phi = \pi/2) = \frac{15v - 3}{2(7 - 5v)}S$$

Using these results, along with the superposition principle, show that maximum stresses for the following cases are given by:

(a) Uniform uniaxial tension loadings of S along x and z directions

$$\sigma_{\max} = \frac{24 - 30v}{2(7 - 5v)}S$$

(b) Tension loading S along z axis and compression loading S along x directions

$$\sigma_{\max} = \frac{15}{7 - 5v}S$$

(c) Tension loadings of S along each Cartesian direction

$$\sigma_{\max} = \frac{3}{2}S$$

Note that part (b) corresponds to far-field pure shear and part (c) coincides with the results found in Exercise 13.23.

13.27[*] Generate plots of the stress concentration factor versus Poisson's ratio (similar to Fig. 13.6) for each case in Exercise 13.26. Compare the results.

13.28[*] There has been some interesting research dealing with materials that have negative values of Poisson's ratio; recall from fundamental theory $-1 \leq v \leq 1/2$. Beginning studies of this concept were done by Lakes (1987) and commonly these types of materials have specialized internal microstructures (e.g., foams and cellular solids) that produce such anomalous behavior. Several interesting consequences occur with $v < 0$, and one such behavior results in decreasing the stress concentration around holes in three-dimensional solids. This can be directly explored by expanding the plot shown in Fig. 13.6 to include the full range of Poisson's ratio. Redevelop Fig. 13.6 for the range $-1 \leq \nu \leq 1/2$ and determine the maximum decrease in the stress concentration factor.

13.29   Using the Morera stress function formulation, define

$$\Phi_{13} = -\frac{1}{2}z\phi_{,1}, \quad \Phi_{23} = -\frac{1}{2}z\phi_{,2}, \quad \Phi_{12,12} = -\frac{\nu}{2}\nabla^2\phi$$

where ϕ is independent of z. Show that this represents plane strain conditions with ϕ equal to the usual Airy stress function.

13.30   Consider the second order stress tensor representation $\sigma_{ij} = \phi_{,kk}\delta_{ij} - \phi_{,ij}$, where ϕ is a proposed stress function. First show that this scheme gives a divergence-less stress; i.e. $\sigma_{ij,j} = 0$. Next using this representation in the general compatibility equations (5.3.3) with no body forces gives the result

$$\phi_{,kkll}\delta_{ij} + \frac{1-\nu}{1+\nu}\phi_{,kkij} = 0$$

Finally show that for the two-dimensional plane stress case, this representation reduces to the ordinary Airy form from Section 7.5.

Nonhomogeneous elasticity

It has been observed that many materials have a spatially varying microstructure that leads to spatial variation in elastic properties and thus requires a nonhomogeneous model. For example, in geomechanics studies, rock and soil material will commonly have depth-dependent properties resulting from the overburden of material lying above a given point. Gradations in microstructure are also commonly found in biological cellular materials such as wood and bone, where biological adaptation has distributed the strongest microstructure in regions that experience the highest stress.

Recently, there has been considerable interest in the development of *graded materials* that have spatial property variations deliberately created to improve mechanical performance (Suresh, 2001; Birman and Byrd, 2007). Such graded properties can be traced back to surface heat treatments for swords, knives, and gear teeth. Various composite materials have been constructed using graded transitions in composition to reduce stress concentrations at interfaces. In the 1990s, interest in graded materials focused on controlling thermal stresses in structures exposed to high-temperature applications and to surface contact damage. This work has led to a new class of engineered materials called *functionally graded materials* (FGMs) that are developed with spatially varying properties to suit particular applications. The graded composition of such materials is commonly established and controlled using advanced manufacturing techniques, including powder metallurgy, chemical vapor deposition, centrifugal casting, solid free-form fabrication, and other schemes.

Our previous developments have been connected to the formulation and solution of isotropic and anisotropic elasticity problems. We now wish to go back and investigate the inhomogeneous isotropic case and explore solutions for a few problems that exist in the literature. We will focus attention on formulation issues that allow the development of tractable boundary value problems and will examine the effect of inhomogeneity on the resulting stress and displacement solution fields. By exploring closed-form solutions to a series of example problems, we will see that in some cases inhomogeneity produces little effect, while in others significant and fundamentally different stress and displacement distributions will occur.

14.1 Basic concepts

For the inhomogeneous model, elastic moduli C_{ijkl} or C_{ij} will now be functions of the spatial coordinates x_m describing the problem; thus, Hooke's law would read

$$\sigma_{ij} = C_{ijkl}(x_m)e_{kl} \qquad (14.1.1)$$

Other than this modification, the structural form of Hooke's law is the same as used previously. Similar to the anisotropic case presented in Chapter 11, the other basic field equations of strain displacement, strain compatibility, and equilibrium will also remain the same. However, it should be recognized that by combining the new form (14.1.1) with these other field equations (e.g., in developing stress or displacement formulations), entirely new and more complicated field equations will be generated. This will create a more complex problem formulation and analytical solutions will of course be more difficult to obtain.

For example, considering the general case with no body forces, the displacement formulation would now yield equilibrium equations in terms of displacement as

$$\sigma_{ij,j} = 0 \Rightarrow \frac{\partial}{\partial x_j}\left[C_{ijkl}\left(u_{k,l} + u_{l,k}\right)\right] = 0 \Rightarrow \frac{\partial}{\partial x_j}\left[C_{ijkl}u_{k,l}\right] = 0 \tag{14.1.2}$$

where we have used the symmetry $C_{ijkl} = C_{ijlk}$. Expanding relation (14.1.2) gives

$$C_{ijkl}u_{k,lj} + C_{ijkl,j}u_{k,l} = 0 \tag{14.1.3}$$

The first term in relation (14.1.3) corresponds to the homogeneous case and for isotropic materials would simply lead to the homogeneous form of Navier's equations (5.4.3) developed in Chapter 5. The second term in (14.1.3) accounts for spatial variation in elastic moduli and includes first-order derivatives of both the elastic moduli and displacements.

Depending on the nature of the material's anisotropy and inhomogeneity, Eq. (14.1.3) could become very complex and thereby limit solution by analytical methods. Only limited studies have included both anisotropy and inhomogeneity (e.g., Lekhnitskii, 1981; Horgan and Miller, 1994; Fraldi and Cowin, 2004; Stampouloglou and Theotokoglou, 2005), and thus most nonhomogeneous analyses have been made under the simplification of material isotropy. For example, using the isotropic assumption, relation (14.1.3) for a two-dimensional plane strain model would reduce to

$$\mu\nabla^2 u + (\lambda + \mu)\frac{\partial}{\partial x}\left(\frac{\partial u}{\partial x} + \frac{\partial v}{\partial y}\right) + \frac{\partial\lambda}{\partial x}\left(\frac{\partial u}{\partial x} + \frac{\partial v}{\partial y}\right) + 2\frac{\partial\mu}{\partial x}\frac{\partial u}{\partial x} + \frac{\partial\mu}{\partial y}\left(\frac{\partial u}{\partial y} + \frac{\partial v}{\partial x}\right) = 0$$
$$\mu\nabla^2 v + (\lambda + \mu)\frac{\partial}{\partial y}\left(\frac{\partial u}{\partial x} + \frac{\partial v}{\partial y}\right) + \frac{\partial\lambda}{\partial y}\left(\frac{\partial u}{\partial x} + \frac{\partial v}{\partial y}\right) + 2\frac{\partial\mu}{\partial y}\frac{\partial v}{\partial y} + \frac{\partial\mu}{\partial x}\left(\frac{\partial u}{\partial y} + \frac{\partial v}{\partial x}\right) = 0 \tag{14.1.4}$$

which clearly simplifies to the homogeneous form (7.1.5) if the elastic moduli λ and μ are constants.

Next, consider the stress formulation for the inhomogeneous but isotropic plane problem with no body forces. Introducing the usual Airy stress function ϕ defined by (8.1.3), the equilibrium equations are again identically satisfied. As before, we look to generate compatibility relations in terms of stress and then incorporate the Airy stress function to develop a single governing field equation. Because the elastic moduli are now functions of spatial coordinates, results for the nonhomogeneous case will differ significantly from our developments in Section 7.5 that lead to a simple biharmonic equation. For the two-dimensional plane strain case, using Hooke's law in the only nonzero compatibility relation (7.1.6) gives the new form for nonhomogeneous materials in terms of the Airy stress function

$$\frac{\partial^2}{\partial x^2}\left(\frac{1-\nu^2}{E}\frac{\partial^2\phi}{\partial x^2} - \frac{\nu(1+\nu)}{E}\frac{\partial^2\phi}{\partial y^2}\right)$$
$$+\frac{\partial^2}{\partial y^2}\left(\frac{1-\nu^2}{E}\frac{\partial^2\phi}{\partial y^2} - \frac{\nu(1+\nu)}{E}\frac{\partial^2\phi}{\partial x^2}\right) + 2\frac{\partial^2}{\partial x\partial y}\left(\frac{1+\nu}{E}\frac{\partial^2\phi}{\partial x\partial y}\right) = 0 \tag{14.1.5}$$

The corresponding relation for the case of plane stress is given by

$$\frac{\partial^2}{\partial x^2}\left(\frac{1}{E}\frac{\partial^2 \phi}{\partial x^2} - \frac{\nu}{E}\frac{\partial^2 \phi}{\partial y^2}\right) + \frac{\partial^2}{\partial y^2}\left(\frac{1}{E}\frac{\partial^2 \phi}{\partial y^2} - \frac{\nu}{E}\frac{\partial^2 \phi}{\partial x^2}\right) + 2\frac{\partial^2}{\partial x \partial y}\left(\frac{1+\nu}{E}\frac{\partial^2 \phi}{\partial x \partial y}\right) = 0 \qquad (14.1.6)$$

Note that results (14.1.5) and (14.1.6) reduce to the biharmonic equation for the case of constant elastic moduli. It should be evident that a biharmonic function will not, in general, satisfy either of these governing equations for nonhomogeneous materials.

In order to formulate tractable problems, the variation in elastic properties is normally taken to be of simple continuous form. For example, in an unbounded domain, the elastic moduli might be chosen to vary in a single direction, as shown schematically in Fig. 14.1, with shading drawn to indicate gradation.

Particular functional forms used to prescribe such nonhomogeneity have commonly used linear, exponential, and power-law variation in elastic moduli of the form

$$\begin{aligned} C_{ij}(x) &= C_{ij}^o(1 + ax) \\ C_{ij}(x) &= C_{ij}^o e^{ax} \\ C_{ij}(x) &= C_{ij}^o x^a \end{aligned} \qquad (14.1.7)$$

where C_{ij}^o and a are prescribed constants. Because experience has indicated that variation in Poisson's ratio normally does not play a significant role in determining magnitudes of stresses and displacements, ν is commonly taken to be constant in many inhomogeneous formulations. Exercises 14.2, 14.3, and 14.7 explore formulation results for some particular elastic moduli variation.

Early work on developing elasticity solutions for inhomogeneous problems began to appear in the literature a half a century ago; see, for example, Du (1961), Ter-Mkrtich'ian (1961), Rostovtsev (1964), and Plevako (1971, 1972). Numerous works followed on refining formulations and developing solutions to problems of engineering interest; see, for example, the beam studies by Sankar (2001). Much of this work was fueled by interest in developing models of functionally graded materials. We now wish to explore specific solutions to some of these problems and compare the results with previously generated homogeneous solutions to find particular differences resulting from the inhomogeneity.

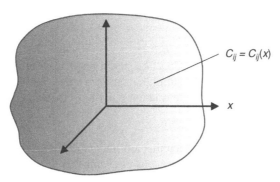

FIG. 14.1 Continuously graded material in a single direction.

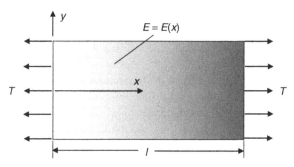

FIG. 14.2 Uniaxial tension of an inhomogeneous sheet.

We first start with a simple example that exploits a scheme to generate nonhomogeneous solutions from corresponding homogeneous solutions. This method has been previously used by Sadd (2010) to determine a variety of plane nonhomogeneous elasticity solutions. This particular technique falls under the more general theoretical framework given by Fraldi and Cowin (2004) who showed that inhomogeneous solutions can be found from an *associated homogeneous problem* providing the problem has a zero eigenvalue stress state, $\det(\sigma_{ij}) = 0$. By definition all plane stress problems satisfy this requirement.

Example 14.1 Uniaxial tension of a graded sheet

Consider a two-dimensional, plane stress problem of a rectangular inhomogeneous sheet under uniform uniaxial tension T, as shown in Fig. 14.2. We will assume that the modulus of elasticity is uniaxially graded such that $E = E(x)$, while Poisson's ratio will be taken to be constant. Recall that this problem was solved for the homogeneous case in Example 8.1.

Based on the boundary conditions, we might guess the same stress field solution as found in the homogeneous case, that is, $\sigma_x = T$, $\sigma_y = \tau_{xy} = 0$, and this field would result from the Airy stress function $\phi = Ty^2/2$. Note that this stress function is biharmonic but, as previously mentioned, it will not identically satisfy the governing equation (14.1.6). Using this stress function along with the prescribed uniaxial gradation $E = E(x)$, the governing relation (14.1.6) would imply that (see Exercise 14.1)

$$\frac{d^2}{dx^2}\left(\frac{1}{E}\right) = 0 \Rightarrow \frac{1}{E} = Ax + B \ \text{ or } \ E = \frac{1}{Ax + B} \tag{14.1.8}$$

where A and B are arbitrary constants. Thus, we find a restriction on the allowable form of the material grading in order to preserve the simplified uniform stress field found in the homogeneous case. It will be more convenient to rewrite relation (14.1.8) in the form

$$E = \frac{E_o}{1 + Kx} \tag{14.1.9}$$

where E_o is the modulus at $x = 0$ and K is another arbitrary constant related to the level of gradation. Note that $K = 0$ corresponds to the homogeneous case with $E = E_o$.

Next we wish to determine the displacement field associated with this stress distribution. This is accomplished by the standard procedure using Hooke's law and the strain displacement relations

$$\frac{\partial u}{\partial x} = e_x = \frac{1}{E}(\sigma_x - \nu\sigma_y) = \frac{T}{E}$$

$$\frac{\partial v}{\partial y} = e_y = \frac{1}{E}(\sigma_y - \nu\sigma_x) = -\nu\frac{T}{E}$$

(14.1.10)

These results are then integrated to get

$$u = \frac{T}{E_o}\left(x + K\left(\frac{x^2}{2} + \nu\frac{y^2}{2}\right)\right)$$

$$v = -\nu\frac{T}{E_o}(1 + Kx)y$$

(14.1.11)

where we have selected zero rigid-body motion terms such that $u(0, 0) = v(0, 0) = \omega_z(0, 0) = 0$. Note the somewhat surprising result that the horizontal displacement u also depends on y, a result coming from the fact that the shear strain must vanish.

The gradation in Young's modulus is shown in Fig. 14.3 for three different gradation cases with $K = -0.5, 0, 5$. These particular parameters give increasing, constant, and decreasing gradation with axial distance x.

The axial displacement behavior for these three gradation cases is shown in Fig. 14.4. As expected, a sheet with material having increasing stiffness (positive gradation, $K = -0.5$) would yield smaller displacements than a corresponding homogeneous sample. The opposite behavior is observed for a sheet with decreasing stiffness ($K = 5$).

Comparison results from this simple example are somewhat limited since the solution scheme started with the assumption that the inhomogeneous stress field coincided with the corresponding homogeneous solution. Thus, differences between the material models only developed in the strain and displacement fields. As we shall see in the coming problems, using a more general problem formulation and solution will produce completely different inhomogeneous stress, strain, and displacement fields.

14.2 Plane problem of a hollow cylindrical domain under uniform pressure

We start our study by re-examining Example 8.6, a plane axisymmetric problem of a hollow cylindrical domain under uniform internal and external pressure loadings, as shown in Fig. 14.5. Following the work of Horgan and Chan (1999a), we choose plane stress conditions and initially allow the modulus of elasticity and Poisson's ratio to be functions of the radial coordinate; that is, $E(r)$ and $\nu(r)$. In polar coordinates, the two nonzero normal stresses can then be expressed in terms of the radial displacement $u(r)$ by

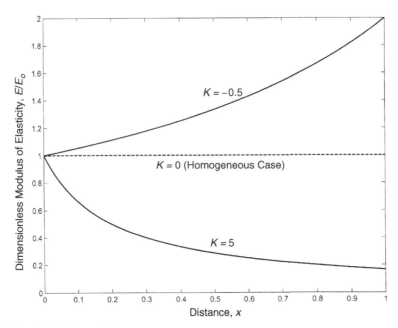

FIG. 14.3 Modulus of elasticity gradation.

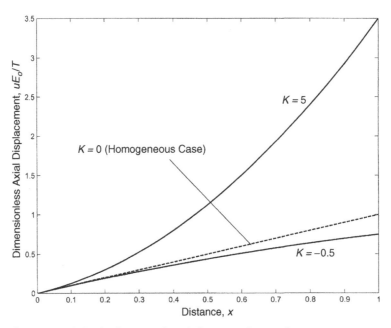

FIG. 14.4 Axial displacement behavior for several gradation cases for $y = 0$.

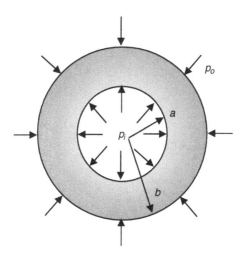

FIG. 14.5 Hollow cylindrical domain under uniform pressure.

$$\sigma_r = \frac{E(r)}{1 - \nu^2(r)} \left[\frac{du}{dr} + \nu(r) \frac{u}{r} \right]$$

$$\sigma_\theta = \frac{E(r)}{1 - \nu^2(r)} \left[\frac{u}{r} + \nu(r) \frac{du}{dr} \right]$$

(14.2.1)

Note that the corresponding plane strain relations can be determined by simple transformation of elastic moduli as given in Table 7.1.

Because it has been shown that variation in Poisson's ratio generally has less effect on the resulting stress than variation in Young's modulus, we now assume that $\nu(r)$ is a constant. Substituting (14.2.1) into the equations of equilibrium then generates the form of Navier's equation for this case

$$\frac{d^2u}{dr^2} + \frac{1}{r}\frac{du}{dr} - \frac{u}{r^2} + \frac{1}{E(r)}\frac{dE(r)}{dr}\left[\frac{du}{dr} + \nu\frac{u}{r}\right] = 0$$

(14.2.2)

This result should be compared to the previously developed relation (8.3.10) for the homogeneous case.

In order to develop a solvable equation, choose the specific power-law variation for Young's modulus

$$E(r) = E_o \left(\frac{r}{a}\right)^n$$

(14.2.3)

where E_o and n are constants and a is the inner boundary radius. Note that E_o has the same units as E and as $n \to 0$ we recover the homogeneous case. In order to gain insight into the relative magnitude of such a gradation, relation (14.2.3) is plotted in Fig. 14.6 for different values of power-law exponent. Note that the case $n = 1$ corresponds to linear variation in Young's modulus. It is observed that values of n greater than 1 produce quite substantial changes in elastic modulus. These particular parameters of the power-law exponent will be used in subsequent comparisons of the stress fields.

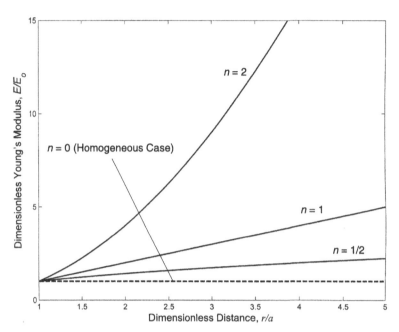

FIG. 14.6 Gradation in Young's modulus for different values of the power-law exponent.

This particular gradation model reduces the Navier equation (14.2.2) to

$$\frac{d^2u}{dr^2} + \frac{(n+1)}{r}\frac{du}{dr} + (n\nu - 1)\frac{u}{r^2} = 0 \tag{14.2.4}$$

The solution to equation (14.2.4) is given by

$$u = Ar^{-(n+k)/2} + Br^{(-n+k)/2} \tag{14.2.5}$$

where A and B are arbitrary constants and $k = \sqrt{n^2 + 4 - 4n\nu} > 0$. Substituting result (14.2.5) into relation (14.2.1) allows determination of the stresses.

Evaluation of the pressure boundary conditions $\sigma_r(a) = -p_i$ and $\sigma_r(b) = -p_o$ allows determination of the arbitrary constants A and B and produces the following stress field

$$\sigma_r = -\frac{a^{-n/2}b^{-n/2}r^{(-2-k+n)/2}}{b^k - a^k}\left[-a^{k+n/2}b^{(2+k)/2}p_o + a^{n/2}b^{(2+k)/2}p_o r^k + a^{(2+k)/2}b^{n/2}p_i\left(b^k - r^k\right)\right]$$

$$\sigma_\theta = \frac{a^{-n/2}b^{-n/2}r^{(-2-k+n)/2}}{b^k - a^k}\left[\frac{\left(a^{(2+k)/2}b^{n/2}p_i - a^{n/2}b^{(2+k)/2}p_o\right)r^k(2 + k\nu - n\nu)}{k - n + 2\nu}\right.$$

$$\left. + \frac{a^{k/2}b^{k/2}\left(-ab^{(k+n)/2}p_i + a^{(k+n)/2)}bp_o\right)(-2 + k\nu + n\nu)}{k + n - 2\nu}\right]$$

$$\tag{14.2.6}$$

As with the homogeneous example, we choose the special case with only internal pressure ($p_o = 0$), which gives the stresses

$$\sigma_r = \frac{p_i a^{(2+k-n)/2}}{b^k - a^k} \left[r^{(-2+k+n)/2} - b^k r^{(-2-k+n)/2} \right]$$

$$\sigma_\theta = \frac{p_i a^{(2+k-n)/2}}{b^k - a^k} \left[\frac{2 + k\nu - n\nu}{k - n + 2\nu} r^{(-2+k+n)/2} + \frac{2 - k\nu - n\nu}{k + n - 2\nu} b^k r^{(-2-k+n)/2} \right]$$

(14.2.7)

The homogeneous solution is found by letting $n \to 0$, which implies that $k \to 2$, giving the result

$$\sigma_r = \frac{p_i a^2}{b^2 - a^2} \left[1 - \frac{b^2}{r^2} \right]$$

$$\sigma_\theta = \frac{p_i a^2}{b^2 - a^2} \left[1 + \frac{b^2}{r^2} \right]$$

(14.2.8)

which matches with the solution shown in Fig. 8.9 for $b/a = 2$.

A plot of the nondimensional stress distributions through the thickness for various gradation cases is shown in Figs. 14.7 and 14.8 for the case of $b/a = 5$ and $\nu = 0.25$. Fig. 14.7 shows the variation in radial stress for different amounts of inhomogeneity reflected by choices of the power-law exponent n. Comparing the homogeneous case ($n = 0$) with increasing gradients of radial inhomogeneity illustrates that the radial stress is not significantly affected by this type of material grading. However, the

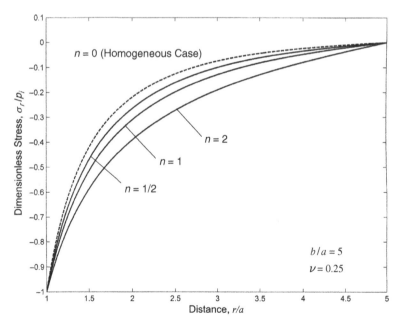

FIG. 14.7 Nondimensional radial stress distribution through domain wall.

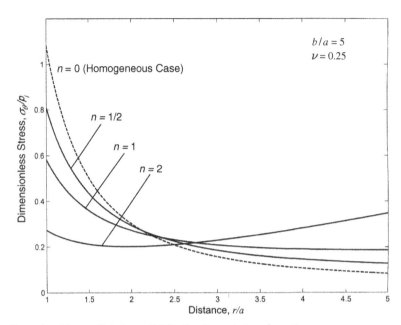

FIG. 14.8 Nondimensional tangential stress distribution through domain wall.

corresponding results for the tangential stress shown in Fig. 14.8 show much more marked differences. While the homogeneous hoop stress is always a monotonically decreasing function of the radial co-ordinate and has its maximum on the inner boundary ($r = a$), the inhomogeneous solid behaves quite differently. The graded material shows behaviors whereby the tangential stress can take a minimum value within the domain ($a < r < b$) and have a maximum value on the outer boundary ($r = b$). More details on these behaviors are given by Horgan and Chan (1999a).

Another interesting special case is that of external pressure loading only ($p_i = 0$). This solution can be easily developed from the general solution (14.2.6) and is given in the following relations (14.2.9)

$$\sigma_r = -\frac{p_o b^{(2+k-n)/2}}{b^k - a^k} \left[r^{(-2+k+n)/2} - a^k r^{(-2-k+n)/2} \right]$$

$$\sigma_\theta = -\frac{p_o b^{(2+k-n)/2}}{b^k - a^k} \left[\frac{2 + kv - nv}{k - n + 2v} r^{(-2+k+n)/2} + \frac{2 - kv - nv}{k + n - 2v} a^k r^{(-2-k+n)/2} \right]$$

(14.2.9)

The homogeneous solution is again found by letting $n \rightarrow 0$, giving the result

$$\sigma_r = -\frac{p_o b^2}{b^2 - a^2} \left[1 - \frac{a^2}{r^2} \right]$$

$$\sigma_\theta = -\frac{p_o b^2}{b^2 - a^2} \left[1 + \frac{a^2}{r^2} \right]$$

(14.2.10)

As in Section 8.4.2, the problem of a stress-free hole in an infinite medium under uniform far-field stress (see Fig. 8.11) can be obtained from (14.2.10) by letting $p_o \to -T$ and $b/a \to \infty$, which gives the result identical to the previous relation (8.4.9)

$$\sigma_r = T\left[1 - \frac{a^2}{r^2}\right], \quad \sigma_\theta = T\left[1 + \frac{a^2}{r^2}\right] \tag{14.2.11}$$

and thus produces the classic stress concentration factor of $K = 2$. The next logical step in our investigation would be to pursue the corresponding concentration effect for the inhomogeneous case using the same limiting process. However, attempting this on relations (14.2.9) *surprisingly fails* to produce satisfactory results because finite stresses for $b/a \to \infty$ require that $(2 + k - n)/2 = k \Rightarrow k = 2 - n$, which is precluded by the original definition $k = \sqrt{n^2 + 4 - 4n\nu}$ unless $n = 0$. Horgan and Chan (1999a) point out that this unexpected result is similar to findings in analogous problems involving certain curvilinear anisotropic materials (e.g., Galmudi and Dvorkin, 1995; Horgan and Baxter, 1996; see also Section 11.7).

Even with this analytical dilemma, we can still pursue the original stress concentration problem of interest (Fig. 8.11) by simply evaluating the general result (14.2.9) for the case with large b/a and $p_o \to -T$. Fig. 14.9 illustrates this case for the tangential stress behavior with $b/a = 20$, $\nu = 0.25$, and $n = -0.2, 0, 0.2, 0.4$, and 0.6. Since the domain includes large variation in the radial coordinate, we restricted the power-law exponent to cases where $n < 1$. It is observed that as n increases, the maximum hoop stress no longer occurs on the inner boundary $r = a$, thus reducing the local stress concentration effect. For negative values of the power-law exponent, the local stress on the hole

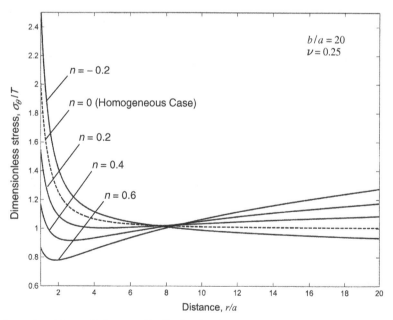

FIG. 14.9 Nondimensional tangential stress distribution for the external loading case with a large b/a.

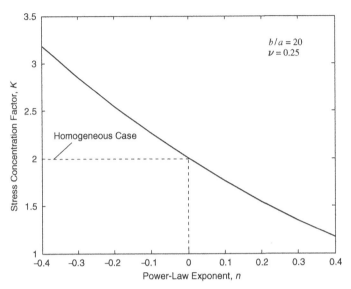

FIG. 14.10 Equivalent stress concentration factor for a small stress-free hole in a large domain under uniform biaxial tension.

boundary will be higher than the homogeneous case, thus creating an increase in the local stress concentration. Fig. 14.10 shows the behavior of the equivalent stress concentration factor, $K = \sigma_\theta(a)/T$, as a function of the power-law exponent over the range $-0.4 \leq n \leq 0.4$. The stress concentration exhibits a decreasing behavior with the gradation parameter. Similar stress-decreasing effects have also been found in studies of anisotropic circular tube problems (Galmudi and Dvorkin, 1995; Horgan and Baxter, 1996).

Clearly, the results in this section indicate that inhomogeneity can significantly alter the elastic stress distribution in such cylindrical domain problems. Stress concentration effects are also changed from corresponding homogeneous values. For this analysis, material inhomogeneity was modeled using a simple radial power-law relation for Young's modulus, and thus solution results were limited to correlations with the power-law exponent for cases with increasing or decreasing modulus.

14.3 Rotating disk problem

The next problem we wish to investigate is that of a solid circular disk (or cylinder) of radius a, rotating with constant angular velocity ω, as shown in Fig. 14.11. The disk is assumed to have zero tractions on its outer boundary, $r = a$. Recall that the solution to the homogeneous problem was developed in Example 8.11.

We follow the inhomogeneous formulation and solution scheme originally presented by Horgan and Chan (1999b). As before, we note that this is an axisymmetric plane problem in which all elastic fields are functions only of the radial coordinate. The rotation produces a centrifugal loading that can

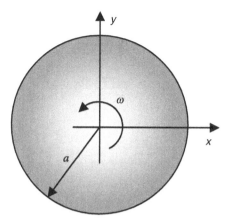

FIG. 14.11 Rotating disk problem.

be handled easily by including a radial body force, $F_r = \rho\omega^2 r$, where ρ is mass density. The only nonzero equilibrium equation then reduces to

$$\frac{d\sigma_r}{dr} + \frac{\sigma_r - \sigma_\theta}{r} + \rho\omega^2 r = 0 \qquad (14.3.1)$$

As in our previous example, we choose plane stress conditions and allow the modulus of elasticity and Poisson's ratio to be functions of the radial coordinate: $E(r)$ and $\nu(r)$. Using Hooke's law in polar coordinates, the two nonzero normal stresses can then again be expressed in terms of the radial displacement $u(r)$ by

$$\sigma_r = \frac{E(r)}{1 - \nu^2(r)}\left[\frac{du}{dr} + \nu(r)\frac{u}{r}\right]$$

$$\sigma_\theta = \frac{E(r)}{1 - \nu^2(r)}\left[\frac{u}{r} + \nu(r)\frac{du}{dr}\right] \qquad (14.3.2)$$

As before, the corresponding plane strain relations can be determined by simple transformation of elastic moduli, as given in Table 7.1.

Again, following similar logic to that in the previous example, variation in Poisson's ratio is of much less significance than Young's modulus, and thus we assume that $\nu(r) = \nu = $ constant. Substituting Eq. (14.3.2) into Eq. (14.3.1) then generates Navier's equation for this case

$$\frac{d^2u}{dr^2} + \frac{1}{r}\frac{du}{dr} - \frac{u}{r^2} + \frac{1}{E(r)}\frac{dE(r)}{dr}\left[\frac{du}{dr} + \nu\frac{u}{r}\right] = -\frac{\rho(1 - \nu^2)\omega^2 r}{E(r)} \qquad (14.3.3)$$

As in our previous pressurized tube example, a power-law distribution for Young's modulus will help reduce the complexity of Navier's equation. On this basis, we again choose a modulus variation of the form (identical to (14.2.3))

$$E(r) = E_o\left(\frac{r}{a}\right)^n \qquad (14.3.4)$$

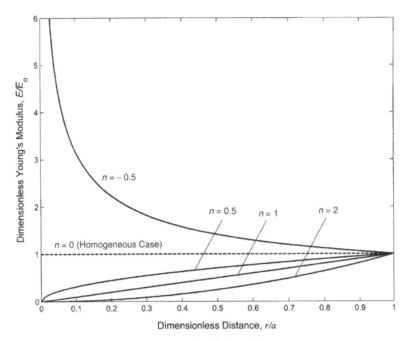

FIG. 14.12 Gradation in Young's modulus for the rotating disk problem.

where E_o and n are constants and a is the outer boundary radius. As before, E_o has the same units as E and as $n \to 0$ we recover the homogeneous case. Various moduli distributions from this form are shown in Fig. 14.12 for different values of the power-law exponent, $n = -0.5, 0, 0.5, 1, 2$. Notice that for cases with $n > 0$, the gradation increases from 0 at the disk center to E_o at the outer boundary. For the case with $n < 0$, the modulus is unbounded at the disk's center, and thus we expect a similar singularity in the stress field for this case.

Using this gradation model, relation (14.3.3) reduces to

$$\frac{d^2u}{dr^2} + \frac{(n+1)}{r}\frac{du}{dr} + (nv-1)\frac{u}{r^2} = -\frac{\rho(1-v^2)\omega^2 r^{1-n}a^n}{E_o} \tag{14.3.5}$$

The general solution to Eq. (14.3.5) follows from standard theory as the sum of homogeneous plus particular solutions. Note that the solution to the homogeneous equation was given by (14.2.5). Combining these results gives

$$u(r) = Ar^{-(n+k)/2} + Br^{(-n+k)/2} - \frac{\rho(1-v^2)\omega^2 a^n}{E_o(nv-3n+8)}r^{3-n} \tag{14.3.6}$$

where A and B are arbitrary constants and $k = \sqrt{(n^2+4-4nv)} > 0$. Solution (14.3.6) requires the following

$$nv - 3n + 8 \neq 0 \quad \text{or} \quad (n+k)/2 \neq 3 \tag{14.3.7}$$

For the case where (14.3.7) does not hold [i.e., (14.3.7) with equality signs], the particular solution must be fundamentally modified, giving the result

$$u(r) = A^* r^{-3} + B^* r^{3-n} - \frac{\rho(1 - \nu^2)\omega^2 a^n}{E_o} \log(r/a) \tag{14.3.8}$$

Relation (14.3.7) with equality signs yields $n = 8/(3 - \nu)$, which implies the interesting fact that for $0 \le \nu \le 1/2$, the gradation power-law exponent must be in the range

$$\frac{8}{3} \le n \le \frac{16}{5} \tag{14.3.9}$$

We now proceed to determine the displacement and stress fields for particular gradation cases invoking bounded solutions at the origin ($r = 0$) and zero tractions ($\sigma_r = 0$) at $r = a$.

Consider first the case with $n > 0$ so that the modulus gradation increases monotonically with radial coordinate. Because now $n + k > 0$, the boundedness condition at the origin requires that $A = 0$, and the third term in solution (14.3.6) implies that $n \le 3$, thus restricting the power-law exponent to the range $0 < n \le 3$. To satisfy the traction-free boundary condition, our solution (14.3.6) must retain the B term, which means that $k - n > 0$, and this will happen only if $1 - n\nu > 0$. Additionally, we must honor the usual range restriction on Poisson's ratio, $0 \le \nu \le 1/2$. Collectively, these conditions place coupled restrictions on the gradation parameter n and Poisson's ratio ν, but these can all be satisfied. Therefore, we now move forward with the solution assuming this is the case.

With $A = 0$ in solution (14.3.6), the radial stress follows from relation (14.3.2)$_1$

$$\sigma_r = \frac{Ba^{-n}E_o}{1 - \nu^2}\left[\nu + \frac{1}{2}(k - n)\right] r^{(n+k-2)/2} - \frac{(3 - n + \nu)\rho\omega^2 r^2}{(n\nu - 3n + 8)} \tag{14.3.10}$$

Applying the zero-traction boundary condition easily determines the constant B, giving the final stress results

$$\sigma_r = \frac{(3 - n + \nu)\rho\omega^2}{(n\nu - 3n + 8)}\left[a^{3-(n+k)/2} r^{(n+k-2)/2} - r^2\right]$$

$$\sigma_\theta = \frac{\rho\omega^2}{(n\nu - 3n + 8)}\left[\frac{(3 - n + \nu)[1 + (\nu/2)(k - n)]}{[\nu + (k - n)/2]} a^{3-(n+k)/2} r^{(n+k-2)/2} - [(3 - n)\nu + 1]r^2\right] \tag{14.3.11}$$

for the case $n\nu - 3n + 8 \ne 0$.

Recall that these results were developed under condition (14.3.7). For the case where this condition is not satisfied, solution form (14.3.8) must be used instead. Similar analysis yields the stress solutions

$$\sigma_r = \rho\omega^2\left[\frac{(n - 4)(n - 2)}{n} r^2 \log(r/a)\right]$$

$$\sigma_\theta = \frac{\rho\omega^2}{n}\left[8r^2 - 3(n - 4)(n - 2)r^2 \log(r/a)\right] \tag{14.3.12}$$

for the case $\frac{8}{3} \le n \le \frac{16}{5}$

The homogeneous case corresponds to $n \to 0$, which means $k \to 2$, and (14.3.11) would reduce to our previous solution (8.4.81).

Returning now to consider the case where $n < 0$ (decreasing radial gradation), we find that the boundedness condition implies that the solution constant A must again be set to 0. However, no restriction is needed on the power of r in the third term of solution (14.3.6). Consideration of the special solution (14.3.8) is no longer needed because $n\nu - 3n + 8 \neq 0$ will always be satisfied. Thus, the solution given by (14.3.11) is also valid for $n < 0$.

To show details on the stress distribution, we will choose the case $\nu = 0$ so that no restriction is placed on the condition that $k - n > 0$, and additionally plane stress results coincide with plane strain values. We also take the case $8 - 3n \neq 0$ and thus relations (14.3.11) are used to determine the stresses. Results for the radial stress distribution are shown in Fig. 14.13 for gradation parameter values $n = -0.5, 0, 0.5, 1, 2$. The results indicate very significant differences in behavior from the homogeneous case $n = 0$, which has its maximum value at the disk center ($r = 0$) and decays to 0 at the outer stress-free boundary. For inhomogeneous cases with positive gradations (e.g., $n = 0.5, 1, 2$), the radial stress actually vanishes at $r = 0$ and at $r = a$, and thus takes on a maximum value at an interior point within the interval $0 < r < a$. It can be shown that this maximum occurs at

$$r = r_m = a\left(\frac{2}{M-1}\right)^{1/(M-3)}, \quad \text{where } M = \frac{1}{2}\left(n + \sqrt{n^2 + 4}\right) \qquad (14.3.13)$$

Note that for the homogeneous case, as $n \to 0, M \to 1$ and $r_m \to 0$, and as $n \to 8/3, M \to 3$ and $r_m \to a$. Also note that for positive gradations ($n > 0$) at fixed r, the radial stress decreases monotonically with the gradation parameter n.

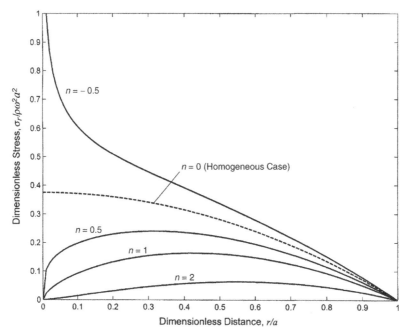

FIG. 14.13 Radial stress distribution in a rotating disk ($\nu = 0$).

For negative values of gradation parameter, the radial stress distribution drastically changes its behavior and is actually unbounded at the origin. The case for $n = -0.5$ is shown in Fig. 14.13, and it is observed that the stress drops rapidly from its singular value at $r = 0$ to 0 at the outer boundary $r = a$. As pointed out by Horgan and Chan (1999b), similar singular behaviors have been found in the analogous problem for homogeneous radially orthotropic materials (Horgan and Baxter, 1996; see Section 11.7) that are due to an *anisotropic focusing* effect at the origin. For the present inhomogeneous case, the singularity corresponds to the unbounded Young's modulus as $r \to 0$ with $n < 0$ (see Fig. 14.12).

It is interesting that the location of maximum stress can actually be controlled by appropriate modulus gradation [see, for example, relation (14.3.13)]. We also find this situation in the torsion problem to be discussed later in the chapter. Also, as discussed by Horgan and Chan (1999b), a design criterion for a disk of uniform stress proposes that

$$\sigma_r = \sigma_\theta, \quad r \in (0, a) \tag{14.3.14}$$

which can be accomplished by a gradation of the form

$$\frac{E(r)}{1 + \nu(r)} = K\rho\omega^2 (a^2 - r^2) \tag{14.3.15}$$

where K is an arbitrary constant.

As shown by Horgan and Chan (1999b), the hoop stress σ_θ for the disk problem has similar behaviors; a detailed plot of this component has been left as an exercise.

14.4 Point force on the free surface of a half-space

By far the most studied inhomogeneous elasticity problem is the half-space domain under point or distributed loadings applied to the free surface. Over the past several decades this problem has received considerable attention; examples include Holl (1940), Lekhnitskii (1961), Gibson (1967), Gibson and Sills (1969), Kassir (1972), Awojobi and Gibson (1973), Carrier and Christian (1973), Calladine and Greenwood (1978), Booker, Balaam, and Davis (1985), Oner (1990), Giannakopoulos and Suresh (1997), and Vrettos (1998). Wang et al. (2003) provide an excellent literature review of previous work. Early applications of these studies were in the field of geomechanics, where the depth variation in the elastic response of soils was investigated.

More current applications involved creating functionally graded materials (FGMs) with depth-dependent properties to provide high surface hardness/stiffness while allowing for softer/tougher core material. Solutions to this type of problem have typically been either for the two-dimensional plane strain/plane stress case or for the three-dimensional axisymmetric geometry. Inhomogeneity has normally included elastic moduli variation with depth coordinates into the elastic half-space using forms similar to relations (14.1.7). Many problems with varying degrees of complexity in either the loading or moduli variation have been solved. We will explore one of the more basic and simple solutions that provide fundamental insight into the effect of inhomogeneity on the stress and displacement fields.

Following the work of Booker et al. (1985), we first explore the two-dimensional plane strain solution of an inhomogeneous half-space with depth-dependent elastic modulus carrying the surface

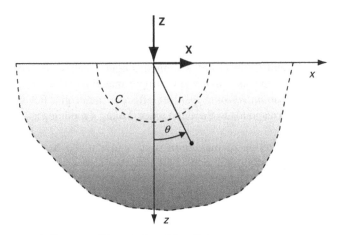

FIG. 14.14 Inhomogeneous half-space with graded modulus E(z).

point loadings, as shown in Fig. 14.14. Of course, this problem is the counterpart of the Flamant solution given for the homogeneous case in Section 8.4.7. Similar to the previous example, the particular variation in Young's modulus is prescribed by the power-law relation

$$E = E_o z^\alpha \qquad (14.4.1)$$

where E_o and α are positive constants. Again, note that E_o has the same units as E and as $\alpha \to 0$ we recover the homogeneous case. The general graded behavior of form (14.4.1) is identical to that shown in Fig. 14.6. As before, stress and displacement fields commonly show little variation with Poisson's ratio, and thus ν is to be kept constant.

The problem is formulated and solved using polar coordinates r and θ, as shown in Fig. 14.14. Reviewing the solution forms from our previous homogeneous case in Section 8.4.7, we follow a semi-inverse solution scheme by proposing a somewhat similar form for the nonhomogeneous stress and displacement fields

$$\sigma_r = \frac{S_r(\theta)}{r}, \quad \sigma_\theta = \frac{S_\theta(\theta)}{r}, \quad \tau_{r\theta} = \frac{S_{r\theta}(\theta)}{r}$$

$$u_r = \frac{U_r(\theta)}{r^\alpha}, \quad u_\theta = \frac{U_\theta(\theta)}{r^\alpha} \qquad (14.4.2)$$

where S_r, S_θ, $S_{r\theta}$, U_r, and U_θ are functions to be determined and α is the power-law exponent from relation (14.4.1). Using these forms, the equilibrium equations (7.6.3) with no body force produce

$$\frac{dS_{r\theta}}{d\theta} - S_\theta = 0$$

$$\frac{dS_\theta}{d\theta} + S_{r\theta} = 0 \qquad (14.4.3)$$

These relations can be easily solved, giving the results

$$S_{r\theta} = A \cos\theta + B \sin\theta$$

$$S_\theta = -A \sin\theta + B \cos\theta \tag{14.4.4}$$

where A and B are arbitrary constants. On the free surface, the stress-free boundary conditions $\sigma_\theta(r, \ 0) = \tau_{r\theta}(r, \ 0) = 0$ and $\sigma_\theta(r, \ \pi) = \tau_{r\theta}(r, \ \pi) = 0$ greatly simplify the solution forms. These conditions imply that A and B vanish and thus $S_{r\theta}$ and S_θ are also zero, indicating that σ_r is the only nonzero stress. This type of stress field is commonly referred to as a *radial stress distribution*. Similar findings were also found in the homogeneous solution.

Combining Hooke's law (plane strain case) with the strain displacement relations gives

$$\frac{\partial u_r}{\partial r} = \frac{1+\nu}{E}[(1-\nu)\sigma_r - \nu\sigma_\theta]$$

$$\frac{1}{r}\left(u_r + \frac{\partial u_\theta}{\partial \theta}\right) = \frac{1+\nu}{E}[(1-\nu)\sigma_\theta - \nu\sigma_r] \tag{14.4.5}$$

$$\frac{1}{r}\frac{\partial u_r}{\partial \theta} + \frac{\partial u_\theta}{\partial r} - \frac{u_\theta}{r} = \frac{2(1+\nu)}{E}\tau_{r\theta}$$

Substituting our assumed forms (14.4.2) into (14.4.5) and using the fact that $S_{r\theta} = S_\theta = 0$ yields

$$-\alpha U_r = \frac{1-\nu^2}{E_o \cos^\alpha\theta} S_r$$

$$\frac{dU_\theta}{d\theta} + U_r = -\frac{(1+\nu)\nu}{E_o \cos^\alpha\theta} S_r \tag{14.4.6}$$

$$\frac{dU_r}{d\theta} - (1+\alpha)U_\theta = 0$$

Relations (14.4.6) represent three linear differential equations for the unknowns S_r, U_r, and U_θ. The system can be reduced to a single equation in terms of a single unknown and then solved. This result may then be back-substituted to determine the remaining unknowns. The final solution results are found to be

$$S_r = -\cos^\alpha\theta\left[C_1 \cos\beta\theta + \frac{1+\alpha}{\beta} C_2 \sin\beta\theta\right]$$

$$U_r = \frac{1-\nu^2}{E_o\alpha}\left[C_1 \cos\beta\theta + \frac{1+\alpha}{\beta} C_2 \sin\beta\theta\right] \tag{14.4.7}$$

$$U_\theta = \frac{1-\nu^2}{E_o\alpha}\left[-\frac{\beta}{1+\alpha} C_1 \sin\beta\theta + C_2 \cos\beta\theta\right]$$

where C_1 and C_2 are arbitrary constants and β is a parameter given by

$$\beta^2 = (1+\alpha)\left[1 - \frac{\alpha\nu}{1-\nu}\right] \tag{14.4.8}$$

As with the homogeneous case, the constants C_1 and C_2 are determined by applying force equilibrium over the semicircular arc C of radius a, as shown in Fig. 14.14. Similar to relations (8.4.33), we may write the force balance as

$$Z = -\int_{-\pi/2}^{\pi/2} \sigma_r(a, \theta) a \cos\theta d\theta = -\int_{-\pi/2}^{\pi/2} S_r(\theta) \cos\theta d\theta = C_1/F_{\alpha\beta}$$

$$X = -\int_{-\pi/2}^{\pi/2} \sigma_r(a, \theta) a \sin\theta d\theta = -\int_{-\pi/2}^{\pi/2} S_r(\theta) \sin\theta d\theta = C_2/F_{\alpha\beta}$$

(14.4.9)

and thus $C_1 = ZF_{\alpha\beta}$ and $C_2 = XF_{\alpha\beta}$, where

$$F_{\alpha\beta} = \frac{2^{(1+\alpha)}(2+\alpha)}{\pi} \frac{\Gamma((3+\alpha+\beta)/2)\Gamma((3+\alpha-\beta)/2)}{\Gamma(3+\alpha)}$$

(14.4.10)

and $\Gamma(\cdot)$ is the *gamma function* defined by $\Gamma(z) = \int_0^\infty e^{-t} t^{z-1} dt$.

The solution is now complete, and the homogeneous case can be extracted by letting $\alpha \to 0$, which yields $\beta = 1$, $F_{\alpha\beta} = 2/\pi$, and the solution reduces to our previous result (8.4.34).

Let us now evaluate the special inhomogeneous case with only normal loading ($X = 0$) and explore the nature of the resulting stress and displacement fields. The solution for this case is given by

$$\sigma_r = -\frac{\cos^\alpha\theta}{r} ZF_{\alpha\beta} \cos\beta\theta$$

$$u_r = \frac{(1-\nu^2)}{E_o \alpha r^\alpha} ZF_{\alpha\beta} \cos\beta\theta$$

(14.4.11)

$$u_\theta = -\frac{(1-\nu^2)\beta}{E_o \alpha r^\alpha (1+\alpha)} ZF_{\alpha\beta} \sin\beta\theta$$

Results (14.4.11) are shown in Figs. 14.15–14.18, with $\nu = 0.25$ for three different values of the power-law exponent including the homogeneous case ($\alpha = 0$). Such results illustrate the effect of inhomogeneity on the resulting elastic fields. Clearly, the degree of nonhomogeneity has a significant influence on these stress and displacement distributions. Figs. 14.15 and 14.16 illustrate that, at a fixed depth into the half-space, both radial stress and displacement directly under the surface loading ($\theta = 0$) will increase with higher values of α. On the other hand, Fig. 14.17 indicates that the magnitude of the tangential displacement will get smaller as the inhomogeneity variation is increased. Radial stress contours are shown in Fig. 14.18 for both the homogeneous ($\alpha = 0$) and nonhomogeneous ($\alpha = 1$) cases. For the inhomogeneous material, the stress contours become elongated and thus are no longer circular, as demonstrated for the homogeneous case.

The corresponding three-dimensional problem shown in Fig. 14.19 has also been developed by Booker et al. (1985). The formulation was constructed using spherical coordinates in much the same manner as the previous two-dimensional problem. Following an axisymmetric, semi-inverse solution scheme, we assume that $u_\theta = 0$, $\tau_{\phi\theta} = \tau_{R\theta} = 0$ and the nonzero stresses and displacements are given by

$$\sigma_R = \frac{S_R(\phi)}{R^2}, \quad \sigma_\phi = \frac{S_\phi(\phi)}{R^2}, \quad \sigma_\theta = \frac{S_\theta(\phi)}{R^2}, \quad \tau_{R\phi} = \frac{S_{R\phi}(\phi)}{R^2}$$

$$U_R = \frac{U_R(\phi)}{R^{1+\alpha}}, \quad U_\phi = \frac{U_\phi(\phi)}{R^{1+\alpha}}$$

(14.4.12)

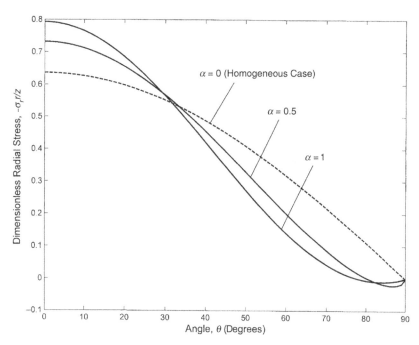

FIG. 14.15 Radial stress distribution in a nonhomogeneous half-space.

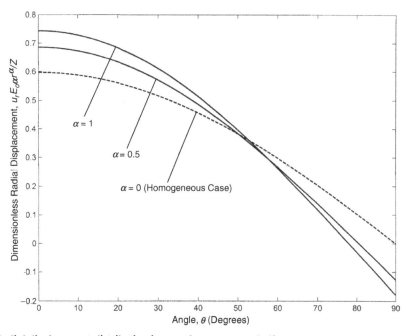

FIG. 14.16 Radial displacement distribution in a nonhomogeneous half-space.

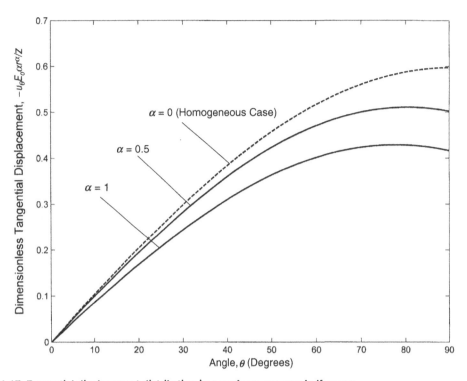

FIG. 14.17 Tangential displacement distribution in a nonhomogeneous half-space.

where S_R, S_ϕ, S_θ, $S_{R\phi}$, U_R, and U_ϕ are functions to be determined and α is the power-law exponent from relation (14.4.1). Using these forms, the equilibrium equations in spherical coordinates (A.6), with no body force, reduce to

$$\frac{dS_{R\phi}}{d\phi} + S_{R\phi} \cot \phi - S_\phi - S_\theta = 0$$

$$\frac{dS_\phi}{d\phi} + (S_\phi - S_\theta) \cot \phi + S_{R\phi} = 0$$

(14.4.13)

Furthermore, by considering the equilibrium of a conical volume bounded by $R_1 \le R \le R_2$, $\phi = \phi_o$, it can be shown that $S_{R\phi}$ and S_ϕ are related and can be expressed by

$$S_{R\phi} = S(\phi)\sin \phi$$

$$S_\phi = S(\phi)\cos \phi$$

(14.4.14)

Combining Hooke's law with the strain−displacement relations again gives a system of relations similar to the previous result (14.4.6) for the two-dimensional case. Unfortunately, the relations for the three-dimensional problem cannot be integrated analytically, and thus final results were generated by Booker et al. (1985) using numerical methods. The radial stress results are shown in Fig. 14.20, with

$\nu = 0.25$ for several cases of power-law exponent, and again the homogeneous case ($\alpha = 0$) corresponds to the Boussinesq solution given in Examples 13.2 and 13.4. Note the similarity of the three-dimensional results with those of the two-dimensional model shown in Fig. 14.15.

Oner (1990) has also provided a solution to the point-force problem shown in Fig. 14.19 but for the power-law gradation in terms of the shear modulus

$$\mu = \mu_o z^n \tag{14.4.15}$$

For this type of inhomogeneity, the resulting Cartesian stress and displacement fields are found to be

$$\sigma_x = \frac{(n+3)P}{2\pi z^2} \cos^{(n+3)}\phi \sin^2\phi \cos^2\theta$$

$$\sigma_y = \frac{(n+3)P}{2\pi z^2} \cos^{(n+3)}\phi \sin^2\phi \sin^2\theta$$

$$\sigma_z = \frac{(n+3)P}{2\pi z^2} \cos^{(n+5)}\phi$$

$$\tau_{xy} = \frac{(n+3)P}{2\pi z^2} \cos^{(n+3)}\phi \sin^2\phi \sin\theta \cos\theta \tag{14.4.16}$$

$$\tau_{yz} = \frac{(n+3)P}{2\pi z^2} \cos^{(n+4)}\phi \sin\phi \sin\theta$$

$$\tau_{zx} = \frac{(n+3)P}{2\pi z^2} \cos^{(n+4)}\phi \sin\phi \cos\theta$$

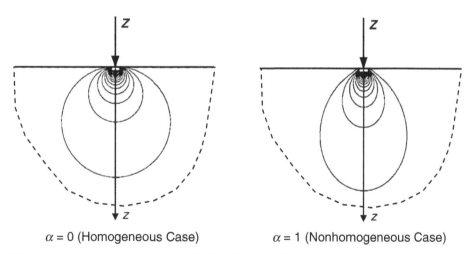

$\alpha = 0$ (Homogeneous Case) $\alpha = 1$ (Nonhomogeneous Case)

FIG. 14.18 Radial stress contour comparisons for the Flamant problem.

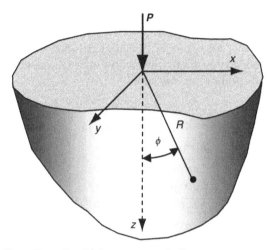

FIG. 14.19 Point load on a three-dimensional inhomogeneous half-space.

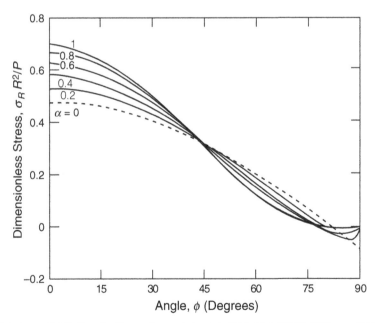

FIG. 14.20 Radial stress distribution for the nonhomogeneous point load problem, with $\nu = 0.25$.

From Booker JR, Balaam NP, Davis EH: The behavior of an elastic non-homogeneous half-space. Part I. Line and point loads, Int J Numer Anal Methods Geomech *9:353–367, 1985; reprinted with permission of John Wiley & Sons.*

$$u = \frac{P}{4\pi\mu_o} \frac{xz}{\left(x^2 + y^2 + z^2\right)^{(n+3)/2}}$$

$$v = \frac{P}{4\pi\mu_o} \frac{yz}{\left(x^2 + y^2 + z^2\right)^{(n+3)/2}}$$

(14.4.17)

$$w = \frac{P}{4\pi\mu_o} \frac{x^2 + y^2 + z^2(n+2)}{(1+n)\left(x^2 + y^2 + z^2\right)^{(n+3)/2}}$$

These results are developed under the power-law exponent restriction $n = (1/\nu) - 2$. Note that the homogeneous case corresponds to $n = 0$, which implies that $\nu = 1/2$ (incompressible case). Under these conditions, the inhomogeneous results (14.4.16) and (14.4.17) reduce to the homogeneous solution developed in Exercise 13.9.

The distribution of the normal stress σ_z is shown in Fig. 14.21 for several cases of the power-law exponent, $n = 0, 1/3, 2/3, 1$. These results indicate behavior similar to that shown in Fig. 14.20 for a

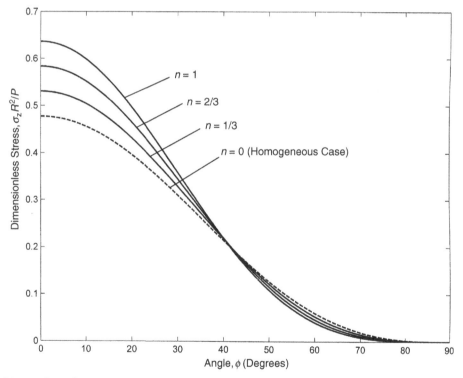

FIG. 14.21 Nondimensional normal stress distribution σ_z for the inhomogeneous point load problem.

See Oner M. Vertical and horizontal deformation of an inhomogeneous elastic half-space, Int J Numer Anal Methods Geomech 14:

613–629, 1990.

comparable inhomogeneous problem. Directly under the loading ($\phi = 0$), the normal stress σ_z increases with increasing inhomogeneity gradient, a result consistent with the findings shown in Figs. 14.15 and 14.20. Oner also presents stress and displacement results for the horizontal point loading (Cerruti problem) shown in Fig. 13.3.

14.5 Antiplane strain problems

Because of its relatively simple formulation, antiplane strain problems have been investigated for nonhomogeneous material. Examples of such work include Clements et al. (1978), Dhaliwal and Singh (1978), Delale (1985), Ang and Clements (1987), Erdogan and Ozturk (1992), Horgan and Miller (1994), Clements, Kusuma, and Ang (1997), and Spencer and Selvadurai (1998). Much of this work has been applied to crack problems related to mode III fracture behaviors. For the homogeneous case, this crack problem was given as Exercise 8.44. The homogeneous formulation of antiplane strain was given in Section 7.4, and we now develop the corresponding inhomogeneous formulation for a particular class of material gradation.

Antiplane strain is based on the existence of only out-of-plane deformation, and thus with respect to a Cartesian coordinate system the assumed displacement field can be written as

$$u = v = 0, \quad w = w(x, y) \tag{14.5.1}$$

This yields the following strains

$$e_x = e_y = e_z = e_{xy} = 0$$

$$e_{xz} = \frac{1}{2}\frac{\partial w}{\partial x}, \quad e_{yz} = \frac{1}{2}\frac{\partial w}{\partial y} \tag{14.5.2}$$

Using Hooke's law, the stresses become

$$\sigma_x = \sigma_y = \sigma_z = \tau_{xy} = 0$$

$$\tau_{xz} = \mu\frac{\partial w}{\partial x}, \quad \tau_{yz} = \mu\frac{\partial w}{\partial y} \tag{14.5.3}$$

Thus, in the absence of body forces, the equilibrium equations reduce to the single equation

$$\frac{\partial}{\partial x}\left(\mu\frac{\partial w}{\partial x}\right) + \frac{\partial}{\partial y}\left(\mu\frac{\partial w}{\partial y}\right) = 0 \tag{14.5.4}$$

where the shear modulus is assumed to be a function of the in-plane coordinates $\mu = \mu(x,y)$.

Following the solution procedure outlined by Dhaliwal and Singh (1978), the transformation

$$w(x, y) = W(x, y)\big/\sqrt{\mu(x, y)} \tag{14.5.5}$$

can be used to reduce relation (14.5.4) to

$$\mu\nabla^2 W + \frac{1}{2}\left(\frac{1}{2\mu}\left(\mu_x^2 + \mu_y^2\right) - \mu_{xx} - \mu_{yy}\right)W = 0 \tag{14.5.6}$$

where $\mu_x = \dfrac{\partial\mu}{\partial x}, \mu_y = \dfrac{\partial\mu}{\partial y}, \mu_{xx} = \dfrac{\partial^2\mu}{\partial x^2}, \mu_{yy} = \dfrac{\partial^2\mu}{\partial y^2}.$

We now assume separable product forms for W and μ such that

$$W(x,y) = X(x)Y(y), \quad \mu(x,y) = \mu_o p(x)q(y) \tag{14.5.7}$$

where μ_o is a constant. Substituting (14.5.7) into (14.5.6) yields a separable relation that can be written as two equations

$$X_{xx} + \left[n^2 + \frac{1}{4}\left(\frac{p_x}{p}\right)^2 - \frac{1}{2}\left(\frac{p_{xx}}{p}\right) \right] X = 0$$

$$Y_{yy} + \left[-n^2 + \frac{1}{4}\left(\frac{q_y}{q}\right)^2 - \frac{1}{2}\left(\frac{q_{yy}}{q}\right) \right] Y = 0 \tag{14.5.8}$$

where n^2 is the separation constant and subscripts indicate partial differentiation as before.

In order to proceed further with an analytical solution, particular choices of the material gradation functions p and q must be made. In particular, we choose the case where

$$\frac{1}{2p}\frac{d^2p}{dx^2} - \frac{1}{4}\left(\frac{1}{p}\frac{dp}{dx}\right)^2 = a_o$$

$$\frac{1}{2q}\frac{d^2q}{dy^2} - \frac{1}{4}\left(\frac{1}{q}\frac{dq}{dy}\right)^2 = b_o \tag{14.5.9}$$

where a_o and b_o are constants; relations (14.5.8) then reduce to

$$X_{xx} + k^2 X = 0$$

$$Y_{yy} - \left(a_o + b_o + k^2\right)Y = 0 \tag{14.5.10}$$

where $k^2 = n^2 - a_o$. Combining the previous results, the general bounded solution to governing equation (14.5.4) for, say, the domain $y \geq 0$, can be written in a Fourier integral form

$$w(x,y) = \frac{1}{\sqrt{\mu}} \int_0^\infty \left[A(\xi)\cos(x\xi) + B(\xi)\sin(x\xi)\right] e^{-\sqrt{a_o + b_o + \xi^2}\, y} d\xi \tag{14.5.11}$$

under the conditions that $k^2 = n^2 - a_o \geq 0$ and $a_o + b_o + k^2 \geq 0$, and where $A(\xi)$ and $B(\xi)$ are arbitrary functions of ξ.

Again, following the work of Dhaliwal and Singh (1978), an application of this solution method can be applied to an unbounded medium containing a crack located at $-1 \leq x \leq 1$, $y = 0$ (see Fig. 14.22). The crack surfaces are to be loaded under self-equilibrated uniform, out-of-plane shear stress S. For this problem the boundary conditions can thus be written as

$$w(x,0) = 0, \quad |x| > 1$$

$$\tau_{yz}(x,0) = S, \quad |x| < 1$$

$$\tau_{yz}(x,y) \to 0 \text{ as } r \to \infty \tag{14.5.12}$$

For the particular material inhomogeneity, we choose $p(x) = e^{\alpha|x|}$ and $q(y) = e^{\beta|y|}$, and thus

$$\mu(x,y) = \mu_o \exp(\alpha|x| + \beta|y|) \tag{14.5.13}$$

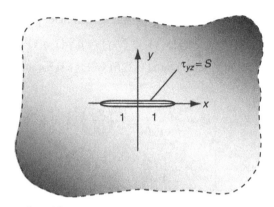

FIG. 14.22 Antiplane strain crack problem.

where α and β are constants. For this choice, $p(x)$ and $q(y)$ satisfy conditions (14.5.9) with $a_o = \alpha^2/4$ and $b_o = \beta^2/4$; our bounded solution scheme then gives

$$w = \frac{1}{\sqrt{\mu}} \int_0^\infty A(\xi)e^{-s(\xi)|y|} \cos(x\xi)d\xi$$

$$\tau_{yz} = -\sqrt{\mu} \int_0^\infty [\beta/2 + s(\xi)]A(\xi)e^{-s(\xi)|y|} \cos(x\xi)d\xi$$

(14.5.14)

where $s(\xi) = \sqrt{\xi^2 + (\alpha^2 + \beta^2)/4}$.

The two boundary conditions $(14.5.12)_{1,2}$ imply that

$$\int_0^\infty A(\xi)\cos(x\xi)d\xi = 0, \quad x > 1$$

$$\int_0^\infty [\beta/2 + s(\xi)]A(\xi)\cos(x\xi)d\xi = -\frac{S}{\sqrt{\mu_o}} e^{-\alpha x/2}, \quad 0 < x < 1$$

(14.5.15)

For some function $F(t)$, we can write an integral representation for $A(\xi)$ in the form

$$A(\xi) = \int_0^1 F(t)J_o(\xi t)dt$$

(14.5.16)

where J_o is the zero-order Bessel function of the first kind. By using the result

$$\int_0^\infty J_o(\xi t)\cos(x\xi)d\xi = \begin{cases} 0, & x > t \\ (t^2 - x^2)^{-1/2}, & 0 \le x < t \end{cases}$$

(14.5.17)

it can be shown that relation $(14.5.15)_1$ will be identically satisfied. The remaining boundary condition $(14.5.15)_2$ can be rewritten as

$$\int_0^\infty [\xi + G(\xi)]A(\xi)\cos(x\xi)d\xi = -\frac{S}{\sqrt{\mu_o}}e^{-\alpha x/2}, \quad 0 < x < 1 \tag{14.5.18}$$

where $G(\xi) = s(\xi) - \xi + \beta/2 = \sqrt{\xi^2 + (\alpha^2 + \beta^2)}\big/4 - \xi + \beta/2$.

Using results (14.5.16) and (14.5.17), relation (14.5.18) can be expressed as the following integral equation

$$F(t) + \int_0^1 F(x)K(x,t)dx = g(t), \quad 0 < t < 1 \tag{14.5.19}$$

with

$$g(t) = -\frac{2tS}{\pi\sqrt{\mu_o}}\int_0^1 \frac{e^{-\alpha x/2}}{\sqrt{t^2 - x^2}}dx = \frac{tS}{\sqrt{\mu_o}}[L_o(\alpha t/2) - I_o(\alpha t/2)]$$

$$K(x,t) = t\int_0^\infty G(\xi)J_o(\xi t)J_o(xt)d\xi \tag{14.5.20}$$

and $I_o(\cdot)$ is the modified Bessel function of the first kind of zero order and $L_o(\cdot)$ is the modified Struve function of zero order (see Abramowitz and Stegun, 1964).

Combining the previous results, the shear stress on the x-axis can be expressed as

$$\tau_{yz}(x,0) = \frac{\sqrt{\mu_o}}{x}e^{-\alpha x/2}\left[\frac{F(1)}{\sqrt{x^2 - 1}} - \int_0^1 \frac{tF'(t)}{\sqrt{x^2 - t^2}}dt\right]$$

$$-\sqrt{\mu_o}e^{-\alpha x/2}\int_0^1 F(t)dt\int_0^\infty G(\xi)J_o(\xi t)\cos(x\xi)d\xi, \quad x > 1 \tag{14.5.21}$$

As mentioned in Chapters 8 and 10, the stress intensity factor plays an important role in fracture mechanics theory. For this out-of-plane deformation case, the stress intensity factor is related to mode III fracture toughness, and its value can be computed by the expression

$$K_{III} = \lim_{x \to 1+}\left(\sqrt{2(x-1)}\,\tau_{yz}(x,0)\right) = \sqrt{\mu_o}e^{\alpha/2}F(1) \tag{14.5.22}$$

To obtain explicit results for the displacement, stress, and stress intensity factor, equation (14.5.19) must be solved to determine the functional behavior of $F(t)$ for various values of α and β. Relation (14.5.19) is a Fredholm integral equation of the second kind and generally requires numerical integration methods to determine the solution. Dhaliwal and Singh (1978) conducted such numerical evaluation, and their results for the stress intensity factor are shown in Fig. 14.23. These results

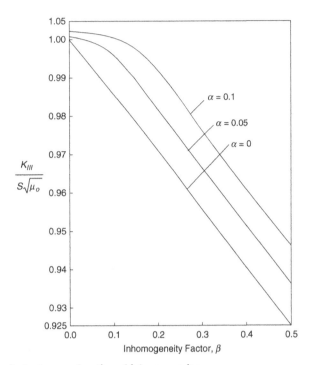

FIG. 14.23 Stress intensity factor as a function of inhomogeneity.

From Dhaliwal RS, Singh BM. On the theory of elasticity of a non-homogeneous medium, J Elasticity 8:211–219, 1978; reprinted with permission from Springer.

generally indicate that the stress intensity factor increases with α but decreases with increasing values of β. Note that for the homogeneous case ($\alpha = \beta = 0$), $K_{III} = S\sqrt{\mu_o}$.

14.6 Torsion problem

We now wish to re-examine the torsion of elastic cylinders for the case where the material is nonhomogeneous. The basic formulation and particular solutions were given in Chapter 9 for the homogeneous case and in Chapter 11 for anisotropic materials. Although a vast amount of work has been devoted to these problems, only a few studies have investigated the corresponding inhomogeneous case. Early work on the torsion of nonhomogeneous cylinders includes Lekhnitskii (1981); later studies were done by Rooney and Ferrari (1995) and Horgan and Chan (1999c). As expected, most closed-form analytical solutions for the inhomogeneous problem are limited to cylinders of revolution, normally with circular cross-sections.

Following the work of Horgan and Chan (1999c), we consider the torsion of a right circular cylinder of radius a, as shown in Fig. 14.24. The cylindrical body is assumed to be isotropic, but with graded shear modulus that is a function only of the radial coordinate $\mu = \mu(r)$. The usual boundary conditions require zero tractions on the lateral boundary S and a resultant pure torque loading T over each end section R.

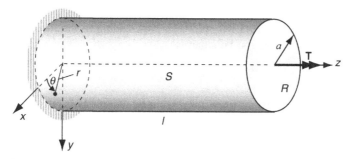

FIG. 14.24 Torsion of a nonhomogeneous circular cylinder.

The beginning formulation steps remain the same as presented previously, and thus the displacements, strains, and stresses are the same as given in Chapter 9

$$u = -\alpha yz$$
$$v = \alpha xz \qquad\qquad (14.6.1)$$
$$w = w(x, y)$$

$$e_x = e_y = e_z = e_{xy} = 0$$

$$e_{xz} = \frac{1}{2}\left(\frac{\partial w}{\partial x} - \alpha y\right)$$

$$\qquad\qquad (14.6.2)$$

$$e_{yz} = \frac{1}{2}\left(\frac{\partial w}{\partial y} + \alpha x\right)$$

$$\sigma_x = \sigma_y = \sigma_z = \tau_{xy} = 0$$

$$\tau_{xz} = \mu\left(\frac{\partial w}{\partial x} - \alpha y\right)$$

$$\qquad\qquad (14.6.3)$$

$$\tau_{yz} = \mu\left(\frac{\partial w}{\partial y} + \alpha x\right)$$

It again becomes useful to introduce the Prandtl stress function, $\phi = \phi(x,y)$

$$\tau_{xz} = \frac{\partial \phi}{\partial y}, \quad \tau_{yz} = -\frac{\partial \phi}{\partial x} \qquad\qquad (14.6.4)$$

so that the equilibrium equations are satisfied identically. We can again generate the compatibility relation among the two nonzero stress components by differentiating and combining relations $(14.6.3)_{2,3}$ to eliminate the displacement terms. Substituting (14.6.4) into that result gives the governing relation in terms of the stress function

$$\frac{\partial}{\partial x}\left(\frac{1}{\mu}\frac{\partial \phi}{\partial x}\right) + \frac{\partial}{\partial y}\left(\frac{1}{\mu}\frac{\partial \phi}{\partial y}\right) = -2\alpha \qquad\qquad (14.6.5)$$

where the shear modulus μ must now be left inside the derivative operations because the material is inhomogeneous. Recall that, for the homogeneous case, relation (14.6.5) reduced to the Poisson equation $\nabla^2 \phi = -2\mu\alpha$.

Incorporation of the boundary condition that tractions vanish on the lateral surface S leads to identical steps as given previously by Eqs. (9.3.10)–(9.3.12), thus leading to the fact that the stress function must be a constant on all cross-section boundaries

$$\frac{d\phi}{ds} = 0 \Rightarrow \phi = \text{constant}, \quad \text{on } S \tag{14.6.6}$$

For simply connected sections, the constant may again be chosen as 0. Invoking the resultant force conditions on the cylinder end planes (domain R), as given by relations (9.3.14), again yields

$$T = \iint_R \left(xT_y^n - yT_x^n \right) dxdy = 2 \iint_R \phi \, dxdy \tag{14.6.7}$$

and the *torsional rigidity* J can again be defined by $J = T/\alpha$.

Because we wish to use a simple radial shear modulus variation, $\mu = \mu(r)$, it is more convenient to use a polar coordinate formulation. For the circular cylinder under study, the problem reduces to an axisymmetric formulation independent of the angular coordinate and the warping displacement vanishes. Under these conditions the displacements, strains, and stresses reduce to

$$u_r = u_z = 0, \quad u_\theta = \alpha rz$$

$$e_r = e_\theta = e_z = e_{rz} = e_{r\theta} = 0, \quad e_{\theta z} = \frac{\alpha r}{2} \tag{14.6.8}$$

$$\sigma_r = \sigma_\theta = \sigma_z = \tau_{rz} = \tau_{r\theta} = 0, \quad \tau_{\theta z} = \alpha \mu r$$

Relations (14.6.4) and (14.6.5) for the stress function formulation then reduce to a system in terms of only the radial coordinate r

$$\tau_{\theta z} = -\frac{d\phi}{dr}$$

$$\frac{1}{r}\frac{d}{dr}\left(\frac{r}{\mu}\frac{d\phi}{dr}\right) = -2\alpha \tag{14.6.9}$$

with boundary condition $\phi(a) = 0$.

The governing differential equation (14.6.9) can be easily integrated to give the general solution

$$\phi(r) = -\alpha \int r\mu(r)dr + C_1 \int \frac{\mu(r)}{r}dr + C_2 \tag{14.6.10}$$

where C_1 and C_2 are arbitrary constants. We require that the solution for ϕ remain bounded as $r \to 0$, thus implying that each integral term in (14.6.10) be finite at the origin. Restricting ourselves to the plausible case where the shear modulus is expected to be nonzero but bounded at the origin, $\lim_{r\to 0}\int r\mu(r)dr \to 0$, while the second integral, $\lim_{r\to 0}\int\left[\frac{\mu(r)}{r}\right]dr$, is singular. Based on these arguments, C_1 must be set to 0. Finally, the boundary condition $\phi(a) = 0$ determines the final constant C_2 and produces the general solution

$$\phi(r) = \alpha \int_r^a \xi\mu(\xi)d\xi \tag{14.6.11}$$

With this result, the shear stress and torsional rigidity then become

$$\tau_{\theta z} = \alpha r \mu(r)$$

$$J = 2\pi \int_0^a r^3 \mu(r) dr$$

(14.6.12)

To explore the effects of inhomogeneity, let us consider some specific gradations in shear modulus. Following some of the examples discussed by Horgan and Chan (1999c), we consider two cases of the following form

$$\mu(r) = \mu_o \left(1 + \frac{n}{a} r\right)^m$$

$$\mu(r) = \mu_o e^{-\frac{n}{a} r}$$

(14.6.13)

where $n \geq 0$ and $\mu_o > 0$ and m are material constants. Note that, for either example, as $n \to 0$ we recover the homogeneous case $\mu(r) = \mu_o$. Also, as $r \to 0$, $\mu \to \mu_o$, and so these material examples all have finite shear modulus at $r = 0$.

Plots of these shear modulus gradations are shown in Fig. 14.25 for various cases of material parameter m with $n = 1$. For the model given by $(14.6.13)_1$, three cases are shown. The $m = 1$ case corresponds to a linearly increasing shear modulus from the central axis of the shaft, while $m = -1$ or -3 gives a nonlinear decreasing gradation in material stiffness. The figure also shows the modulus variation for the exponential graded model given by $(14.6.13)_2$ for the case $n = 1$. All gradation forms

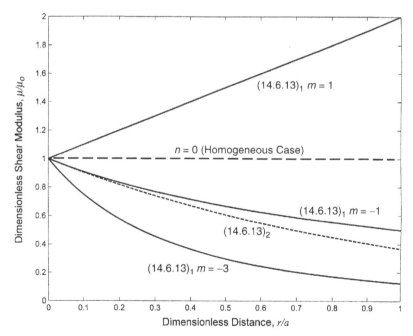

FIG. 14.25 Shear modulus behavior for torsion problems ($n = 1$).

(14.6.13) allow simple solutions to be generated for the stress function, shear stress, and torsional rigidity.

Solutions for the gradation model given by $(14.6.13)_1$ are found to be

$$\phi(r) = \begin{cases} \dfrac{\mu_o\alpha}{2}(a^2 - r^2) + \dfrac{\mu_o\alpha n}{3}\left(a^2 - \dfrac{r^3}{a}\right), & m = 1 \\[3ex] -\mu_o\alpha\left[\dfrac{a}{n}r - \left(\dfrac{a}{n}\right)^2 \log\left|1 + \dfrac{n}{a}r\right|\right] + \mu_o\alpha\left[\dfrac{a^2}{n} - \left(\dfrac{a}{n}\right)^2 \log|1 + n|\right], & m = -1 \end{cases}$$

(14.6.14)

$$\tau_{\theta z} = \mu_o\alpha r\left(1 + \dfrac{n}{a}r\right)^m$$

(14.6.15)

Note that the solution for the stress function requires integration through relation (14.6.11), and thus closed-form solutions can only be determined for integer and other special values of the parameter m. From relation (14.6.15), it can be shown that if $m \geq -1$, the maximum shear stress always occurs at the boundary $r = a$. Recall that this result was found to be true in general for all homogeneous cylinders of any cross-section geometry (see Exercise 9.5). However, for the inhomogeneous case when $m < -1$, the situation changes and the location of maximum shear stress can occur in the cylinder's interior.

Horgan and Chan (1999c) have shown that for the case with $n > 0$, the choice of $m < -1 - 1/n$ produces a maximum shear stress $\tau_{\theta z}$ at $r = -a/n(1 + m)$. These results imply that modulus gradation can be adjusted to allow control of the location of $(\tau_{\theta z})_{\max}$. Dimensionless shear stress distributions for model $(14.6.13)_1$ are shown in Fig. 14.26 for various cases of material parameters m and n. As expected, higher stresses occur for a gradation with increasing shear modulus. For the homogeneous case, the shear stress distribution will be linear, as predicted from both elasticity theory and mechanics of materials. For the nonhomogeneous cases with $n = 1$ and $m = \pm 1$, it is noted that the maximum shear stress occurs on the boundary of the shaft. However, for the case shown with $n = 1$ and $m = -3$, the maximum stress occurs interior at $r = a/2$ according to our previous discussion.

Considering next the solutions for the gradation model $(14.6.13)_2$, relations (14.6.11) and (14.6.12) give

$$\phi(r) = \mu_o\alpha e^{-\frac{n}{a}r}\left(\frac{ar}{n} + \left(\frac{a}{n}\right)^2\right) - \mu_o\alpha a^2\left(\frac{1}{n} + \frac{1}{n^2}\right)e^{-n}$$

(14.6.16)

$$\tau_{\theta z} = \mu_o\alpha r e^{-\frac{n}{a}r}$$

It can easily be shown that if $n \leq 1$ the maximum shear stress will exist on the outer boundary, while if $n > 1$ the maximum moves to an interior location within the shaft. Nondimensional shear stress distributions for this exponential gradation are shown in Fig. 14.27 for several values of the parameter n. As observed in the previous model, for the case with decreasing radial gradation ($n > 0$), the shear stress will always be less than the corresponding homogeneous distribution. With $n < 0$, we have an increasing radial gradation that results in stresses larger than the homogeneous values.

The discussion of the location of maximum shear stress can be generalized for the radial gradation case by going back to the general shear stress solution $(14.6.12)_1$. Using this relation, it can be shown

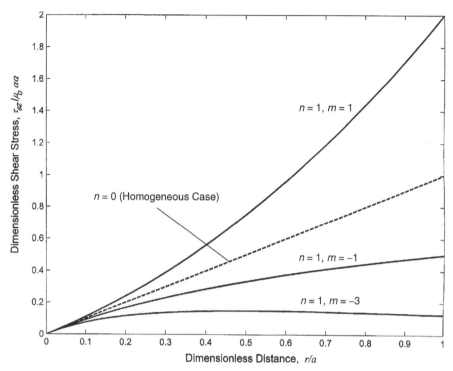

FIG. 14.26 Shear stress distribution for the torsion problem with modulus given by model $(14.6.13)_1$.

that the necessary and sufficient condition on $\mu(r)$ for an extremum of $\tau_{\theta z}$ at an interior point $(r < a)$ is given by

$$\mu'(r) < -\frac{\mu(r)}{a}, \quad r \in (0, a) \tag{14.6.17}$$

where the prime indicates differentiation. Furthermore the location r_0 of this extremum is specified by

$$r_0 = -\frac{\mu(r_0)}{\mu'(r_0)} \tag{14.6.18}$$

Further analysis to determine the torsional rigidities can also be carried out (see, for example, Exercise 14.26). Horgan and Chan (1999c) have explored torsional rigidities in detail and have developed several general results including upper and lower bounding theorems.

It should be noted that the general field equation for the torsion problem (14.6.5) is quite similar to the corresponding field equation for the antiplane strain problem (14.5.4) discussed previously. Thus, a transformation scheme similar to (14.5.5) may also help reduce the torsion field equation into a more tractable relation. Following this concept, we use the transformation

$$\phi(x, y) = \Phi(x, y)\sqrt{\mu(x, y)} \tag{14.6.19}$$

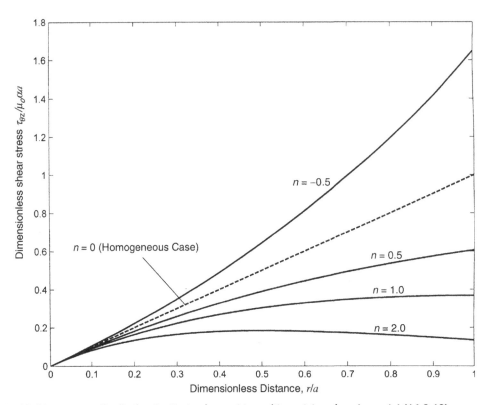

FIG. 14.27 Shear stress distribution for the torsion problem with modulus given by model (14.6.13)$_2$.

into (14.6.5) and find the following reduction

$$\nabla^2 \Phi + F(\mu)\Phi = -2\alpha\sqrt{\mu} \tag{14.6.20}$$

where

$$F(\mu) = \frac{1}{2\mu}\left[\nabla^2\mu - \frac{3}{2\mu}|\nabla\mu|^2\right] = -\sqrt{\mu}\,\nabla^2\left(\mu^{-1/2}\right) \tag{14.6.21}$$

Assuming that $\mu > 0$ in the domain, boundary condition (14.6.6) implies that

$$\Phi = 0, \quad \text{on} \ \ S \tag{14.6.22}$$

This formulation leads to significant simplification for the case where $F(\mu) = 0$, which corresponds to the situation where $\mu^{-1/2}$ is harmonic. For such a case, the governing equation reduces to the Poisson equation

$$\nabla^2\Phi = -2\alpha\sqrt{\mu} \tag{14.6.23}$$

and standard solution techniques for solving such equations can then be applied.

Another noteworthy area of significant research dealing with graded materials includes studies of static and dynamic fracture mechanics; see, for example, Parameswaran and Shukla (1999, 2002). These and other studies have investigated material gradation effects on crack tip stresses around stationary and moving cracks.

This concludes our exploration into nonhomogeneous elasticity solutions. The examples given illustrate some of the interesting effects caused by spatial variation of elastic moduli. As should be evident, problem formulation and solution are more challenging. Solutions to such problems commonly indicate significant differences in the elastic stress and displacement fields when compared with corresponding homogeneous solutions. In some cases, material gradation will reduce maximum stresses and change the spatial location where such maxima occur. This provides the possibility of tailoring material variation to achieve desired stresses in a structure and thus of functionally grading the material. The difficult part of this concept is how to develop manufacturing techniques that will produce the desired continuous modulus variation in realistic materials used in engineering applications.

References

Abramowitz M, Stegun IA: Handbook of Mathematical Functions with Formulas, Graphs and Mathematical Tables, *National Bureau of Standards*, 1964.

Ang WT, Clements DL: On some crack problems for inhomogeneous elastic materials, *Int J Solids Struct* 23: 1089–1104, 1987.

Awojobi AO, Gibson RE: Plane strain and axially symmetric problems of a linearly nonhomogeneous elastic half-space, *Quart J Mech Appl Math* 21:285–302, 1973.

Birman V, Byrd LW: Modeling and analysis of functionally graded materials and structures, *Appl Mech Rev* 60: 195–216, 2007.

Booker JR, Balaam NP, Davis EH: The behavior of an elastic non-homogeneous half-space. Part I. Line and point loads, *Int J Numer Anal Methods Geomech* 9:353–367, 1985.

Calladine CR, Greenwood JA: Line and point loads on a non-homogeneous incompressible elastic half-space, *J Mech Appl Math* 31:507–529, 1978.

Carrier WD, Christian JT: Analysis of an inhomogeneous elastic half-space, *J Soil Mech Foundations Engng* 99: 301–306, 1973.

Clements DL, Atkinson C, Rogers C: Antiplane crack problems for an inhomogeneous elastic material, *Acta Mech* 29:199–211, 1978.

Clements DL, Kusuma J, Ang WT: A note on antiplane deformations of inhomogeneous elastic materials, *Int J Eng Sci* 35:593–601, 1997.

Delale F: Mode III fracture of bonded non-homogeneous materials, *Eng Fract Mech* 22:213–226, 1985.

Dhaliwal RS, Singh BM: On the theory of elasticity of a non-homogeneous medium, *J Elasticity* 8:211–219, 1978.

Du CH: *The two-dimensional problems of the nonhomogeneous isotropic medium, problems of continuum mechanics: contributions in honor of the 70th birthday of I. Muskehlishvili*, Noordhoff, 1961, SIAM, pp 104–108.

Erdogan F, Ozturk M: Diffusion problems in bonded nonhomogeneous materials with an interface cut, *Int J Eng Sci* 30:1507–1523, 1992.

Fraldi M, Cowin SC: Inhomogeneous elastostatic problem solutions constructed from stress-associated homogeneous solutions, *J Mech Phys Solids* 52:2207–2233, 2004.

Galmudi D, Dvorkin J: Stresses in anisotropic cylinders, *Mech Res Commun* 22:109−113, 1995.

Giannakopoulos AE, Suresh S: Indentation of solids with gradients in elastic properties: Part I. Point force, *Int J Solids Struct* 34:2357−2392, 1997.

Gibson RE: Some results concerning displacements and stresses in a non-homogeneous elastic half space, *Geotechnique* 17:58−67, 1967.

Gibson RE, Sills GC: On the loaded elastic half space with a depth varying Poisson's ratio, *ZAMP* 20:691−695, 1969.

Holl DL: Stress transmission in earths, *Proc Highway Res Board* 20:709−721, 1940.

Horgan CO, Baxter SC: Effects of curvilinear anisotropy of radially symmetric stresses in anisotropic linearly elastic solids, *J Elasticity* 42:31−48, 1996.

Horgan CO, Chan AM: The pressurized hollow cylinder or disk problem for functionally graded isotropic linearly elastic materials, *J Elasticity* 55:43−59, 1999a.

Horgan CO, Chan AM: The stress response of functionally graded isotropic linearly elastic rotating disks, *J Elasticity* 55:219−230, 1999b.

Horgan CO, Chan AM: Torsion of functionally graded isotropic linearly elastic bars, *J Elasticity* 52:181−199, 1999c.

Horgan CO, Miller KL: Antiplane shear deformations for homogeneous and inhomogeneous anisotropic linearly elastic solids, *J Appl Mech* 61:23−29, 1994.

Kassir MK: Boussinesq problems for nonhomogeneous solids, *J Eng Mech* 98:457−470, 1972.

Lekhnitskii SG: Radial distribution of stresses in a wedge and in a half-plane with variable modulus of elasticity, *PMM* 26:146−151, 1961.

Lekhnitskii SG: *Theory of elasticity of an anisotropic body*, Moscow, 1981, Mir Publishers.

Oner M: Vertical and horizontal deformation of an inhomogeneous elastic half-space, *Int J Numer Anal Methods Geomech* 14:613−629, 1990.

Parameswaran V, Shukla A: Crack-tip stress fields for dynamic fracture in functionally gradient materials, *Mech Mater* 31:579−596, 1999.

Parameswaran V, Shukla A: Asymptotic stress fields for stationary cracks along the gradient in functionally graded materials, *J Appl Mech* 69:240−243, 2002.

Plevako VP: On the theory of elasticity of inhomogeneous media, *PMM* 35:853−860, 1971.

Plevako VP: On the possibility of using harmonic functions for solving problems of the theory of elasticity of nonhomogeneous media, *PMM* 36:886−894, 1972.

Rooney FJ, Ferrari M: Torsion and flexure of inhomogeneous materials, *Comput Eng* 5:901−911, 1995.

Rostovtsev NA: On the theory of elasticity of a nonhomogeneous medium, *PMM* 28:601−611, 1964.

Sadd MH: Some simple Cartesian solutions to plane nonhomogeneous elasticity problems, *Mech Res Commun* 37:22−27, 2010.

Sankar BV: An elasticity solution for functionally graded beams, *Compos Sci Technol* 61:689−696, 2001.

Spencer AJM, Selvadurai APS: Some generalized anti-plane strain problems for an inhomogeneous elastic half space, *J Eng Math* 34:403−416, 1998.

Stampouloglou IH, Theotokoglou EE: The anisotropic and angularly inhomogeneous elastic wedge under a monomial load distribution, *Arch Appl Mech* 75:1−17, 2005.

Suresh S: Graded materials for resistance to contact deformation and damage, *Science* 292:2447−2451, 2001.

Ter-Mkrtich'ian LN: Some problems in the theory of elasticity of nonhomogeneous elastic media, *PMM* 25:1120−1125, 1961.

Vrettos C: The Boussinesq problem for soils with bounded non-homogeneity, *Int J Numer Anal Methods Geomech* 22:655−669, 1998.

Wang CD, Tzeng CS, Pan E, Liao JJ: Displacements and stresses due to a vertical point load in an inhomogeneous transversely isotropic half-space, *Int J Rock Mech Min Sci* 40:667−685, 2003.

Exercises

14.1 Show that, for the case of plane stress, the governing compatibility relation in terms of the Airy stress function is given by (14.1.6)

$$\frac{\partial^2}{\partial x^2}\left(\frac{1}{E}\frac{\partial^2\phi}{\partial x^2} - \frac{v}{E}\frac{\partial^2\phi}{\partial y^2}\right) + \frac{\partial^2}{\partial y^2}\left(\frac{1}{E}\frac{\partial^2\phi}{\partial y^2} - \frac{v}{E}\frac{\partial^2\phi}{\partial x^2}\right) + 2\frac{\partial^2}{\partial x\partial y}\left(\frac{1+v}{E}\frac{\partial^2\phi}{\partial x\partial y}\right) = 0$$

Next determine the reduced form of this equation for the special case $E = E(x)$ and $v = $ constant.

14.2 Consider a special case of Eq. (14.1.4). Parameswaran and Shukla (1999) presented a two-dimensional study where the shear modulus and Lamé's constant varied as $\mu(x) = \mu_o(1 + ax)$ and $\lambda(x) = k\mu(x)$, where μ_o, a, and k are constants. For such a material, show that in the absence of body forces the two-dimensional Navier's equations become

$$\mu_o(1 + ax)\left(k\frac{\partial\vartheta}{\partial x} + 2\frac{\partial^2 u}{\partial x^2} + \frac{\partial^2 u}{\partial y^2} + \frac{\partial^2 v}{\partial x\partial y}\right) + \mu_o a\left(k\vartheta + 2\frac{\partial u}{\partial x}\right) = 0$$

$$\mu_o(1 + ax)\left(k\frac{\partial\vartheta}{\partial y} + 2\frac{\partial^2 v}{\partial y^2} + \frac{\partial^2 v}{\partial x^2} + \frac{\partial^2 u}{\partial x\partial y}\right) + \mu_o a\left(\frac{\partial u}{\partial y} + \frac{\partial v}{\partial x}\right) = 0$$

where $\vartheta = \dfrac{\partial u}{\partial x} + \dfrac{\partial v}{\partial y}$.

14.3 Parameswaran and Shukla (2002) recently presented a fracture mechanics study of nonhomogeneous material behavior related to functionally graded materials. They investigated a two-dimensional plane stress problem where Poisson's ratio remained constant but Young's modulus varied as $E(x) = E_o e^{ax}$, where E_o and a are constants. For this case, with zero body forces, show that the governing Airy stress function equation is given by

$$\nabla^4\phi - 2a\frac{\partial}{\partial x}\left(\nabla^2\phi\right) + a^2\nabla^2\phi - a^2(1+v)\frac{\partial^2\phi}{\partial y^2} = 0$$

This result may be compared with the more general case given by relation (14.1.6). Note that, when $a = 0$, this result reduces to the homogeneous form $\nabla^4\phi = 0$. The nonhomogeneous result is a challenging equation, and its solution was developed for the limited case near the tip of a crack using asymptotic analysis.

14.4 Explicitly integrate relations (14.1.10) and develop the displacements relations (14.1.11).

14.5* Plot horizontal and vertical displacement contours of relations (14.1.11) in the unit domain $0 \le (x, y) \le 1$ for cases $K = -0.5$ and 5. Normalize values with respect to T/E_o and take $v = 0.3$. Qualitatively compare these results with those expected for the homogeneous case $K = 0$.

14.6 Following the scheme used in Example 14.1, consider the same stress field case $\phi = Ty^2/2$ but with modulus variation in the y-direction, $E = E(y)$. First show that the required modulus variation is given by $E = E_o/(1 + Ky)$. Next determine the resulting displacement field assuming as before zero displacements and rotation at the origin, to get

$$u = \frac{T}{E_o}(1 + Ky)x, \quad v = -v\frac{T}{E_o}\left(y + K\left(\frac{y^2}{2} + \frac{x^2}{2v}\right)\right)$$

14.7 Consider the plane stress inhomogeneous case with only variation in elastic modulus given by $E = E(y) = 1/(Ay + B)$. Further assume that the Airy function depends only on y, $\phi = \phi(y)$. Show that governing relation (14.1.6) reduces to the ordinary differential equation

$$(Ay + B)\frac{d^4\phi}{dy^4} + 2A\frac{d^3\phi}{dy^3} = 0$$

Solve this relation to find that $d^3\phi/dy^3 = C_1/(y + B/A)^2$, thus giving the stress forms

$$\sigma_x = \frac{d^2\phi}{dy^2} = -\frac{C_1}{y + B/A} + C_2, \quad \sigma_y = \tau_{xy} = 0$$

where C_1 and C_2 are arbitrary constants of integration.

14.8 The stress state $\sigma_x = \sigma_y = 0$, $\tau_{xy} = T = $ constant corresponds to the Airy stress function $\phi = -Txy$. For this case with constant Poisson's ratio, show that the general equation (14.1.6) greatly reduces and gives the result that the allowable Young's modulus must be of the form $1/E = f(x) + g(y)$, where $f(x)$ and $g(y)$ are arbitrary functions.

14.9 Consider a stress function formulation for the axisymmetric problem discussed in Section 14.2. The appropriate compatibility relation for this case has been previously developed in Example 8.11; see (8.4.76). Using the plane stress Hooke's law, express this compatibility relation in terms of stress and then in terms of the Airy function to get the result

$$\frac{d}{dr}\left[\frac{1}{E(r)}\left(r\frac{d^2\phi}{dr^2} - v(r)\frac{d\phi}{dr}\right)\right] - \frac{1}{E(r)}\left(\frac{1}{r}\frac{d\phi}{dr} - v(r)\frac{d^2\phi}{dr^2}\right) = 0$$

14.10 For the hollow cylinder problem shown in Fig. 14.5, use the given boundary conditions to explicitly determine the arbitrary constants A and B and the stress relations (14.2.6).

14.11 For the problem in Fig. 14.5 with only internal pressure, show that the general stress field (14.2.6) reduces to relations (14.2.7).

14.12 For the hollow cylinder problem illustrated in Fig. 14.5, show that the usual restrictions on Poisson's ratio, $0 \leq v \leq 1/2$, and $n > 0$ imply that

$$-2 + k + n \geq 0, \quad -2 - k + n \leq 0$$

$$\frac{2 + kv - nv}{k - n + 2v} \geq 1, \quad \frac{2 - kv - nv}{k + n - 2v} \leq 1$$

Using these results, develop arguments to justify that the stresses in solution (14.2.7) must satisfy $\sigma_r < 0$ and $\sigma_\theta > 0$ in the cylinder's domain. Thus, stresses in the nonhomogeneous problem have behavior similar to those of the ungraded case.

14.13* Following procedures similar to those for the homogeneous problem (see Section 8.4.1), develop the following stress field for a pressurized hole in an infinite nonhomogeneous medium with moduli variation given by (14.2.3)

$$\sigma_r = -p_i\left(\frac{a}{r}\right)^{(2+k-n)/2}$$

$$\sigma_\theta = p_i\frac{2 - kv - nv}{k + n - 2v}\left(\frac{a}{r}\right)^{(2+k-n)/2}$$

Plot the dimensionless stress fields for this case using the same parameters ν and n used in Figs. 14.7 and 14.8.

14.14 For the inhomogeneous rotating disk problem with $\nu = 0$, explore the solution for the special case of $n = 3$ and show that the stresses reduce to

$$\sigma_r = 0, \quad \sigma_\theta = \rho\omega^2 r^2$$

Also determine the displacement solution and show the surprising result that $u(0) > 0$. This strange behavior has been referred to as a *cavitation* at the disk's center.

14.15* Using MATLAB® or similar software, make a plot similar to Fig. 14.13 showing the behavior of the hoop stress σ_θ for the case with $\nu = 0$ and $n = -0.5, 0, 0.5, 1, 2$. Discuss your results.

14.16 Using the design criterion (14.3.14), incorporate the fundamental equations for the inhomogeneous rotating disk problem to explicitly develop the required gradation given by relation (14.3.15).

14.17 Rather than using polar coordinates to formulate the inhomogeneous half-space problem of Section 14.4, some researchers have used Cartesian coordinates instead. Using the x, z-coordinates as shown in Fig. 14.14, consider the plane strain case with inhomogeneity only in the shear modulus such that $\mu = \mu(z)$ and $\nu = $ constant. First show that combining the strain displacement relations with Hooke's law gives

$$\sigma_x = 2\mu(z)\left(\frac{\partial u}{\partial x} + \nu^*\vartheta\right)$$

$$\sigma_z = 2\mu(z)\left(\frac{\partial w}{\partial z} + \nu^*\vartheta\right)$$

$$\tau_{xz} = \mu(z)\left(\frac{\partial u}{\partial z} + \frac{\partial w}{\partial x}\right)$$

where $\vartheta = e_{kk} = \frac{\partial u}{\partial x} + \frac{\partial w}{\partial z}$ is the dilatation and $\nu^* = \frac{\nu}{(1-2\nu)}$. Next show that the equilibrium equations yield

$$\nabla^2 u + (1 + 2\nu^*)\frac{\partial\vartheta}{\partial x} + h(z)\left(\frac{\partial u}{\partial z} + \frac{\partial w}{\partial x}\right) = 0$$

$$\nabla^2 w + (1 + 2\nu^*)\frac{\partial\vartheta}{\partial z} + 2h(z)\left(\frac{\partial w}{\partial z} + \nu^*\vartheta\right) = 0$$

where $h(z) = \left(\frac{1}{\mu}\right)\left(\frac{d\mu}{dz}\right)$. Explore the simplification of these equations for the special inhomogeneous case in which $\mu = \mu_o e^{\alpha z}$, where μ_o and α are constants.

14.18 Consider the axisymmetric half-space problem shown in Fig. 14.19 for the case where only Poisson's ratio is allowed to vary with depth coordinate $v = v(z)$. Using cylindrical coordinates (r, θ, z) to formulate the problem, first show that the stress field can be expressed by

$$\sigma_r = 2\mu\left(\frac{\partial u_r}{\partial r} + v^*(z)\vartheta\right)$$

$$\sigma_z = 2\mu\left(\frac{\partial u_z}{\partial z} + v^*(z)\vartheta\right)$$

$$\sigma_\theta = 2\mu\left(\frac{u_r}{r} + v^*(z)\vartheta\right)$$

$$\tau_{rz} = \mu\left(\frac{\partial u_r}{\partial z} + \frac{\partial u_z}{\partial r}\right)$$

where $\vartheta = \dfrac{\partial u_r}{\partial r} + \dfrac{u_r}{r} + \dfrac{\partial u_z}{\partial z}$ is the dilatation and $v^*(z) = \dfrac{v(z)}{[1 - 2v(z)]}$. Next show that the equilibrium equations reduce to

$$\left(\nabla^2 - \frac{1}{r}\right)u_r + \frac{\partial}{\partial r}((1 + 2v^*)\vartheta) = 0$$

$$\nabla^2 u_z + \frac{\partial}{\partial z}((1 + 2v^*)\vartheta) = 0$$

14.19 For the antiplane strain problem, verify that transformation (14.5.5) will reduce the governing equilibrium equation to relation (14.5.6). Next show that the separation of variables scheme defined by (14.5.7) will lead to the two equations (14.5.8).

14.20 Explicitly show that the inhomogeneity functions $p(x) = e^{-\alpha|x|}$ and $q(y) = e^{-\beta|y|}$, that were used for the antiplane crack problem, do in fact satisfy relations (14.5.9) with $a_o = \alpha^2/4$ and $b_o = \beta^2/4$.

14.21 For the torsion problem discussed in Section 14.6, explicitly justify the reductions in polar coordinates summarized by relations (14.6.8) and (14.6.9).

14.22 Verify that the general solutions to equations (14.6.9) are given by (14.6.11) and (14.6.12) for the torsion problem.

14.23 For the torsion problem, verify the solutions (14.6.14) and (14.6.15) for the gradation model (14.6.13)₁.

14.24 For the torsion problem, verify the solutions (14.6.16) for the gradation model (14.6.13)₂.

14.25 Investigate the issue of finding the location of $(\tau_{\theta z})_{max}$ for the torsion problem. First verify the results given by (14.6.17) and (14.6.18). Next apply these general relations to the specific gradation models given by (14.6.13)₁,₂, and develop explicit results for locations of the extrema.

14.26 Show that the torsional rigidity for the exponential gradation case $(14.6.13)_2$ is given by

$$J = 2\pi a^4 \mu_o \left[\frac{6}{n^4} - e^{-n} \left(\frac{1}{n} + \frac{3}{n^2} + \frac{6}{n^3} + \frac{6}{n^4} \right) \right]$$

Next use a Taylor series expansion for small n to show that

$$J \approx \frac{\pi a^4 \mu_o}{2} \left[1 - \frac{4n}{5} + O(n^2) \right].$$

Compare this result with the homogeneous case.

Micromechanics applications

15

In recent years, considerable interest has developed in micromechanical modeling of solids. This interest has been fueled by the realization that many materials have heterogeneous microstructures that play a dominant role in determining macro deformational behavior. Materials where this occurs include multiphase fiber and particulate composites, soil, rock, concrete, and various granular materials. These materials have microstructures that occur at a variety of length scales from meters to nanometers, and general interest lies with the case where the length scale is smaller than other characteristic lengths in the problem. The response of such heterogeneous solids shows strong dependence on the micromechanical behaviors between different material phases. Classical theories of continuum mechanics have limited ability to predict such behaviors, and this has lead to the development of many new micromechanical theories of solids.

Work in this area, initiated over a century ago by Volterra (1907), began with studies of elastic stress and displacement fields around dislocations and other imperfections. More recently, using continuum mechanics principles, theories have been developed in which the material response depends on particular microscale length parameters connected with the existence of inner degrees of freedom and *nonlocal* continuum behavior. By nonlocal behavior we mean that the stress at a point depends not only on the strain at that point but also on the strains of neighboring points. Mindlin (1964) developed a general *linear elasticity theory with microstructure* that allowed the stress to depend on both the strain and an additional *kinematic microdeformation* tensor. Related research has led to the development of *micropolar* and *couple-stress theories*; see Eringen (1968). These approaches allow material deformation to include additional independent *microrotational* degrees of freedom. Elastic continuum theories using *higher-order gradients* have also been developed to model micromechanical behavior of solids; see, for example, Mindlin (1965), Chang and Gao (1995), Aifantis (1999), and Li, Miskioglu, and Altan (2004). A general review of modeling heterogeneous elastic solids has been provided by Nemat-Nasser and Hori (1993). Along similar lines, Cowin and Nunziato (1983) developed a theory of elastic materials with *voids* including an independent *volume fraction* in the constitutive relations.

Another interesting theory called *doublet mechanics* (Ferrari et al., 1997) represents heterogeneous solids in a discrete fashion as arrays of points or particles that interact through prescribed micromechanical laws. Other related work has investigated elastic materials with *distributed cracks*; see, for example, Budianski and O'Connell (1976) and Kachanov (1994). Originally developed by Biot (reference collection, 1992), *poroelasticity* allows for the coupling action between a porous elastic solid and the contained pore fluid. The coupled diffusion—deformation mechanisms provide useful applications in many geomechanics problems. Some work has approached the heterogeneous problem using statistical and probabilistic methods to develop models with random variation in micromechanical properties; see, for example, Ostoja-Starzewski and Wang (1989, 1990). The monograph

Elasticity. https://doi.org/10.1016/B978-0-12-815987-3.00015-3

477

by Mura (1987) provides considerable elastic modeling of dislocations, inclusions, cracks, and other inhomogeneities using the *eigenstrain* technique.

We now present an introduction to some of these particular modeling schemes, including dislocations, singular stress states, elastic materials with distributed cracks, micropolar/couple-stress theory, elastic voids, doublet mechanics and higher gradient elasticity. Our brief coverage focuses on only the linear elastic response of a given theory, generally including one or two example applications. Of course, many other theories have been developed, and the choice of topics to be presented is based on their appropriateness for the educational goals of the text. This review provides a good foundation for further study and pursuit of additional theories that may be more appropriate for a given material.

15.1 Dislocation modeling

Deformations of an elastic solid may depend not only on the action of the external loadings, but also on *internal microstructural defects* that may be present in the material. In crystalline materials, such internal defects are commonly associated with imperfections in the atomic lattice and are referred to as *dislocations*. The particular type of defect depends on the basic atomic lattice structure of the crystal, and an example imperfection is shown in Fig. 15.1 for the case of a *simple cubic* packing geometry. This imperfection is associated with the insertion of an extra plane of atoms (indicated by the dotted line) and is referred to as an *edge dislocation*. Other examples exist, and we now investigate the elastic stress and displacement fields of two particular dislocation types. As previously mentioned, studies on dislocation modeling began over a century ago by Volterra (1907) and detailed summaries of this work have been given by Weertman and Weertman (1964), Lardner (1974), and Landau and Lifshitz (1986).

The two most common types of imperfections are the *edge* and *screw dislocations*, and these are shown for a *simple cubic* crystal in Fig. 15.2. As mentioned, the edge dislocation occurs when an extra plane(s) of atoms is inserted into the regular crystal as shown, while the screw dislocation is associated

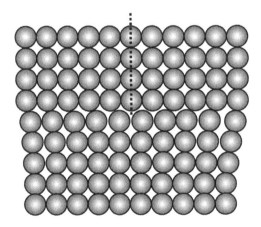

FIG. 15.1 Edge dislocation.

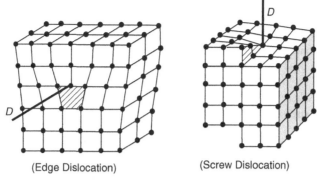

(Edge Dislocation)   (Screw Dislocation)

FIG. 15.2 Schematics of edge and screw dislocations.

with a shearing deformational shift along a regular plane. The effect of such dislocations is to produce a local stress and displacement field in the vicinity of the imperfection. For such cases, the local stress field will exhibit singular but single-valued behavior, while the displacements will be finite and multivalued. This displacement discontinuity can be measured by evaluating the cyclic property around a closed contour C that encloses the dislocation line D shown in Fig. 15.2. The value of this discontinuity is called the *Burgers vector* b and is given by the following integral relation

$$b_i = \oint_C du_i = \oint_C \frac{\partial u_i}{\partial x_j} dx_j \tag{15.1.1}$$

Note that for the cases shown in Fig. 15.2, the magnitude of the Burgers vector will be one atomic spacing.

In order to determine the elastic stress and displacement fields around edge and screw dislocations, we consider idealized elastic models. The edge dislocation model is shown in Fig. 15.3. For this case

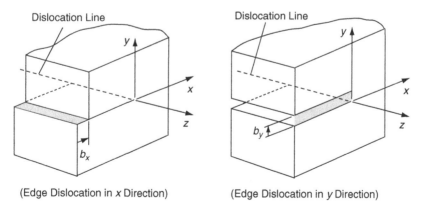

(Edge Dislocation in x Direction)   (Edge Dislocation in y Direction)

FIG. 15.3 Edge dislocation models.

the medium has been cut in the x,z-plane for $x \le 0$, and the dislocation line coincides with the z-axis. Two cases can be considered that include displacement discontinuities in either the x or y directions. The action of these discontinuities produces a local stress and strain field that we wish to determine. For the edge dislocation, a plane strain displacement field in the x,y-plane can be chosen. The Burgers vector for the general case with both in-plane discontinuities would read $b = (b_x, b_y, 0)$. This type of problem can be solved by complex variable methods using the cyclic displacement condition $[u + iv]_C = b_x + i b_y$, where the contour C lies in the x,y-plane and encloses the origin. We expect in this problem singular stresses at the origin.

Example 15.1 Edge dislocation in the x direction

We first consider in detail an edge dislocation where $b_x = b$, $b_y = 0$. The appropriate displacement field must give rise to the required multivaluedness, and this can be accomplished through a field of the form

$$u = \frac{b}{2\pi}\left[\tan^{-1}\frac{y}{x} + \frac{1}{2(1-\nu)}\frac{xy}{x^2+y^2}\right]$$

$$v = -\frac{b}{2\pi}\left[\frac{1-2\nu}{4(1-\nu)}\log(x^2+y^2) - \frac{1}{2(1-\nu)}\frac{y^2}{x^2+y^2}\right]$$

(15.1.2)

The stresses associated with these displacements are found to be

$$\sigma_x = -bB\frac{y(3x^2+y^2)}{(x^2+y^2)^2}$$

$$\sigma_y = bB\frac{y(x^2-y^2)}{(x^2+y^2)^2}$$

$$\tau_{xy} = bB\frac{x(x^2-y^2)}{(x^2+y^2)^2}$$

(15.1.3)

$$\sigma_z = \nu(\sigma_x + \sigma_y)$$

$$\tau_{xz} = \tau_{yz} = 0$$

where $B = \mu/2\pi(1-\nu)$. In cylindrical coordinates, the stresses are

$$\sigma_r = \sigma_\theta = -\frac{bB}{r}\sin\theta$$

$$\tau_{r\theta} = \frac{bB}{r}\cos\theta$$

(15.1.4)

It should be noted that this solution actually follows from a portion of the general Michell solution (8.3.6), $\phi = b_{12}r\log r\sin\theta$.

A similar set of field functions can be determined for the edge dislocation case of a y discontinuity with $b_x = 0$, $b_y = b$.

Example 15.2 Screw dislocation in the *z* direction

Next consider the screw dislocation case, as shown in Fig. 15.4. For this problem the material is again cut in the *x,z*-plane for $x \leq 0$, and the dislocation line coincides with the *z*-axis. The displacement discontinuity is now taken in the *z* direction as shown, and thus the Burgers vector becomes $b_x = b_y = 0$, $b_z = b$.

This case can be easily solved with the following displacement field

$$u = v = 0$$

$$w = \frac{b}{2\pi} \tan^{-1} \frac{y}{x} \qquad (15.1.5)$$

Clearly, this field satisfies the required cyclic displacement discontinuity. As previously discussed, fields of the form (15.1.5) are commonly called *antiplane strain* elasticity (see the discussion in Section 7.4). The resulting stresses for this case are

$$\tau_{xz} = -\frac{\mu b}{2\pi} \frac{y}{x^2 + y^2}$$

$$\tau_{yz} = \frac{\mu b}{2\pi} \frac{x}{x^2 + y^2} \qquad (15.1.6)$$

$$\sigma_x = \sigma_y = \sigma_z = \tau_{xy} = 0$$

In cylindrical coordinates, these stresses can be expressed in simpler form as

$$\tau_{\theta z} = \frac{\mu b}{2\pi r}$$

$$\sigma_r = \sigma_\theta = \sigma_z = \tau_{r\theta} = \tau_{rz} = 0 \qquad (15.1.7)$$

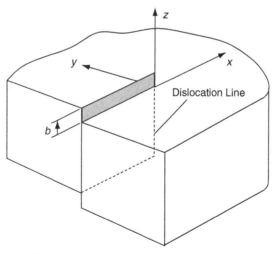

FIG. 15.4 Screw dislocation model.

Notice that the edge and screw dislocation stress fields are singular at the origin. This is expected because of the nature of the displacement discontinuities associated with each problem. Other aspects of dislocation modeling include determination of the associated strain energy (see Exercises), effect of external force fields, dislocation interaction, and movement. These and other modeling issues can be found in Weertman and Weertman (1964), Lardner (1974), and Landau and Lifshitz (1986).

15.2 Singular stress states

As discussed in the previous section, elasticity theory can be used to model defects in solids. Such studies may involve modeling of imperfections that are not simple edge or screw dislocations. For example, other defects may include voids and inclusions of arbitrary shape and distribution. In some cases these defects can produce localized, self-equilibrated residual stress fields from, say, trapped gases, thermal mismatch associated with an inclusion, and so forth. For many such problems, elasticity models can be developed by using solutions from a particular solution class sometimes referred to as *singular stress states*. These stress states include a variety of concentrated force and moment systems yielding stress, strain, and displacement fields that have singular behaviors at particular points in the domain. Such cases commonly include concentrated forces as developed in the solution of the Kelvin problem (see Example 13.1). Combinations and superposition of this fundamental solution are normally made to generate more complex and applicable solutions. We now develop some basic singular stress states and investigate their fundamental features.

Define a *regular elastic state* in a domain D as the set

$$S(x) = \{u, e, \sigma\} \tag{15.2.1}$$

where the displacement vector u and stress and strain tensors σ and e satisfy the elasticity field equations in D.

We use the Papkovich−Neuber solution scheme from Section 13.4 with redefined scalar and vector potential functions to allow the displacement solution to be expressed as

$$2\mu u = \nabla(\phi + R \cdot \psi) - 4(1 - \nu)\psi \tag{15.2.2}$$

where ϕ is the scalar potential and ψ is the vector potential satisfying the equations

$$\nabla^2 \phi = -\frac{R \cdot F}{2(1 - \nu)}, \quad \nabla^2 \psi = \frac{F}{2(1 - \nu)} \tag{15.2.3}$$

with body force F. Using elements of potential theory (Kellogg, 1953), a particular solution to equations (15.2.3) in a bounded domain D can be written as

$$\phi(x) = \frac{1}{8\pi(1 - \nu)} \int_D \frac{\xi \cdot F(\xi)}{\widehat{R}(x, \xi)} \, dV(\xi)$$

$$\psi(x) = -\frac{1}{8\pi(1 - \nu)} \int_D \frac{F(\xi)}{\widehat{R}(x, \xi)} \, dV(\xi) \tag{15.2.4}$$

where $\widehat{R} = |x - \xi|$.

Useful relations for the dilatation, strains, and stresses are given by

$$e_{kk} = -\frac{1-2v}{\mu}\psi_{k,k}$$

$$e_{ij} = \frac{1}{2\mu}\left[\phi_{,ij} - (1-2v)(\psi_{i,j}+\psi_{j,i}) + x_k\psi_{k,ij}\right] \qquad (15.2.5)$$

$$\sigma_{ij} = \phi_{,ij} - 2v\delta_{ij}\psi_{k,k} - (1-2v)(\psi_{i,j}+\psi_{j,i}) + x_k\psi_{k,ij}$$

Let us now investigate a series of example singular states of interest. Zero body forces will be chosen for these examples.

Example 15.3 Concentrated force in an infinite medium (Kelvin problem)

Consider first the simplest singular state problem of a concentrated force acting in an infinite medium, as shown in Fig. 15.5. Recall this was referred to as the *Kelvin problem* and was solved previously in Example 13.1. The solution to this problem is given by the Papkovich potentials

$$\phi = 0, \quad \psi = -\frac{1}{8\pi(1-v)}\frac{P}{R} \qquad (15.2.6)$$

where $R = |x| = \sqrt{x^2+y^2+z^2}$.

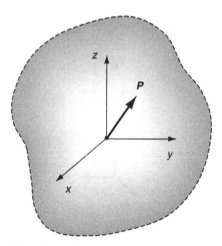

FIG. 15.5 Concentrated force singular state problem.

Example 15.4 Kelvin state with unit loads in coordinate directions

Consider next the combined Kelvin problem with unit loads a_α ($\alpha = 1, 2, 3$) acting along each of three coordinate directions, as shown in Fig. 15.6. This singular state is denoted by $S^\alpha(x)$ and is given by the potentials

$$\phi^\alpha = 0, \quad \psi_i^\alpha = -C\frac{\delta_{i\alpha}}{R}, \quad \text{where } C = \frac{1}{8\pi(1-\nu)} \tag{15.2.7}$$

The displacements and stresses corresponding to this case become

$$u_i^\alpha = \frac{C}{2\mu R}\left[\frac{x_\alpha x_i}{R^2} + (3 - 4\nu)\delta_{\alpha i}\right]$$

$$\sigma_{ij}^\alpha = -\frac{C}{R^3}\left[\frac{3x_\alpha x_i x_j}{R^2} + (1 - 2\nu)(\delta_{\alpha i}x_j + \delta_{\alpha j}x_i - \delta_{ij}x_\alpha)\right] \tag{15.2.8}$$

As a special case of this state, consider $a_\alpha = [0, 0, 1]$, which would be state $S^z(x)$ with potentials

$$\phi^z = \psi_x^z = \psi_y^z = 0, \quad \psi_z^z = -\frac{C}{R} \tag{15.2.9}$$

and in spherical coordinates (R,ϕ,θ) (see Fig. 1.6) yields the following displacements and stresses

$$u_R = \frac{2C(1-\nu)}{\mu}\frac{\cos\phi}{R}, \quad u_\phi = -\frac{C(3-4\nu)}{2\mu}\frac{\sin\phi}{R}, \quad u_\theta = 0$$

$$\sigma_R = -2C(2-\nu)\frac{\cos\phi}{R^2}, \quad \sigma_\theta = \sigma_\phi = C(1-2\nu)\frac{\cos\phi}{R^2} \tag{15.2.10}$$

$$\tau_{R\phi} = C(1-2\nu)\frac{\sin\phi}{R^2}, \quad \tau_{R\theta} = \tau_{\phi\theta} = 0$$

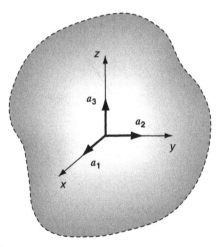

FIG. 15.6 Unit concentrated loadings.

For the case with the force in the x direction, that is, the state $S^x(x)$, we get the following fields

$$u_R = \frac{2C(1-2\nu)}{\mu R}\sin\phi\ \cos\theta$$

$$u_\phi = \frac{C(3-4\nu)}{2\mu R}\cos\phi\ \cos\theta$$

$$u_\theta = -\frac{C(3-4\nu)}{2\mu R}\sin\theta$$

$$\sigma_R = -\frac{2C(2-\nu)}{R^2}\sin\phi\ \cos\theta \qquad (15.2.11)$$

$$\sigma_\theta = \sigma_\phi = \frac{C(1-2\nu)}{R^2}\sin\phi\ \cos\theta$$

$$\tau_{R\phi} = \frac{C(2\nu-1)}{R^2}\cos\phi\ \cos\theta$$

$$\tau_{R\theta} = \frac{C(1-2\nu)}{R^2}\sin\theta,\ \tau_{\theta\phi} = 0$$

Notice that for the Kelvin state the displacements are of order $O(1/R)$, while the stresses are $O(1/R^2)$, and that

$$\int_S T^n\, ds = P, \quad \int_S R\times T^n\, ds = 0$$

for any closed surface S enclosing the origin.

Using the basic Kelvin problem, many *related singular states* can be generated. For example, define $S'(x) = S_{,\alpha} = \{u_{,\alpha}, \sigma_{,\alpha}, e_{,\alpha}\}$, where $\alpha = 1, 2, 3$. Now, if the state S is generated by the Papkovich potentials $\phi(x)$ and $\psi(x)$, then S' is generated by

$$\phi'(x) = \phi_{,\alpha} + \psi_\alpha$$
$$\psi'(x) = \psi_{i,\alpha}\, e_i \qquad (15.2.12)$$

Further, define the Kelvin state $S^\alpha(x;\xi)$ as that corresponding to a unit load applied in the x_α direction at point ξ, as shown in Fig. 15.7. Note that $S^\alpha(x;\xi) = S^\alpha(x-\xi)$. Also define the set of nine states $S^{\alpha\beta}(x)$ by the relation

$$S^{\alpha\beta}(x) = S^\alpha_{,\beta}(x) \qquad (15.2.13)$$

or equivalently

$$\hat{S}^{\alpha\beta}(x) = \frac{S^\alpha(x_1,x_2,x_3) - S^\alpha(x_1 - \delta_{\beta1}h, x_2 - \delta_{\beta2}h, x_3 - \delta_{\beta3}h)}{h}$$
$$= \frac{S^\alpha(x_1,x_2,x_3) - S^\alpha(x_1,x_2,x_3; \delta_{\beta1}h, \delta_{\beta2}h, \delta_{\beta3}h)}{h} \qquad (15.2.14)$$

and thus

$$S^{\alpha\beta} = \lim_{h\to0}\hat{S}^{\alpha\beta} \qquad (15.2.15)$$

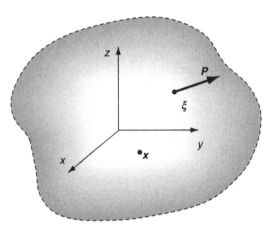

FIG. 15.7 Generalized Kelvin state.

Example 15.5 Force doublet

Consider the case of two concentrated forces acting along a common line of action but in opposite directions, as shown in Fig. 15.8. The magnitude of each force is specified as $1/h$, where h is the spacing distance between the two forces. We then wish to take the limit as $h \to 0$, and this type of system is called a *force doublet*. Recall that this problem was first defined in Chapter 13; see Exercise 13.20.

From our previous constructions, the elastic state for this case is given by $S\alpha\alpha(x)$ with no sum over α. This form matches the suggested solution scheme presented in Exercise 13.20.

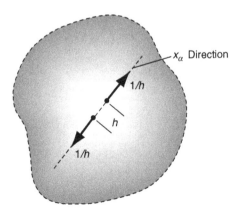

FIG. 15.8 Force doublet state.

Example 15.6 Force doublet with a moment (about γ-axis)

Consider again the case of a double-force system with equal and opposite forces but acting along different lines of action, as shown in Fig. 15.9. For this situation the two forces produce a moment about an axis perpendicular to the plane of the forces. Again, the magnitudes of the forces are taken to be $1/h$, where h is the spacing between the lines of action, and the limit is to be taken as $h \to 0$.

The elastic state for this case is specified by $\mathcal{S}^{\alpha\beta}(\boldsymbol{x})$, where $\alpha \neq \beta$, and the resulting moment acts along the γ-axis defined by the unit vector $\boldsymbol{e}_\gamma = \boldsymbol{e}_\alpha \times \boldsymbol{e}_\beta$. It can be observed from Fig. 15.9 that $\mathcal{S}^{\alpha\beta} = -\mathcal{S}^{\beta\alpha}$. From the previous Eqs. (15.2.7), (15.2.12), and (15.2.13), the Papkovich potentials for state $\mathcal{S}^{\alpha\beta}(\boldsymbol{x})$ are given by

$$\phi^{\alpha\beta} = -C\frac{\delta_{\alpha\beta}}{R}, \quad \psi_i^{\alpha\beta} = C\delta_{\alpha i}\frac{x_\beta}{R^3}, \quad C = \frac{1}{8\pi(1-v)} \tag{15.2.16}$$

and this yields the following displacements and stresses

$$u_i^{\alpha\beta} = -\frac{C}{2\mu R^3}\left(\frac{3x_\alpha x_\beta x_i}{R^2} + (3-4v)\delta_{\alpha i}x_\beta - \delta_{\alpha\beta}x_i - \delta_{\beta i}x_\alpha\right) \tag{15.2.17}$$

$$\sigma_{ij}^{\alpha\beta} = \frac{C}{R^3}\left(\frac{15x_\alpha x_\beta x_i x_j}{R^4} + \frac{3(1-2v)}{R^2}\left(\delta_{\alpha i}x_\beta x_j + \delta_{\alpha j}x_\beta x_i - \delta_{ij}x_\alpha x_\beta\right)\right.$$
$$\left. -\frac{3}{R^2}\left(\delta_{\beta i}x_\alpha x_j + \delta_{\beta j}x_\alpha x_i + \delta_{\alpha\beta}x_i x_j\right) - (1-2v)\left(\delta_{\alpha i}\delta_{\beta j} + \delta_{\alpha j}\delta_{\beta i} + \delta_{ij}\delta_{\alpha\beta}\right)\right) \tag{15.2.18}$$

Note the properties of state $\mathcal{S}^{\alpha\beta} = \{\boldsymbol{u}^{\alpha\beta}, \boldsymbol{\sigma}^{\alpha\beta}, \boldsymbol{e}^{\alpha\beta}\}$: $\boldsymbol{u}^{\alpha\beta} = O(R^{-2})$, $\boldsymbol{\sigma}^{\alpha\beta} = O(R^{-3})$, and

$$\int_S \boldsymbol{T}^{\alpha\beta}\,dS = 0, \quad \int_S \boldsymbol{R}\times\boldsymbol{T}^{\alpha\beta}\,dS = \varepsilon_{\gamma\alpha\beta}\boldsymbol{e}_\gamma$$

for any closed surface S enclosing the origin.

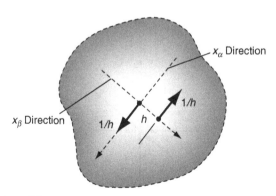

FIG. 15.9 Double-force system with a moment.

Example 15.7 Center of compression/dilatation

A *center of compression* (or dilatation) is constructed by the superposition of three mutually perpendicular force doublets, as shown in Fig. 15.10. The problem was introduced previously in Exercise 13.21. The elastic state for this force system is given by

$$S^o(x) = \frac{1}{2(1-2\nu)C} S^{\alpha\alpha}(x) \tag{15.2.19}$$

with summation over $\alpha = 1, 2, 3$. This state is then associated with the following potentials

$$\phi^o = \frac{-3}{2(1-2\nu)} \frac{1}{R}, \quad \psi_i^o = \frac{x_i}{2(1-2\nu)} \frac{1}{R^3} \tag{15.2.20}$$

and these yield the displacements and stresses

$$u_i^o = -\frac{x_i}{2\mu R^3}$$

$$\sigma_{ij}^o = \frac{1}{R^3}\left(\frac{3x_i x_j}{R^2} - \delta_{ij}\right) \tag{15.2.21}$$

Note that this elastic state has zero dilatation and rotation. In spherical coordinates the displacements and stresses are given by

$$u_R^o = -\frac{1}{2\mu R^2}, \quad u_\theta^o = u_\phi^o = 0$$

$$\sigma_R^o = \frac{2}{R^3}, \quad \sigma_\theta^o = \sigma_\phi^o = -\frac{1}{R^3}, \quad \tau_{R\theta}^o = \tau_{R\phi}^o = \tau_{\theta\phi}^o = 0 \tag{15.2.22}$$

A *center of dilatation* follows directly from the center of compression with a simple sign reversal and thus can be specified by $-S^o(x)$.

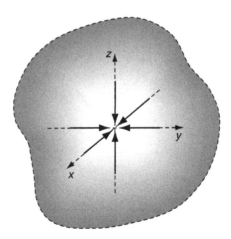

FIG. 15.10 Center of compression.

Example 15.8 Center of rotation

Using the cross-product representation, a *center of rotation* about the α-axis can be expressed by the state

$$^\alpha S(x) = \frac{1}{2}\epsilon_{\alpha\beta\gamma}S^{\beta\gamma}(x) \qquad (15.2.23)$$

where summation over β and γ is implied. Thus, centers of rotation about the coordinate axes can be written as

$$^1 S(x) = \frac{1}{2}\left(S^{23} - S^{32}\right)$$

$$^2 S(x) = \frac{1}{2}\left(S^{31} - S^{13}\right) \qquad (15.2.24)$$

$$^3 S(x) = \frac{1}{2}\left(S^{12} - S^{21}\right)$$

Using the solution (15.2.16), the potentials for this state become

$$^\alpha \phi = 0, \quad ^\alpha \psi_i = \frac{C}{2R^3}\epsilon_{\alpha ij}x_j \qquad (15.2.25)$$

with the constant C defined in relation (15.2.16). The corresponding displacements and stresses follow as

$$^\alpha u_i = -\frac{1}{8\pi\mu R^3}\,\epsilon_{\alpha ij}x_j$$

$$^\alpha \sigma_{ij} = -\frac{3}{8\pi R^5}\left(\epsilon_{\alpha ik}x_k x_j + \epsilon_{\alpha jk}x_k x_i\right) \qquad (15.2.26)$$

This state has the following properties: $\int_S {}^\alpha T\,dS = 0$, $\int_S R \times {}^\alpha T\,dS = \delta_{\alpha i}e_i$, where the integration is taken over any closed surface enclosing the origin.

In order to develop additional singular states that might be used to model distributed singularities, consider the following property.

Definition: Let $S(x;\lambda) = \{u(x;\lambda),\, \sigma(x;\lambda),\, e(x;\lambda)\}$ be a regular elastic state for each parameter $\lambda \in$ [a, b] with zero body forces. Then the state S^* defined by

$$S^*(x) = \int_a^b S(x;\lambda)d\lambda \qquad (15.2.27)$$

is also a regular elastic state. This statement is just another form of the superposition principle, and it allows the construction of integrated combinations of singular elastic states as shown in the next example.

Example 15.9 Half line of dilatation

A *line of dilatation* may be created through the superposition relation (15.2.27) by combining centers of dilatation. Consider the case shown in Fig. 15.11 that illustrates a line of dilatation over the negative x_3-axis. Let $S^o(x;\lambda)$ be a center of compression located at $(0, 0, -\lambda)$ for all $\lambda \in [0, \infty)$. From our previous definitions, it follows that

$$^zS^o(x) = -\int_0^\infty S^o(x;\lambda)d\lambda \tag{15.2.28}$$

where $S^o(x;\lambda) = S^o(x_1, x_2, x_3 + \lambda)$

will represent the state for a half line of dilatation along the negative x_3-axis, that is, $x_3 \in [0, \infty)$.

Using the displacement solution for the center of compression (15.2.21) in (15.2.28) yields the following displacement field for the problem

$$^zu_1^o = \frac{x_1}{2\mu}\int_0^\infty \frac{d\lambda}{\widehat{R}^3}$$

$$^zu_2^o = \frac{x_2}{2\mu}\int_0^\infty \frac{d\lambda}{\widehat{R}^3} \tag{15.2.29}$$

$$^zu_3^o = \frac{1}{2\mu}\int_0^\infty \frac{(x_3 + \lambda)d\lambda}{\widehat{R}^3}$$

which can be expressed in vector form as

$$^zu^o = -\frac{1}{2\mu}\int_0^\infty \nabla\left(\frac{1}{\widehat{R}}\right)d\lambda = \frac{1}{2\mu}\int_0^\infty \frac{d\lambda}{\widehat{R}} = \nabla\log(R + x_3) \tag{15.2.30}$$

The potentials for this state can be written as

$$^z\phi^o = \log(R + x_3), {}^z\psi_i^o = 0 \tag{15.2.31}$$

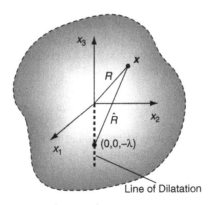

FIG. 15.11 Line of dilatation along the negative x_3-axis.

Notice the singularity at $R = -x_3$, and of course this behavior is expected along the negative x_3-axis because of the presence of the distributed centers of dilatation.

In spherical coordinates the displacement and stress fields become

$$^z u_R^o = \frac{1}{2\mu R}, \quad ^z u_\phi^o = \frac{-\sin\ \phi}{2\mu R(1 + \cos\ \phi)}, \quad ^z u_\theta^o = 0$$

$$^z \sigma_R^o = -\frac{1}{R^2}, \quad ^z \sigma_\phi^o = \frac{\cos\ \phi}{R^2(1 + \cos\ \phi)}, \quad ^z \sigma_\theta^o = \frac{1}{R^2(1 + \cos\ \phi)} \qquad (15.2.32)$$

$$^z \tau_{R\phi}^o = \frac{\sin\ \phi}{R^2(1 + \cos\ \phi)}, \quad ^z \tau_{R\theta}^o = {}^z \tau_{\theta\phi}^o = 0$$

15.3 Elasticity theory with distributed cracks

Many brittle solids such as rock, glass, ceramics, and concretes contain microcracks. It is generally accepted that the tensile and compressive strength of these materials is determined by the coalescence of these flaws into macrocracks, thus leading to overall fracture. The need to appropriately model such behaviors has led to many studies dealing with the elastic response of materials with distributed cracks. Some studies have simply developed moduli for elastic solids containing distributed cracks; see, for example, Budiansky and O'Connell (1976), Hoenig (1979), and Hori and Nemat-Nasser (1983). Other work (Kachanov, 1994) has investigated the strength of cracked solids by determining local crack interaction and propagation behaviors. Reviews by Kachanov (1994) and Chau, Wong, and Wang (1995) provide good summaries of work in this field.

We now wish to present some brief results of studies that have determined the elastic constants of microcracked solids, as shown in Fig. 15.12. It is assumed that a locally isotropic elastic material contains a distribution of planar elliptical cracks as shown. Some studies have assumed a random crack distribution, thus implying an overall isotropic response; other investigators have considered preferred

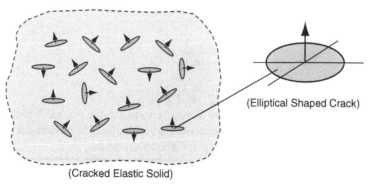

(Elliptical Shaped Crack)

(Cracked Elastic Solid)

FIG. 15.12 Elastic solid containing a distribution of cracks.

crack orientations, giving rise to anisotropic behaviors. Initial research assumed that the crack density is *dilute so that crack interaction effects can be neglected.* Later studies include crack interaction using the well-established *self-consistent approach.* In general, the effective moduli are found to depend on a *crack density parameter*, defined by the following

$$\varepsilon = \frac{2N}{\pi}\left\langle\frac{A^2}{P}\right\rangle \tag{15.3.1}$$

where N is the number of cracks per unit volume, A is the crack face area, P is the crack perimeter, and the angle brackets indicate the average value. Space limitations prevent us going into details of the various analyses, and thus only effective moduli results are given. Three particular examples are presented, and all cases assume no crack closure.

Example 15.10 Isotropic dilute crack distribution

Consider first the special case of a random dilute distribution of circular cracks of radius a. Note for the circular crack case the crack density parameter defined by (15.3.1) reduces to $\varepsilon = N\langle a^3\rangle$. Results for *the effective Young's modulus $\bar{E}$, shear modulus $\bar{\mu}$, and Poisson's ratio $\bar{\nu}$* are given by

$$\frac{\bar{E}}{E} = \frac{45(2-\nu)}{45(2-\nu)+16(1-\nu^2)(10-3\nu)\varepsilon}$$

$$\frac{\bar{\mu}}{\mu} = \frac{45(2-\nu)}{45(2-\nu)+32(1-\nu)(5-\nu)\varepsilon} \tag{15.3.2}$$

$$\varepsilon = \frac{45(\nu-\bar{\nu})(2-\nu)}{16(1-\nu^2)(10\bar{\nu}-3\nu\bar{\nu}-\nu)}$$

where E, μ, and ν are the moduli for the uncracked material.

Example 15.11 Planar transverse isotropic dilute crack distribution

Next consider the case of a dilute distribution of cracks arranged randomly but with all crack normals oriented along a common direction, as shown in Fig. 15.13. For this case results for the effective moduli are as follows

$$\frac{\bar{E}}{E} = \frac{3}{3+16(1-\nu^2)\varepsilon}$$

$$\frac{\bar{\mu}}{\mu} = \frac{3(2-\nu)}{3(2-\nu)+16(1-\nu)\varepsilon} \tag{15.3.3}$$

where $\bar{E}$ and $\bar{\mu}$ are the effective moduli in the direction normal to the cracks. A plot of this behavior for $\nu=0.25$ is shown in Fig. 15.14. It is observed that both effective moduli decrease with crack density, and the decrease is more pronounced for Young's modulus.

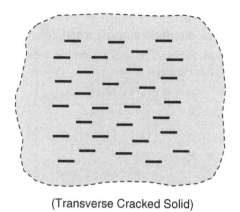

(Transverse Cracked Solid)

FIG. 15.13 Cracked elastic solid with a common crack orientation.

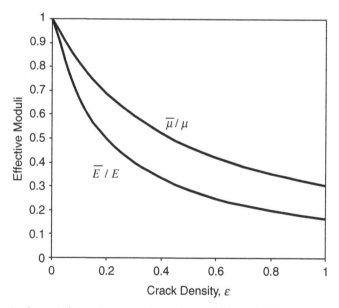

FIG. 15.14 Effective elastic moduli for a transversely cracked solid ($\nu = 0.25$).

Example 15.12 Isotropic crack distribution using a self-consistent model

Using the self-consistent method, effective moduli for the random distribution case can be developed. The results for this case are given by

$$\frac{\bar{E}}{E} = 1 - \frac{16(1-\bar{\nu}^2)(10-3\bar{\nu})\varepsilon}{45(2-\bar{\nu})}$$

$$\frac{\bar{\mu}}{\mu} = 1 - \frac{32(1-\bar{\nu})(5-\bar{\nu})\varepsilon}{45(2-\bar{\nu})} \qquad (15.3.4)$$

$$\varepsilon = \frac{45(\nu-\bar{\nu})(2-\bar{\nu})}{16(1-\bar{\nu}^2)(10\nu-3\nu\bar{\nu}-\bar{\nu})}$$

It is interesting to note that as $\varepsilon \to 9/16$, all effective moduli decrease to zero. This can be interpreted as a critical crack density where the material will lose its coherence. Although it would be expected that such a critical crack density would exist, the accuracy of this particular value is subject to the assumptions of the modeling and is unlikely to match universally with all materials.

In the search for appropriate models of brittle microcracking solids, there has been a desire to find a correlation between failure mechanisms (fracture) and effective elastic moduli. However, it has been pointed out (Kachanov, 1990, 1994) that such a correlation appears to be unlikely because failure-related properties such as stress intensity factors are correlated to local behavior, while the effective elastic moduli are determined by volume average procedures. External loadings on cracked solids can close some cracks and possibly produce frictional sliding, thereby affecting the overall moduli. This interesting process creates induced anisotropic behavior as a result of the applied loading. In addition to these studies of cracked solids, there also exists a large volume of work on determining effective elastic moduli for heterogeneous materials containing particulate and/or fiber phases—that is, distributed inclusions. A review of these studies has been given by Hashin (1983). Unfortunately, space does not permit a detailed review of this work.

15.4 Micropolar/couple-stress elasticity

As previously mentioned, the response of many heterogeneous materials has indicated dependency on microscale length parameters and on additional microstructural degrees of freedom. Solids exhibiting such behavior include a large variety of cemented particulate materials such as particulate composites, ceramics, and various concretes. This concept can be qualitatively illustrated by considering a simple lattice model of such materials, as shown in Fig. 15.15. Using such a scheme, the macro load transfer within the heterogeneous particulate solid is modeled using the microforces and moments between adjacent particles (see Chang and Ma, 1991; Sadd et al., 1992, 2004). Depending on the micro-structural packing geometry (sometimes referred to as fabric), this method establishes a lattice network that can be thought of as an interconnected series of elastic bar or beam elements interconnected at particle centers. Thus, the network represents in some way the material microstructure and brings into the model microstructural dimensions such as the grid size. Furthermore, the elastic network establishes internal bending moments and forces, which depend on internal degrees of freedom

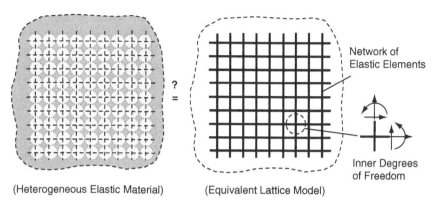

(Heterogeneous Elastic Material)   (Equivalent Lattice Model)

FIG. 15.15 Heterogeneous materials with microstructure.

(e.g., rotations) at each connecting point in the microstructure as shown. These internal rotations would be, in a sense, independent of the overall macro deformations.

This concept then suggests that an elastic continuum theory including an independent rotation field with concentrated pointwise moments might be suitable for modeling heterogeneous materials. Such approaches have been formulated under the names *Cosserat continuum; oriented media; asymmetric elasticity;* and *micropolar, micromorphic,* or *couple-stress theories.* The Cosserat continuum, developed in 1909, was historically one of the first models of this category. However, over the next 50 years very little activity occurred in this field. Renewed interest began during the 1960s, and numerous articles on theoretical refinements and particular analytical and computational applications were produced. The texts and articles by Eringen (1968, 1999) and Kunin (1983) provide detailed background on much of this work, while Nowacki (1986) presents a comprehensive account on dynamic and thermoelastic applications of such theories.

Micropolar theory incorporates an additional internal degree of freedom called the microrotation and allows for the existence of body and surface couples. For this approach, the new kinematic strain–deformation relation is expressed as

$$e_{ij} = u_{j,i} - \varepsilon_{ijl}\phi_l \tag{15.4.1}$$

where e_{ij} is the usual strain tensor, u_i is the displacement vector, and ϕ_i is the *microrotation vector.* Note that this new kinematic variable ϕ_i is independent of the displacement u_i, and thus is not in general the same as the usual macrorotation vector

$$\omega_i = \frac{1}{2}\,\varepsilon_{ijk}u_{k,j} \neq \phi_i \tag{15.4.2}$$

Later in our discussion we relax this restriction and develop a more specialized theory that normally allows simpler analytical problem solution.

The body and surface couples (moments) included in the new theory introduce additional terms in the equilibrium equations. Skipping the derivation details, the linear and angular equilibrium equations thus become

$$\sigma_{ji,j} + F_i = 0$$
$$m_{ji,j} + \varepsilon_{ijk}\sigma_{jk} + C_i = 0 \tag{15.4.3}$$

where σ_{ij} is the usual stress tensor, F_i is the body force, m_{ij} is the surface moment tensor normally referred to as the *couple-stress tensor*, and C_i is the body couple per unit volume. Notice that as a consequence of including couple stresses and body couples, *the stress tensor σ_{ij} is no longer symmetric*. For linear elastic isotropic materials, the constitutive relations for a micropolar material are given by

$$\sigma_{ij} = \lambda e_{kk}\delta_{ij} + (\mu + \kappa)e_{ij} + \mu e_{ji}$$
$$m_{ij} = \alpha\phi_{k,k}\delta_{ij} + \beta\phi_{i,j} + \gamma\phi_{j,i} \qquad (15.4.4)$$

where $\lambda, \mu, \kappa, \alpha, \beta, \gamma$ are the micropolar elastic moduli. Note that classical elasticity relations correspond to the case where $\kappa = \alpha = \beta = \gamma = 0$. The requirement of a positive definite strain energy function puts the following restrictions on these moduli

$$0 \le 3\lambda + 2\mu + \kappa, \ \ 0 \le 2\mu + \kappa, \ \ 0 \le \kappa$$
$$0 \le 3\alpha + \beta + \gamma, \ \ -\gamma \le \beta \le \gamma, \ \ 0 \le \gamma \qquad (15.4.5)$$

Lakes (2016) has provided some physical meaning of these elastic constants along with additional information on constants for the void case presented in Section 15.5.

Relations (15.4.1) and (15.4.4) can be substituted into the equilibrium equations (15.4.3) to establish two sets of governing field equations in terms of the displacements and microrotations. Appropriate boundary conditions to accompany these field equations are more problematic. For example, it is not completely clear how to specify the microrotation ϕ_i and/or couple-stress m_{ij} on domain boundaries. Some developments on this subject have determined particular field combinations whose boundary specification guarantees a unique solution to the problem.

15.4.1 Two-dimensional couple-stress theory

Rather than continuing on with the general three-dimensional equations, we now move directly into two-dimensional problems under plane strain conditions. In addition to the usual assumption $u = u(x,y)$, $v = v(x,y)$, $w = 0$, we include the restrictions on the microrotation, $\phi_x = \phi_y = 0$, $\phi_z = \phi_z(x,y)$. Furthermore, relation (15.4.2) is relaxed and *the microrotation is allowed to coincide with the macrorotation*

$$\phi_i = \omega_i = \frac{1}{2}\epsilon_{ijk}u_{k,j} \qquad (15.4.6)$$

This particular theory is then a special case of micropolar elasticity and is commonly referred to as couple-stress theory. Eringen (1968) refers to this theory as indeterminate because the antisymmetric part of the stress tensor is not determined solely by the constitutive relations.

Stresses on a typical in-plane element are shown in Fig. 15.16. Notice the similarity of this force system to the microstructural system illustrated previously in Fig. 15.15. For the two-dimensional case with no body forces or body couples, the equilibrium equations (15.4.3) reduce to

$$\frac{\partial\sigma_x}{\partial x} + \frac{\partial\tau_{yx}}{\partial y} = 0$$

$$\frac{\partial\tau_{xy}}{\partial x} + \frac{\partial\sigma_y}{\partial y} = 0 \qquad (15.4.7)$$

$$\frac{\partial m_{xz}}{\partial x} + \frac{\partial m_{yz}}{\partial y} + \tau_{xy} - \tau_{yx} = 0$$

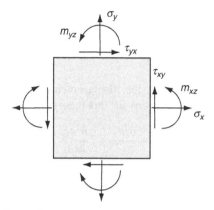

FIG. 15.16 Couple stresses on a planar element.

The in-plane strains can be expressed as

$$e_x = \frac{\partial u}{\partial x}, \quad e_y = \frac{\partial v}{\partial y}$$

$$e_{xy} = \frac{\partial v}{\partial x} - \phi_z, \quad e_{yx} = \frac{\partial u}{\partial y} + \phi_z \qquad (15.4.8)$$

while using (15.4.6) gives

$$\phi_z = \frac{1}{2}\left(\frac{\partial v}{\partial x} - \frac{\partial u}{\partial y}\right) \qquad (15.4.9)$$

Notice that substituting (15.4.9) into (15.4.8)$_2$ gives the result $e_{xy} = e_{yx}$.

The constitutive equations (15.4.4) yield the following forms for the stress components

$$\sigma_x = \lambda(e_x + e_y) + (2\mu + \kappa)e_x$$

$$\sigma_y = \lambda(e_x + e_y) + (2\mu + \kappa)e_y$$

$$\tau_{xy} = (2\mu + \kappa)e_{xy} = \tau_{yx} \qquad (15.4.10)$$

$$m_{xz} = \gamma\frac{\partial \phi_z}{\partial x}, \quad m_{yz} = \gamma\frac{\partial \phi_z}{\partial y}$$

In regard to the last pair of equations of this set, some authors (Mindlin, 1963; Boresi and Chong, 2011) define the gradients of the rotation ϕ_z as the curvatures. Thus, they establish a linear constitutive law between the couple stresses and curvatures. This approach is completely equivalent to the current method. It is to be noted from Eq. (15.4.10) that under the assumptions of couple-stress theory we find the unpleasant situation that the antisymmetric part of the stress tensor disappears from the constitutive relations. In order to remedy this, we can solve for the antisymmetric stress term from the moment equilibrium equation (15.4.7)$_3$ to get

$$\tau_{[xy]} = \frac{1}{2}(\tau_{xy} - \tau_{yx}) = -\frac{1}{2}\left(\frac{\partial m_{xz}}{\partial x} + \frac{\partial m_{yz}}{\partial y}\right)$$

$$= -\frac{\gamma}{2}\nabla^2\phi_z$$

(15.4.11)

By cross-differentiation we can eliminate the displacements from Eqs. (15.4.8) and (15.4.9) and develop the particular compatibility equations for this theory; i.e

$$\frac{\partial^2 e_x}{\partial y^2} + \frac{\partial^2 e_y}{\partial x^2} = 2\frac{\partial^2 e_{xy}}{\partial x \partial y}$$

$$\frac{\partial^2 \phi_z}{\partial x \partial y} = \frac{\partial^2 \phi_z}{\partial y \partial x}$$

(15.4.12)

$$\frac{\partial \phi_z}{\partial x} = \frac{\partial e_{xy}}{\partial x} - \frac{\partial e_x}{\partial y}$$

$$\frac{\partial \phi_z}{\partial y} = \frac{\partial e_y}{\partial x} - \frac{\partial e_{xy}}{\partial y}$$

Using the constitutive forms (15.4.10), these relations may be expressed in terms of the stresses as

$$\frac{\partial^2 \sigma_x}{\partial y^2} + \frac{\partial^2 \sigma_y}{\partial x^2} - \nu\nabla^2(\sigma_x + \sigma_y) = \frac{\partial^2}{\partial x \partial y}(\tau_{xy} + \tau_{yx})$$

$$\frac{\partial m_{xz}}{\partial y} = \frac{\partial m_{yz}}{\partial x}$$

(15.4.13)

$$m_{xz} = l^2 \frac{\partial}{\partial x}(\tau_{xy} + \tau_{yx}) - 2l^2 \frac{\partial}{\partial y}[\sigma_x - \nu(\sigma_x + \sigma_y)]$$

$$m_{yz} = 2l^2 \frac{\partial}{\partial x}[\sigma_y - \nu(\sigma_x + \sigma_y)] - l^2 \frac{\partial}{\partial y}(\tau_{xy} + \tau_{yx})$$

where $\nu = \lambda/(2\lambda + 2\mu + \kappa)$ and $l = \sqrt{\gamma/(4\mu + 2\kappa)}$ is a material constant with units of length. Notice that this result then introduces a length scale into the problem. If $l = 0$, the couple-stress effects are eliminated and the problem reduces to classical elasticity. It should also be pointed out that only three of the four equations in set (15.4.13) are independent because the second relation can be established from the other equations.

Proceeding along similar lines as classical elasticity, we introduce a stress function approach (Carlson, 1966) to solve (15.4.13). A self-equilibrated form satisfying (15.4.7) identically can be written as

$$\sigma_x = \frac{\partial^2 \Phi}{\partial y^2} - \frac{\partial^2 \Psi}{\partial x \partial y}, \qquad \sigma_y = \frac{\partial^2 \Phi}{\partial x^2} + \frac{\partial^2 \Psi}{\partial x \partial y}$$

$$\tau_{xy} = -\frac{\partial^2 \Phi}{\partial x \partial y} - \frac{\partial^2 \Psi}{\partial y^2}, \qquad \tau_{yx} = \frac{\partial^2 \Phi}{\partial x \partial y} + \frac{\partial^2 \Psi}{\partial x^2}$$

(15.4.14)

$$m_{xz} = \frac{\partial \Psi}{\partial x}, \qquad m_{yz} = \frac{\partial \Psi}{\partial y}$$

where $\Phi = \Phi(x,y)$ and $\Psi = \Psi(x,y)$ are the stress functions for this case. If Ψ is taken to be zero, the representation reduces to the usual Airy form with no couple stresses. Using form (15.4.14) in the compatibility equations (15.4.13) produces

$$\nabla^4 \Phi = 0$$

$$\frac{\partial}{\partial x}\left(\Psi - l^2\nabla^2\Psi\right) = -2(1-\nu)l^2\frac{\partial}{\partial y}\left(\nabla^2\Phi\right) \tag{15.4.15}$$

$$\frac{\partial}{\partial y}\left(\Psi - l^2\nabla^2\Psi\right) = 2(1-\nu)l^2\frac{\partial}{\partial x}\left(\nabla^2\Phi\right)$$

Differentiating the second equation with respect to x and the third with respect to y and adding eliminates Φ and gives the following result

$$\nabla^2\Psi - l^2\nabla^4\Psi = 0 \tag{15.4.16}$$

Thus, the two stress functions satisfy governing equations $(15.4.15)_1$ and (15.4.16). Now we consider a specific application of this theory for the following stress concentration problem.

Example 15.13 Stress concentration around a circular hole: micropolar elasticity

We now wish to investigate the effects of couple-stress theory on the two-dimensional stress distribution around a circular hole in an infinite medium under uniform tension at infinity. Recall that this problem was previously solved for the nonpolar case in Example 8.7 and the problem geometry is shown in Fig. 8.12. The hole is to have radius a, and the far-field stress is directed along the x direction as shown. The solution for this case is first developed for the micropolar model and then the additional simplification for couple-stress theory is incorporated. This solution was first presented by Kaloni and Ariman (1967) and later by Eringen (1968).

As expected for this problem the plane strain formulation and solution are best done in polar coordinates (r,θ). For this system, the equilibrium equations become

$$\frac{\partial \sigma_r}{\partial r} + \frac{1}{r}\frac{\partial \tau_{\theta r}}{\partial \theta} + \frac{\sigma_r - \sigma_\theta}{r} = 0$$

$$\frac{\partial \tau_{r\theta}}{\partial r} + \frac{1}{r}\frac{\partial \sigma_\theta}{\partial \theta} + \frac{\tau_{r\theta} - \tau_{\theta r}}{r} = 0 \tag{15.4.17}$$

$$\frac{\partial m_{rz}}{\partial r} + \frac{1}{r}\frac{\partial m_{\theta z}}{\partial \theta} + \frac{m_{rz}}{r} + \tau_{r\theta} - \tau_{\theta r} = 0$$

while the strain—deformation relations are

$$e_r = \frac{\partial u_r}{\partial r}, \quad e_\theta = \frac{1}{r}\left(\frac{\partial u_\theta}{\partial \theta} + u_r\right)$$

$$e_{r\theta} = \frac{\partial u_\theta}{\partial r} - \phi_z, \quad e_{\theta r} = \frac{1}{r}\left(\frac{\partial u_r}{\partial \theta} - u_\theta\right) + \phi_z \tag{15.4.18}$$

Continued

The constitutive equations in polar coordinates read as

$$\sigma_r = \lambda(e_r + e_\theta) + (2\mu + k)e_r$$

$$\sigma_\theta = \lambda(e_r + e_\theta) + (2\mu + k)e_\theta$$

$$\tau_{r\theta} = (\mu + k)e_{r\theta} + \mu e_{\theta r}, \quad \tau_{\theta r} = (\mu + k)e_{\theta r} + \mu e_{r\theta} \qquad (15.4.19)$$

$$m_{rz} = \gamma \frac{\partial \phi_z}{\partial r}, \quad m_{\theta z} = \gamma \frac{1}{r}\frac{\partial \phi_z}{\partial \theta}$$

and the strain—compatibility relations take the form

$$\frac{\partial e_{\theta r}}{\partial r} - \frac{1}{r}\frac{\partial e_r}{\partial \theta} + \frac{e_{\theta r} - e_{e\theta}}{r} - \frac{\partial \phi_z}{\partial r} = 0$$

$$\frac{\partial e_\theta}{\partial r} - \frac{1}{r}\frac{\partial e_{r\theta}}{\partial \theta} + \frac{e_\theta - e_r}{r} - \frac{1}{r}\frac{\partial \phi_z}{\partial \theta} = 0 \qquad (15.4.20)$$

$$\frac{\partial m_{\theta z}}{\partial r} - \frac{1}{r}\frac{\partial m_{rz}}{\partial \theta} + \frac{m_{\theta z}}{r} = 0$$

For the polar coordinate case, the stress—stress function relations become

$$\sigma_r = \frac{1}{r}\frac{\partial \Phi}{\partial r} + \frac{1}{r^2}\frac{\partial^2 \Phi}{\partial \theta^2} - \frac{1}{r}\frac{\partial^2 \Psi}{\partial r \partial \theta} + \frac{1}{r^2}\frac{\partial \Psi}{\partial \theta}$$

$$\sigma_\theta = \frac{1}{r^2}\frac{\partial^2 \Phi}{\partial r^2} + \frac{1}{r}\frac{\partial^2 \Psi}{r \partial r \partial \theta} - \frac{1}{r^2}\frac{\partial \Psi}{\partial \theta}$$

$$\tau_{r\theta} = -\frac{1}{r}\frac{\partial^2 \Phi}{\partial r \partial \theta} + \frac{1}{r^2}\frac{\partial \Phi}{\partial \theta} - \frac{1}{r}\frac{\partial \Psi}{\partial r} - \frac{1}{r^2}\frac{\partial^2 \Psi}{\partial \theta^2} \qquad (15.4.21)$$

$$\tau_{\theta r} = -\frac{1}{r}\frac{\partial^2 \Phi}{\partial r \partial \theta} + \frac{1}{r^2}\frac{\partial \Phi}{\partial \theta} + \frac{\partial^2 \Psi}{\partial r^2}$$

$$m_{rz} = \frac{\partial \Psi}{\partial r}, \quad m_{\theta z} = \frac{1}{r}\frac{\partial \Psi}{\partial \theta}$$

Using constitutive relations (15.4.19), the compatibility equations (15.4.20) can be expressed in terms of stresses, and combining this result with (15.4.21) will yield the governing equations for the stress functions in polar coordinates

$$\frac{\partial}{\partial r}\left(\Psi - l_1^2 \nabla^2 \Psi\right) = -2(1-\nu)l_2^2 \frac{1}{r}\frac{\partial}{\partial \theta}\left(\nabla^2 \Phi\right)$$

$$\frac{1}{r}\frac{\partial}{\partial \theta}\left(\Psi - l_1^2 \nabla^2 \Psi\right) = 2(1-\nu)l_2^2 \frac{\partial}{\partial r}\left(\nabla^2 \Phi\right) \qquad (15.4.22)$$

where

$$l_1^2 = \frac{\gamma(\mu + k)}{k(2\mu + k)}, \quad l_2^2 = \frac{\gamma}{2(2\mu + \kappa)}$$

$$\nabla^2 = \frac{\partial^2}{\partial r^2} + \frac{1}{r}\frac{\partial}{\partial r} + \frac{1}{r^2}\frac{\partial^2}{\partial \theta^2} \qquad (15.4.23)$$

Note that for the micropolar case, two length parameters l_1 and l_2 appear in the theory.

The appropriate solutions to equations (15.4.22) for the problem under study are given by

$$\Phi = \frac{T}{4}r^2(1 - \cos\ 2\theta) + A_1\ \log\ r + \left(\frac{A_2}{r^2} + A_3\right)\cos\ 2\theta$$

$$\Psi = \left(\frac{A_4}{r^2} + A_5K_2(r/l_1)\right)\sin\ 2\theta$$

(15.4.24)

where K_n is the *modified Bessel function of the second kind or order n* and A_i are constants to be determined with $A_4 = 8(1 - \nu)l_1^2 A_3$. Using this stress function solution, the components of the stress and couple stress then follow from Eq. (15.4.21) to be

$$\sigma_r = \frac{T}{2}(1 + \cos2\theta) + \frac{A_1}{r^2} - \left(\frac{6A_2}{r^2} + \frac{4A_3}{r^2} - \frac{6A_4}{r^4}\right)\cos2\theta$$

$$+ \frac{2A_5}{l_1 r}\left[\frac{3l_1}{r}K_o(r/l_1) + \left(1 + \frac{6l_1^2}{r^2}\right)K_1(r/l_1)\right]\cos2\theta$$

$$\sigma_\theta = \frac{T}{2}(1 - \cos2\theta) + \frac{A_1}{r^2} - \left(\frac{6A_2}{r^4} - \frac{6A_4}{r^4}\right)\cos2\theta$$

$$- \frac{2A_5}{l_1 r}\left[\frac{3l_1}{r}K_o(r/l_1) + \left(1 + \frac{6l_1^2}{r^2}\right)K_1(r/l_1)\right]\cos2\theta$$

$$\tau_{r\theta} = -\left(\frac{T}{2} + \frac{6A_2}{r^4} + \frac{2A_3}{r^2} - \frac{6A_4}{r^4}\right)\sin2\theta$$

(15.4.25)

$$+ \frac{A_5}{l_1 r}\left[\frac{6l_1}{r}K_o(r/l_1) + \left(1 + \frac{12l_1^2}{r^2}\right)K_1(r/l_1)\right]\sin2\theta$$

$$\tau_{\theta r} = -\left(\frac{T}{2} + \frac{6A_2}{r^4} + \frac{2A_3}{r^2} - \frac{6A_4}{r^4}\right)\sin2\theta$$

$$+ \frac{A_5}{l_1^2}\left[\left(1 + \frac{6l_1^2}{r^2}\right)K_o(r/l_1) + \left(\frac{3l_1}{r} - \frac{12l_1^3}{r^3}\right)K_1(r/l_1)\right]\sin2\theta$$

$$m_{rz} = -\left\{\frac{2A_4}{r^3} + \frac{A_5}{l_1}\left[\frac{2l_1}{r}K_o(r/l_1) + \left(1 + \frac{4l_1^2}{r^2}\right)K_1(r/l_1)\right]\right\}\sin2\theta$$

$$m_{\theta z} = \left\{\frac{2A_4}{r^3} + \frac{2A_5}{r}\left[K_o(r/l_1) + \frac{2l_1}{r}K_1(r/l_1)\right]\right\}\cos2\theta$$

For boundary conditions we use the usual forms for the nonpolar variables, while the couple stress m_{rz} is taken to vanish on the hole boundary and at infinity

$$\sigma_r(a,\theta) = \tau_{r\theta}(a,\theta) = m_{rz}(a,\theta) = 0$$

$$\sigma_r(\infty,\theta) = \frac{T}{2}(1 + \cos2\theta)$$

(15.4.26)

$$\tau_{r\theta}(\infty,\theta) = -\frac{T}{2}\sin2\theta$$

$$m_{rz}(\infty,\theta) = 0$$

Continued

Using these conditions, sufficient relations can be developed to determine the arbitrary constants A_i, giving the results

$$A_1 = -\frac{T}{2}a^2, A_2 = -\frac{Ta^4(1-F)}{4(1+F)}$$

$$A_3 = \frac{Ta^2}{2(1+F)}, A_4 = \frac{4T(1-\nu)a^2 l_2^2}{1+F} \qquad (15.4.27)$$

$$A_5 = -\frac{Tal_1 F}{(1+F)K_1(a/l_1)}$$

where

$$F = 8(1-\nu)\frac{l_2^2}{l_1^2}\left[4 + \frac{a^2}{l_1^2} + \frac{2a}{l_1}\frac{K_o(a/l_1)}{K_1(a/l_1)}\right]^{-1} \qquad (15.4.28)$$

This then completes the solution to the problem.

Let us now investigate the maximum stress and discuss the nature of the concentration behavior in the vicinity of the hole. As in the previous nonpolar case, the circumferential stress σ_θ on the hole boundary will be the maximum stress. From the previous solution

$$\sigma_\theta(a, \theta) = T\left(1 - \frac{2\cos 2\theta}{1+F}\right) \qquad (15.4.29)$$

As expected, the maximum value of this quantity occurs at $\theta = \pm\pi/2$, and thus the stress concentration factor for the micropolar stress problem is given by

$$K = \frac{(\sigma_\theta)_{\max}}{T} = \frac{3+F}{1+F} \qquad (15.4.30)$$

Notice that for micropolar theory, the stress concentration depends on the material parameters and on the *size of the hole*.

This problem has also been solved by Mindlin (1963) for couple-stress theory, and this result may be found from the current solution by letting $l_1 = l_2 = l$. Fig. 15.17 illustrates the behavior of the stress concentration factor as a function of a/l_1 for several cases of length ratio l_2/l_1 with $\nu = 0$. It is observed that the micropolar/couple-stress concentration factors are less than that predicted by classical theory ($K = 3$), and differences between the theories depend on the ratio of the hole size to the microstructural length parameter l_1 (or l). If the length parameter is small in comparison to the hole size, very small differences in the stress concentration predictions occur. For the case $l_1 = l_2 = l = 0$, it can be shown that $F \to 0$, thus giving $K = 3$, which matches with the classical result. Mindlin (1963) also investigated other far-field loading conditions for this problem. He showed that for the case of equal biaxial loading, the couple-stress effects disappear completely, while for pure shear loading couple-stress effects produce a significant reduction in the stress concentration when compared to classical theory.

Originally, it was hoped that this solution could be used to explain the observed reduction in stress concentration factors for small holes in regions of high stress gradients. Unfortunately, it has been pointed out by several authors (Schijve, 1966; Ellis and Smith, 1967; Kaloni and Ariman, 1967), that for typical metals the reduction in the stress concentration for small holes cannot be accurately accounted for using couple-stress theory.

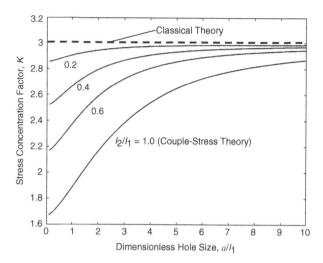

FIG. 15.17 Stress concentration behavior for the micropolar theory ($\nu = 0$).

Additional similar solutions for stress concentrations around circular inclusions have been developed by Weitsman (1965) and Hartranft and Sih (1965). More recent studies have had success in applying micropolar/couple-stress theory to fiber-reinforced composites (Sun and Yang, 1975) and granular materials (Chang and Ma, 1991). With respect to computational methods, micropolar finite element techniques have been developed by Kennedy and Kim (1987) and Kennedy (1999).

15.5 Elasticity theory with voids

A micromechanics model has been proposed for materials with distributed voids. The linear theory was originally developed by Cowin and Nunziato (1983) and a series of application papers followed, including Cowin and Puri (1983) and Cowin (1984a, b, 1985). The theory is intended for elastic materials containing a uniform distribution of small voids, as shown in Fig. 15.18. When the void volume vanishes, the material behavior reduces to classical elasticity theory. The primary new feature of this theory is the introduction of a volume fraction (related to void volume), which is taken as an independent kinematic variable. The other variables of displacement and strain are retained in their

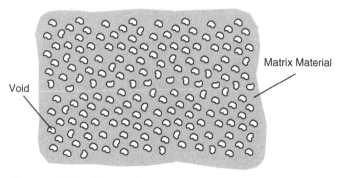

FIG. 15.18 Elastic continuum with distributed voids.

usual form. The inclusion of the new variable requires additional microforces to provide proper equilibrium of the micropore volume.

The theory begins by expressing the material mass density as the following product

$$\rho = \gamma v \tag{15.5.1}$$

where ρ is the bulk (overall) mass density, γ is the mass density of the matrix material, and v is the matrix volume fraction or volume distribution function. This function describes the way the medium is distributed in space and is taken to be an independent variable, thus introducing an additional kinematic degree of freedom in the theory. The linear theory with voids deals with small changes from a stress- and strain-free reference configuration. In this configuration relation (15.5.1) can be written as $\rho_R = \gamma_R v_R$. The independent kinematical variables of this theory are the usual displacements u_i and the change in volume fraction from the reference configuration expressed by

$$\phi = v - v_R \tag{15.5.2}$$

The strain–displacement relations are those of classical elasticity

$$e_{ij} = \frac{1}{2}\left(u_{i,j} + u_{j,i}\right) \tag{15.5.3}$$

and likewise for the equilibrium equations (with no body forces)

$$\sigma_{ij,j} = 0 \tag{15.5.4}$$

The general development of this theory includes external body forces and dynamic inertial terms. However, our brief presentation does not include these complexities.

In order to develop the microequilibrium of the void volume, new micromechanics theory involving the balance of equilibrated force is introduced. Details of this development are beyond the scope of our presentation, and we give only the final results

$$h_{i,i} + g = 0 \tag{15.5.5}$$

where h_i is the *equilibrated stress vector* and g is the *intrinsic equilibrated body force*. Simple physical meanings of these variables have proved difficult to provide. However, Cowin and Nunziato (1983) have indicated that these variables can be related to particular self-equilibrated singular-force systems as previously discussed in Section 15.2. In particular, h_i and g can be associated with double-force systems as presented in Example 15.5, and the expression $h_{i,i}$ can be related to centers of dilatation; see Example 15.7.

The constitutive equations for linear isotropic elastic materials with voids provide relations for the stress tensor, equilibrated stress vector, and intrinsic body force of the form

$$\sigma_{ij} = \lambda e_{kk}\delta_{ij} + 2\mu e_{ij} + \beta\phi\delta_{ij}$$
$$h_i = \alpha\phi_{,i} \tag{15.5.6}$$
$$g = -\omega\dot{\phi} - \xi\phi - \beta e_{kk}$$

where the material constants λ, μ, α, β, ξ, ω all depend on the reference fraction v_R and satisfy the inequalities

$$\mu \geq 0, \quad \alpha \geq 0, \quad \xi \geq 0, \quad \omega \geq 0, \quad 3\lambda + 2\mu \geq 0, \quad M = \frac{3\lambda + 2\mu}{\beta^2}\xi \geq 3 \tag{15.5.7}$$

Note that even though we have dropped dynamic inertial terms, constitutive relation $(15.5.6)_3$ includes a *time-dependent response* in the volume fraction. This fact indicates that the theory will predict a *viscoelastic* type of behavior (Cowin, 1985) even for problems with time-independent boundary conditions and homogeneous deformations.

For this theory, the boundary conditions on stress and displacement are the same as those of classical elasticity. The boundary conditions on the self-equilibrated stress vector are taken to have a vanishing normal component; that is, $h_i n_i = 0$, where n_i is the surface unit normal vector. Using this with the constitutive statement $(15.5.6)_2$ develops the boundary specification on the volume fraction

$$\phi_{,i} n_i = 0 \tag{15.5.8}$$

This completes our brief general presentation of the theory, and we will now discuss the solution to the stress concentration problem around a circular hole discussed previously in Example 15.13.

Example 15.14 Stress concentration around a circular hole—elasticity with voids

Consider again the stress concentration problem of a stress-free circular hole of radius a in an infinite plane under uniform tension, as shown in Fig. 8.12. We now outline the solution given by Cowin (1984b) and compare the results with the micropolar, couple-stress, and classical solutions. The problem is formulated under the usual plane stress conditions

$$\sigma_x = \sigma_x(x,y), \quad \sigma_y = \sigma_y(x,y), \quad \tau_{xy} = \tau_{xy}(x,y), \quad \sigma_z = \tau_{xz} = \tau_{yz} = 0$$

For this two-dimensional case the constitutive relations reduce to

$$\sigma_{ij} = \frac{2\mu}{\lambda + 2\mu}(\lambda e_{kk} + \beta\phi)\delta_{ij} + 2\mu e_{ij}$$

$$g = -\omega\dot\phi - \left(\xi - \frac{\beta^2}{\lambda + 2\mu}\right)\phi - \frac{2\mu\beta}{\lambda + 2\mu}e_{kk} \tag{15.5.9}$$

where all indices are taken over the limited range 1, 2. Using a stress formulation, the single nonzero compatibility relation becomes

$$\sigma_{kk,mm} - \frac{\mu\beta}{\lambda + \mu}\phi_{,mm} = 0 \tag{15.5.10}$$

Introducing the usual Airy stress function, denoted here by ψ, allows this relation to be written as

$$\nabla^4 \psi - \frac{\mu\beta}{\lambda + \mu}\nabla^2\phi = 0 \tag{15.5.11}$$

For this case, relation (15.5.5) for balance of equilibrated forces reduces to

$$\alpha\nabla^2\phi - \frac{\alpha}{h^2}\phi - \omega\dot\phi = \frac{\beta}{3\lambda + 2\mu}\left(\nabla^2\psi - \frac{\mu\beta}{\lambda + \mu}\phi\right) \tag{15.5.12}$$

The parameter h is defined by

$$\frac{\alpha}{h^2} = \xi - \frac{\beta^2}{\lambda + \mu} \tag{15.5.13}$$

Continued

and has units of length, and thus can be taken as a microstructural length measure for this particular theory.

Relations (15.5.11) and (15.5.12) now form the governing equations for the plane stress problem. The presence of the time-dependent derivative term in (15.5.12) requires some additional analysis. Using Laplace transform theory, Cowin (1984b) shows that under steady boundary conditions, the solutions ϕ and ψ can be determined from the limiting case where $t \to \infty$, which is related to taking $\omega = 0$. Thus, taking the Laplace transform of (15.5.11) and (15.5.12) gives the following

$$\nabla^4 \bar{\psi} - \frac{\mu\beta}{\lambda+\mu}\nabla^2\bar{\phi} = 0$$

$$\alpha\nabla^2\bar{\phi} - \frac{\alpha}{\bar{h}^2}\bar{\phi} = \frac{\beta}{3\lambda+2\mu}\left(\nabla^2\bar{\psi} - \frac{\mu\beta}{\lambda+\mu}\bar{\phi}\right)$$

(15.5.14)

where $\bar{\phi} = \bar{\phi}(s), \bar{\psi} = \bar{\psi}(s)$ are the standard Laplace transforms of ϕ, ψ, and s is the Laplace transform variable. The basic definition of this transform is given by

$$\bar{f}(s) = \int_0^\infty f(t)e^{-st}dt$$

and the parameter $\bar{h}$ is defined by $\frac{\alpha}{\bar{h}^2} = \frac{\alpha}{h^2} + \omega s$. Boundary conditions on the problem follow from our previous discussions to be

$$\sigma_r = \tau_{r\theta} = \frac{\partial\bar{\phi}}{\partial r} = 0 \quad \text{on} \quad r = a$$

For the circular hole problem, the solution to system (15.5.14) is developed in polar coordinates. Guided by the results from classical elasticity, we look for solutions of the form $f(r) + g(r)\cos2\theta$, where f and g are arbitrary functions of the radial coordinate. Employing this scheme, the properly bounded solution satisfying the boundary condition $\frac{\partial\bar{\phi}}{\partial r} = 0$ at $r = a$ is found to be

$$\bar{\phi} = \frac{-\xi\bar{p}}{M\beta\omega s + \beta\xi(M-3)} + \frac{A_3(\lambda+\mu)}{\mu\bar{h}^2\beta}[\bar{F}(r) - 1]\cos2\theta$$

$$\bar{\psi} = \frac{\mu\beta}{\lambda+\mu}\bar{h}^2\bar{\phi} + \left[\frac{\bar{p}r^2}{4} + A_1\log\, r\left(\frac{A_2}{r^2} + A_3 - \frac{\bar{p}r^2}{4}\right)\cos2\theta\right]$$

(15.5.15)

where $\bar{F}$ is given by

$$\bar{F}(r) = 1 + \frac{4\mu\xi\bar{h}^2}{(\lambda+\mu)M\omega s + 4\mu\xi N}\left[\frac{1}{r^2} + \frac{2\bar{h}K_2(r/\bar{h})}{a^3 K_2'(a/\bar{h})}\right]$$

(15.5.16)

and $\bar{p}$ is the Laplace transform of the uniaxial stress at infinity, K_2 is the modified Bessel function of the second kind of order 2

$$N = \frac{\lambda+\mu}{4\mu}(M-3) \geq 0$$

and the constants A_1, A_2, A_3 are determined from the stress-free boundary conditions as

$$A_1 = -\frac{1}{2}\bar{p}a^2, \quad A_2 = -\frac{1}{4}\bar{p}a^4, \quad A_3 = \frac{1}{2}\bar{p}a^2\bar{F}(a)$$

Note in relation (15.5.16) the bar on F indicates the dependency on the Laplace transform parameter s, and the bar is to be removed for the case where $s \to 0$ and $\overline{h}(s)$ is replaced by h.

This completes the solution for the Laplace-transformed volume fraction and Airy stress function. The transformed stress components can now be obtained from the Airy function using the usual relations

$$\overline{\sigma}_r = \frac{1}{r}\frac{\partial \overline{\psi}}{\partial r} + \frac{1}{r^2}\frac{\partial^2 \overline{\psi}}{\partial \theta^2}, \quad \overline{\sigma}_\theta = \frac{\partial^2 \overline{\psi}}{\partial r^2}, \quad \overline{\tau}_{r\theta} = \frac{1}{r^2}\frac{\partial \overline{\psi}}{\partial \theta} - \frac{1}{r}\frac{\partial^2 \overline{\psi}}{\partial r\partial \theta} \tag{15.5.17}$$

We now consider the case where the far-field tension T is a constant in time, and thus $\overline{p} = T/s$. Rather than formally inverting (inverse Laplace transformation) the resulting stress components generated from relations (15.5.17), Cowin develops results for the cases of $t = 0$ and $t \to \infty$. It turns out that for the initial condition at $t = 0$, the stresses match those found from classical elasticity (see Example 8.7). However, for the final-value case ($t \to \infty$), which implies ($s \to 0$), the stresses are different than predictions from classical theory.

Focusing our attention on only the hoop stress, the elasticity with voids solution for the final-value case is determined as

$$\sigma_\theta = \frac{T}{2}\left\{\left(1 + \frac{a^2}{r^2}\right) + \cos2\theta\left[a^2\frac{F''(r)}{F(a)} - \left(1 + 3\frac{a^4}{r^4}\right)\right]\right\} \tag{15.5.18}$$

The maximum value of this stress is again found at $r = a$ and $\theta = \pm\pi/2$ and is given by

$$(\sigma_\theta)_{max} = \sigma_\theta(a, \pm\pi/2) = T\left(3 - \frac{a^2}{2}\frac{F''(a)}{F(a)}\right)$$

$$= T\left[3 + \left(2N + [1 + (4 + L^2)N]\frac{K_1(L)}{LK_o(L)}\right)^{-1}\right] \tag{15.5.19}$$

where $L = a/h$. It is observed from this relation that the stress concentration factor $K = (\sigma_\theta)_{max}/T$ will always be greater than or equal to 3. Thus, the elasticity theory with voids predicts an *elevation* of the stress concentration when compared to the classical result. The behavior of the stress concentration factor as a function of the dimensionless hole size L is shown in Fig. 15.19. It can be seen that the concentration factor reduces to the classical result as L approaches zero or infinity. For a particular value of the material parameter N, the stress concentration takes on a maximum value at a finite intermediate value of L.

It is interesting to compare these results with our previous studies of the same stress concentration problem using some of the other micromechanical theories discussed in this chapter. Recall in Example 15.13 we solved the identical problem for micropolar and couple-stress theories, and results were given in Fig. 15.17. Figs. 15.17 and 15.19 both illustrate the stress concentration behavior as a function of a nondimensional ratio of hole radius divided by a microstructural length parameter. Although the current model with voids indicates an elevation of stress concentration, the micropolar and couple-stress results show a decrease in this factor. Micropolar/couple-stress theory also predicts that the largest difference from the classical result occurs at a dimensionless hole size ratio of zero. However, for elasticity with voids this difference occurs at a finite hole size ratio approximately between 2 and 3. It is apparent that micropolar theory (allowing independent microrotational deformation) gives fundamentally different results than the current void theory, which allows for an independent microvolumetric deformation.

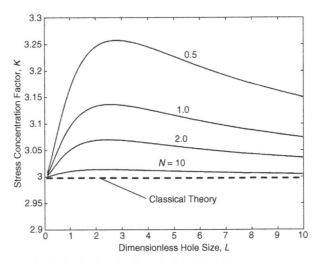

FIG. 15.19 Stress concentration behavior for elastic material with voids.

15.6 Doublet mechanics

We now wish to investigate an elastic micromechanical theory that has demonstrated applications for particulate materials in predicting observed behaviors that cannot be shown using classical continuum mechanics. The theory known as *doublet mechanics* (DM) was originally developed by Granik (1978). It has been applied to granular materials by Granik and Ferrari (1993) and Ferrari et al. (1997) and to asphalt concrete materials by Sadd and Dai (2004). Doublet mechanics is a micromechanical theory based on a discrete material model whereby solids are represented as arrays of points or nodes at finite distances. A pair of such nodes is referred to as a *doublet,* and the nodal spacing distances introduce *length scales* into the theory. Current applications of this scheme have normally used regular arrays of nodal spacing, thus generating a regular lattice microstructure with similarities to the micropolar model shown in Fig. 15.15. Each node in the array is allowed to have a translation and rotation, and increments of these variables are expanded in a Taylor series about the nodal point. The order at which the series is truncated defines the degree of approximation employed. The lowest-order case that uses only a single term in the series will not contain any length scales, while using additional terms results in a multi-length-scale theory. The allowable kinematics develop microstrains of elongation, shear, and torsion (about the doublet axis). Through appropriate constitutive assumptions, these microstrains can be related to corresponding elongational, shear, and torsional microstresses.

Although not necessary, *a granular interpretation* of doublet mechanics is commonly employed, in which the material is viewed as an assembly of circular or spherical particles. A pair of such particles represents a doublet, as shown in Fig. 15.20. Corresponding to the doublet *(A,B)* there exists a *doublet or branch vector* ζ_α connecting the adjacent particle centers and defining the doublet axis α. The magnitude of this vector $\eta_\alpha = |\zeta_\alpha|$ is simply the sum of the two radii for particles in contact. However, in general the particles need not be in contact, and the length scale η_α could be used to represent a more general microstructural feature. As mentioned, the kinematics allow relative elongational, shearing,

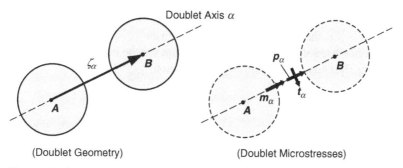

FIG. 15.20 Doublet mechanics geometry.

and torsional motions between the particles, and this is used to develop *elongational microstress* p_α, *shear microstress* t_α, *and torsional microstress* m_α, as shown in Fig. 15.20. It should be pointed out that these microstresses are not second-order tensors in the usual continuum mechanics sense. Rather, they are vector quantities that represent the elastic microforces and microcouples of interaction between doublet particles. Their directions are dependent on the doublet axes that are determined by the material microstructure. Also, these microstresses are not continuously distributed but rather exist only at particular points in the medium being simulated by DM theory.

If $u(x,t)$ is the displacement field coinciding with a particle displacement, then the increment function can be written as

$$\Delta u_\alpha = u(x + \zeta_\alpha, t) - u(x, t) \qquad (15.6.1)$$

where $\alpha = 1, \ldots, n$, and n is referred to as the *valence* of the lattice. Considering only the case where the doublet interactions are symmetric, it can be shown that the shear and torsional microdeformations and stresses vanish, and thus only extensional strains and stresses exist. For this case the extensional microstrain ε_α (representing the elongational deformation of the doublet vector) is defined by

$$\varepsilon_\alpha = \frac{q_\alpha \cdot \Delta u_\alpha}{\eta_\alpha} \qquad (15.6.2)$$

where $q_\alpha = \zeta_\alpha/\eta_\alpha$ is the unit vector in the α direction. The increment function (15.6.1) can be expanded in a Taylor series as

$$\Delta u_\alpha = \sum_{m=1}^{M} \frac{(\eta_\alpha)^m}{m!} (q_\alpha \cdot \nabla)^m u(x, t) \qquad (15.6.3)$$

Using this result in relation (15.6.2) develops the series expansion for the extensional microstrain

$$\varepsilon_\alpha = q_{\alpha i} \sum_{m=1}^{M} \frac{(\eta_\alpha)^{m-1}}{m!} q_{\alpha k_1} \cdots q_{\alpha k_m} \frac{\partial^m u_i}{\partial x_{k_1} \cdots \partial x_{k_m}} \qquad (15.6.4)$$

where $q_{\alpha k}$ are the direction cosines of the doublet directions with respect to the coordinate system. As mentioned, the number of terms used in the series expansion of the local deformation field determines the order of approximation in DM theory. For the first-order case ($m = 1$), the scaling parameter η_α drops from the formulation, and the elongational microstrain is reduced to

$$\varepsilon_\alpha = q_{\alpha i} q_{\alpha j} e_{ij} \tag{15.6.5}$$

where $e_{ij} = \frac{1}{2}(u_{i,j} + u_{j,i})$ is the usual continuum strain tensor.

For this case, it has been shown that the DM solution can be calculated directly from the corresponding continuum elasticity solution through the relation

$$\sigma_{ij} = \sum_{\alpha=1}^{n} q_{\alpha i} q_{\alpha j} p_\alpha \tag{15.6.6}$$

This result can be expressed in matrix form

$$\{\boldsymbol{\sigma}\} = [\boldsymbol{Q}]\{\boldsymbol{p}\} \Rightarrow \{\boldsymbol{p}\} = [\boldsymbol{Q}]^{-1}\{\boldsymbol{\sigma}\} \tag{15.6.7}$$

where for the two-dimensional case, $\{\boldsymbol{\sigma}\} = \{\sigma_x \sigma_y \tau_{xy}\}^T$ is the continuum elastic stress vector in Cartesian coordinates, $\{\boldsymbol{p}\}$ is the microstress vector, and $[\boldsymbol{Q}]$ is a transformation matrix. For plane problems, this transformation matrix can be written as

$$[\boldsymbol{Q}] = \begin{bmatrix} (q_{11})^2 & (q_{21})^2 & (q_{31})^2 \\ (q_{12})^2 & (q_{22})^2 & (q_{32})^2 \\ q_{11}q_{12} & q_{21}q_{22} & q_{31}q_{32} \end{bmatrix} \tag{15.6.8}$$

This result allows a straightforward development of first-order DM solutions for many problems of engineering interest; see Ferrari et al. (1997).

Example 15.15 Doublet mechanics solution of the Flamant problem

We now wish to investigate a specific application of the doublet mechanics model for a two-dimensional problem with regular particle packing microstructure. The case of interest is the Flamant problem of a concentrated force acting on the free surface of a semi-infinite solid, as shown in Fig. 15.21. The classical elasticity solution to this problem was originally developed in Section 8.4.7, and the Cartesian stress distribution was given by

$$\sigma_x = -\frac{2Px^2y}{\pi(x^2+y^2)^2}$$

$$\sigma_y = -\frac{2Py^3}{\pi(x^2+y^2)^2} \tag{15.6.9}$$

$$\tau_{xy} = -\frac{2Pxy^2}{\pi(x^2+y^2)^2}$$

This continuum mechanics solution specifies that the normal stresses are *everywhere compressive* in the half space, and a plot of the distribution of normal and shear stresses on a surface $y = $ constant was shown in Fig. 8.20.

The doublet mechanics model of this problem is established by choosing a regular two-dimensional hexagonal packing, as shown in Fig. 15.21. This geometrical microstructure establishes three doublet axes at angles $\gamma = 60°$ as shown. Using only first-order approximation,

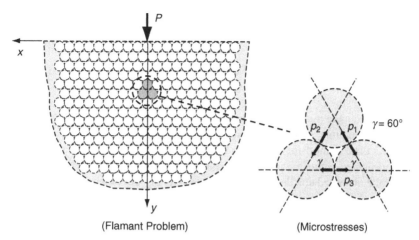

(Flamant Problem)   (Microstresses)

FIG. 15.21 Flamant problem for the doublet mechanics model.

DM shear and torsional microstresses vanish, leaving only elongational microstress components (p_1, p_2, p_3) as shown. Positive elongational components correspond to tensile forces between particles.

For this fabric geometry the transformation matrix (15.6.8) becomes

$$[Q] = \begin{bmatrix} \cos^2\gamma & \cos^2\gamma & 1 \\ \sin^2\gamma & \sin^2\gamma & 0 \\ -\cos\gamma\sin\gamma & \cos\gamma\sin\gamma & 0 \end{bmatrix} \tag{15.6.10}$$

Using this transformation in relation (15.6.7) produces the following microstresses

$$p_1 = -\frac{4Py^2\left(\sqrt{3}x+y\right)}{3\pi\left(x^2+y^2\right)^2}$$

$$p_2 = -\frac{4Py^2\left(\sqrt{3}x-y\right)}{3\pi\left(x^2+y^2\right)^2} \tag{15.6.11}$$

$$p_3 = -\frac{2Py\left(3x^2-y^2\right)}{3\pi\left(x^2+y^2\right)^2}$$

Although these DM microstresses actually exist only at discrete points and in specific directions as shown in Fig. 15.21, we use these results to make continuous contour plots over the half-space domain under study. In this fashion we can compare DM predictions with the corresponding classical elasticity results. Reviewing the stress fields given by (15.6.9) and (15.6.11), we can directly compare only the horizontal elasticity component σ_x with the doublet mechanics micro-

Continued

stress p_3. The other stress components act in different directions and thus do not allow a simple direct comparison.

Fig. 15.22 illustrates contour plots of the elasticity σ_x and DM p_3 stress components. As mentioned previously, the classical elasticity results predict a totally compressive stress field as shown. Note, however, the difference in predictions from doublet mechanics theory. There exists a symmetric region of tensile microstress below the loading point in the region $y \geq \sqrt{3}|x|$. It has been pointed out in the literature that there exists experimental evidence of such tensile behavior in granular and particulate composite materials under similar surface loading, and Ferrari et al. (1997) refer to this issue as *Flamant's paradox*. It would appear that micromechanical effects are the mechanisms for the observed tensile behaviors, and DM theory offers a possible approach to predict this phenomenon. Additional anomalous elastic behaviors have been reported for other plane elasticity problems; see Ferrari et al. (1997) and Sadd and Dai (2004).

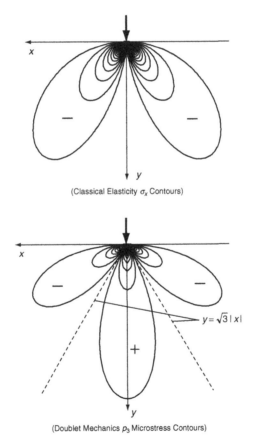

(Classical Elasticity σ_x Contours)

(Doublet Mechanics p_3 Microstress Contours)

FIG. 15.22 Comparison of horizontal stress fields from classical elasticity and doublet mechanics.

15.7 Higher gradient elasticity theories

As pointed out in Chapter 6, for classical elasticity the strain energy can be expressed as a quadratic function of the infinitesimal strain tensor $U = U(e_{ij})$, and thus the stress $\sigma_{ij} = \partial U/\partial e_{ij}$ will be a linear function of the strain. With interest in micromechanical behavior and especially focusing on finding better ways to model singularities, localization and size effects, researchers began to explore new expanded elasticity theories that would also include higher gradients of strain in the constitutive relation. These approaches have become known as *higher grade continua theories*. In addition to his work on micropolar elasticity, Mindlin (1965) was one of the originators of higher gradient theories, and he developed a linear elasticity theory that included the strain and its first and second derivatives in the constitutive relation for stress. It was anticipated that such theories would possibly remove various singularities in the solution field variables and thereby produce a more acceptable model of real materials. Starting in the 1980's Afantis and coworkers began extensive research in this field now called *gradient elasticity theory*. Countless papers have been published, and Aifantis (2003) and Askes and Aifantis (2011) provide reviews on many of the more recent developments and applications of this higher grade continua research.

For the isotropic case Mindlin (1965) expressed the strain energy in terms of the usual strain and also included a sizeable list of first and second derivatives of the strain

$$U = \frac{1}{2}\lambda e_{ii}e_{jj} + \mu e_{ij}e_{ij} + a_1 e_{ij,j}e_{ik,k}\cdots + b_1 e_{ii,jj}e_{kk,ll} + \cdots \tag{15.7.1}$$

Using the usual hyperelastic relation, the stress can be computed, and a typical example might look like

$$\sigma_{ij} = \frac{\partial U}{\partial e_{ij}} = \lambda e_{kk}\delta_{ij} + 2\mu e_{ij} + c_1 e_{kk,ll}\delta_{ij} + c_2 e_{ij,kk} + \frac{1}{2}c_3\left(e_{kk,ij} + e_{kk,ji}\right) \tag{15.7.2}$$

where λ and μ are the usual elastic constants and c_i being new material constants. Aifantis and others have generally simplified this constitutive form into

$$\sigma_{ij} = \lambda e_{kk}\delta_{ij} + 2\mu e_{ij} - c[\lambda e_{kk}\delta_{ij} + 2\mu e_{ij}]_{,ll} \tag{15.7.3}$$

where c is the remaining single gradient material constant which has units of length squared, and hence $\sqrt{c}$ would represent a *length measure*. We thus now have a linear elastic theory that contains the usual strain terms *and second order gradients in the strain tensor*. Over the years, other different and often more complicated constitutive forms have been developed. However, these are beyond the level of our brief study, and thus we will limit further investigation by only using constitutive form (15.7.3). The remaining elasticity field equations of strain-displacement, compatibility and equilibrium remain the same.

Following Ru and Aifantis (1993), we can establish a simple relationship between the gradient theory (15.7.3) and classical elasticity. First noting that constitutive law (15.7.3) can be expressed in direct notational and operator form

$$\boldsymbol{\sigma} = \left(1 - c\nabla^2\right)\boldsymbol{\sigma}^0, \quad \boldsymbol{\sigma}^0 = (\lambda \boldsymbol{I}tr + 2\mu)\boldsymbol{e} \tag{15.7.4}$$

where fields marked by $(\ )^0$ correspond to classical elasticity. Thus in absence of body forces the equilibrium equations can be expressed by

$$\nabla \cdot \boldsymbol{\sigma} = \left(1 - c\nabla^2\right)\nabla \cdot \boldsymbol{\sigma}^0 = 0 \qquad (15.7.5)$$

Considering the displacement form of the equilibrium equations (5.4.4), we can then write

$$\left(1 - c\nabla^2\right)\boldsymbol{L}\boldsymbol{u} = 0, \quad \boldsymbol{L} = \mu\nabla^2 + (\lambda + \mu)\nabla\nabla\cdot \qquad (15.7.6)$$

and it is noted that $\boldsymbol{L}\boldsymbol{u}^0 = 0$. Combining these results together generates the relation

$$\left(1 - c\nabla^2\right)\boldsymbol{u} = \boldsymbol{u}^0 \qquad (15.7.7)$$

and thus this particular gradient elasticity displacement solution is related to the classical result.

Special boundary conditions for this gradient theory are somewhat involved, and we will avoid a detailed development of this issue. Commonly an extra boundary condition for this theory is to also specify that the second normal derivative of the displacement vanishes; i.e. $\partial^2 u/\partial n^2 = 0$.

Numerous solutions using gradient elastic constitutive laws have been developed. We now will explore a couple of these and compare solutions with the corresponding classical elasticity case. This will allow us to examine the differences and to determine some of the effects of gradient theory.

Example 15.16 Gradient elasticity solution to screw dislocation in z-direction

Consider the elasticity problem of a screw dislocation in the z-direction as previously presented for the classical elasticity case in Example 15.2. For a screw dislocation the displacement field has the special form of anti-plane strain where

$$u = v = 0, \quad w = w(x, y) \qquad (15.7.8)$$

The problem geometry and jump in displacement were shown in Fig. 15.4.

We follow the gradient elasticity solution that has been given by Gutkin and Aifantis (1996). For the anti-plane strain displacement field, the only two non-zero strains are given by

$$e_{xz} = \frac{1}{2}\frac{\partial w}{\partial x}, \quad e_{yz} = \frac{1}{2}\frac{\partial w}{\partial y} \qquad (15.7.9)$$

Using the gradient constitutive law (15.7.3), the only two non-zero stresses are

$$\tau_{xz} = \mu\frac{\partial}{\partial x}\left[w - c\left(\frac{\partial^2 w}{\partial x^2} + \frac{\partial^2 w}{\partial y^2}\right)\right], \quad \tau_{yz} = \mu\frac{\partial}{\partial y}\left[w - c\left(\frac{\partial^2 w}{\partial x^2} + \frac{\partial^2 w}{\partial y^2}\right)\right] \qquad (15.7.10)$$

The equilibrium equations reduce to the single relation

$$\frac{\partial\tau_{xz}}{\partial x} + \frac{\partial\tau_{yz}}{\partial y} = 0 \qquad (15.7.11)$$

and using (15.7.10) in (15.7.11) gives the governing equation for the w-displacement

$$\frac{\partial^2 w}{\partial x^2} + \frac{\partial^2 w}{\partial y^2} - c\left(\frac{\partial^4 w}{\partial x^4} + 2\frac{\partial^4 w}{\partial x^2\partial y^2} + \frac{\partial^4 w}{\partial y^4}\right) = 0 \qquad (15.7.12)$$

This equation can be solved using Fourier transforms. We omit the details of such an analysis and only quote the final result

$$w = \frac{b}{2\pi} \left[\tan^{-1}\left(\frac{y}{x}\right) + \text{sgn}(y) \int_0^\infty \frac{\xi \sin \xi x}{\xi^2 + 1/c} e^{-|y|\sqrt{\xi^2 + 1/c}} \, d\xi \right] \tag{15.7.13}$$

where b denotes the *Burgers vector* specifying the magnitude of the discontinuous displacement, see Fig. 15.4. This displacement gives the following strain components

$$e_{xz} = \frac{b}{4\pi} \left[-\frac{y}{r^2} + \frac{y}{r\sqrt{c}} \, K_1\left(\frac{r}{\sqrt{c}}\right) \right]$$

$$e_{yz} = \frac{b}{4\pi} \left[\frac{x}{r^2} + \frac{x}{r\sqrt{c}} \, K_1\left(\frac{r}{\sqrt{c}}\right) \right] \tag{15.7.14}$$

with $r = \sqrt{x^2 + y^2}$ and K_1 being the modified Bessel function of the second kind of order 1. Note that the classical elasticity solution was given in Example 15.2

$$w = \frac{b}{2\pi} \, \tan^{-1}\left(\frac{y}{x}\right), \quad e_{xz} = -\frac{b}{4\pi} \frac{y}{r^2}, \quad e_{yz} = \frac{b}{4\pi} \frac{x}{r^2} \tag{15.7.15}$$

From the structure of the displacement, strain and stress solution fields, it is apparent that the gradient solution contains the classical solution plus an additive term due to the new model. This result is consistent with the findings of Ru and Aifantis (1993) for general boundary value problems of this theory. Using MATLAB Code C-11, Fig. 15.23 illustrates the spatial behaviors of the strain component e_{xz} for each theory. It is clearly seen that the gradient theory removes the singular behavior for the strain components (and also for the stresses). Note that the gradient strain and stress predications go to zero at the dislocation interface ($r = 0$). The removal of singular behaviors from classical elasticity is a common occurrence when using gradient theories.

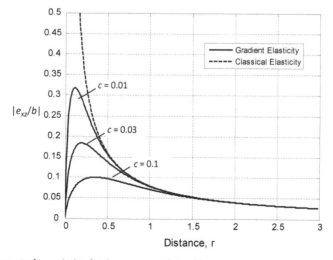

FIG. 15.23 Local shear strain e_{xz} behavior for a screw dislocation.

Example 15.17 Gradient elasticity solution to Flamant problem

Consider again the Flamant problem (Fig. 8.21) which represents a point force or line load P acting normal to the free surface of an elastic half-space. We wish to determine the gradient elasticity solution and compare particular features with the classical solution previously given in Section 8.4.7.

For this problem we follow the solution developed by Li et al. (2004). They formulated the problem in two-dimensional *plane strain* with

$$u = u(x, y), \quad v = v(x, y), \quad w = 0 \tag{15.7.16}$$

which gives the strain field

$$e_x = \frac{\partial u}{\partial x}, \quad e_y = \frac{\partial v}{\partial y}, \quad e_{xy} = \frac{1}{2}\left(\frac{\partial u}{\partial y} + \frac{\partial v}{\partial x}\right), \quad e_z = e_{xz} = e_{yz} = 0 \tag{15.7.17}$$

For this deformation, the stresses follow from constitutive law (15.7.3)

$$\sigma_x = \lambda(e_x + e_y) + 2\mu e_x - c[\lambda(e_x + e_y) + 2\mu e_x]_{,mm}$$
$$\sigma_y = \lambda(e_x + e_y) + 2\mu e_y - c[\lambda(e_x + e_y) + 2\mu e_y]_{,mm}$$
$$\sigma_z = \lambda(e_x + e_y) - c[\lambda(e_x + e_y)]_{,mm} \tag{15.7.18}$$
$$\tau_{xy} = 2\mu e_{xy} - 2\mu c[\lambda(e_x + e_y)]_{,mm}$$
$$\tau_{yz} = \tau_{xz} = 0$$

Combining (15.7.17) into (15.7.18), and then inserting this result into the equilibrium equations gives

$$\mu\nabla^2 u + (\lambda + \mu)\frac{\partial}{\partial x}\left(\frac{\partial u}{\partial x} + \frac{\partial v}{\partial y}\right) + c\nabla^2\left[\mu\nabla^2 u + (\lambda + \mu)\frac{\partial}{\partial x}\left(\frac{\partial u}{\partial x} + \frac{\partial v}{\partial y}\right)\right] = 0$$
$$\mu\nabla^2 v + (\lambda + \mu)\frac{\partial}{\partial y}\left(\frac{\partial u}{\partial x} + \frac{\partial v}{\partial y}\right) + c\nabla^2\left[\mu\nabla^2 v + (\lambda + \mu)\frac{\partial}{\partial y}\left(\frac{\partial u}{\partial x} + \frac{\partial v}{\partial y}\right)\right] = 0 \tag{15.7.19}$$

This system of equations can again be solved by Fourier transforms, and details are provided in Li et al. (2004). Extracting the solution for the in-plane stresses

$$\sigma_x = -\frac{P}{\pi}\int_0^\infty (1 - \xi y)e^{-\xi y}\cos(\xi x)d\xi = -\frac{2Px^2 y}{\pi(x^2 + y^2)^2}$$

$$\sigma_y = -\frac{P}{\pi}\int_0^\infty (1 + \xi y)e^{-\xi y}\cos(\xi x)d\xi = -\frac{2Py^3}{\pi(x^2 + y^2)^2} \tag{15.7.20}$$

$$\tau_{xy} = -\frac{P}{\pi}\int_0^\infty \xi y e^{-\xi y}\sin(\xi x)d\xi = -\frac{2Pxy^2}{\pi(x^2 + y^2)^2}$$

which surprisingly turn out to be the same as the classical elasticity results, see equations (15.6.9). Ru and Aifantis (1993) have shown that gradient problems of this type and boundary condition have this general feature of stress field solutions matching the classical prediction.

It should be pointed out that a more recent study of this problem by Lazar and Maugin (2006) using a different set of boundary conditions claims that *both* the gradient elastic stresses and displacements are free of singularities. Since Li et al. (2004) did not provide a finalized detailed displacement solution, we will choose the equivalent results from Lazar and Maugin (2006). They give the following relation for the vertical displacement

$$v(x,y) = \frac{P}{2\pi\mu}\left[2(1-\nu)(\log\ r + K_0(r/\sqrt{c}\ ) - \frac{x^2}{r^2} + \frac{x^2-y^2}{r^2}\left(\frac{2c}{r^2} - K_2(r/\sqrt{c})\right)\right] \qquad (15.7.21)$$

where again $K_{0,2}$ are the modified Bessel functions of the second kind of orders 0 and 2. We now wish to further explore this result on free surface $y = 0$. The corresponding prediction from classical elasticity is given by

$$v(x,0) = \frac{P(1-\nu)}{\pi\mu}\log x + v_o \qquad (15.7.22)$$

where v_o is an arbitrary rigid-body vertical displacement, and in fact the gradient solution (15.7.21) should also include such a term. Note the logarithmic singularity in the classical prediction.

The vertical surface displacements from gradient and classical theories are shown in Fig. 15.24 for the case $\nu = 0.3$ and $c = 0.1$. The displacements are normalized with respect to the factor $\frac{P(1-\nu)}{\pi\mu}$. Unlike the stresses, displacement results indicate that the gradient solution does not match with the classical prediction. More importantly the logarithmic singularity at the loading point in classical elasticity is not found, and the gradient theory predicts a finite value of the displacement directly under the loading point. Again we see that gradient elasticity eliminates a singular behavior found in classical theory.

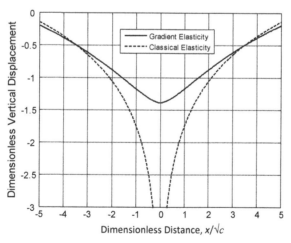

FIG. 15.24 Dimensionless vertical surface displacement comparison for the Flamant problem with $\nu = 0.3$ and $c = 0.1$.

Examination of published literature on gradient elasticity indicates a variety of constitutive choices which lead to different results for a given problem. In some cases, singular behaviors are removed from the classical predictions while in other cases they are not. Many additional problems in gradient elasticity have appeared in the literature and no doubt new developments of this theory will continue. With respect to other continuum mechanics areas, gradient theories have also had many applications in plasticity, viscoplasticity and localization problems.

Many other micromechanical theories of solids have been developed and reported in the literature. Our brief study has been able to discuss only a few of the more common modeling approaches within the context of linear elastic behavior. This has been and will continue to be a very challenging and interesting area in solid mechanics research.

References

Aifantis EC: Gradient deformation models at nano, micro, and macro scales, *J Eng Mater Technol* 121:189–202, 1999.

Aifantis EC: Update on a class of gradient theories, *Mech Mater* 35:259–280, 2003.

Askes H, Aifantis EC: Gradient elasticity in statics and dynamics: an overview of formulations, length scale identification procedures, finite element implementations and new results, *Int J Solids Struct* 48:1962–1990, 2011.

Biot MA: Twenty-one papers by M. A. Biot. In Tolstoy I, editor: *Acoustic Society of America*, New York, 1992, Woodbury.

Boresi AP, Chong KP: *Elasticity in engineering mechanics*, New York, 2011, John Wiley.

Budiansky B, O'Connell RJ: Elastic moduli of a cracked solid, *Int J Solids Struct* 12:81–97, 1976.

Carlson DE: Stress functions for plane problems with couple stresses, *J Appl Math Phys* 17:789–792, 1966.

Chang CS, Gao J: Second-gradient constitutive theory for granular material with random packing structure, *Int J Solids Struct* 32:2279–2293, 1995.

Chang CS, Ma LA: Micromechanical-base micro-polar theory for deformation of granular solids, *Int J Solids Struct* 28:67–86, 1991.

Chau KT, Wong RHC, Wang P: *Validation of microcrack models by experiments on real and replicated rocks, mechanics of materials with discontinuities and heterogeneities*, ASME Publication, AMD 201:133146.

Cowin SC: A note on the problem of pure bending for a linear elastic material with voids, *J Elast* 14:227–233, 1984a.

Cowin SC: The stresses around a hole in a linear elastic material with voids, *Quart J Mech Appl Math* 37:441–465, 1984b.

Cowin SC: The viscoelastic behavior of linear elastic materials with voids, *J Elast* 15:185–191, 1985.

Cowin SC, Nunziato JW: Linear elastic materials with voids, *J Elast* 13:125–147, 1983.

Cowin SC, Puri P: The classical pressure vessel problems for linear elastic materials with voids, *J Elast* 13:157–163, 1983.

Ellis RW, Smith CW: A thin-plate analysis and experimental evaluation of couple-stress effects, *Exp Mech* 7:372–380, 1967.

Eringen AC: Theory of micropolar elasticity. In , New York, 1968, Academic Press, pp 662–729. Liebowitz H, editor: *Fracture*, vol. II. New York, 1968, Academic Press, pp 662–729.

Eringen AC: *Microcontinuum field theories, I foundations and solids*, New York, 1999, Springer.

Ferrari M, Granik VT, Imam A, Nadeau J: *Advances in doublet mechanics*, Berlin, 1997, Springer.

Granik VT: *Microstructural mechanics of granular media*, Technical Report IM/MGU 78–241, Inst. Mech. of Moscow State University (in Russian).

Granik VT, Ferrari M: Microstructural mechanics of granular media, *Mech Mater* 15:301–322, 1993.

Gutkin MY, Aifantis EC: Screw dislocation in gradient elasticity, *Scripta Matl* 35:1353–1358, 1996.

Hartranft RJ, Sih GC: The effect of couple-stress on the stress concentration of a circular inclusion, *J Appl Mech* 32:429–431, 1965.

Hashin Z: Analysis of composite materials—a survey, *J Appl Mech* 50:481–504, 1983.

Hoenig A: Elastic moduli on a non-random cracked body, *Int J Solids Struct* 15:137–154, 1979.

Hori H, Nemat-Nasser S: Overall moduli of solids with microcracks: load induced anisotropy, *J Mech Phys Solids* 31:155–171, 1983.

Kachanov M: On the relationship between fracturing of brittle microcracking solid and its effective elastic properties. In Ju JW, Krajcinovic D, Schreyer HL, editors: *Damage mechanics in engineering materials*, ASME Publication, pp 11–15. AMD 109.

Kachanov M: Elastic solids with many cracks and related problems, *Adv Appl Mech* 30:259–445, 1994.

Kaloni PN, Ariman T: Stress concentration effects in micropolar elasticity, *J Appl Math Phys* 18:136–141, 1967.

Kellogg OD: *Foundations of potential theory*, New York, 1953, Dover.

Kennedy TC: Modeling failure in notched plates with micropolar strain softening, *Compos Struct* 44:71–79, 1999.

Kennedy TC, Kim JB: Finite element analysis of a crack in a micropolar elastic material. In Raghavan R, CoKonis TJ, editors: *Computers in engineering*, 1987, pp 439–444. ASME 3.

Kunin IA: *Elastic media with microstructure II three-dimensional models*, Berlin, 1983, Springer.

Landau LD, Lifshitz EM: *Theory of elasticity*, London, 1986, Pergamon Press.

Lakes RS: Physical meaning of elastic constants in Cosserat, void and microstretch elasticity, *J Mech Mater Struct* 11:217–229, 2016.

Lardner RW: *Mathematical theory of dislocations and fracture*, Toronto, 1974, University of Toronto Press.

Lazar M, Maugin GA: A note of line forces in gradient elasticity, *Mech Res Commun* 33:674–680, 2006.

Li S, Miskioglu I, Altan BS: Solution to line loading of a semi-infinite solid in gradient elasticity, *Int J Solids Struct* 41:3395–3410, 2004.

Mindlin RD: Influence of couple-stress on stress concentrations, *Exp Mech* 3(1–7), 1963.

Mindlin RD: Microstructure in linear elasticity, *Arch Ration Mech Anal* 16:51–78, 1964.

Mindlin RD: Second gradient of strain and surface-tension in linear elasticity, *Int J Solids Struct* 1:417–438, 1965.

Mura T: *Micromechanics of defects in solids*, Dordrecht, 1987, Martinus Nijhoff.

Nemat-Nasser S, Hori H: Micromechanics: overall properties of heterogeneous materials. In , Amsterdam, 1993, North-Holland, pp . Achenbach JD, Budiansky B, Lauwerier HA, Saffman PG, Van Wijngaarden L, Willlis JR, editors: *Adv Appl Mech*, vol. 37. Amsterdam, 1993, North-Holland.

Nowacki W: *Theory of asymmetric elasticity*, Oxford, UK, 1986, Pergamon Press.

Ostoja-Starzewski M, Wang C: Linear elasticity of planar Delaunay networks: random field characterization of effective moduli, *Acta Mech* 80:61–80, 1989.

Ostoja-Starzewski M, Wang C: Linear elasticity of planar Delaunay networks part II: Voigt and Reuss bounds, and modification for centroids, *Acta Mech* 84:47–61, 1990.

Ru CQ, Aifantis EC: A simple approach to solve boundary-value problems in gradient elasticity, *Acta Mech* 101: 59–68, 1993.

Sadd MH, Dai Q: A comparison of micromechanical modeling of asphalt materials using finite elements and doublet mechanics, *Mech Mater* 37:641–662, 2004.

Sadd MH, Dai Q, Parameswaran V, Shukla A: Microstructural simulation of asphalt materials: modeling and experimental studies, *J Mater Civ Eng* 16:107–115, 2004.

Sadd MH, Qiu L, Boardman WG, Shukla A: Modelling wave propagation in granular media using elastic networks, *Int J Rock Mech Min Sci Geomech* 29:161–170, 1992.

Schijve J: Note of couple stresses, *J Mech Phys Solids* 14:113–120, 1966.

Sun CT, Yang TY: A couple-stress theory for gridwork-reinforced media, *J Elast* 5:45–58, 1975.

Volterra V: Sur l'equilibre des corps elastiques multiplement connexes, *Ann Ec Norm (Paris)* 24(Series 3): 401–517, 1907.

Weertman J, Weertman JR: *Elementary dislocation theory*, New York, 1964, Macmillan.

Weitsman Y: Couple-stress effects on stress concentration around a cylindrical inclusion in a field of uniaxial tension, *J Appl Mech* 32:424–428, 1965.

Exercises

15.1 Show that the general plane strain edge dislocation problem shown in Fig. 15.3 can be solved using methods of Chapter 10 with the two complex potentials

$$\gamma(z) = \frac{i\mu b}{4\pi(1-v)}\log\, z,\ \ \psi(z) = -\frac{i\mu\bar{b}}{4\pi(1-v)}\log\, z$$

where $b = b_x + i b_y$. In particular, verify the cyclic property $[u + iv]_C = -b$, where C is any circuit in the x,y-plane around the dislocation line. Also determine the general stress and displacement field.

15.2 Justify that the edge dislocation solution (15.1.2) provides the required multivalued behavior for the displacement field. Explicitly develop the resulting stress fields given by (15.1.3) and (15.1.4).

15.3 Show that the screw dislocation displacement field (15.1.5) gives the stresses (15.1.6) and (15.1.7).

15.4 For the edge dislocation model, consider a cylinder of finite radius with axis along the dislocation line (z-axis). Show that although the stress solution gives rise to tractions on this cylindrical surface, the resultant forces in the x and y directions will vanish.

15.5 The stress field (15.1.7) for the screw dislocation produces no tangential or normal forces on a cylinder of finite radius with axis along the dislocation line (z-axis). However, show that if the cylinder is of finite length, the stress $\tau_{z\theta}$ on the ends will not necessarily be zero and will give rise to a resultant couple.

15.6 Show that the strain energy (per unit length) associated with the screw dislocation model of Example 15.2 is given by

$$W_{screw} = \frac{\mu b^2}{4\pi}\log\frac{R_o}{R_c}$$

where R_o is the outer radius of the crystal and R_c is the core radius of the dislocation. This quantity is sometimes referred to as the *self-energy*. The radial dimensions are somewhat arbitrary, although R_c is sometimes taken as five times the magnitude of the Burgers vector.

15.7 Using similar notation as Exercise 15.6, show that the strain energy associated with the edge dislocation model of Example 15.1 can be expressed by

$$W_{edge} = \frac{\mu b^2}{4\pi(1-v)}\log\frac{R_o}{R_c}$$

Note that this energy is larger than the value developed for the screw dislocation in Exercise 15.6. Evaluate the difference between these energies for the special case of $v = 1/3$.

15.8 For the Kelvin state as considered in Example 15.4, explicitly justify the displacement and stress results given in relations (15.2.8) and (15.2.10).

15.9 Verify that the displacements and stresses for the center of compression are given by (15.2.21) and (15.2.22).

15.10 A fiber discontinuity is to be modeled using a line of centers of dilatation along the x_1-axis from 0 to a. Show that the displacement field for this problem is given by

$$u_1 = \frac{1}{2\mu} \left(\frac{1}{\widehat{R}} - \frac{1}{R} \right)$$

$$u_2 = \frac{1}{2\mu} \left(\frac{1}{R} \frac{x_1 x_2}{x_2^2 + x_3^2} - \frac{1}{\widehat{R}} \frac{(x_1 - a)x_2}{x_2^2 + x_3^2} \right)$$

$$u_3 = \frac{1}{2\mu} \left(\frac{1}{R} \frac{x_1 x_3}{x_2^2 + x_3^2} - \frac{1}{\widehat{R}} \frac{(x_1 - a)x_3}{x_2^2 + x_3^2} \right)$$

where $\widehat{R} = \sqrt{(x_1 - a)^2 + x_2^2 + x_3^2}$ and R is identical to that illustrated in Fig. 15.11.

15.11* For the isotropic self-consistent crack distribution case in Example 15.12, show that for the case $\nu = 0.5$, relation (15.3.4)$_3$ reduces to

$$\varepsilon = \frac{9}{16} \left(\frac{1 - 2\bar{\nu}}{1 - \bar{\nu}^2} \right)$$

Verify the total loss of moduli at $\varepsilon = \frac{9}{16}$. Using these results, develop plots of the effective moduli ratios $\bar{\nu}/\nu, \bar{E}/E, \bar{\mu}/\mu$ versus the crack density. Compare these results with the corresponding values from the dilute case given in Example 15.10.

15.12 Develop the compatibility relations for couple-stress theory given by (15.4.12). Next, using the constitutive relations, eliminate the strains and rotations, and express these relations in terms of the stresses, thus verifying equations (15.4.13).

15.13 Explicitly justify that the stress–stress function relations (15.4.14) are a self-equilibrated form.

15.14 For the couple-stress theory, show that the two stress functions satisfy

$$\nabla^4 \Phi = 0, \quad \nabla^2 \Psi - l^2 \nabla^4 \Psi = 0$$

15.15 Using the general stress relations (15.4.25) for the stress concentration problem of Example 15.13, show that the circumferential stress on the boundary of the hole is given by

$$\sigma_\theta(a, \theta) = T \left(1 - \frac{2 \cos 2\theta}{1 + F} \right)$$

Verify that this expression gives a maximum at $\theta = \pm\pi/2$, and explicitly show that this value will reduce to the classical case of $3T$ by choosing $l_1 = l_2 = l = 0$.

15.16 Starting with the general relations (15.5.6), verify that the two-dimensional plane stress constitutive equations for elastic materials with voids are given by (15.5.9).

15.17 For elastic materials with voids, using the single strain–compatibility equation, develop the stress and stress function compatibility forms (15.5.10) and (15.5.11).

15.18* Compare the hoop stress $\sigma_\theta(r, \pi/2)$ predictions from elasticity with voids given by relation (15.5.18) with the corresponding results from classical theory. Choosing $N = \frac{1}{2}$ and $L = 2$, for the elastic material with voids, make a comparative plot of $\sigma_\theta(r, \pi/2)/T$ versus r/a for these two theories.

15.19* For the doublet mechanics Flamant solution in Example 15.15, develop contour plots (similar to Fig. 15.22) for the microstresses p_1 and p_2. Are there zones where these microstresses are tensile?

15.20 Consider the gradient elasticity problem under a one-dimensional deformation field of the form $u = u(x), v = w = 0$. Using constitutive form (15.7.3), determine the stress components. Next show that equilibrium equations reduce to $\dfrac{d\sigma_x}{dx} = 0$, and this will lead to the equation

$$\frac{d^4u}{dx^4} - \frac{1}{c}\frac{d^2u}{dx^2} = 0$$

Finally show that the solution for the displacement is given by

$$u = c_1 + c_2 x + c_3 \ \sinh\left(\frac{x}{\sqrt{c}}\right) + c_4 \ \cosh\left(\frac{x}{\sqrt{c}}\right)$$

15.21 Starting with the given displacement form (15.7.8), explicitly justify relations (15.7.9) − (15.7.12) in the gradient dislocation Example 15.16.

15.22 Using integral tables verify that the gradient Flamant stress integral solution in Example 15.17 reduces to the classical elasticity form as per equations (15.7.20).

Numerical finite and boundary element methods

Reviewing the previous chapters would indicate that analytical solutions to elasticity problems are normally accomplished for regions and loadings with relatively simple geometry. For example, many solutions can be developed for two-dimensional problems, while only a limited number exist for three dimensions. Solutions are commonly available for problems with simple shapes such as those having boundaries coinciding with Cartesian, cylindrical, and spherical coordinate surfaces. Unfortunately, problems with more general boundary shape and loading are commonly intractable or require very extensive mathematical analysis and numerical evaluation. Because most real-world problems involve structures with complicated shape and loading, a gap was formed between what was needed in applications and what can be solved by analytical closed-form methods.

Over the years, this need to determine deformation and stresses in complex problems has led to the development of many approximate and numerical solution methods (see brief discussion in Section 5.7). Approximate methods based on energy techniques were outlined in Section 6.7, but it was pointed out that these schemes have limited success in developing solutions for problems of complex shape. Methods of numerical stress analysis normally recast the mathematical elasticity boundary value problem into a direct numerical routine. One such early scheme is the *finite difference method* (FDM) in which derivatives of the governing field equations are replaced by algebraic difference equations. This method generates a system of algebraic equations at various computational grid points in the body, and the solution to the system determines the unknown variable at each grid point. Although simple in concept, FDM has not been able to provide a useful and accurate scheme to handle general problems with geometric and loading complexity. Over the past few decades, two methods have emerged that provide necessary accuracy, general applicability, and ease of use. This has led to their acceptance by the stress analysis community and has resulted in the development of many private and commercial computer codes implementing each numerical scheme.

The first of these techniques is known as the *finite element method* (FEM) and involves dividing the body under study into a number of pieces or subdomains called *elements*. The solution is then approximated over each element and is quantified in terms of values at special locations within the element called the *nodes*. The discretization process establishes an algebraic system of equations for unknown nodal values, which approximate the continuous solution. Because element size, shape, and approximating scheme can be varied to suit the problem, the method can accurately simulate solutions to problems of complex geometry and loading. The FEM has thus become a primary tool for practical stress analysis and is also used extensively in many other fields of engineering and science.

The second numerical scheme, called the *boundary element method* (BEM), is based on an integral statement of elasticity [see relation (6.4.7)]. This statement may be cast into a form with unknowns only over the boundary of the domain under study. The boundary integral equation is then solved using finite element concepts where the boundary is divided into elements and the solution is approximated over each element using appropriate interpolation functions. This method again produces an algebraic

Elasticity. https://doi.org/10.1016/B978-0-12-815987-3.00016-5

system of equations to solve for unknown nodal values that approximate the solution. Similar to FEM techniques, the BEM also allows variations in element size, shape, and approximating scheme to suit the application, and thus the method can accurately solve a large variety of problems.

Generally, an entire course is required to present sufficient finite and boundary element theory to prepare properly for the techniques' numerical/computational applications. Thus, the brief presentation in this chapter provides only an overview of each method, focusing on narrow applications for two-dimensional elasticity problems. The primary goal is to establish a basic level of understanding that will allow a quick look at applications and enable connections to be made between numerical solutions (simulations) and those developed analytically in the previous chapters. This brief introduction provides the groundwork for future and more detailed study in these important areas of computational solid mechanics.

16.1 Basics of the finite element method

Finite element procedures evolved out of matrix methods used by the structural mechanics community during the 1950s and 1960s. Over subsequent years, extensive research has clearly established and tested numerous FEM formulations, and the method has spread to applications in many fields of engineering and science. FEM techniques have been created for discrete and continuous problems including static and dynamic behavior with both linear and nonlinear responses. The method can be applied to one-, two-, or three-dimensional problems using a large variety of standard element types. We, however, limit our discussion to only two-dimensional, linear isotropic elastostatic problems. Numerous texts have been generated that are devoted exclusively to this subject; for example, Reddy (2006), Bathe (1982), Zienkiewicz and Taylor (2005), Fung and Tong (2001), and Cook et al. (2002).

As mentioned, the method discretizes the domain under study by dividing the region into subdomains called elements. In order to simplify formulation and application procedures, elements are normally chosen to be simple geometric shapes, and for two-dimensional problems these would be polygons including triangles and quadrilaterals. A two-dimensional example of a rectangular plate with a circular hole divided into triangular elements is shown in Fig. 16.1. Two different meshes (discretizations) of the same problem are illustrated, and even at this early stage in our discussion, it is apparent that improvement of the representation is found using the *finer mesh* with a larger number of smaller elements. Within each element, an approximate solution is developed, and this is quantified at particular locations called the nodes. Using a linear approximation, these nodes are located at the vertices of the triangular element as shown in the figure. Other higher-order approximations (quadratic, cubic, etc.) can also be used, resulting in additional nodes located in other positions. We present only a finite element formulation using linear, two-dimensional triangular elements.

Typical basic steps in a linear, static finite element analysis include the following

1. Discretize the body into a finite number of element subdomains.
2. Develop approximate solution over each element in terms of nodal values.
3. Based on system connectivity, assemble elements and apply all continuity and boundary conditions to develop an algebraic system of equations among nodal values.
4. Solve assembled system for nodal values; post process solution to determine additional variables of interest if necessary.

The basic formulation of the method lies in developing the element equation that approximately represents the elastic behavior of the element. This development is done for the generic case, thus creating a

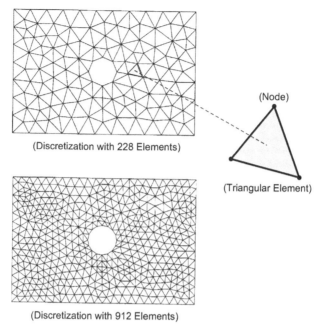

(Discretization with 228 Elements)

(Node)

(Triangular Element)

(Discretization with 912 Elements)

FIG. 16.1 Finite element discretization using triangular elements.

model applicable to all elements in the mesh. As pointed out in Chapter 6, energy methods offer schemes to develop approximate solutions to elasticity problems, and although these schemes were not practical for domains of complex shape, they can be easily applied over an element domain of simple geometry (i.e., triangle). Therefore, methods of virtual work leading to a Ritz approximation prove to be very useful in developing element equations for FEM elasticity applications. Another related scheme to develop the desired element equation uses a more mathematical approach known as the *method of weighted residuals*. This second technique starts with the governing differential equations, and through appropriate mathematical manipulations, a so-called *weak form* of the system is developed. Using a Ritz/Galerkin scheme, an approximate solution to the weak form is constructed, and this result is identical to the method based on energy and virtual work. Before developing the element equations, we first discuss the necessary procedures to create approximate solutions over an element in the system.

16.2 Approximating functions for two-dimensional linear triangular elements

Limiting our discussion to the two-dimensional case with triangular elements, we wish to investigate procedures necessary to develop a linear approximation of a scalar variable $u(x,y)$ over an element. Fig. 16.2 illustrates a typical triangular element denoted by Ω_e in the x,y-plane. Looking for a linear approximation, the variable is represented as

$$u(x,y) = c_1 + c_2 x + c_3 y \tag{16.2.1}$$

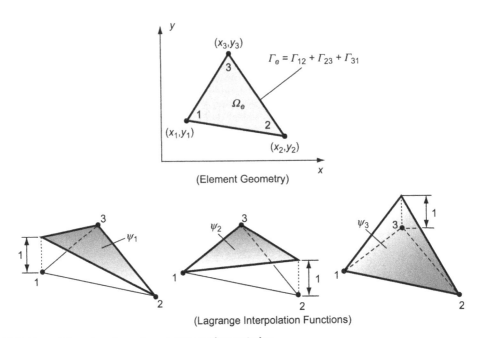

(Element Geometry)

(Lagrange Interpolation Functions)

FIG. 16.2 Linear triangular element geometry and interpolation.

where c_i are constants. It should be kept in mind that in general the solution variable is expected to have nonlinear behavior over the entire domain and our linear (planar) approximation is only proposed over the element. We therefore are using a piecewise linear approximation to represent the general nonlinear solution over the entire body. This approach usually gives sufficient accuracy if a large number of elements are used to represent the solution field. Other higher-order approximations including quadratic, cubic, and specialized nonlinear forms can also be used to improve the accuracy of the representation.

It is normally desired to express the representation (16.2.1) in terms of the nodal values of the solution variable. This can be accomplished by first evaluating the variable at each of the three nodes

$$u(x_1, y_1) = u_1 = c_1 + c_2 x_1 + c_3 y_1$$
$$u(x_2, y_2) = u_2 = c_1 + c_2 x_2 + c_3 y_2 \qquad (16.2.2)$$
$$u(x_3, y_3) = u_3 = c_1 + c_2 x_3 + c_3 y_3$$

Solving this system of algebraic equations, the constants c_i can be expressed in terms of the nodal values u_i, and the general results are given by

$$c_1 = \frac{1}{2A_e}(\alpha_1 u_1 + \alpha_2 u_2 + \alpha_3 u_3)$$

$$c_2 = \frac{1}{2A_e}(\beta_1 u_1 + \beta_2 u_2 + \beta_3 u_3) \qquad (16.2.3)$$

$$c_3 = \frac{1}{2A_e}(\gamma_1 u_1 + \gamma_2 u_2 + \gamma_3 u_3)$$

where A_e is the area of the element, and $\alpha_i = x_j y_k - x_k y_j$, $\beta_i = y_j - y_k$, $\gamma_i = x_k - x_j$, where $i \neq j \neq k$ and i, j, k permute in natural order. Substituting for c_i in (16.2.1) gives

$$u(x,y) = \frac{1}{2A_e} \left[(\alpha_1 u_1 + \alpha_2 u_2 + \alpha_3 u_3) \right.$$

$$+ (\beta_1 u_1 + \beta_2 u_2 + \beta_3 u_3)x$$

$$\left. + (\gamma_1 u_1 + \gamma_2 u_2 + \gamma_3 u_3)y \right]$$

$$= \sum_{i=1}^{3} u_i \psi_i(x,y)$$

(16.2.4)

where ψ_i are the *interpolation functions* for the triangular element given by

$$\psi_i(x,y) = \frac{1}{2A_e} (\alpha_i + \beta_i x + \gamma_i y)$$

(16.2.5)

It is noted that the form of the interpolation functions depends on the initial approximation assumption and on the shape of the element. Each of the three interpolation functions represents a planar surface as shown in Fig. 16.2, and it is observed that they will satisfy the following conditions

$$\psi_i(x_j, y_j) = \delta_{ij}, \quad \sum_{i=1}^{3} \psi_i = 1$$

(16.2.6)

Functions satisfying such conditions are referred to as *Lagrange interpolation functions*.

This method of using interpolation functions to represent the approximate solution over an element quantifies the approximation in terms of nodal values. In this fashion, the continuous solution over the entire problem domain is represented by discrete values at particular nodal locations. This discrete representation can be used to determine the solution at other points in the region using various other interpolation schemes. With these representation concepts established, we now pursue a brief development of the plane elasticity element equations using the virtual work formulation.

16.3 Virtual work formulation for plane elasticity

The principle of virtual work developed in Section 6.5 can be stated over a finite element volume V_e with boundary S_e as

$$\int_{V_e} \sigma_{ij} \delta e_{ij} dV = \int_{S_e} T_i^n \delta u_i dS + \int_{V_e} F_i \delta u_i dV$$

(16.3.1)

For plane elasticity with an element of uniform thickness h_e, $V_e = h_e \Omega_e$ and $S_e = h_e \Gamma_e$, and the previous relation can be reduced to the two-dimensional form

$$h_e \int_{\Omega_e} (\sigma_x \delta e_x + \sigma_y \delta e_y + 2\tau_{xy} \delta e_{xy}) \, dxdy$$

$$- h_e \int_{\Gamma_e} \left(T_x^n \delta u + T_y^n \delta v \right) dS - h_e \int_{\Omega_e} (F_x \delta u + F_y \delta v) dxdy = 0$$

(16.3.2)

Using matrix notation, this relation can be written as

$$
h_e \int_{\Omega_e} \left(\left\{ \begin{matrix} \delta e_x \\ \delta e_y \\ 2\delta e_{xy} \end{matrix} \right\}^T \left\{ \begin{matrix} \sigma_x \\ \sigma_y \\ \tau_{xy} \end{matrix} \right\} \right) dxdy
$$

$$
- h_e \int_{\Gamma_e} \left(\left\{ \begin{matrix} \delta u \\ \delta v \end{matrix} \right\}^T \left\{ \begin{matrix} T_x^n \\ T_y^n \end{matrix} \right\} \right) dS - h_e \int_{\Omega_e} \left(\left\{ \begin{matrix} \delta u \\ \delta v \end{matrix} \right\}^T \left\{ \begin{matrix} F_x \\ F_y \end{matrix} \right\} \right) dxdy = 0 \tag{16.3.3}
$$

We now proceed to develop an element formulation in terms of the displacements and choose a linear approximation for each component

$$
u(x,y) = \sum_{i=1}^{3} u_i \psi_i(x,y)
$$

$$
v(x,y) = \sum_{i=1}^{3} v_i \psi_i(x,y) \tag{16.3.4}
$$

where $\psi_i(x,y)$ are the Lagrange interpolation functions given by (16.2.5). Using this scheme there will be two unknowns or *degrees of freedom* at each node, resulting in a total of six degrees of freedom for the entire linear triangular element. Because the strains are related to displacement gradients, this interpolation choice results in a *constant strain element* (CST), and of course the stresses will also be element-wise constant. Relation (16.3.4) can be expressed in matrix form

$$
\left\{ \begin{matrix} u \\ v \end{matrix} \right\} = \begin{bmatrix} \psi_1 & 0 & \psi_2 & 0 & \psi_3 & 0 \\ 0 & \psi_1 & 0 & \psi_2 & 0 & \psi_3 \end{bmatrix} \left\{ \begin{matrix} u_1 \\ v_1 \\ u_2 \\ v_2 \\ u_3 \\ v_3 \end{matrix} \right\} = [\psi]\{\Delta\} \tag{16.3.5}
$$

The strains can then be written as

$$
\{e\} = \left\{ \begin{matrix} e_x \\ e_y \\ 2e_{xy} \end{matrix} \right\} = \begin{bmatrix} \partial/\partial x & 0 \\ 0 & \partial/\partial y \\ \partial/\partial y & \partial/\partial x \end{bmatrix} \left\{ \begin{matrix} u \\ v \end{matrix} \right\}
$$

$$
= \begin{bmatrix} \partial/\partial x & 0 \\ 0 & \partial/\partial y \\ \partial/\partial y & \partial/\partial x \end{bmatrix} \{\psi\}\{\Delta\} = [B]\{\Delta\} \tag{16.3.6}
$$

where

$$[B] = \begin{bmatrix} \dfrac{\partial\psi_1}{\partial x} & 0 & \dfrac{\partial\psi_2}{\partial x} & 0 & \dfrac{\partial\psi_3}{\partial x} & 0 \\[2ex] 0 & \dfrac{\partial\psi_1}{\partial y} & 0 & \dfrac{\partial\psi_2}{\partial y} & 0 & \dfrac{\partial\psi_3}{\partial y} \\[2ex] \dfrac{\partial\psi_1}{\partial y} & \dfrac{\partial\psi_1}{\partial x} & \dfrac{\partial\psi_2}{\partial y} & \dfrac{\partial\psi_2}{\partial x} & \dfrac{\partial\psi_3}{\partial y} & \dfrac{\partial\psi_3}{\partial x} \end{bmatrix} \tag{16.3.7}$$

$$= \frac{1}{2A_e} \begin{bmatrix} \beta_1 & 0 & \beta_2 & 0 & \beta_3 & 0 \\ 0 & \gamma_1 & 0 & \gamma_2 & 0 & \gamma_3 \\ \gamma_1 & \beta_1 & \gamma_2 & \beta_2 & \gamma_3 & \beta_3 \end{bmatrix}$$

Hooke's law then takes the form

$$\{\sigma\} = [C]\{e\} = [C][B]\{\Delta\} \tag{16.3.8}$$

where $[C]$ is the elasticity matrix that can be generalized to the orthotropic case (see Section 11.2) by

$$[C] = \begin{bmatrix} C_{11} & C_{12} & 0 \\ C_{12} & C_{22} & 0 \\ 0 & 0 & C_{66} \end{bmatrix} \tag{16.3.9}$$

For isotropic materials

$$C_{11} = C_{22} = \begin{cases} \dfrac{E}{1-v^2} & \dots \text{ plane stress} \\[3ex] \dfrac{E(1-v)}{(1+v)(1-2v)} & \dots \text{ plane strain} \end{cases}$$

$$C_{12} = \begin{cases} \dfrac{Ev}{1-v^2} & \dots \text{ plane stress} \\[3ex] \dfrac{Ev}{(1+v)(1-2v)} & \dots \text{ plane strain} \end{cases} \tag{16.3.10}$$

$$C_{66} = \mu = \frac{E}{2(1+v)} \quad \dots \text{ plane stress and plane strain}$$

Using results (16.3.5), (16.3.6), and (16.3.8) in the virtual work statement (16.3.3) gives

$$h_e \int_{\Omega_e} \{\delta\Delta\}^T \left([B]^T[C][B]\right)\{\Delta\}dxdy$$

$$-h_e \int_{\Omega_e} \{\delta\Delta\}^T [\psi]^T \begin{Bmatrix} F_x \\ F_y \end{Bmatrix} dxdy - h_e \int_{\Gamma_e} \{\delta\Delta\}^T [\psi]^T \begin{Bmatrix} T_x^n \\ T_y^n \end{Bmatrix} dS = 0 \qquad (16.3.11)$$

which can be written in compact form

$$\{\delta\Delta\}^T \left([K]\{\Delta\} - \{F\} - \{Q\}\right) = 0 \qquad (16.3.12)$$

Because this relation is to hold for arbitrary variations $\{\delta\Delta\}^T$, the expression in parentheses must vanish, giving the finite element equation

$$[K]\{\Delta\} = \{F\} + \{Q\} \qquad (16.3.13)$$

The equation matrices are defined as follows

$$[K] = h_e \int_{\Omega_e} [B]^T[C][B]dxdy \quad \cdots \text{ stiffness matrix}$$

$$\{F\} = h_e \int_{\Omega_e} [\psi]^T \begin{Bmatrix} F_x \\ F_y \end{Bmatrix} dxdy \quad \cdots \text{ body force vector} \qquad (16.3.14)$$

$$\{Q\} = h_e \int_{\Gamma_e} [\psi]^T \begin{Bmatrix} T_x^n \\ T_y^n \end{Bmatrix} dS \quad \cdots \text{ loading vector}$$

Using the specific interpolation functions for the constant strain triangular element, the $[B]$ matrix had constant components given by (16.3.7). If we assume that the elasticity matrix also does not vary over the element, then the stiffness matrix is given by

$$[K] = h_e A_e [B]^T[C][B] \qquad (16.3.15)$$

and multiplying out the matrices gives the specific form

$$[K] = \frac{h_e}{4A_e} \begin{bmatrix} \beta_1^2 C_{11} + \gamma_1^2 C_{66} & \beta_1\gamma_1 C_{12} + \beta_1\gamma_1 C_{66} & \beta_1\beta_2 C_{11} + \gamma_1\gamma_2 C_{66} & \beta_1\gamma_2 C_{12} + \beta_2\gamma_1 C_{66} & \beta_1\beta_3 C_{11} + \gamma_1\gamma_3 C_{66} & \beta_1\gamma_3 C_{12} + \beta_3\gamma_1 C_{66} \\ . & \gamma_1^2 C_{22} + \beta_1^2 C_{66} & \beta_2\gamma_1 C_{12} + \beta_1\gamma_2 C_{66} & \gamma_1\gamma_2 C_{22} + \beta_1\beta_2 C_{66} & \beta_3\gamma_1 C_{12} + \beta_1\gamma_3 C_{66} & \gamma_1\gamma_3 C_{22} + \beta_1\beta_3 C_{66} \\ . & . & \beta_2^2 C_{11} + \gamma_2^2 C_{66} & \beta_2\gamma_2 C_{12} + \beta_2\gamma_2 C_{66} & \beta_2\beta_3 C_{11} + \gamma_2\gamma_3 C_{66} & \beta_2\gamma_3 C_{12} + \beta_3\gamma_2 C_{66} \\ . & . & . & \gamma_2^2 C_{22} + \beta_2^2 C_{66} & \beta_3\gamma_2 C_{12} + \beta_2\gamma_3 C_{66} & \gamma_2\gamma_3 C_{22} + \beta_2\beta_3 C_{66} \\ . & . & . & . & \beta_3^2 C_{11} + \gamma_3^2 C_{66} & \beta_3\gamma_3 C_{12} + \beta_3\gamma_3 C_{66} \\ . & . & . & . & . & \gamma_3^2 C_{22} + \beta_3^2 C_{66} \end{bmatrix}$$

$$(16.3.16)$$

Note that the stiffness matrix is always symmetric, and thus only the top right (or bottom left) portion need be explicitly written out. If we also choose body forces that are element-wise constant, the body force vector $\{F\}$ can be integrated to give

$$\{F\} = \frac{h_e A_e}{3} \{ F_x \ F_y \ F_x \ F_y \ F_x \ F_y \}^T \qquad (16.3.17)$$

The $\{Q\}$ matrix involves integration of the tractions around the element boundary, and its evaluation depends on whether an element side falls on the boundary of the domain or is located in the region's interior. The evaluation also requires a modeling decision on the assumed traction variation on the element sides. Most problems can be adequately modeled using constant, linear, or quadratic variation in the element boundary tractions. For the typical triangular element shown in Fig. 16.2, the $\{Q\}$ matrix may be written as

$$
\begin{aligned}
\{Q\} &= h_e \int_{\Gamma_e} [\psi]^T \begin{Bmatrix} T_x^n \\ T_y^n \end{Bmatrix} dS \\
&= h_e \int_{\Gamma_{12}} [\psi]^T \begin{Bmatrix} T_x^n \\ T_y^n \end{Bmatrix} dS + h_e \int_{\Gamma_{23}} [\psi]^T \begin{Bmatrix} T_x^n \\ T_y^n \end{Bmatrix} dS + h_e \int_{\Gamma_{31}} [\psi]^T \begin{Bmatrix} T_x^n \\ T_y^n \end{Bmatrix} dS
\end{aligned}
\tag{16.3.18}
$$

Wishing to keep our study brief in theory, we take the simplest case of element-wise constant boundary tractions, which allows explicit calculation of the boundary integrals. For this case, the integral over element side Γ_{12} is given by

$$
h_e \int_{\Gamma_{12}} [\psi]^T \begin{Bmatrix} T_x^n \\ T_y^n \end{Bmatrix} dS = h_e \int_{\Gamma_{12}} \begin{Bmatrix} \psi_1 T_x^n \\ \psi_1 T_y^n \\ \psi_2 T_x^n \\ \psi_2 T_y^n \\ \psi_3 T_x^n \\ \psi_3 T_y^n \end{Bmatrix} dS = \frac{h_e L_{12}}{2} \begin{Bmatrix} T_x^n \\ T_y^n \\ T_x^n \\ T_y^n \\ 0 \\ 0 \end{Bmatrix}_{12}
\tag{16.3.19}
$$

where L_{12} is the length of side Γ_{12}. Note that we have used the fact that alongside Γ_{12}, ψ_1 and ψ_2 vary linearly and $\psi_3 = 0$. Following similar analysis, the boundary integrals along sides Γ_{23} and Γ_{31} are found to be

$$
h_e \int_{\Gamma_{23}} [\psi]^T \begin{Bmatrix} T_x^n \\ T_y^n \end{Bmatrix} dS = \frac{h_e L_{23}}{2} \begin{Bmatrix} 0 \\ 0 \\ T_x^n \\ T_y^n \\ T_x^n \\ T_y^n \end{Bmatrix}_{23} , \quad h_e \int_{\Gamma_{23}} [\psi]^T \begin{Bmatrix} T_x^n \\ T_y^n \end{Bmatrix} dS = \frac{h_e L_{31}}{2} \begin{Bmatrix} T_x^n \\ T_y^n \\ 0 \\ 0 \\ T_x^n \\ T_y^n \end{Bmatrix}_{31}
\tag{16.3.20}
$$

It should be noted that for element sides that lie in the region's interior, values of the boundary tractions will not be known before the solution is found; therefore, the previous relations cannot be used to evaluate the contributions of the $\{Q\}$ matrix explicitly. However, for this situation, the stresses and tractions are in internal equilibrium, and thus the integrated result from one element will cancel that from the opposite adjacent element when the finite element system is assembled. For element sides that coincide with the region's boundary, any applied boundary tractions are then incorporated into the results given by relations (16.3.19) and (16.3.20). Our simplifications of choosing element-wise constant values for the elastic moduli, body forces, and tractions were made only for convenience of the current abbreviated presentation. Normally, FEM modeling allows considerably more generality in these choices, and integrals in the basic element equation (16.3.14) are then evaluated numerically for such applications.

16.4 FEM problem application

Applications using the linear triangular element discretize the domain into a connected set of such elements (see, for example, Fig. 16.1). The mesh geometry establishes which elements are interconnected and identifies those on the boundary of the domain. Using computer implementation, each element in the mesh is mapped or transformed onto a master element in a local coordinate system where all calculations are done. The overall problem is then modeled by assembling the entire set of elements through a process of invoking equilibrium at each node in the mesh. This procedure creates a global assembled matrix system equation of similar form as (16.3.13). Boundary conditions are then incorporated into this global system to reduce the problem to a solvable set of algebraic equations for the unknown nodal displacements. We do not pursue the theoretical and operational details in these procedures, but rather focus attention on a particular example to illustrate some of the key steps in the process.

Example 16.1 Elastic plate under uniform tension

Consider the plane stress problem of an isotropic elastic plate under uniform tension with zero body forces, as shown in Fig. 16.3. For convenience, the plate is taken with unit dimensions and thickness and is discretized into two triangular elements as shown. This simple problem is chosen in order to demonstrate some of the basic FEM solution procedures previously presented. More complex examples are discussed in the next section to illustrate the general power and utility of the numerical technique.

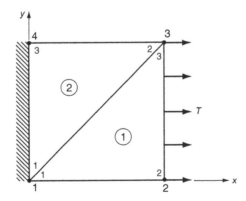

FIG. 16.3 FEM analysis of an elastic plate under uniform tension.

The element mesh is labeled as shown with local node numbers within each element and global node numbers (1−4) for the entire problem. We start by developing the equation for each element and then assemble the two elements to model the entire plate. For element 1, the geometric parameters are $\beta_1 = -1$, $\beta_2 = 1$, $\beta_3 = 0$, $\gamma_1 = 0$, $\gamma_2 = -1$, $\gamma_3 = 1$, and $A_1 = 1/2$. For the isotropic plane stress case, the element equation follows from our previous work

$$
\frac{E}{2(1-\nu^2)}
\begin{bmatrix}
1 & 0 & -1 & \nu & 0 & -\nu \\
& \dfrac{1-\nu}{2} & \dfrac{1-\nu}{2} & -\dfrac{1-\nu}{2} & -\dfrac{1-\nu}{2} & 0 \\
& & \dfrac{3-\nu}{2} & -\dfrac{1+\nu}{2} & -\dfrac{1-\nu}{2} & \nu \\
& & & \dfrac{3-\nu}{2} & \dfrac{1-\nu}{2} & -1 \\
& & & & \dfrac{1-\nu}{2} & 0 \\
& & & & & 1
\end{bmatrix}
\begin{Bmatrix}
u_1^{(1)} \\
v_1^{(1)} \\
u_2^{(1)} \\
v_2^{(1)} \\
u_3^{(1)} \\
v_3^{(1)}
\end{Bmatrix}
=
\begin{Bmatrix}
T_{1x}^{(1)} \\
T_{1y}^{(1)} \\
T_{2x}^{(1)} \\
T_{2y}^{(1)} \\
T_{3x}^{(1)} \\
T_{3y}^{(1)}
\end{Bmatrix}
\qquad (16.4.1)
$$

In similar fashion for element 2, $\beta_1 = 0$, $\beta_2 = 1$, $\beta_3 = -1$, $\gamma_1 = -1$, $\gamma_2 = 0$, $\gamma_3 = 1$, $A_1 = 1/2$, and the element equation becomes

$$
\frac{E}{2(1-\nu^2)}
\begin{bmatrix}
\dfrac{1-\nu}{2} & 0 & 0 & -\dfrac{1-\nu}{2} & -\dfrac{1-\nu}{2} & \dfrac{1-\nu}{2} \\
& 1 & -\nu & 0 & \nu & -1 \\
& & 1 & 0 & -1 & \nu \\
& & & \dfrac{1-\nu}{2} & \dfrac{1-\nu}{2} & -\dfrac{1-\nu}{2} \\
& & & & \dfrac{3-\nu}{2} & -\dfrac{1-\nu}{2} \\
& & & & & \dfrac{3-\nu}{2}
\end{bmatrix}
\begin{Bmatrix}
u_1^{(2)} \\
v_1^{(2)} \\
u_2^{(2)} \\
v_2^{(2)} \\
u_3^{(2)} \\
v_3^{(2)}
\end{Bmatrix}
=
\begin{Bmatrix}
T_{1x}^{(2)} \\
T_{1y}^{(2)} \\
T_{2x}^{(2)} \\
T_{2y}^{(2)} \\
T_{3x}^{(2)} \\
T_{3y}^{(2)}
\end{Bmatrix}
\qquad (16.4.2)
$$

These individual element equations are to be assembled to model the plate, and this is carried out using the global node numbering format by enforcing x and y equilibrium at each node. The final result is given by the assembled global system

Continued

$$
\begin{bmatrix}
K_{11}^{(1)}+K_{11}^{(2)} & K_{12}^{(1)}+K_{12}^{(2)} & K_{13}^{(1)} & K_{14}^{(1)} & K_{15}^{(1)}+K_{13}^{(2)} & K_{16}^{(1)}+K_{14}^{(2)} & K_{15}^{(1)} & K_{16}^{(1)} \\
\cdot & K_{22}^{(1)}+K_{22}^{(2)} & K_{23}^{(1)} & K_{24}^{(1)} & K_{25}^{(1)}+K_{23}^{(2)} & K_{26}^{(1)}+K_{24}^{(2)} & K_{25}^{(1)} & K_{26}^{(1)} \\
\cdot & \cdot & K_{33}^{(1)} & K_{34}^{(1)} & K_{35}^{(1)} & K_{36}^{(1)} & 0 & 0 \\
\cdot & \cdot & \cdot & K_{44}^{(1)} & K_{45}^{(1)} & K_{46}^{(1)} & 0 & 0 \\
\cdot & \cdot & \cdot & \cdot & K_{55}^{(1)}+K_{33}^{(2)} & K_{56}^{(1)}+K_{34}^{(2)} & K_{35}^{(2)} & K_{36}^{(2)} \\
\cdot & \cdot & \cdot & \cdot & \cdot & K_{66}^{(1)}+K_{44}^{(2)} & K_{45}^{(2)} & K_{46}^{(2)} \\
\cdot & \cdot & \cdot & \cdot & \cdot & \cdot & K_{55}^{(2)} & K_{56}^{(2)} \\
\cdot & \cdot & \cdot & \cdot & \cdot & \cdot & \cdot & K_{66}^{(2)}
\end{bmatrix}
$$

$$
\begin{Bmatrix}
U_1 \\ V_1 \\ U_2 \\ V_2 \\ U_3 \\ V_3 \\ U_4 \\ V_4
\end{Bmatrix}
=
\begin{Bmatrix}
T_{1x}^{(1)}+T_{1x}^{(2)} \\
T_{1y}^{(1)}+T_{1y}^{(2)} \\
T_{2x}^{(1)} \\
T_{2y}^{(1)} \\
T_{3x}^{(1)}+T_{2x}^{(2)} \\
T_{3y}^{(1)}+T_{2y}^{(2)} \\
T_{3x}^{(2)} \\
T_{3y}^{(2)}
\end{Bmatrix}
\tag{16.4.3}
$$

where U_i and V_i are the global x and y nodal displacements, and $K_{ij}^{(1)}$ and $K_{ij}^{(2)}$ are the local stiffness components for elements 1 and 2 as given in relations (16.4.1) and (16.4.2).

The next step is to use the problem boundary conditions to reduce this global system. Because the plate is fixed along its left edge, $U_1 = V_1 = U_4 = V_4 = 0$. Using the scheme presented in Eqs. (16.3.18)–(16.3.20), the tractions on the right edge are modeled by choosing $T_{2x}^{(1)} = \text{T}/2$, $T_{2y}^{(1)} = 0$, $T_{3x}^{(1)} + T_{2x}^{(2)} = \text{T}/2$, $T_{3y}^{(1)} + T_{2y}^{(2)} = 0$. These conditions reduce the global system to

$$
\begin{bmatrix}
K_{33}^{(1)} & K_{34}^{(1)} & K_{35}^{(1)} & K_{36}^{(1)} \\
\cdot & K_{44}^{(1)} & K_{45}^{(1)} & K_{46}^{(1)} \\
\cdot & \cdot & K_{55}^{(1)}+K_{33}^{(2)} & K_{56}^{(1)}+K_{34}^{(2)} \\
\cdot & \cdot & \cdot & K_{66}^{(1)}+K_{44}^{(2)}
\end{bmatrix}
\begin{Bmatrix}
U_2 \\ V_2 \\ U_3 \\ V_3
\end{Bmatrix}
=
\begin{Bmatrix}
\text{T}/2 \\ 0 \\ \text{T}/2 \\ 0
\end{Bmatrix}
\tag{16.4.4}
$$

This result can then be solved for the nodal unknowns, and for the case of material with properties $E = 207$ GPa and $\nu = 0.25$, the solution is found to be

$$\left\{ \begin{array}{c} U_2 \\ V_2 \\ U_3 \\ V_3 \end{array} \right\} = \left\{ \begin{array}{c} 0.492 \\ 0.081 \\ 0.441 \\ -0.030 \end{array} \right\} T \times 10^{-11} m \qquad (16.4.5)$$

Note that the FEM displacements are not symmetric as expected from analytical theory. This is caused by the fact that our simple two-element discretization eliminated the symmetry in the original problem. If another symmetric mesh were used, the displacements at nodes 2 and 3 would then be more symmetric. As a postprocessing step, the forces at nodes 1 and 4 could now be computed by back-substituting solution (16.4.5) into the general Eq. (16.4.3). Many of the basic steps in an FEM solution are demonstrated in this hand-calculation example. However, the importance of the numerical method lies in its computer implementation, and examples of this are now discussed.

16.5 FEM code applications

The power and utility of the finite element method lies in the use of computer codes that implement the numerical method for problems of general shape and loading. A very large number of both private and commercial FEM computer codes have been developed over the past several decades. Many of these codes (e.g., ABAQUS, ANSYS, ALGOR, NASTRAN, ADINA) offer very extensive element libraries and can handle linear and nonlinear problems under either static or dynamic conditions. However, the use of such general codes requires considerable study and practice and would not suit the needs of this chapter. Therefore, rather than attempting to use a general code, we follow our numerical theme of employing MATLAB® software, which offers a simple FEM package appropriate for our limited needs. The MATLAB® code is called the *PDE Toolbox* and is one of the many toolboxes distributed with the basic software.

    This software package provides an FEM code that can solve two-dimensional elasticity problems using linear triangular elements. Additional problems governed by other partial differential equations can also be handled, and this allows the software to be used for the torsion problem as well. The PDE Toolbox is very easy to use, and its simple graphical user interface and automeshing features allow the user to create problem geometry quickly and appropriately mesh the domain. Some additional user details on this MATLAB® package are provided in Appendix C (Example Code C12). We now present some example FEM solutions to problems developed by analytical methods in earlier chapters.

Example 16.2 Circular and elliptical holes in a plate under uniform tension

We wish to investigate the numerical finite element solution to the two-dimensional problem of an elastic plate under uniform tension that contains a circular or elliptical hole. These problems were previously solved for the case of an unbounded plane domain; the circular hole (see Fig. 8.12) was

Continued

developed in Example 8.7, while the elliptical hole (Fig. 10.17) was solved in Example 10.7. It is noted that for a standard FEM solution we must have a finite size domain to discretize into a mesh. Because of problem symmetry, only one-quarter of the domain need be analyzed; however, due to the simple problem geometry we do not use this fact and the entire domain is discretized.

The circular hole example is shown in Fig. 16.4. The code allows many different meshes to be generated, and the particular case shown is a fine mesh with 3648 elements and 1912 nodes. Using this software, various types of FEM results can be plotted, and the particular graphic shows contours of the horizontal normal stress σ_x. The concentration effect around the hole is clearly evident, and FEM results give a stress concentration factor $K \approx 2.9$ (based on nominal stress applied at the boundaries). Recall that our theoretical result for an infinite plane gave $K = 3$, and results for the finite-width plate can be found from Peterson (1974), giving $K \approx 3.2$ for this geometry (width/diameter ≈ 4.23). Thus, the FEM result is slightly less than that predicted from theory and Peterson, indicating that the numerical model has some difficulty in properly capturing the high stress gradient in the vicinity of the hole. Using a finer mesh or higher-order elements would result in a value closer to the theoretical/experimental prediction.

A similar problem with an elliptical hole of aspect ratio $b/a = 2$ is shown in Fig. 16.5. The mesh for this case has 3488 elements with 1832 nodes. Again, FEM results are illustrated with contours of horizontal stress σ_x. The concentration effect is reflected by the high stress values at the top and bottom of the ellipse, and the stress concentration factor was found to be $K \approx 3.3$. Using Fig. 10.18 from our previous analytical solution in Example 10.7, an aspect ratio of 2 would result in $K = 5$. Thus, we again experience a lower concentration prediction from the finite element

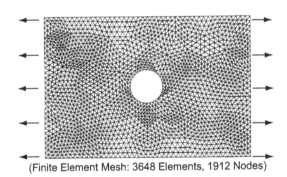

(Finite Element Mesh: 3648 Elements, 1912 Nodes)

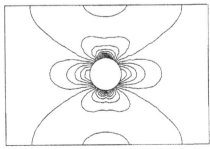

(Contours of Horizontal Stress σ_x)

FIG. 16.4 FEM solution of a plate under uniform tension containing a circular hole.

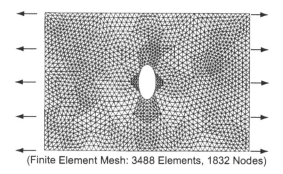

(Finite Element Mesh: 3488 Elements, 1832 Nodes)

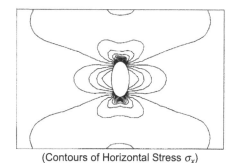

(Contours of Horizontal Stress σ_x)

FIG. 16.5 FEM solution of a plate under uniform tension containing an elliptical hole.

model, and the difference between FEM and theory is larger than in the previous example with the circular hole. The lower FEM concentration value is again attributable to the numerical model's inability to simulate the high stress gradient near the top and bottom of the ellipse boundary. Again, a finer mesh and/or higher-order elements would result in better FEM predictions.

Example 16.3 Circular disk under diametrical compression

Consider next the problem of a circular disk under diametrical compression, as originally discussed in Example 8.10. The problem was solved for the case of concentrated loadings as shown in Fig. 8.35, and contours of the maximum shear stress were compared with photoelastic data in Fig. 8.36. Recall that the photoelastic contours result from an actual experiment in which the loading is distributed over a small portion of the top and bottom of the disk. This distributed loading case was solved using the MATLAB® PDE Toolbox, and the results are shown in Fig. 16.6. The figure illustrates two FEM models with different meshes along with contours of maximum shear stress. With the loading distributed over a small portion of the disk boundary, the maximum stresses occur slightly interior to the loading surfaces. As expected, the finer mesh produces better results that more closely compare with analytical and photoelastic predictions shown in Fig. 8.36.

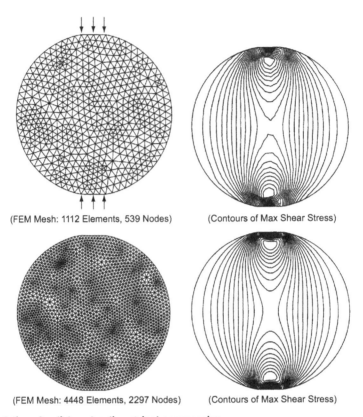

(FEM Mesh: 1112 Elements, 539 Nodes)   (Contours of Max Shear Stress)

(FEM Mesh: 4448 Elements, 2297 Nodes)   (Contours of Max Shear Stress)

FIG. 16.6 FEM solution of a disk under diametrical compression.

Example 16.4 Torsion problem examples

Recall that in Chapter 9 we formulated the torsion problem in terms of the Prandtl stress function ϕ, which satisfies the Poisson equation $\nabla^2\phi = -2\mu\alpha$ in the cross-section. For simply connected sections, $\phi = 0$ on the boundary, while for multiply connected sections the function could be set to zero on the outer boundary but must be a different constant on each inner boundary. This two-dimensional problem is easily solved using finite elements and in particular using the PDE Toolbox. Fig. 16.7 illustrates three sections that have been solved using the MATLAB® software with linear triangular elements. The first problem is that of a circular section with a circular keyway, and this problem was originally presented in Exercises 9.23 and 9.24. FEM results show stress function contours over the section, and the slope of these contours gives the shear stress in the perpendicular direction. It is readily apparent that the maximum shear stress occurs at the root of the keyway acting tangent to the boundary at point A. The second example shown is a square section with a square keyway. The stress function contour lines indicate high-stress regions at the two re-entrant corners of the keyway. The final example is a multiply connected section with a square outer boundary and a triangular inner hole. Contours for this case show three high-stress regions at each vertex of the triangular cutout. Countless other torsion examples can be quickly analyzed using this simple FEM code, and quantitative stress results can also be generated (see Exercise 16.10).

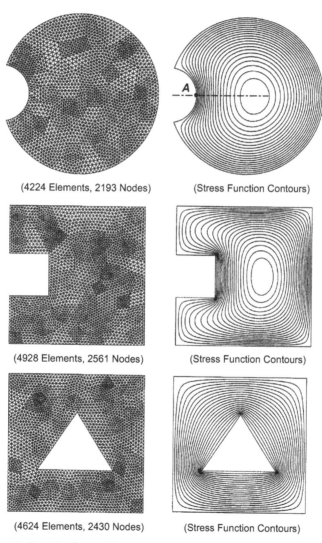

(4224 Elements, 2193 Nodes)   (Stress Function Contours)

(4928 Elements, 2561 Nodes)   (Stress Function Contours)

(4624 Elements, 2430 Nodes)   (Stress Function Contours)

FIG. 16.7 FEM solutions to three torsion problems.

16.6 Boundary element formulation

A second numerical method has recently emerged that provides good computational abilities and has some particular advantages when compared to FEM. The technique, known as the *boundary element method* (BEM), has been widely used by computational mechanics investigators, leading to the development of many private and commercial codes. Similar to the finite element method, BEM can analyze many different problems in engineering science including those in thermal sciences and fluid mechanics. Although the method is not limited to elastic stress analysis, this brief presentation discusses only this particular case. Many texts have been written that provide additional details on this subject (see, for example, Banerjee and Butterfield, 1981; Brebbia and Dominguez, 1992).

The formulation of BEM is based on an integral statement of elasticity, and this can be cast into a relation involving unknowns only over the boundary of the domain under study. This originally led to the *boundary integral equation* (BIE) method, and early work in the field was reported by Rizzo (1967) and Cruse (1969). Subsequent research realized that finite element methods could be used to solve the boundary integral equation by dividing the boundary into elements over which the solution is approximated using appropriate interpolation functions. This process generates an algebraic system of equations to solve for the unknown nodal values that approximate the boundary solution. A procedure to calculate the solution at interior domain points can also be determined from the original boundary integral equation. This scheme also allows variation in element size, shape, and approximating scheme to suit the application, thus providing similar advantages as FEM.

By discretizing only the boundary of the domain, BEM has particular advantages over FEM. The first issue is that the resulting boundary element equation system is generally much smaller than that generated by finite elements. It has been pointed out in the literature that boundary discretization is somewhat easier to interface with *computer-aided design* (CAD) codes that create the original problem geometry. A two-dimensional comparison of equivalent FEM and BEM meshes for a rectangular plate with a central circular hole is shown in Fig. 16.8. It is apparent that a significant reduction in the number of elements (by a factor of 5) is realized in the BEM mesh. It should be pointed out, however, that the BEM scheme does not automatically compute the solution at interior points, and thus additional computational effort is required to find such information.

Some studies have indicated that BEM more accurately determines stress concentration effects. Problems of infinite extent (e.g., full-space or half-space domains) create some difficulty in developing appropriate FEM meshes, whereas particular BEM schemes can automatically handle the infinite nature of the problem and only require limited boundary meshing. There exist several additional advantages and disadvantages related to each method; however, we will not pursue further comparison and debate. For linear elasticity, both methods offer considerable utility to solve very complex problems numerically. We now proceed with a brief development of the boundary element method for two-dimensional elasticity problems.

The integral statement of elasticity was developed in Section 6.4.3 as Somigliana's identity. Using the reciprocal theorem, one elastic state was selected as the fundamental solution, while the other state was chosen as the desired solution field. For a region V with boundary S, this led to the following result

$$cu_j(\xi) = \int_S [T_i(x)G_{ij}(x,\xi) - u_i T_{ikj}(x,\xi)n_k]dS + \int_V F_i G_{ij}(x,\xi)dV \qquad (16.6.1)$$

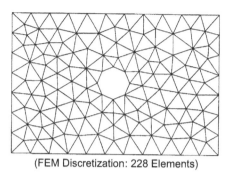

(FEM Discretization: 228 Elements)

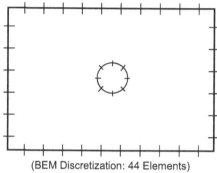

(BEM Discretization: 44 Elements)

FIG. 16.8 Comparison of typical FEM and BEM meshes.

where the coefficient c is given by

$$c = \begin{cases} 1, \xi \ in \ V \\ \dfrac{1}{2}, \xi \ on \ S \\ 0, \xi \ outside \ V \end{cases} \tag{16.6.2}$$

$G_{ij}(x, \xi)$ is the *displacement Green's function* that comes from the fundamental solution to the elasticity equations and corresponds to the solution of the displacement field at point x produced by a unit concentrated body force e located at point ξ

$$u_i^G(x) = G_{ij}(x, \xi)e_j(\xi) \tag{16.6.3}$$

The stresses associated with this state are specified by

$$\sigma_{ij}^G = T_{ijk}(x, \xi)e_k = \left[\lambda G_{lk,l}\delta_{ij} + \mu\left(G_{ik,j} + G_{jk,i}\right) \right] e_k \tag{16.6.4}$$

and the tractions follow to be

$$T_i^G = \sigma_{ij}^G n_j = T_{ijk}n_j e_k = p_{ik}e_k \tag{16.6.5}$$

with $p_{ik} = T_{ijk}n_j$. Relation (16.6.1) gives the displacement of a given observational point ξ in terms of boundary and volume integrals. If point ξ is chosen to lie on boundary S, then the expression will contain unknowns (displacements and tractions) only on the boundary. For this case (ξ on S), relation (16.6.2) indicates $c = 1/2$, but this is true only if the boundary has a continuous tangent (i.e., is smooth). Slight modifications are necessary for the case of nonsmooth boundaries (see Brebbia and Dominguez, 1992).

Restricting our attention to only the two-dimensional plane strain case, the Green's function becomes (Brebbia and Dominguez, 1992)

$$G_{ij} = \frac{1}{8\pi\mu(1-v)}\left[(3-4v)\ln\left(\frac{1}{r}\right)\delta_{ij} + r_{,i}r_{,j}\right] \tag{16.6.6}$$

where $r = |x - \xi|$ is the distance between points x and ξ. Relation (16.6.5) can be used to determine the traction p_{ij} associated with this specific Green's function, giving the result

$$p_{ij} = T_{ikj}n_k = -\frac{1}{4\pi(1-v)r}\left[(1-2v)\left(\frac{\partial r}{\partial n}\delta_{ij} + r_{,i}n_j - r_{,j}n_i\right) + 2\frac{\partial r}{\partial n}r_{,i}r_{,j}\right] \tag{16.6.7}$$

It is convenient to use matrix notation in the subsequent formulation and thus define

$$G = \begin{bmatrix} G_{11} & G_{12} \\ G_{21} & G_{22} \end{bmatrix}, \quad p = \begin{bmatrix} p_{11} & p_{12} \\ p_{21} & p_{22} \end{bmatrix}$$
$$u = \begin{Bmatrix} u_1 \\ u_2 \end{Bmatrix}, \quad T = \begin{Bmatrix} T_1 \\ T_2 \end{Bmatrix}, \quad F = \begin{Bmatrix} F_1 \\ F_2 \end{Bmatrix} \tag{16.6.8}$$

The boundary integral Eq. (16.6.1) can then be expressed in two-dimensional form as

$$c^i u^i = \int_\Gamma [GT - pu]dS + \int_R GFdV \tag{16.6.9}$$

It is noted that by allowing point ξ to be on the boundary, this relation will contain unknown displacements or tractions only over Γ. We now wish to apply numerical finite element concepts to solve (16.6.9) by discretizing the boundary Γ and region R into subdomains over which the solution will be approximated. Only the simplest case is presented here in which the approximating scheme assumes piecewise constant values for the unknowns.

Referring to Fig. 16.9, a typical boundary Γ is discretized into N elements. The unknown boundary displacements and tractions are assumed to be constant over each element and equal to the value at each midnode. Subdivision of the interior into cells would also be required in order to compute integration of the body force term over R. However, such interior integrals can be reformulated in terms of boundary integrals, thereby maintaining efficiency of the basic boundary techniques. This reformulation is not discussed here, and we now formally drop body force contributions from further consideration. Using this discretization scheme, relation (16.6.9) can be written as

$$c^i u^i + \sum_{j=1}^N \left(\int_{\Gamma_j} pdS\right)u^j = \sum_{j=1}^N \left(\int_{\Gamma_j} GdS\right)T^j \tag{16.6.10}$$

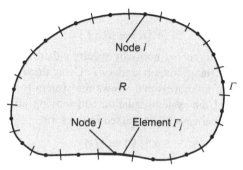

FIG. 16.9 Boundary discretization using elements with constant approximation.

where index i corresponds to a particular node where the Green's function concentrated force is applied, and index j is related to each of the boundary elements including the case $j = i$. Notice that for the choice of constant approximation over the element, there is no formal interpolation function required, and nodal values are simply brought outside of the element integrations.

Reviewing the previous expressions (16.6.6) and (16.6.7), the integral terms $\int_{\Gamma_j} G dS$ and $\int_{\Gamma_j} p dS$ relate node i to node j and are sometimes referred to as *influence functions*. Each of these terms generates 2×2 matrices that can be defined by

$$\hat{A}^{ij} = \int_{\Gamma_j} p dS$$
$$B^{ij} = \int_{\Gamma_j} G dS$$

(16.6.11)

For the constant element case, some of the integrations in (16.6.11) can be carried out analytically, while other cases use numerical integration commonly employing Gauss quadrature. It should be noted that the $i = j$ case generates a singularity in the integration, and special methods are normally used to handle this problem.

Relation (16.6.10) can thus be written as

$$c^i u^i + \sum_{j=1}^{N} \hat{A}^{ij} u^j = \sum_{j=1}^{N} B^{ij} T^j$$

(16.6.12)

and this result specifies the value of u at node i in terms of values of u and T at all other nodes on the boundary. If the boundary is smooth, $c^i = \frac{1}{2}$ at all nodes. By defining

$$A^{ij} = \begin{cases} \hat{A}^{ij}, & i \neq j \\ \hat{A}^{ij} + c^i, & i = j \end{cases}$$

(16.6.13)

Eq. (16.6.12) can be written in compact form as

$$\sum_{j=1}^{N} A^{ij} u^j = \sum_{j=1}^{N} B^{ij} T^j$$

(16.6.14)

or in matrix form

$$[A]\{u\} = [B]\{T\} \tag{16.6.15}$$

Boundary conditions from elasticity theory normally specify either the displacements or tractions or a mixed combination of the two variables over boundary Γ. Using these specified values in (16.6.14) or (16.6.15) reduces the number of unknowns and allows the system to be rearranged. Placing all unknowns on the left-hand side of the system equation and moving all known variables to the right generates a final system that can always be expressed in the form

$$[C]\{X\} = \{D\} \tag{16.6.16}$$

where all unknown boundary displacements and tractions are located in the column matrix $\{X\}$ and all known boundary data have been multiplied by the appropriate influence function and moved into $\{D\}$. Relation (16.6.16) represents a system of linear algebraic equations that can be solved for the desired unknown boundary information. This BEM system is generally much smaller than that from a corresponding FEM model. However, unlike the FEM system, the $[C]$ matrix from (16.6.16) is not in general symmetric, thereby requiring more computational effort to solve for nodal unknowns. Using modern computing systems, usually this added computational effort is not a significant factor in the solution of linear elasticity problems.

Once this solution is complete, all boundary data is known and the solution at any desired interior point can be calculated reusing the basic governing boundary integral equation. For example, at an interior point relation (16.6.9) with no body forces will take the form

$$u^i = \int_\Gamma GT ds - \int_\Gamma pu dS \tag{16.6.17}$$

Following our previous constant element approximation, this expression can be discretized as

$$
\begin{aligned}
u^i &= \sum_{j=1}^N \left(\int_{\Gamma_j} G dS \right) T^j - \sum_{j=1}^N \left(\int_{\Gamma_j} p dS \right) u^j \\
&= \sum_{j=1}^N B^{ij} T^j - \sum_{j=1}^N \hat{A}^{ij} u^j
\end{aligned}
\tag{16.6.18}
$$

and the interior displacement can then be determined using standard computational evaluation of the influence functions $\hat{A}^{ij}$ and B^{ij}. Internal values of strain and stress can also be computed using (16.6.17) in the strain-displacement relations and Hooke's law, thereby generating expressions similar to relation (16.6.18); see Brebbia and Dominguez (1992).

Example 16.5 BEM solution of a circular hole in a plate under uniform tension

Consider again the problem of Example 16.2 of a plate under uniform tension that contains a stress-free circular hole. The finite element solution was shown in Fig. 16.4 for a very fine FEM mesh with 3648 triangular elements. A simple BEM FORTRAN code using constant and quadratic elements is provided in the text by Brebbia and Dominguez (1992), and this was used to develop the numerical solution. This simple BEM code does not have drawing or automeshing capabilities, and thus problem data was input by hand. The boundary element solution was generated using two different models that incorporated problem symmetry to analyze half of the domain, as shown in Fig. 16.10. One model used 32 constant elements, while the second case used 14 three-noded quadratic elements.

The constant element model is limited to having nodes located at the midpoint of each element (see Fig. 16.9), and thus does not allow direct determination of the highest stress at the edge of the hole. For this case using the stress value at node A in Fig. 16.10, the stress concentration factor is found to be $K \approx 2.75$. The quadratic element case uses three nodes per element, including nodes located at the element boundaries. For this case node B in Fig. 16.10 is used to determine the highest stress, and this gives a stress concentration factor of $K \approx 3.02$. The particular model has a width-to-diameter ratio of 10, and for such geometry, results from Peterson (1974) would predict a stress concentration of about 3. As expected, the BEM results using constant elements were not as good as the predictions using quadratic interpolation. Comparing this BEM analysis with the finite element results in Example 16.2 indicates that the quadratic boundary element results appear to give a more accurate estimate of the actual stress concentration using much fewer elements. However, this conclusion is based on the particular element models for each analysis, and using other element types and meshes could produce somewhat different results and comparisons.

Many additional FEM and BEM examples can be developed and compared to illustrate other interesting features of these computational stress analysis methods. Unfortunately, such an excursion would require developments outside the current scope of the text, and thus we will not pursue such material. The interested reader is encouraged to consult the chapter references for additional study.

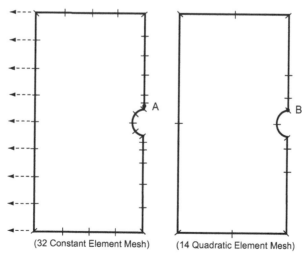

(32 Constant Element Mesh)   (14 Quadratic Element Mesh)

FIG. 16.10 BEM solution of a plate under uniform tension containing a circular hole.

References

Banerjee PK, Butterfield R: *Boundary element methods in engineering science*, New York, 1981, McGraw-Hill.

Bathe KJ: *Finite element procedures in engineering analysis*, Englewood Cliffs, NJ, 1982, Prentice Hall.

Brebbia CA, Dominguez J: *Boundary elements: an introductory course*, Southampton, UK, 1992, WIT Press.

Cook RD, Malkus DS, Witt R, Plesha ME: *Concepts and applications of finite elements in engineering*, New York, 2002, John Wiley.

Cruse TA: Numerical solutions in three-dimensional elastostatics, *Int J Solids Struct* 5:1259–1274, 1969.

Fung YC, Tong P: *Classical and computational solid mechanics*, Singapore, 2001, World Scientific.

Peterson RE: *Stress concentration factors*, New York, 1974, John Wiley.

Reddy JN: *An introduction to the finite element method*, New York, 2006, McGraw-Hill.

Rizzo RJ: An integral equation approach to boundary value problems of classical elastostatics, *Q Appl Math* 25: 83–95, 1967.

Zienkiewicz OC, Taylor RL: *The finite element method*, New York, 2005, McGraw-Hill.

Exercises

16.1 Starting with the general linear form (16.2.1), verify the interpolation relations (16.2.4) and (16.2.5).

16.2 For the constant strain triangular element, show that the stiffness matrix is given by (16.3.16).

16.3 For the case of element-wise constant body forces, verify that the body force vector is given by relation (16.3.17) for the linear triangular element.

16.4 Verify boundary relation (16.3.19) for the linear triangular element with constant boundary tractions T_x^n and T_y^n.

16.5 For Example 16.1, show that the element stiffness equations for the isotropic case are given by relations (16.4.1) and (16.4.2).

16.6 Verify the nodal displacement solution given by (16.4.5) in Example 16.1.

16.7* Using the MATLAB® PDE Toolbox (or equivalent), develop an FEM solution for the stress concentration problem under biaxial loading given in Exercise 8.25. Compare the stress concentration factor from the numerical results with the corresponding analytical predictions.

16.8* Using the MATLAB® PDE Toolbox (or equivalent), develop an FEM solution for the stress concentration problem under shear loading given in Exercise 8.26. Compare the stress concentration factor from the numerical results with the corresponding analytical predictions.

16.9* Using the MATLAB® PDE Toolbox (or equivalent), develop an FEM solution for the curve beam problem shown in Fig. 8.32. At the fixed section, compare numerical stress results (σ_x) with analytical predictions (σ_θ) given by Eqs. (8.4.65).

16.10* Using the MATLAB® PDE Toolbox (or equivalent), develop an FEM solution for the torsion of a cylinder of circular section with circular keyway as shown in Exercise 9.23. Verify the result of Exercise 9.24, that the maximum shear stress on the keyway is approximately twice that found on a solid shaft. In order to investigate the shear stress, use the Toolbox plot selection window to plot contours of the variable $abs(grad(u))$.

16.11 Verify the traction relation (16.6.7) for the plane strain case.

Basic field equations in Cartesian, cylindrical, and spherical coordinates

For convenience, the basic three-dimensional field equations of elasticity are listed here for Cartesian, cylindrical, and spherical coordinate systems. This will eliminate searching for these results in various chapters of the text. Cylindrical and spherical coordinates are related to the basic Cartesian system, as shown in Fig. A.1.

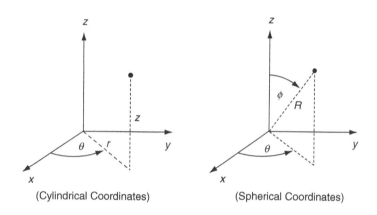

(Cylindrical Coordinates)   (Spherical Coordinates)

FIG. A.1 Cylindrical and spherical coordinate systems.

Strain—displacement relations
Cartesian coordinates

$$e_x = \frac{\partial u}{\partial x} , \quad e_y = \frac{\partial v}{\partial y} , \quad e_z = \frac{\partial w}{\partial z}$$

$$e_{xy} = \frac{1}{2}\left(\frac{\partial u}{\partial y} + \frac{\partial v}{\partial x}\right)$$

$$e_{yz} = \frac{1}{2}\left(\frac{\partial v}{\partial z} + \frac{\partial w}{\partial y}\right)$$

$$e_{zx} = \frac{1}{2}\left(\frac{\partial w}{\partial x} + \frac{\partial u}{\partial z}\right)$$

(A.1)

Cylindrical coordinates

$$e_r = \frac{\partial u_r}{\partial r}, \quad e_\theta = \frac{1}{r}\left(u_r + \frac{\partial u_\theta}{\partial \theta}\right), \quad e_z = \frac{\partial u_z}{\partial z}$$

$$e_{r\theta} = \frac{1}{2}\left(\frac{1}{r}\frac{\partial u_r}{\partial \theta} + \frac{\partial u_\theta}{\partial r} - \frac{u_\theta}{r}\right)$$

$$e_{\theta z} = \frac{1}{2}\left(\frac{\partial u_\theta}{\partial z} + \frac{1}{r}\frac{\partial u_z}{\partial \theta}\right)$$

$$e_{zr} = \frac{1}{2}\left(\frac{\partial u_r}{\partial z} + \frac{\partial u_z}{\partial r}\right)$$

(A.2)

Spherical coordinates

$$e_R = \frac{\partial u_R}{\partial R}, \quad e_\phi = \frac{1}{R}\left(u_R + \frac{\partial u_\phi}{\partial \phi}\right)$$

$$e_\theta = \frac{1}{R\sin\phi}\left(\frac{\partial u_\theta}{\partial \theta} + \sin\phi\, u_R + \cos\phi\, u_\phi\right)$$

$$e_{R\phi} = \frac{1}{2}\left(\frac{1}{R}\frac{\partial u_R}{\partial \phi} + \frac{\partial u_\phi}{\partial R} - \frac{u_\phi}{R}\right)$$

$$e_{\phi\theta} = \frac{1}{2R}\left(\frac{1}{\sin\phi}\frac{\partial u_\phi}{\partial \theta} + \frac{\partial u_\theta}{\partial \phi} - \cot\phi\, u_\theta\right)$$

$$e_{\theta R} = \frac{1}{2}\left(\frac{1}{R\sin\phi}\frac{\partial u_R}{\partial \theta} + \frac{\partial u_\theta}{\partial R} - \frac{u_\theta}{R}\right)$$

(A.3)

Equilibrium equations
Cartesian coordinates

$$\frac{\partial \sigma_x}{\partial x} + \frac{\partial \tau_{yx}}{\partial y} + \frac{\partial \tau_{zx}}{\partial z} + F_x = 0$$

$$\frac{\partial \tau_{xy}}{\partial x} + \frac{\partial \sigma_y}{\partial y} + \frac{\partial \tau_{zy}}{\partial z} + F_y = 0$$

$$\frac{\partial \tau_{xz}}{\partial x} + \frac{\partial \tau_{yz}}{\partial y} + \frac{\partial \sigma_z}{\partial z} + F_z = 0$$

(A.4)

Cylindrical coordinates

$$\frac{\partial \sigma_r}{\partial r} + \frac{1}{r}\frac{\partial \tau_{r\theta}}{\partial \theta} + \frac{\partial \tau_{rz}}{\partial z} + \frac{1}{r}(\sigma_r - \sigma_\theta) + F_r = 0$$

$$\frac{\partial \tau_{r\theta}}{\partial r} + \frac{1}{r}\frac{\partial \sigma_\theta}{\partial \theta} + \frac{\partial \tau_{\theta z}}{\partial z} + \frac{2}{r}\tau_{r\theta} + F_\theta = 0 \qquad\text{(A.5)}$$

$$\frac{\partial \tau_{rz}}{\partial r} + \frac{1}{r}\frac{\partial \tau_{\theta z}}{\partial \theta} + \frac{\partial \sigma_z}{\partial z} + \frac{1}{r}\tau_{rz} + F_z = 0$$

Spherical coordinates

$$\frac{\partial \sigma_R}{\partial R} + \frac{1}{R}\frac{\partial \tau_{R\phi}}{\partial \phi} + \frac{1}{R\sin\phi}\frac{\partial \tau_{R\theta}}{\partial \theta} + \frac{1}{R}\left(2\sigma_R - \sigma_\phi - \sigma_\theta + \tau_{R\phi}\cot\phi\right) + F_R = 0$$

$$\frac{\partial \tau_{R\phi}}{\partial R} + \frac{1}{R}\frac{\partial \sigma_\phi}{\partial \phi} + \frac{1}{R\sin\phi}\frac{\partial \tau_{\phi\theta}}{\partial \theta} + \frac{1}{R}\left[(\sigma_\phi - \sigma_\theta)\cot\phi + 3\tau_{R\phi}\right] + F_\phi = 0 \qquad\text{(A.6)}$$

$$\frac{\partial \tau_{R\theta}}{\partial R} + \frac{1}{R}\frac{\partial \tau_{\phi\theta}}{\partial \phi} + \frac{1}{R\sin\phi}\frac{\partial \sigma_\theta}{\partial \theta} + \frac{1}{R}\left(2\tau_{\phi\theta}\cot\phi + 3\tau_{R\theta}\right) + F_\theta = 0$$

Hooke's law
Cartesian coordinates

$$e_x = \frac{1}{E}[\sigma_x - \nu(\sigma_y + \sigma_z)]$$

$$\sigma_x = \lambda(e_x + e_y + e_z) + 2\mu e_x \qquad\qquad e_y = \frac{1}{E}[\sigma_y - \nu(\sigma_z + \sigma_x)]$$

$$\sigma_y = \lambda(e_x + e_y + e_z) + 2\mu e_y$$

$$\sigma_z = \lambda(e_x + e_y + e_z) + 2\mu e_z \qquad\qquad e_z = \frac{1}{E}[\sigma_z - \nu(\sigma_x + \sigma_y)]$$

$$\tau_{xy} = 2\mu e_{xy} \qquad\qquad\qquad\qquad\qquad e_{xy} = \frac{1+\nu}{E}\tau_{xy} \qquad\text{(A.7)}$$

$$\tau_{yz} = 2\mu e_{yz} \qquad\qquad\qquad\qquad\qquad e_{yz} = \frac{1+\nu}{E}\tau_{yz}$$

$$\tau_{zx} = 2\mu e_{zx} \qquad\qquad\qquad\qquad\qquad e_{zx} = \frac{1+\nu}{E}\tau_{zx}$$

Cylindrical coordinates

$$\sigma_r = \lambda(e_r + e_\theta + e_z) + 2\mu e_r$$
$$\sigma_\theta = \lambda(e_r + e_\theta + e_z) + 2\mu e_\theta$$
$$\sigma_z = \lambda(e_r + e_\theta + e_z) + 2\mu e_z$$
$$\tau_{r\theta} = 2\mu e_{r\theta}$$
$$\tau_{\theta z} = 2\mu e_{\theta z}$$
$$\tau_{zr} = 2\mu e_{zr}$$

$$e_r = \frac{1}{E}[\sigma_r - \nu(\sigma_\theta + \sigma_z)]$$
$$e_\theta = \frac{1}{E}[\sigma_\theta - \nu(\sigma_z + \sigma_r)]$$
$$e_z = \frac{1}{E}[\sigma_z - \nu(\sigma_r + \sigma_\theta)]$$
$$e_{r\theta} = \frac{1+\nu}{E}\tau_{r\theta}$$
$$e_{\theta z} = \frac{1+\nu}{E}\tau_{\theta z}$$
$$e_{zr} = \frac{1+\nu}{E}\tau_{zr}$$

(A.8)

Spherical coordinates

$$\sigma_R = \lambda(e_R + e_\phi + e_\theta) + 2\mu e_R$$
$$\sigma_\phi = \lambda(e_R + e_\phi + e_\theta) + 2\mu e_\phi$$
$$\sigma_\theta = \lambda(e_R + e_\phi + e_\theta) + 2\mu e_\theta$$
$$\tau_{R\phi} = 2\mu e_{R\phi}$$
$$\tau_{\phi\theta} = 2\mu e_{\phi\theta}$$
$$\tau_{\theta R} = 2\mu e_{\theta R}$$

$$e_R = \frac{1}{E}\left[\sigma_R - \nu(\sigma_\phi + \sigma_\theta)\right]$$
$$e_\phi = \frac{1}{E}\left[\sigma_\phi - \nu(\sigma_\theta + \sigma_R)\right]$$
$$e_\theta = \frac{1}{E}\left[\sigma_\theta - \nu(\sigma_R + \sigma_\phi)\right]$$
$$e_{R\phi} = \frac{1+\nu}{E}\tau_{R\phi}$$
$$e_{\phi\theta} = \frac{1+\nu}{E}\tau_{\phi\theta}$$
$$e_{\theta R} = \frac{1+\nu}{E}\tau_{\theta R}$$

(A.9)

Equilibrium equations in terms of displacements (Navier's equations)
Cartesian coordinates

$$\mu\nabla^2 u + (\lambda + \mu)\frac{\partial}{\partial x}\left(\frac{\partial u}{\partial x} + \frac{\partial v}{\partial y} + \frac{\partial w}{\partial z}\right) + F_x = 0$$

$$\mu\nabla^2 v + (\lambda + \mu)\frac{\partial}{\partial y}\left(\frac{\partial u}{\partial x} + \frac{\partial v}{\partial y} + \frac{\partial w}{\partial z}\right) + F_y = 0 \qquad \text{(A.10)}$$

$$\mu\nabla^2 w + (\lambda + \mu)\frac{\partial}{\partial z}\left(\frac{\partial u}{\partial x} + \frac{\partial v}{\partial y} + \frac{\partial w}{\partial z}\right) + F_z = 0$$

Cylindrical coordinates

$$\mu\left(\nabla^2 u_r - \frac{u_r}{r^2} - \frac{2}{r^2}\frac{\partial u_\theta}{\partial \theta}\right) + (\lambda + \mu)\frac{\partial}{\partial r}\left(\frac{1}{r}\frac{\partial}{\partial r}(ru_r) + \frac{1}{r}\frac{\partial u_\theta}{\partial \theta} + \frac{\partial u_z}{\partial z}\right) + F_r = 0$$

$$\mu\left(\nabla^2 u_\theta - \frac{u_\theta}{r^2} + \frac{2}{r^2}\frac{\partial u_r}{\partial \theta}\right) + (\lambda + \mu)\frac{1}{r}\frac{\partial}{\partial \theta}\left(\frac{1}{r}\frac{\partial}{\partial r}(ru_r) + \frac{1}{r}\frac{\partial u_\theta}{\partial \theta} + \frac{\partial u_z}{\partial z}\right) + F_\theta = 0 \qquad \text{(A.11)}$$

$$\mu\nabla^2 u_z + (\lambda + \mu)\frac{\partial}{\partial z}\left(\frac{1}{r}\frac{\partial}{\partial r}(ru_r) + \frac{1}{r}\frac{\partial u_\theta}{\partial \theta} + \frac{\partial u_z}{\partial z}\right) + F_z = 0$$

Spherical coordinates

$$\mu\left(\nabla^2 u_R - \frac{2u_R}{R^2} - \frac{2}{R^2}\frac{\partial u_\phi}{\partial \phi} - \frac{2u_\phi \cot \phi}{R^2} - \frac{2}{R^2 \sin \phi}\frac{\partial u_\theta}{\partial \theta}\right)$$

$$+ (\lambda + \mu)\frac{\partial}{\partial R}\left(\frac{1}{R^2}\frac{\partial}{\partial R}(R^2 u_R) + \frac{1}{R \sin \phi}\frac{\partial}{\partial \phi}(u_\phi \sin \phi) + \frac{1}{R \sin \phi}\frac{\partial u_\theta}{\partial \theta}\right) + F_R = 0$$

$$\mu\left(\nabla^2 u_\phi + \frac{2}{R^2}\frac{\partial u_R}{\partial \phi} - \frac{u_\phi}{R^2 \sin^2 \phi} - \frac{2\cos \phi}{R^2 \sin^2 \phi}\frac{\partial u_\theta}{\partial \theta}\right)$$

$$+ (\lambda + \mu)\frac{1}{R}\frac{\partial}{\partial \phi}\left(\frac{1}{R^2}\frac{\partial}{\partial R}(R^2 u_R) + \frac{1}{R \sin \phi}\frac{\partial}{\partial \phi}(u_\phi \sin \phi) + \frac{1}{R \sin \phi}\frac{\partial u_\theta}{\partial \theta}\right) + F_\phi = 0 \qquad \text{(A12)}$$

$$\mu\left(\nabla^2 u_\theta - \frac{u_\theta}{R^2 \sin^2 \phi} + \frac{2}{R^2 \sin^2 \phi}\frac{\partial u_R}{\partial \theta} + \frac{2\cos \phi}{R^2 \sin^2 \phi}\frac{\partial u_\phi}{\partial \theta}\right)$$

$$+ (\lambda + \mu)\frac{1}{R \sin \phi}\frac{\partial}{\partial \theta}\left(\frac{1}{R^2}\frac{\partial}{\partial R}(R^2 u_R) + \frac{1}{R \sin \phi}\frac{\partial}{\partial \phi}(u_\phi \sin \phi) + \frac{1}{R \sin \phi}\frac{\partial u_\theta}{\partial \theta}\right) + F_\theta = 0$$

Transformation of field variables between Cartesian, cylindrical, and spherical components

B

This appendix contains some three-dimensional transformation relations between displacement and stress components in Cartesian, cylindrical, and spherical coordinates. The coordinate systems are shown in Fig. A.1 in, Appendix A, and the related stress components are illustrated in Fig. B.1. These results follow from the general transformation laws, (1.5.1) and, (3.3.3). Note that the stress results can also be applied for strain transformation.

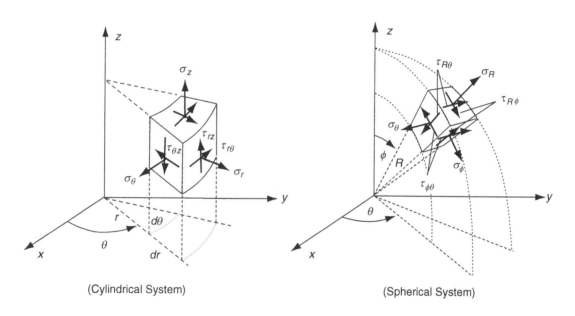

(Cylindrical System)   (Spherical System)

FIG. B.1 Stress components in cylindrical and spherical coordinates.

Cylindrical components from Cartesian

The transformation matrix for this case is given by

$$[Q] = \begin{bmatrix} \cos\theta & \sin\theta & 0 \\ -\sin\theta & \cos\theta & 0 \\ 0 & 0 & 1 \end{bmatrix} \tag{B.1}$$

Displacement transformation

$$
\begin{aligned}
u_r &= u\cos\theta + v\sin\theta \\
u_\theta &= -u\sin\theta + v\cos\theta \\
u_z &= w
\end{aligned}
\tag{B.2}
$$

Stress transformation

$$
\begin{aligned}
\sigma_r &= \sigma_x \cos^2\theta + \sigma_y \sin^2\theta + 2\tau_{xy}\sin\theta\cos\theta \\
\sigma_\theta &= \sigma_x \sin^2\theta + \sigma_y \cos^2\theta - 2\tau_{xy}\sin\theta\cos\theta \\
\sigma_z &= \sigma_z \\
\tau_{r\theta} &= -\sigma_x \sin\theta\cos\theta + \sigma_y \sin\theta\cos\theta + \tau_{xy}\left(\cos^2\theta - \sin^2\theta\right) \\
\tau_{\theta z} &= \tau_{yz}\cos\theta - \tau_{zx}\sin\theta \\
\tau_{zr} &= \tau_{yz}\sin\theta + \tau_{zx}\cos\theta
\end{aligned}
\tag{B.3}
$$

Spherical components from cylindrical

The transformation matrix from cylindrical to spherical coordinates is given by

$$[Q] = \begin{bmatrix} \sin\phi & 0 & \cos\phi \\ \cos\phi & 0 & -\sin\phi \\ 0 & 1 & 0 \end{bmatrix} \tag{B.4}$$

Displacement transformation

$$u_R = u_r \sin\phi + u_z \cos\phi$$
$$u_\phi = u_r \cos\phi - u_z \sin\phi \qquad \text{(B.5)}$$
$$u_\theta = u_\theta$$

Stress transformation

$$\sigma_R = \sigma_r \sin^2\phi + \sigma_z \cos^2\phi + 2\tau_{rz} \sin\phi \cos\phi$$
$$\sigma_\phi = \sigma_r \cos^2\phi + \sigma_z \sin^2\phi - 2\tau_{rz} \sin\phi \cos\phi$$
$$\sigma_\theta = \sigma_\theta$$
$$\tau_{R\phi} = (\sigma_r - \sigma_z)\sin\phi \cos\phi - \tau_{rz}\left(\sin^2\phi - \cos^2\phi\right) \qquad \text{(B.6)}$$
$$\tau_{\phi\theta} = \tau_{r\theta} \cos\phi - \tau_{\theta z} \sin\phi$$
$$\tau_{\theta R} = \tau_{r\theta} \sin\phi + \tau_{\theta z} \cos\phi$$

Spherical components from Cartesian

The transformation matrix from Cartesian to spherical coordinates can be obtained by combining the previous transformations given by (B.1) and (B.4). Tracing back through tensor transformation theory, this is accomplished by the simple matrix multiplication

$$[Q] = \begin{bmatrix} \sin\phi & 0 & \cos\phi \\ \cos\phi & 0 & -\sin\phi \\ 0 & 1 & 0 \end{bmatrix} \begin{bmatrix} \cos\theta & \sin\theta & 0 \\ -\sin\theta & \cos\theta & 0 \\ 0 & 0 & 1 \end{bmatrix}$$

$$= \begin{bmatrix} \sin\phi \cos\theta & \sin\phi \sin\theta & \cos\phi \\ \cos\phi \cos\theta & \cos\phi \sin\theta & -\sin\phi \\ -\sin\theta & \cos\theta & 0 \end{bmatrix} \qquad \text{(B.7)}$$

Displacement transformation

$$u_R = u \sin\phi \cos\theta + v \sin\phi \sin\theta + w \cos\phi$$
$$u_\phi = u \cos\phi \cos\theta + v \cos\phi \sin\theta - w \sin\phi \qquad \text{(B.8)}$$
$$u_\theta = -u \sin\theta + v \cos\theta$$

Stress transformation

$$\sigma_R = \sigma_x \sin^2\phi \cos^2\theta + \sigma_y \sin^2\phi \sin^2\theta + \sigma_z \cos^2\phi$$
$$+ 2\tau_{xy} \sin^2\phi \sin\theta \cos\theta + 2\tau_{yz} \sin\phi \cos\phi \sin\theta + 2\tau_{zx} \sin\phi \cos\phi \cos\theta$$

$$\sigma_\phi = \sigma_x \cos^2\phi \cos^2\theta + \sigma_y \cos^2\phi \sin^2\theta + \sigma_z \sin^2\phi$$
$$+ 2\tau_{xy} \cos^2\phi \sin\theta \cos\theta - 2\tau_{yz} \sin\phi \cos\phi \sin\theta - 2\tau_{zx} \sin\phi \cos\phi \cos\theta$$

$$\sigma_\theta = \sigma_x \sin^2\theta + \sigma_y \cos^2\theta - 2\tau_{xy} \sin\theta \cos\theta$$

$$\tau_{R\phi} = \sigma_x \sin\phi \cos\phi \cos^2\theta + \sigma_y \sin\phi \cos\phi \sin^2\theta - \sigma_z \sin\phi \cos\phi$$
$$+ 2\tau_{xy} \sin\phi \cos\phi \sin\theta \cos\theta - \tau_{yz}\left(\sin^2\phi - \cos^2\phi\right)\sin\theta$$
$$- \tau_{zx}\left(\sin^2\phi - \cos^2\phi\right)\cos\theta$$

$$\tau_{\phi\theta} = -\sigma_x \cos\phi \sin\theta \cos\theta + \sigma_y \cos\phi \sin\theta \cos\theta + \tau_{xy} \cos\phi\left(\cos^2\theta - \sin^2\theta\right)$$
$$- \tau_{yz} \sin\phi \cos\theta + \tau_{zx} \sin\phi \sin\theta$$

$$\tau_{\theta R} = -\sigma_x \sin\phi \sin\theta \cos\theta + \sigma_y \sin\phi \sin\theta \cos\theta + \tau_{xy} \sin\phi\left(\cos^2\theta - \sin^2\theta\right)$$
$$+ \tau_{yz} \cos\phi \cos\theta - \tau_{zx} \cos\phi \sin\theta$$

(B.9)

Inverse transformations of these results can be computed by formally inverting the system equations or redeveloping the results using tensor transformation theory.

MATLAB® Primer

Many locations in the text used numerical methods to calculate and plot solutions to a variety of elasticity problems. Although other options (e.g., MAPLE, Mathematica) are available, the author has found MATLAB software well suited to conduct this numerical work. This particular software has all of the necessary computational and plotting tools to enable very efficient and simple applications for elasticity. MATLAB is a professional engineering and scientific software package developed and marketed by MathWorks, Inc. In recent years, it has achieved widespread and enthusiastic acceptance throughout the engineering community. Many engineering schools now require and/or use MATLAB as one of their primary computing tools, and it is expected that it will continue to replace older structured programming methods. Its popularity is due to a long history of well-developed and tested products, to its ease of use by students, and to its compatibility across many different computer platforms (e.g., PC, Mac, and UNIX). The purpose of this appendix is to present a few MATLAB basics to help students apply particular software applications for the needs of the text. The software package itself contains an excellent *Help* module that provides extensive information on various commands and procedures. Also, many books are available on the software package; see for example Palm (2008) or Gilat (2017). It is assumed that the reader has some prior background and experience with either MATLAB or another programming language such as FORTRAN, BASIC or C, and thus has a fundamental understanding of programming techniques.

C.1 Getting started

MATLAB is both a computer programming language and a software environment for using the language. Under the MS Windows Operating System, the MATLAB window will appear as shown in Fig. C.1. It is from this window that the Help menu can be accessed and this provides extensive information on most topics. In this Command window, the user can type instructions after the prompt "≫". However, it is much more efficient to create and save application programs within the Editor/Debugger window. This window is activated by going to the Editor menu above the Command window and selecting either New to start a new creation or Open to open an existing file. MATLAB files are called m-files and have the extension "*.m". Within the Editor/Debugger window, a new application code can be created or an existing one can be modified. In either case, the resulting file can then be saved for later use, and the current file can be run from this window. An example program appearing in the Editor/Debugger window is shown in Fig. C.2. Since MATLAB is continually being updated, it should be noted that all windows shown in the text will be for a 2009 version of the code.

C.2 Examples

Rather than attempting a step-by-step explanation of various MATLAB commands, we instead pursue a *learn-by-example* approach. In this fashion most of the needed procedures will be demonstrated

559

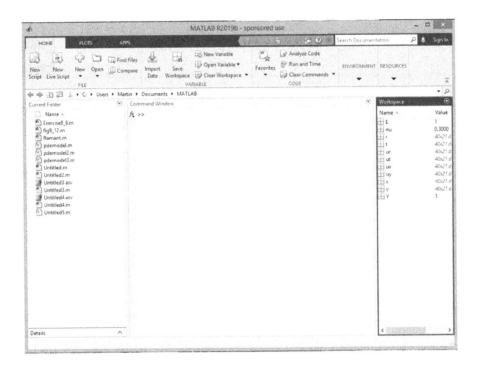

FIG. C.1 MATLAB Command window.

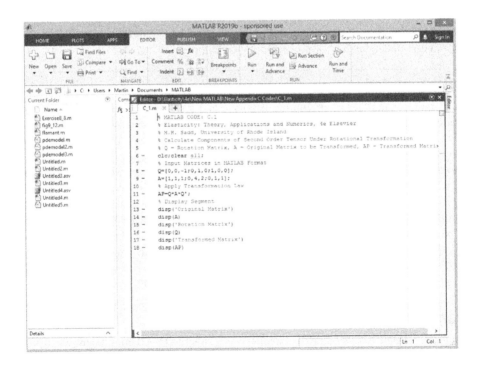

FIG. C.2 MATLAB Editor/Debugger window.

through presentation and discussion of several example codes that have been previously used in the text. Key features to be learned from these examples include:

- input of problem data
- generation of spatial variables
- calculation procedures
- plotting and display techniques

Readers with previous programming experience should be able to quickly review these examples and use them (with selected support from the Help menu) to develop their own codes.

Example C.1 Transformation of second-order tensors

Our first example is a simple MATLAB code to conduct the transformation of second-order tensors. The transformation rule was given by relation $(1.5.1)_3$ and this has been incorporated into the code shown in the frame below. Code lines preceded with a percent (%) symbol will not be executed and are used for comments to explain the coding. A semicolon ending a code line will suppress screen-printing that particular calculation. The rotation tensor $[Q]$ and tensor to be transformed $[A]$ are input from within the program, and thus the code must be modified if either of these matrices are changed. MATLAB's automatic abilities with matrix multiplication allow very simple coding. The disp command displays text and each of the matrices to the screen after the code is run.

```
% MATLAB CODE: C.1
% Elasticity: Theory, Applications and Numerics, 4e Elsevier
% M.H. Sadd, University of Rhode Island
% Calculate Components of Second Order Tensor Under Rotational Transformation
% Q = Rotation Matrix, A = Original Matrix to be Transformed, AP = Transformed Matrix
clc;clear all;
% Input Matrices in MATLAB Format
Q=[0,0,-1;0,1,0;1,0,0];
A=[1,1,1;0,4,2;0,1,1];
% Apply Transformation Law
AP=Q*A*Q';
% Display Segment
disp('Original Matrix')
disp(A)
disp('Rotation Matrix')
disp(Q)
disp('Transformed Matrix')
disp(AP)
```

Continued

Screen output created from this code is given by

```
>> Original Matrix
    1    1    1
    0    4    2
    0    1    1
Rotation Matrix
    0    0   -1
    0    1    0
    1    0    0
Transformed Matrix
    1   -1    0
   -2    4    0
   -1    1    1
```

Example C.2 Calculation of invariants, eigenvalues, and eigenvectors of a matrix

MATLAB provides very simple and efficient methods to determine the invariants, principal values, and directions of second-order tensors (matrices). The basic tool for the eigenvalue problem is the command eig( ), which will generate both the eigenvalues and eigenvectors. The simple code below illustrates the basics for the matrix originally given in Example 1.3.

```
% MATLAB CODE: C.2
% Elasticity: Theory, Applications and Numerics, 4e Elsevier
% M.H. Sadd, University of Rhode Island
% Program to Calculate Matrix Invariants, Principal Values & Directions
% Program Uses Matrix from Example 1-3
clc;clear all;
% Input Matrix
A=[2,0,0;0,3,4;0,4,-3]
% Calculate Invariants
invariants=[trace(A),(trace(A)^2-trace(A*A))/2,det(A)]
[V,L]=eig(A);
% Principal Values are the Diagonal Elements of the L Matix
principal_values=[L(1,1),L(2,2),L(3,3)]
% Principal Directions are the Columns of the V Matrix
principal_directions=[V(:,1),V(:,2),V(:,3)]
```

Screen output created from this code is given by
```
A =
   2   0    0
   0   3    4
   0   4   -3
invariants =
   2  -25  -50
principal_values =
  -5   2   5
principal_directions =
        0     1.0000        0
   0.4472          0  -0.8944
  -0.8944          0  -0.4472
```

Example C.3 Stress Contour Plotting from Fig. 3.8

MATLAB provides several useful techniques to make contour plots of variables. The following code uses the basic contour(. . .) command to handle scalar variables in two-dimensions. It is necessary to first generate a mesh grid of variable values in the domain of interest. For this case, the mesh grid was initially developed in polar coordinates and then translated to Cartesian for use with stress equations (3.6.3). Also included is a routine to draw the outer boundary of the disk.

```
% MATLAB CODE: C.3
% Elasticity: Theory, Applications and Numerics, 4e Elsevier
% M.H. Sadd, University of Rhode Island
% Plot Tau-Max Stress Contours in a Diametrically Loaded Disk
clc;clear all;clf
% Generate Mesh
[t,r]=meshgrid(0:pi/50:2*pi,0:.01:1);
[x,y]=pol2cart(t,r);
% Set Size
R=1;D=2*R;
% Calculate Stress Fields
r1=sqrt(((x.^2)+((R-y).^2)));
r2=sqrt(((x.^2)+((R+y).^2)));
sx=(-2/pi)*((((R-y).*(x.^2))./(r1.^4))+(((R+y).*(x.^2))./(r2.^4))-(1/D));
sy=(-2/pi)*((((R-y).^3)./(r1.^4))+(((R+y).^3)./(r2.^4))-(1/D));
txy=(2/pi)*(((((R-y).^2).*x)./(r1.^4))-((((R+y).^2).*x)./(r2.^4)));
tmax=sqrt((((sx-sy)./2).^2)+(txy.^2));
% Draw Outer Boundary
tt=0:.01:2*pi;
xx=cos(tt);yy=sin(tt);
plot(xx,yy,'k','linewidth',2.5)
hold on;axis off;axis equal
% Plot Tau-Max Contours with Particular Values
contour(x,y,tmax,[0,0.1,0.2,0.4,0.6,0.7,0.8,1,1.5,2,3],'k','linewidth',2);
```

Example C.4 XY plotting of stresses in Fig. 8.9

The MATLAB code shown in the frame below was developed to calculate and plot the radial and circumferential stresses in a thick-walled cylinder under internal pressure loading from Example 8.6. The theoretical expressions for the stresses were given by Eqs. (8.4.3) and the plot is shown in Fig. 8.9. The nondimensional radial coordinate r/r_2 is conveniently generated from 0.5 to 1, and the length(r) expression gives the number of terms in the r-array which is used as the limiter on the looping index. This simple code illustrates the for-end looping and calculation procedure and the basic XY plot call used to draw the two stress curves. Note the plot formatting used to label the axes. Additional format control is available through the Help menu and the plot can also be edited within the generated plot window. As shown in the next example, this looping scheme can actually be avoided by using MATLAB's automatic handling of array mathematics.

```
% MATLAB CODE: C.4
% Elasticity: Theory, Applications and Numerics, 4e Elsevier
% M.H. Sadd, University of Rhode Island
% Calculate and Plot Stresses in Thick-Walled Cylinder Problem Example 8.6
% Internal Pressure Loading Case with r1/r2=0.5; Generate Figure 8-9
clc;clear all;
% Generate Non-Dimensional Radial Coordinate Space: r/r2
r=[0.5:0.01:1];
% Calculation Loop for Stresses
for i=1:length(r)
    sr(i)=(1/3)*(1-(1/r(i))^2);
    st(i)=(1/3)*(1+(1/r(i))^2);
end
% Plotting Call
plot(r,sr,r,st)
xlabel('Dimensionless Distance, r/r_2')
ylabel('Dimensionless Stress')
```

Example C.5 Polar contour plotting of hoop stress around circular hole in Fig. 8.13

This example illustrates a code used to plot the hoop stresses on the boundary of a circular hole in an infinite plane under uniform far-field tension (see problem geometry in Fig. 8.12). The hoop stress was given by Eq. (8.4.15)$_2$, and the plot on the boundary of the hole is shown in Fig. 8.13. Similar to the previous example, this code calculates the necessary stress values and then displays them in a polar plot. The radial coordinate is specified to be 1, while the angular coordinate is generated from 0 to 2π. For this code the use of the for-loop is dropped since MATLAB can automatically handle calculations with matrix/vector arguments. The plotting call used in this example is the polar command that generates Fig. 8.13. Additional formatting can be applied to the polar plot call to control line type, thickness, etc., and these can also be edited in the plot window.

```
% MATLAB CODE: C.5
% Elasticity: Theory, Applications and Numerics, 4e Elsevier
% M.H. Sadd, University of Rhode Island
% Calculate and Plot Normalized Hoop Stress on Circular Hole
% In Infinite Plane Under Uniform Tension at Infinity: Example 8.7
% Non-Dimensional Plot Generates Figure 8-13
clc;clear all;
% Input (r/a)- Variable and Generate Angular Coordinate Space
r=1;
t=[0:0.01:2*pi];
% Calculation Loop
st=0.5*(1+(1/r)^2)-0.5*(1+3*(1/r)^4)*cos(2*t);
% Plotting Call
polar(t,st)
title('Non-Dimensional Hoop-stress Around Hole')
```

Example C.6 Displacement vector distribution plotting in Fig. 8.22

Consider next the example of plotting the displacement vectors for the Flamant problem of Example 8.8 shown in Fig. 8.19. For the normal loading case, the displacement field was given by relations (8.4.43) and these were plotted in Fig. 8.22 for the near-field case ($0 < r < 0.5$) with a Poission's ratio of 0.3 and $Y/E = 1$. Radial coordinates are generated by the logspace command as explained in the comment line, and this range can easily be changed to investigate other regions of the half-plane. Within the calculation, the Cartesian coordinates and displacement components are changed to reflect the system used in Chapter 8. The plotting is done using the quiver command, which draws two-dimensional vectors with components ux and uy at locations x and y. Note the use of the meshgrid which creates a mesh of points in the two-dimensional field to do our calculations and plotting. Also note the use of a period in front of products of arrays to specify that the operation is to be done on an element-by-element basis. Additional details on this type of plotting can be found in the Help menu.

Continued

```
% MATLAB CODE: C.6
% Elasticity: Theory, Applications and Numerics, 4e Elsevier
% M.H. Sadd, University of Rhode Island
% Displacement Vector Distribution Plot for Flamant Problem - Figure 8-19
% Normal Loading Case (X=0), Generates Figure 8-22
clc;clear all;
% Input Parameters
Y=1; E=1; nu=0.3;
% Input Radial Coordinates: logspace(M,N,*) Generates Region 10^M < r < 10^N
r=logspace(-3,-0.5,40);
% Input Theta Coordinates
t=[0:pi/20:pi];
% Create Mesh
[t,r]=meshgrid(t,r);
ur=Y/(pi*E)*((1-nu)*(t-pi/2).*cos(t)-2*log(r).*sin(t));
ut=Y/(pi*E)*(-(1-nu)*(t-pi/2).*sin(t)-2*log(r).*cos(t)-(1+nu)*cos(t));
% Calculate Cartesian Displacement Components - Flip y-Component
ux=cos(t).*ur-sin(t).*ut;
uy=-(sin(t).*ur+cos(t).*ut);
% Covert to Cartesian Coordinate Mesh
[x,y]=pol2cart(-t,r);
% Plotting Call for Vector Distribution
quiver(x,y,ux,uy)
axis equal;
```

Example C.7 Plotting of maximum shear stresses below flat punch problem in Fig. 8.41

The solution to some contact mechanics problems were developed in Section 8.5. For the particular case of a flat rigid indenter, the half-space stresses were given by relations (8.5.9) and the maximum shearing stress distribution below the punch was shown in Fig. 8.41. The following code does the numerical evaluation of the complicated integrals in (8.5.9), calculates pointwise values of τ_{max}, and then uses the contour command to plot the contours shown in Fig. 8.41. Numerical evaluation of the integrals is determined by using the quadv command, which requires the use of function programming to properly pass the integrands into the computation. This fact requires that the code must start with a function name. See the Help menu for more information on function programming and the use of numerical integration commands.

```
% MATLAB CODE: C.7
% Elasticity: Theory, Applications and Numerics, 4e Elsevier
% M.H. Sadd, University of Rhode Island
% Calculate and Plot Tau Stress Contours Under Flat Punch Loading (P=1)
% Numerically Evaluate Integrals in Solution (8.5.9)Using quadv(.)
```

```
% Singularity at s=a is Handled using Integral Range Limiter e
function flat_punch
clc;clear all;clf
a=1; %input loading width
e=0.0001; %input range limiter
[x,y]=meshgrid(-2*a:0.1:2*a,0.05:0.1:4*a+0.1);
for i=1:length(x)
    for j=1:length(y)
        sx(i,j)=quadv(@(s)Irx(x(i,j),y(i,j),s,a),-(a-e),a-e);
        sy(i,j)=quadv(@(s)Iry(x(i,j),y(i,j),s,a),-(a-e),a-e);
        txy(i,j)=quadv(@(s)Irxy(x(i,j),y(i,j),s,a),-(a-e),a-e);
        smax(i,j)=sqrt((((sx(i,j)-sy(i,j))/2)^2)+(txy(i,j)^2));
    end
end
% Draw half-space boundary line
plot([-2*a,2*a],[0,0],'k','linewidth',3)
hold on
contour(x,-y,smax,40,'k','linewidth',1.4)
axis off;hold off;
title('\tau_m_a_x Contours: Frictionless Rigid Punch Loading on a Half-Space')
function I=Irx(x,y,s,a)
I=-(2*y/pi)*(1/(pi*sqrt(a^2-s^2)))*(x-s)^2/(((x-s)^2+y^2)^2);
function I=Iry(x,y,s,a)
I=-(2*y^3/pi)*(1/(pi*sqrt(a^2-s^2)))/(((x-s)^2+y^2)^2);
function I=Irxy(x,y,s,a)
I=-(2*y^2/pi)*(1/(pi*sqrt(a^2-s^2)))*(x-s)/(((x-s)^2+y^2)^2);
```

Example C.8 Plotting of warping displacement contours in Fig. 9.8

The following MATLAB code has been developed to calculate and plot the warping displacements for the torsion of a cylinder with the elliptical section shown in Fig. 9.7. This code uses many of the same commands from the previous examples to input parameters, generate the variable grid space, and conduct the calculations to determine the warping displacement array. The plotting call uses the contour command that generates contours of constant w within the grid space that was created to lie inside the elliptical boundary. The code generates the displacement contours shown in Fig. 9.8. Again, additional formatting details on this plotting command can be found in the Help menu.

Continued

```
% MATLAB CODE: C.8
% Elasticity: Theory, Applications and Numerics, 4e Elsevier
% M.H. Sadd, University of Rhode Island
% Generates 2-D Warping Displacement Contours for
% Elliptical Section Under Torsion - Figure 9-8
clc, clear all
% Input Geometry and Angle of Twist
a=1.0; b=0.5; alpha=1.0;
% Input Grid Space
[t,r]=meshgrid(0:pi/20:2*pi,0:0.05:1);
% Generate Contour Data
K=alpha*(b2-a2)/(a^2+b2);
x=a*r.*cos(t);
y=b*r.*sin(t);
w=K*x.*y;
% Plotting Call with 20 Contours
contour(x,y,w,20,'k-');
axis equal
```

Example C.9 Plotting three-dimensional warping displacement surface in Fig. 9.8

This example illustrates how MATLAB can make a three-dimensional surface plot of the warping displacement for the torsion of an elliptical section. Fig. 9.8 shows the surface plot created from this code. After generating appropriate x and y location arrays, the warping displacement values are calculated and stored in the array w. Using the surf command, the warping surface is then generated in a three-dimensional system. Numerous viewing parameters can be edited to suit the desired view.

```
% MATLAB CODE: C.9
% Elasticity: Theory, Applications and Numerics, 4e Elsevier
% M.H. Sadd, University of Rhode Island
% Three Dimensional Plot of Warping Displacement Surface
% for Elliptical Section Under Torsion
clc;clear all;
a=1;b=0.5;
[t,r]=meshgrid(0:pi/20:2*pi,0:0.05:1);
x=a*r.*cos(t);
y=b*r.*sin(t);
w=-x.*y;
surfc(x,y,w);
```

```
colormap gray
h=findobj(gcf,'type','patch');
set(h,'LineWidth',1.0, 'EdgeColor','k');
axis([-1 1 -1 1])
axis square
view(20,10)
```

Example C.10 Determination of roots for orthotropic materials

In Chapter 11 anisotropic solutions to plane elasticity problems required the roots of the characteristic equation (11.5.7). As indicated in Exercise 11.15, for orthotropic materials this characteristic equation reduces to a quadratic with roots $\beta_{1,2}$. The MATLAB code shown will calculate these two roots and write the results to the command window screen. This code illustrates some of the basic formatting issues related to inputting names and data, and printing calculated information to the screen.

```
% MATLAB CODE: C.10
% Elasticity: Theory, Applications and Numerics, 4e Elsevier
% M.H. Sadd, University of Rhode Island
% Calculate Beta Parameters for Orthotropic E-Glass Material
clc; clear all;
% Input Number of Materials, Names and Stiffness Moduli
N=1;
name(1,:)='E-Glass/Epoxy';
e1(1)=38.6; e2(1)=8.3; nu12(1)=0.26; mu12(1)=4.2;
% Calulate Compliances
for i=1:N
s11(i)=1/e1(i); s22(i)=1/e2(i); s12(i)=-nu12(i)/e1(i); s66(i)=1/mu12(i);
% Calculate Beta Values
b1(i)=sqrt(-(1/(2*s11(i)))*(-(2*s12(i)+s66(i))+sqrt((2*s12(i)+s66(i))^2-4*s11(i)
*s22(i))));
b2(i)=sqrt(-(1/(2*s11(i)))*(-(2*s12(i)+s66(i))-sqrt((2*s12(i)+s66(i))^2-4*s11(i)
*s22(i))));
% Print Results to Screen
fprintf(1,'\n ')
disp(name(i,:))
fprintf(1,' beta(1)=%5.3f',b1(i))
fprintf(1,' beta(2)=%5.3f',b2(i))
end
```

Continued

Screen output from this particular case is given by
```
>>
  E-Glass/Epoxy
  beta(1)=0.758 beta(2)=2.845
```

Example C.11 Plotting of screw dislocation strains for gradient elasticity solution

This MATLAB code plots the strain distribution (15.7.15) for the gradient elasticity solution. No new commands are introduced in this code.

```
% MATLAB CODE: C.11
% Elasticity: Theory, Applications and Numerics, 4e Elsevier
% M.H. Sadd, University of Rhode Island
% Calculation & Plot Gradient Elasticity Dislocation Example 15.16
clc;clear all;clf
for n=[1,3,10]
c=.01*n;
r=0.001:0.01:3;
ege=abs((1/(4*pi))*(-(1./r)+(1/sqrt(c)).*besselk(1,r/sqrt(c))));
ece=abs((1/(4*pi))*(-(1./r)));
plot(r,ege,'k-','linewidth',2)
hold on
plot(r,ece,'k--','linewidth',2)
axis([0,3,0,0.5]);grid on
xlabel('Distance, r'); ylabel('| e_x_z / b |')
title('Strain Field Near Screw Dislocation')
legend('Gradient Elasticity','Classical Elasticity')
end
```

Example C.12 : PDE Toolbox—finite element application of circular hole in a plate under uniform tension from Example 16.2

In Chapter 16 the PDE Toolbox was presented as a MATLAB application software that could conduct two-dimensional finite element analysis. We will now present some of the basic steps in generating the solution to the problem of a circular hole in a plate under uniform tension as originally discussed in Example 16.2. Once properly installed, the toolbox is activated by typing the command *pdetool* in the MATLAB Command window or clicking on the *APPS* tab to find the *PDE* tab. This will bring up the Graphical User Interface (GUI) window shown in Fig. C.3. Within this window, the first step is to select the type of problem by choosing one of the items from the pull-down menu as shown in Fig. C.4. The figure illustrates the selection of "Structural Mech., Plane Stress". Other useful choices would include "Structural Mech., Plane Strain" for plane strain analysis and "Generic Scalar", which can be used to find numerical solutions to the torsion problem (see Example 16.4

and Fig. 16.7). After selecting the problem type, a click on the "PDE" button will bring down a window to input the desired values of elastic moduli and body forces.

Once the problem type and input parameters have been chosen, problem geometry can be created using the simple drawing package within the GUI window. This feature is activated by selecting "Draw Mode" from the Draw menu. The problem of a rectangular plate with a central circular hole is created by drawing a rectangle and circle, and then subtracting the circular area from the rectangle as shown in the drawing mode window in Fig. C.5.

After completing the problem geometry, the next step is to input the appropriate boundary conditions. This step is done in the boundary mode screen that is selected from the Boundary menu pull-down. Fig. C.6 shows this screen for the current example, and the software has automatically divided the rectangular and circular boundaries into four segments. Unfortunately, this simple code

Continued

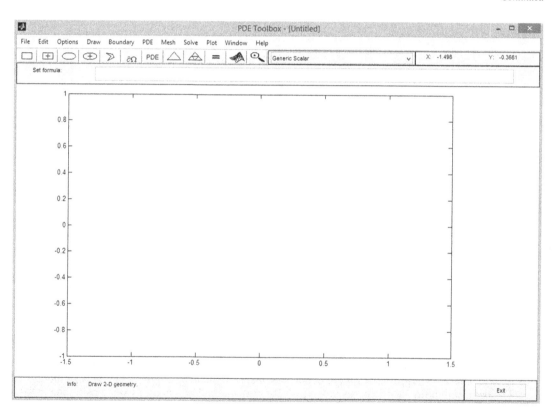

FIG. C.3 Graphical user interface for the PDE Toolbox.

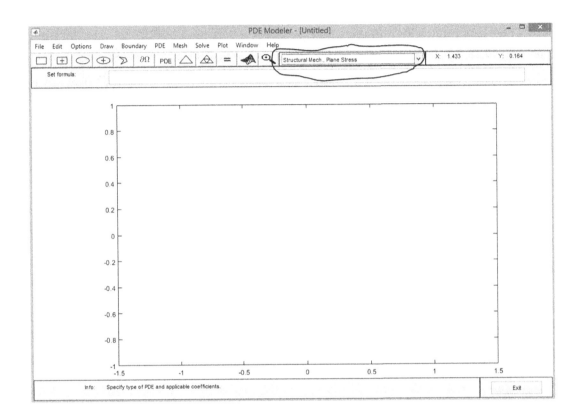

FIG. C.4 Selection of problem type in PDE Toolbox.

does not provide selection options in defining the boundary segments, and this places limits on boundary condition specification. In any event once in the boundary mode screen, one or more boundary segments can be selected through a simple mouse point-and-click. Boundary condition specification can then be made on these selected segments by going to "Specify Boundary Conditions..." found in the Boundary menu pull-down. This will bring forward the Boundary Condition window shown in Fig. C.7. Through appropriate selection of the various parameters defined in the window, displacement (Dirichlet), traction (Neumann), and mixed conditions can be specified. For displacement specification, $u = \text{r1}$ and $v = \text{r2}$, while traction conditions correspond to $T_x = \text{g1}$ and $T_y = \text{g2}$. In this fashion, conditions on each boundary segment can be specified.

The next step in the FEA process is to mesh the domain, and this is easily done using the auto-meshing features of the toolbox. After completing the boundary conditions, an initial coarse mesh can be generated by simply clicking on the toolbar button denoted with a triangle. Normally this coarse mesh will not be appropriate for final use, and a finer mesh can be generated by clicking

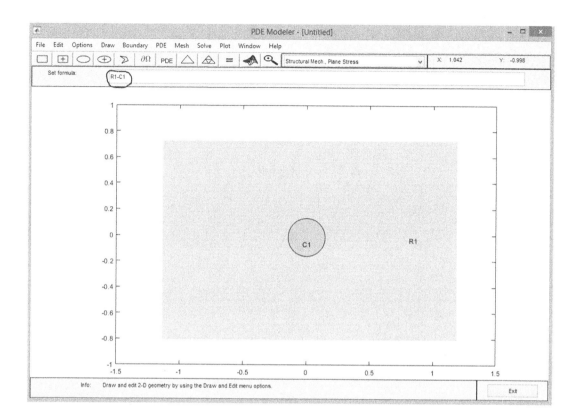

FIG. C.5 Draw mode screen for rectangle with circular hole.

on the toolbar button with the finer triangles. Continued use of this button will generate meshes with increasing numbers of elements. Normally only one or two mesh refinements are necessary to create a useful mesh. These and other meshing procedures can also be found under the Mesh tab in the main menu. Fig. 16.4 illustrates a reasonably fine mesh for this problem.

Having finished the creation of problem geometry, boundary conditions, and finite element meshing, the solution is now ready to be completed and the results displayed. However, before running the solver, selection of the desired graphical solution output should be made. The choice of results to be displayed is found on the Parameters tab under the Plot menu pull-down. Selecting this tab will activate the Plot Selection window as shown in Fig. C.8. Numerous choices are available on the variables to be plotted and the type of plot to be made. Selecting the variable x stress and choosing a contour plot type produces contours of σ_x as shown in Fig. 16.4. Many other graphical displays can be sequentially generated and saved for later use. Different stages of the finite element solution can be revisited after completion of the final solution. However, going

Continued

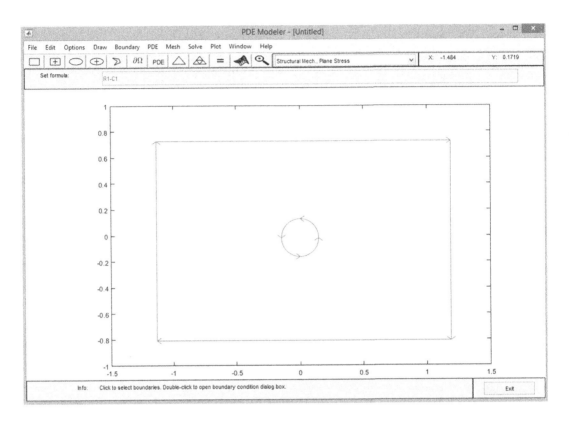

FIG. C.6 Boundary mode screen for rectangle with circular hole.

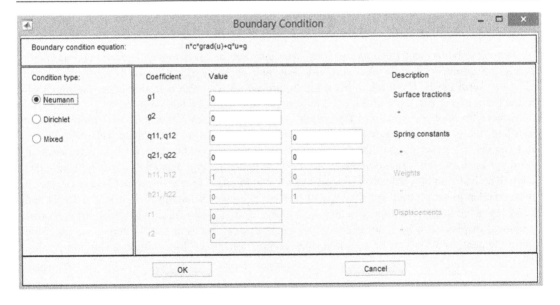

FIG. C.7 Boundary specification window.

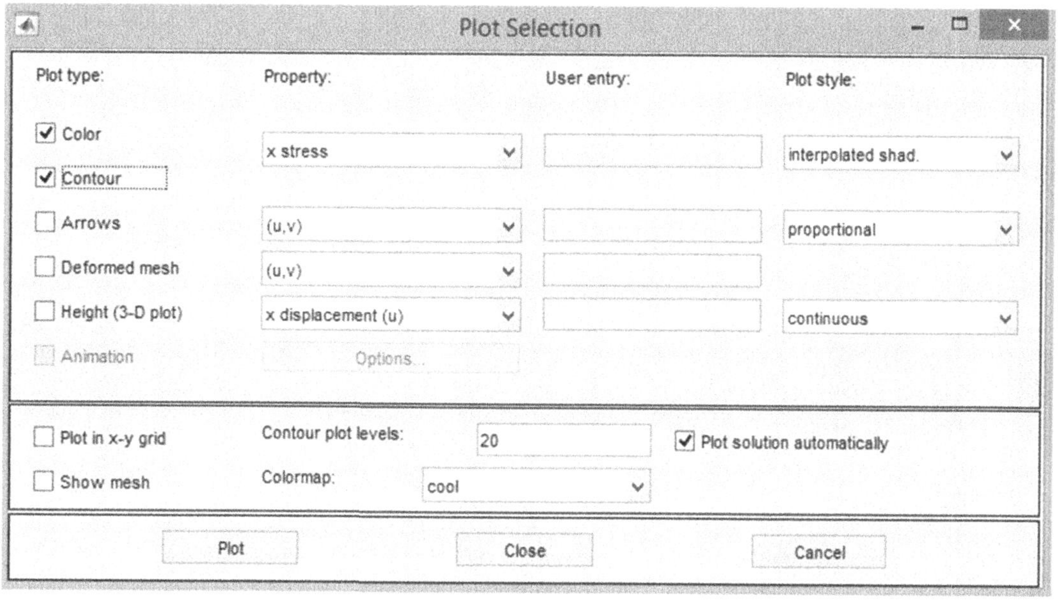

FIG. C.8 Plot Selection window.

back to an earlier stage will require that all subsequent solution steps be redone; for example, going back to the drawing stage will require re-input of boundary conditions and meshing data. Because the software is so easy to use, redoing certain steps is normally not a difficult task. This brief discussion only presented some of the basics of the PDE Toolbox. Further and more detailed information can be found under the Help menu, and an entire User Manual (pdf format) is available for reference and/or printout.

References

Palm WJ: *A concise introduction to MATLAB*, New York, 2008, McGraw-Hill.
Gilat A: *MATLAB: An Introduction with Applications*, 6th Ed, Hoboken, NJ, 2017, John Wiley & Sons.

Review of mechanics of materials

Beginning undergraduate studies of the mechanics of deformable solids are normally taught in a course called *mechanics of materials* or *strength of materials*. Based on very restrictive assumptions, this study develops stress, strain, and displacement field solutions for a very limited class of elastic solids with simple geometry. Strength of materials theory commonly makes use of assumptions on the geometry of the deformation (e.g., *plane sections remain plane*) and thus assumes the distribution of displacements and strains. Further simplification is also sometimes made on the stress field. Because of the level of approximation, strength of materials is often referred to as the *elementary theory* when compared to the more exact elasticity model. Nevertheless, decades of application have shown that mechanics of materials provides reasonable estimates for many practical stress analysis problems. Furthermore, strength of materials solutions have provided guidance for the development of particular elasticity solutions.

We now pursue a brief review of the basic strength of materials solutions of extension, torsion, and bending/shear of elastic rods and beams, as shown in Fig. D.1. Rod and beam structures are normally defined as prismatic solids with a length dimension much larger than the other two dimensions located within the cross-section. General loadings to such structures commonly include axial force P, shear force V, torque T, and bending moment M. Mechanics of materials theory develops an approximate solution for each of these four loading types.

Commonly these solutions will be restricted to cases with particular cross-sectional shapes that are related to section properties of A = area, J = polar moment of inertia, and I = rectangular moment of

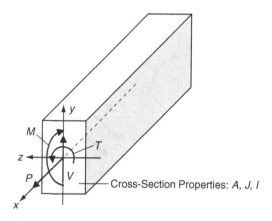

FIG. D.1 Extension, torsion, and bending/shear deformation of beam-type structures.

inertia. Because these solutions are useful to compare with related elasticity models, we now briefly review their development. In addition to these problems, we review curved beams and cylindrical pressure vessels.

D.1 Extensional deformation of rods and beams

We begin with the simplest case concerning the extensional deformation of an elastic rod or beam under purely axial loading P, as shown in Fig. D.2. For this case the cross-section can be of general shape, but the resultant loading must pass through the section's centroid so as not to produce bending effects. The fundamental deformation assumption is that all points in the cross-section displace uniformly in the axial direction (x-direction), thus making the problem one-dimensional.

Under this assumption, the only nonzero stress component considered is the normal component $\sigma = \sigma_x$ and it is assumed to be uniformly distributed over the section, as shown in Fig. D.2. A simple force balance will give $P = \sigma A$, where A is the cross-sectional area. This result then generates the simple stress relation

$$\sigma = \frac{P}{A} \tag{D.1}$$

Because the problem is one-dimensional, Hooke's law reduces to $\sigma = E\varepsilon$ and the single strain component is given by $\varepsilon = du/dx$. Combining these results with relation (D.1) produces the simple displacement or deformation relation

$$u = \int \frac{P}{AE}dx$$
$$= \frac{PL}{AE}\,(\text{constant loading}) \tag{D.2}$$

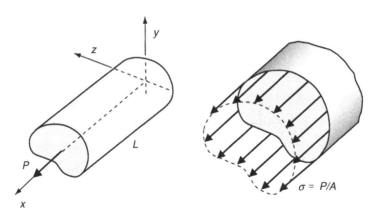

FIG. D.2 Extensional deformation problem.

D.2 Torsion of circular rods

The next loading case concerns the torsional loading and deformation of rods, as shown in Fig. D.3. For this case, the cross-section must be circular or hollow circular, thereby simplifying the section deformation. The deformation assumption for this problem is that points within the cross-section displace only tangentially and in proportion to the distance from the section's center. Thus, cross-sections perpendicular to the rod's axis remain plane during the deformation and no section warping will occur.

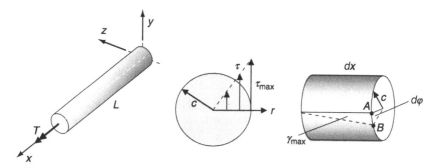

FIG. D.3 Torsional deformation problem.

Under such deformation, the section shear strain $\gamma = \gamma_{r\theta}$ will vary linearly from the center. Typical of the elementary theory, only one nonzero stress component will be considered; this is the shear stress, $\tau = \tau_{r\theta}$, lying in the cross-sectional plane. Because the strain component varies linearly, the section shear stress also behaves in the same manner as shown in Fig. D.3. Applying equilibrium between the applied loading T and the assumed shear stress distribution gives

$$T = \int_A \left(\frac{\tau_{max}}{c} r\right) r dA = \frac{\tau_{max}}{c} \int_A r^2 dA = \frac{\tau_{max}}{c} J$$

where $J = \int_A r^2 dA$ is known as the *polar moment of inertia* of the cross-section and for a solid circular section of radius c, $J = \dfrac{\pi c^4}{2}$. Rearranging the previous expression gives the standard stress relationship for the torsion problem

$$\tau_{max} = \frac{Tc}{J} \tag{D.3}$$

and relation (D.3) can be used to calculate the shear stress at any radial distance.

To determine the angle of twist φ, consider a rod element of length dx as shown in Fig. D.3. Under small torsional deformation, the outer fiber arc AB can be expressed in two ways, thus giving the relation $\gamma_{max} dx = d\varphi c$. This result then implies

$$\frac{d\varphi}{dx} = \frac{\gamma_{max}}{c} = \frac{T}{J\mu}$$

where μ is the shear modulus. Integrating this result gives the familiar relation

$$\varphi = \int \frac{T}{J\mu} dx$$

$$= \frac{TL}{J\mu} \text{(constant loading)}$$

(D.4)

D.3 Bending deformation of beams under moments and shear forces

The application of transverse external loadings will introduce internal bending moments M and shear forces V in beam type structures, as shown in Fig. D.4. Each of these internal forces will generate stresses within the structure, and mechanics of materials theory has developed approximate relations to calculate them. Beam deflection relations have also been formulated to determine the resulting deformation of the beam's centroidal axis (x-axis).

Before heading into these stress and deflection analysis relations, we first explore the standard methods of determining bending moment and shear force distributions in beams. These distributions will be needed for stress and deflection calculations. Typically this procedure involves a static equilibrium analysis of the beam, taking into account the particular support conditions and the nature of the applied loadings. Although other, more complicated conditions can be modeled, most beam problems involve three types of idealized supports: pinned, roller, and fixed, as shown in Fig. D.5. Such support conditions involve particular constraints on the deformation and these can be translated into particular support reaction forces.

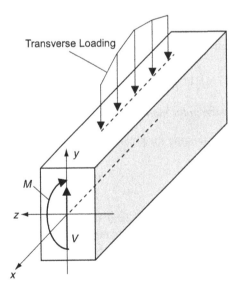

FIG. D.4 Bending and shear loadings on beam structures.

Pinned Support:
No Horizontal or
Vertical Movement

Resulting Reactions:
Horizontal and Vertical
Forces

Roller Support:
No Vertical Movement

Resulting Reactions:
Vertical Force and
Zero Moment

Fixed Support:
No Horizontal or
Vertical Movement
and No Rotation

Resulting Reactions:
Horizontal and Vertical
Forces and Moment

FIG. D.5 Common supports for beam problems.

We will now explore the typical procedures to determine the internal bending moment and shear force distribution for a particular beam problem, with the understanding that other problems with different loadings and support conditions can be handled in the same fundamental manner. Consider the problem of a simply supported beam (pinned and roller supports) carrying a single concentrated loading of P, as shown in Fig. D.6.

We wish to determine the moment and shear distribution as a function of coordinate x. This is easily done by constructing one or more sections through the beam in locations where the distributions are continuous, and given by a single unique relation. For the problem under study, two such regions exist: $0 \leq x \leq a$ and $a \leq x \leq L$. After making the appropriate sections, a free-body diagram of each portion of the beam can then be constructed, as shown in Fig. D.7. Note that the vertical reaction R from the left support has been calculated from overall equilibrium analysis of the entire beam, and the shear force V and bending moment M have been included at the cut location x. Normal positive sign conventions for the shear and moment are as drawn in the figure.

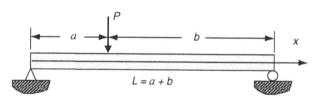

$$L = a + b$$

FIG. D.6 Simply supported beam example.

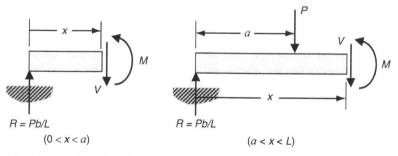

$R = Pb/L$

$(0 < x < a)$

$R = Pb/L$

$(a < x < L)$

FIG. D.7 Free-body diagrams of sectioned beam segments.

Applying equilibrium analysis (balance of vertical forces and moments) yields the following results for the shear force and bending moment in each portion of the beam

$$V(x) = \begin{cases} Pb/L, & 0 \le x \le a \\ -Pa/L, & a < x \le L \end{cases}$$

$$M(x) = \begin{cases} Pbx/L, & 0 \le x \le a \\ Pa(L-x)/L, & a \le x \le L \end{cases}$$

(D.5)

These results are plotted in Fig. D.8, and the maximum values can then be easily determined; for example, $M_{\max} = M|_{x=a} = Pba/L$. Note that a general relation between the moment and shear, $V = \dfrac{dM}{dx}$, can be developed by an equilibrium analysis of a differential beam element. Solutions to other problems follow using the same basic procedures.

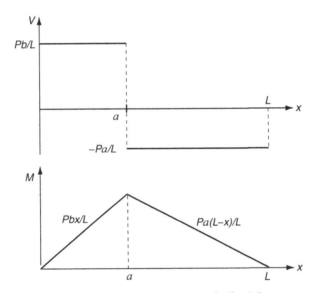

FIG. D.8 Shear and bending moment diagrams of the beam problem in Fig. D.6.

Referring to Fig. D.9, the fundamental mechanics of materials assumption for beam theory is that plane sections perpendicular to the beam axis before deformation remain plane after deformation. Recall that the beam axis is the line that goes through the centroid of each cross-sectional area. This assumption leads to the result that the extensional strains due to bending vary linearly from the beam axis. Neglecting all other normal strain and stress components, the bending stress, $\sigma = \sigma_x$, also varies linearly, and thus $\sigma = Ky$, where K is some constant. Applying equilibrium between the applied loading M and the assumed bending stress distribution gives

$$M = -\int_A Ky^2 dA = -K\int_A y^2 dA = -KI$$

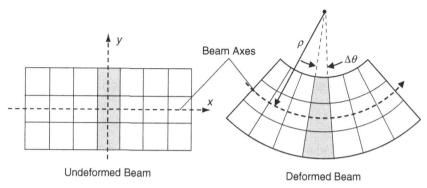

FIG. D.9 Assumed deformation within beams.

where $I = \int_A y^2 dA$ is the *moment of inertia* of section area A about the neutral axis (z-axis in Fig. D.4) that goes through the centroid of the cross-section. This result establishes the value for the constant K, and thus the stress relation is now given by the familiar relation

$$\sigma = -\frac{My}{I} \tag{D.6}$$

This simple linear relation predicts maximum stresses at either the top or the bottom of the section depending on the location of the centroid; this is illustrated in Fig. D.10 for the case of a centrally located centroid. Note that the positive moment produces compression at the top of the section and tension at the bottom.

The fundamental hypothesis that plane sections remain plane during deformation provides the basis to determine the theory for beam deflection analysis. As shown in Fig. D.9, the beam axis is bent into a locally circular shape with a radius of curvature ρ. Denoting $\Delta\theta$ as the included angle between nearby sections, the longitudinal strain can be expressed as

$$\varepsilon = \frac{(\rho - y)\Delta\theta - \rho\Delta\theta}{\rho\Delta\theta} = -\frac{y}{\rho}$$

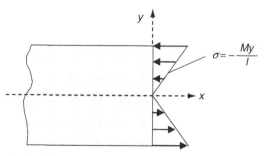

FIG. D.10 Bending stress distribution in the beam section.

where y represents the coordinate measure as shown. Using Hooke's law and relation (D.6), the strain can also be written as

$$\varepsilon = \frac{\sigma}{E} = -\frac{My}{EI}$$

Combining these two results gives the Euler–Bernoulli curvature–flexure relation

$$\frac{1}{\rho} = \frac{M}{EI} \tag{D.7}$$

where $1/\rho$ is the *curvature*. From geometry, the curvature of any two-dimensional space curve $v(x)$ is given by

$$\frac{1}{\rho} = \frac{\dfrac{d^2v}{dx^2}}{\left[1 + \left(\dfrac{dv}{dx}\right)^2\right]^{3/2}} \approx \frac{d^2v}{dx^2}$$

where we have assumed small deformations and small slopes. Now interpreting $v(x)$ as the vertical deflection of the beam axis (positive upward), we can write the equation of the elastic curve as

$$\frac{d^2v}{dx^2} = \frac{M}{EI} \tag{D.8}$$

Note that for the case where deflection is positive downward, the right-hand side of relation (D.8) picks up a minus sign. Once the moment distribution $M(x)$ has been determined, relation (D.8) can then be integrated to determine the transverse beam deflection—that is, the elastic deflection curve $v(x)$. This solution scheme requires the use of particular boundary conditions to evaluate the arbitrary constants of integration that are generated during the integrations.

Consider the simple cantilever beam example shown in Fig. D.11. The beam is fixed at $x = L$ and carries a single concentrated force at the free end, $x = 0$. Taking a single section at any location x, the shear and moment distributions are easily found to be $V = -P$ and $M = -Px$. Using Eq. (D.8) and integrating twice yields

$$EI\frac{d^2v}{dx^2} = -Px$$

$$EI\frac{dv}{dx} = -\frac{Px^2}{2} + C_1$$

$$EIv = -\frac{Px^3}{6} + C_1x + C_2$$

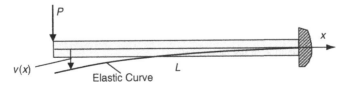

FIG. D.11 Cantilever beam example.

Boundary conditions at $x = L$ require zero deflection and zero slope, and thus lead to relations

$$EI\frac{dv(L)}{dx} = -\frac{PL^2}{2} + C_1 = 0$$

$$EIv(L) = -\frac{PL^3}{6} + C_1L + C_2 = 0$$

which can be solved to determine the constants $C_1 = PL^2/2$ and $C_2 = -PL^3/3$. Combining these results gives the final form for the beam deflection relation

$$v = \frac{P}{6EI}\left(-x^3 + 3L^2x - 2L^3\right)$$

From this relation the maximum deflection is found at the free end

$$v_{max} = v(0) = -\frac{PL^3}{3EI}$$

The results of another example beam deflection problem are shown in Fig. D.12.

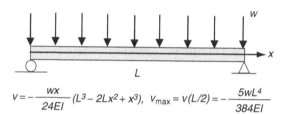

$$V = -\frac{wx}{24EI}(L^3 - 2Lx^2 + x^3), \quad v_{max} = v(L/2) = -\frac{5wL^4}{384EI}$$

FIG. D.12 Uniformly loaded beam deflection problem.

The final step in our review of beam problems concerns the effect of the shear force V. Although not explicitly stated, the previous discussion of beam deflection was concerned only with the bending moment loading. It has been shown that shear effects on beam deflections are only important for very short beams whose length to section dimension ratio is less than 10. Generally, then, mechanics of materials theory neglects shear force effects when calculating beam deflections. However, in regard to beam stresses, the internal shear force must give rise to a resulting shear stress distribution over the cross-section. For this case, no simple assumption exists for the deformation or strain distribution, and thus we must make some modification from our previous stress analysis developments.

Fig. D.13 illustrates a typical beam element for the general case where the moment will be changing with location x. Thus, the resulting bending stress distribution on the left-hand side of the element will not be identical to the stress on the right-hand side. This fact will create an imbalance of forces and will generate a shear stress τ on a horizontal plane, as shown in the sectioned element in Fig. D.13. Note that the identical shear stress will also exist on the vertical plane at the same location y'. Assuming that this shear stress is uniformly distributed over the differential beam element length dx, we apply a simple equilibrium force balance in the x-direction to get

$$\int_{A'}\left(\frac{M + dM}{I}\right)ydA - \int_{A'}\left(\frac{M}{I}\right)ydA - \tau(tdx) = 0 \Rightarrow \tau = \frac{dM}{dx}\left(\frac{1}{It}\right)\int_{A'}ydA$$

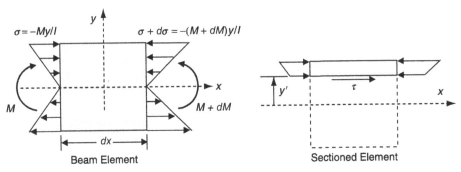

FIG. D.13 Loadings on a beam element.

where t is the thickness of the section at $y = y'$, and A' is the partial section area above the level $y = y'$. Now, as previously mentioned, $\dfrac{dM}{dx} = V$, and if we let $Q = \int_{A'} y \, dA$, our force balance reduces to

$$\tau = \frac{VQ}{It} \tag{D.9}$$

This is the mechanics of materials formula for the shear stress distribution in beam-bending problems. It should be observed that Q is the first area moment of section A' about the neutral axis, and it varies as a function of the vertical coordinate measure y', vanishing when y' corresponds to the top or bottom of the section. This parameter can also be expressed by

$$Q = \int_{A'} y \, dA = \bar{y}' A'$$

where $\bar{y}'$ is the vertical distance to the centroid of the partial area A'.

To explore shear stress variation across an example beam section, consider the rectangular section shown in Fig. D.14. Recall that for a cross-section of rectangular shape of height h and width b, the moment of inertia is given by the relation $I = \dfrac{1}{12} bh^3$. Taking $y = y'$, the relation for Q can be written as

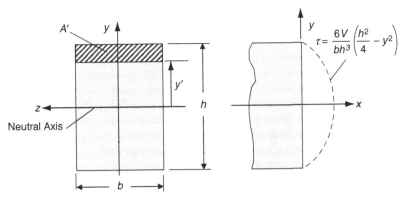

FIG. D.14 Rectangular section shear stress distribution analysis.

$$Q = \bar{y}'A' = \left[y + \frac{1}{2}\left(\frac{h}{2} - y\right)\right]\left(\frac{h}{2} - y\right)b = \frac{1}{2}\left(\frac{h^2}{4} - y^2\right)b \qquad \text{(D.10)}$$

Using these results in the shear stress formula (D.9) gives

$$\tau = \frac{VQ}{It} = \frac{6V}{bh^3}\left(\frac{h^2}{4} - y^2\right) \qquad \text{(D.11)}$$

which predicts a parabolic shear stress distribution over the section (see Fig. D.14) that vanishes at the top and bottom and takes on maximum $\tau_{max} = \dfrac{3V}{2A}$ at the neutral axis ($y = 0$).

This concludes our brief review of the standard four stress and deflection analyses for extension, torsion, and bending and shear of beams.

D.4 Curved beams

We now discuss the mechanics of materials analysis of curved beams. This topic is often omitted in the first undergraduate course and is sometimes covered in later courses on advanced mechanics of materials, machine design, or structures. The analysis is concerned with the bending deformation of a prismatic beam (constant cross-section) that is in a circular shape, as shown in Fig. D.15. These structures commonly occur in various machine parts such as hooks and links. Clearly this structure is not modeled well using straight beam theory, and thus strength of materials develops a suitable curved beam analysis.

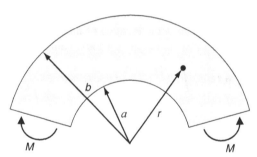

FIG. D.15 Curved beam geometry.

It has been shown that for curved beams the normal strain no longer varies linearly from the neutral axis. We again consider only cross-sections that have an axis of symmetry perpendicular to the moment axis, as shown in Fig. D.16. Consistent with mechanics of materials theory, we assume again that cross-sections remain plane after the deformation. Using the isolated beam segment illustrated in Fig. D.16, this assumption allows simple calculation of the strain distribution.

Considering a beam fiber located a distance r from the center of curvature, the original fiber length is $r d\theta$ and the change in length is given by $(R - r)d\psi$, where each section rotates an amount $d\psi/2$ because of the applied moment. Using classical definition, the fiber strain is given by

$$\varepsilon_\theta = \varepsilon = \frac{(R - r)d\psi}{r d\theta} = k\frac{(R - r)}{r} \qquad \text{(D.12)}$$

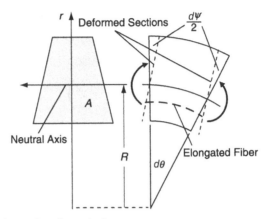

FIG. D.16 Curved beam section and strain analysis.

where $k = d\psi/d\theta$ is a constant parameter for a given element. It is observed that, unlike straight beams, the strain varies in a nonlinear hyperbolic fashion. Under the usual assumption that one component of stress and strain exists, Hooke's law gives the bending stress

$$\sigma_\theta = \sigma = E\varepsilon = Ek\frac{(R-r)}{r} \qquad (D.13)$$

With the results just given, the location of the neutral axis and the stress–moment relation can be determined. Similar to straight beam theory, the location of the neutral axis is found by requiring that the resultant force normal to the cross-sectional area A must vanish, and thus

$$\int_A \sigma dA = \int_A Ek\frac{(R-r)}{r}dA = 0 \Rightarrow R = \frac{A}{\int_A \frac{dA}{r}} \qquad (D.14)$$

The location dimension R is then a function only of section properties and does not correspond to the section centroid as found in straight beam theory.

This location can be easily calculated for particular geometric shapes. For example, using the inner and outer radial dimensions shown in Fig. D.15, the location for a rectangular section is $R = (b-a)/\log(b/a)$. Note that this result indicates that, even for a rectangular section with two axes of symmetry, the neutral axis is not located at the centroid (geometric center).

The stress–moment relation is found by the usual equilibrium statement that balances the applied section moment to the resulting stress field

$$M = \int_A (R-r)\sigma dA = \int_A Ek\frac{(R-r)^2}{r}dA$$

$$= Ek\left(R^2 \int_A \frac{dA}{r} - 2R \int_A dA + \int_A rdA \right) \qquad (D.15)$$

$$= EkA(\bar{r} - R)$$

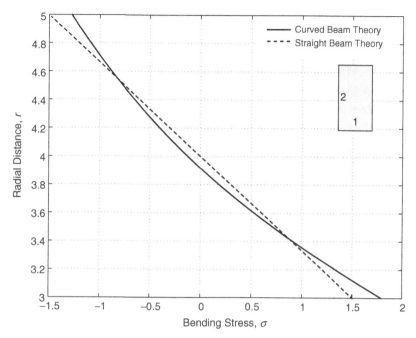

FIG. D.17 Comparison of curved and straight beam theory for rectangular section.

where $\bar{r}$ is the location of the section centroid measured from the center of curvature. Combining this result with relation (D.13) gives the desired relation

$$\sigma = \frac{M(R - r)}{Ar(\bar{r} - R)} \qquad (D.16)$$

A specific comparative example is shown in Fig. D.17 for a curved beam of rectangular section of unit thickness with properties $a = 3$, $b = 5$, and $M = 1$ (using suitable consistent units). The results compare bending stresses predicted from both straight and curved beam theory. The nonlinear distribution from curved beam theory is clearly evident; however, results from straight beam theory compare reasonably well. For beams with a relatively large radius of curvature, the two theories are in good agreement, while for cases with small values of $\bar{r}/(b - a)$, the differences become significant.

D.5 Thin-walled cylindrical pressure vessels

We conclude our review of mechanics of materials with a discussion of the analysis of thin-walled cylindrical pressure vessels. The elementary theory is concerned only with the uniform stresses developed in the side walls away from any concentration effects at the ends. The only loadings on the vessel are due to application of a uniform internal pressure p. It is further assumed that the vessel thickness t is much smaller than the mean radius r of the side wall.

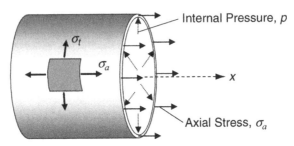

FIG. D.18 Thin-walled cylindrical vessel under internal pressure.

As shown in Fig. D.18, under these conditions an axial stress σ_a and a hoop stress σ_t are generated at all points on the lateral sides of the cylinder. Because the vessel is assumed to have thin walls, variation of the stress through the wall thickness can be neglected. The resulting state of stress is then assumed to be biaxial, under the condition that the pressure loading on the inside surface is normally much smaller than the axial and hoop stresses.

To determine these two side-wall stresses in terms of vessel geometry and pressure loading, a simple equilibrium analysis is done. The cylindrical vessel is first sectioned to isolate a semicircular strip of width dx, as shown in Fig. D.19. Then an equilibrium analysis in the horizontal direction is conducted on the segment to give

$$2\sigma_t(tdx) - p(2rdx) = 0 \Rightarrow$$

$$\sigma_t = \frac{pr}{t} \tag{D.17}$$

Note that because the vessel is thin, our analysis makes no distinction between inner, outer, and mean vessel radius.

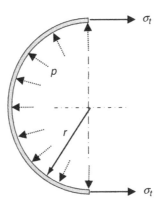

FIG. D.19 Cylindrical vessel section.

To determine the axial stress, we conduct an axial force balance of a sectioned half-vessel (similar to Fig. D.18 with a left end cap) to get

$$\sigma_a(2\pi rt) - p(\pi r^2) = 0 \Rightarrow$$
$$\sigma_a = \frac{pr}{2t}$$

(D.18)

Relations (D.17) and (D.18) provide the mechanics of materials predictions for the stresses in the pressure vessel structure. Note that the hoop stress is twice the magnitude of the axial stress. These forms indicate that for $r/t \gg 1$ the two side-wall stresses are much larger than p.

Index

A

Airy stress function, 154–156, 163, 298–299, 348, 424, 434
Alternating symbol, 7
Analytical solution procedures, 111–112
  complex variable method, 112
  Fourier method, 112
  integral transform method, 112
  power series method, 112
Angle of twist, 243–244, 273, 343, 579–580
Anisotropic elasticity. *See also* Nonhomogeneous elasticity
  curvilinear anisotropic problems, 364–372
  fracture mechanics applications, 360–364
  material microstructures, 332f
  material symmetry, 333–339
    axis of symmetry, 336–337
    complete symmetry, 338–339
    hydrostatic compression of monoclinic cube,
      338b–339b
    perpendicular planes of symmetry, 335–336
    plane of symmetry, 334–335
  plane deformation problems, 347–360
    concentrated force system in an infinite plane,
      352b–353b
    concentrated force system on surface of half-plane,
      354b–355b
    infinite plate with an elliptical hole, 355b–357b
    uniform pressure loading case, 357–360
    uniform tension of an anisotropic sheet, 351b–352b
  restrictions on elastic moduli, 339–341, 340t
  torsion of solid possessing plane of material symmetry,
    341–347
    displacement formulation, 344–345
    general solution to governing equation, 345–347
    stress formulation, 342–344
Anisotropic/anisotropy, 434
  behavior, 331
  circular tube problems, 443–444
  focusing effect, 449
  materials, 86, 127, 339
  model, 86
Annular plate problem, 395b–396b
Annular region, 295
Antiplane elasticity, 481b
Antiplane strain, 153–154. *See also* Plane strain
  problems, 458–462, 460f
Antisymmetric tensor, 33

Approximate models, 145, 523
Approximate solution procedures, 112–113
  Ritz method, 113
Approximating functions for 2D linear triangular elements,
    525–527, 526f
Argument, 291
Associated homogeneous problem, 436–437
Averaging operator, 151
Axial displacement, 436b–437b
Axial stress–strain data, 84
Axisymmetric/axisymmetry, 145
  plane problem, 444–445
  solution in polar coordinates, 183–184

B

BASIC, 559
Beam
  bending by uniform transverse loading, 169b–173b
  curved beam problems, 207b
  pure bending of, 167b–168b
  subject to transverse sinusoidal loading, 175b–177b
  uniaxial tension of, 165b–166b
Beltrami representation, 423
Beltrami stress function, 423
Beltrami–Michell equation, 147, 150, 421–423
  compatibility equations, 104, 245
BEM. *See* Boundary element method (BEM)
Bessel function, 461
Betti/Rayleigh reciprocal theorem, 130–131
Biaxial loading, 191–192
BIE. *See* Boundary integral equation (BIE)
Biharmonic
  equation, 155, 163, 298–299, 434
  functions, 155
  operator, 155
Biological adaptation, 433
Biological cellular materials, 433
Body forces, 57–58, 58f, 123
Boundary conditions, 97, 101f, 532–535
  and fundamental problem classifications, 98–103
    boundary stress components on coordinate surfaces,
      100f
    composite elastic continuum, 102f
    line of symmetry boundary condition, 99f
    typical boundary conditions, 99f
    typical composite bodies, 101f

Boundary element method (BEM), 113−114, 131−132, 523−524. *See also* Finite element method (FEM)
  formulation, 540−545
    BEM solution of plate under uniform tension containing circular hole, 545f
    boundary discretization using elements with constant approximation, 543f
    comparison of typical FEM and BEM meshes, 541f
Boundary equation, torsion solutions from, 253−258
  elliptical section, 254b−255b
  equilateral triangular section, 256b−257b
  higher-order boundary polynomials, 258b
Boundary integral equation (BIE), 131−132, 523−524, 540
Boundary tractions, 123
Boundary value problems, 433
Boundary-loading function, 308b−309b
Boundary-value problem, 145
Bounded solution scheme, 459
Boussinesq potentials, 416
Boussinesq solution, 454−455
Boussinesq's problem, 413f, 413b
  revisiting, 416b−417b
Branch cut, 296
Branch point, 296
Brazilian test, 211b−214b
Bulk modulus of elasticity, 90
Burgers vector, 478−479

C

C language, 559
CAD. *See* Computer-aided design (CAD)
Calculus of residues, 295
Cartesian coordinates, 24−25, 145, 411, 549
  equilibrium equations, 550
    in displacements, 553
  Hooke's law, 551
  solutions
    using Fourier methods, 174−182
    using polynomials, 163−173
  strain−displacement relations in, 549
Cartesian systems, 21
Cartesian tensors, 10−12
  calculus of, 17−21
    divergence or Gauss theorem, 20
    Green's theorem in plane, 20
    Stokes theorem, 20
    zero-value or localization theorem, 21
Cast iron, 84−85
Castigiliano's theorems, 136−138
Cauchy integral
  formula, 294, 307
  theorem, 294

Cauchy Principal Value, 219
Cauchy stress tensor, 61−62
Cauchy−Euler differential equation, 370
Cauchy−Riemann
  equations, 293, 394
  relations, 294
Center of compression, 488f, 488b
Center of rotation, 489b
Central crack problem, 324
Centrifugal loading, 444−445
Cerruti's problem, 414f, 414b
Cesàro integral, 46−47
Characteristic equation, 13
Circular disk under diametrical compression, 537b
Circular domain examples, 305−310
  circular disk problem, 306f
  circular plate with edge loading, 309f
  disk under uniform compression, 308f
Circular plate problems, 391b−393b
Circumferentially orthotropic, 365−366
Clapeyron's theorem, 130
Classical Hertz theory, 223
Coefficient of thermal expansion, 92, 380
  tensor, 92
Comma notation, 17
Comparing relations, 332−333
Compatibility equations, 42−43, 83
Complementary virtual
  strain energy, 134
  work, 134
Complete solution, 423
Complete symmetry, 338−339
Completeness, 407
Complex boundary-loading function, 306
Complex conjugate, 291
  roots, 346
Complex plane, 291
Complex potentials, 312b−313b
  finite multiply connected domains, 304
  finite simply connected domains, 303
  infinite domains, 305
  structure of, 303−305
Complex variable function, 292
Complex variable method, 112
  for plane problems, 394−401
    annular plate problem, 395b−396b
    circular hole in infinite plane under uniform heat flow, 397b, 398f, 398b
    elliptical hole in infinite plane under uniform heat flow, 399f, 399b−401b
Complex variable theory, 291
  applications

using conformal mapping method, 315–319
  to fracture mechanics, 320–323
circular domain examples, 305–310
complex formulation of plane elasticity problem, 298–301
complex plane, 292f
  contour in, 294f
conformal mapping, 297f–298f
multiply connected domain, 296f
plane and half-plane problems, 310–315
resultant boundary conditions, 301–302, 302f
review, 291–298
Westergaard method for crack analysis, 323–324
Compliance tensor, 332
Computer-aided design (CAD), 540
Concentrated edge loading, 309b–310b
Conformal mapping method, 297, 315–319
  general mapping for infinite plane with interior hole, 316f
  infinite plane with elliptical hole, 318f
  mappings for infinite plane with circular and elliptical holes, 317f
  stress concentration factor for elliptical hole problem, 319f
Conformal transformation. See Conformal mapping method
Conical shaft, 267–270, 269f
  torsion of variable diameter, 267–270
Conjugate functions, 249, 293–294
Conservation of energy principle, 379–380
Constant element model, 544b–545b
Constant strain element (CST), 528
Constant stress state example, 300b–301b
Constitutive equations, 83–84
Contact mechanics, 217–218
  problems, 202–203
Continuum energy equation, 127
Continuum mechanics, 57
  principles, 477
  theory, 83–84
Contour line, 252
  stress distributions and, 68–69
  for torsion-membrane analogy, 253f
Contraction, 5
Convex regions, 415
Coordinate transformations, 8–9, 11b, 12f, 12b
Cosserat continuum, 495
Couple-stress
  tensor, 495–496
  theory, 58, 477, 482, 496
Crack problems in elasticity, 320–323
Cross product, 16
CST. See Constant strain element (CST)
Cubic anisotropy, 337
Curved beams, 587–589
Curved cantilever under end loading, 208–217

disk under diametrical compression, 211b–214b
Curvilinear anisotropic model, 364–365
Curvilinear anisotropic problems, 364–372
  2D polar-orthotropic problem, 365–368
    idealized orthogonal curvilinear microstructure, 365f
  3D spherical-orthotropic problem, 368–372
Curvilinear coordinate systems, 410
Curvilinear cylindrical, 48–49
  coordinates, 73–76
Curvilinear system, 90
Cyclic function, 304
Cylindrical body, 241–242
Cylindrical components from Cartesian, 556
Cylindrical coordinates, 411, 550
  equilibrium equations in, 551
  equilibrium equations in displacements, 553
  Hooke's law, 552
  strain–displacement relations in, 550
  system, 48
Cylindrical equation development, 75
Cylindrical systems, 21

D

Deformation, 31–34, 523
  curvilinear cylindrical and spherical coordinates, 48–49
  field, 145
  geometric construction of small deformation theory, 34–39
  principal strains, 41–42
  process, 123
  spherical and deviatoric strains, 42
  strain compatibility, 42–47
  strain transformation, 34–39
  theory, 83
  2D deformation, 32f
  between two neighboring points, 32f
Degrees of freedom, 528
Del operator, 18
Deviatoric strains, 42
  tensors, 42
Deviatoric stress tensor, 66–68
Differential element illustrated, 75
Dilatation. See Center of compression
Dirac delta function, 309b–310b
Directional derivative, 17
Discretization process, 523
Disk under diametrical compression, 211b–214b
Dislocation modeling, 478–482
  edge dislocation, 478f, 480b
    models, 479f
    schematics of edge and screw dislocations, 479f
  screw dislocation, 481f, 481b

Displacement
    equilibrium equations in
        Cartesian coordinates, 553
        cylindrical coordinates, 553
        spherical coordinates, 553
    fields, 450
    form, 153−154
    formulation, 104−106
    gradient matrix, 37
    Green's function, 131, 541
    problem, 102
    and strain, 31
Displacement potential. *See also* Stress functions
    functions, 105−106
    Galerkin vector representation, 409−414
    Helmholtz displacement vector representation, 407−408
    Lamé's strain potential, 408−409
    Papkovich−Neuber representation, 414−418
    solution, 384−385
    spherical coordinate formulations, 418−421
    stress functions, 421−425
Displacement transformation
    cylindrical components from Cartesian, 556
    spherical components from Cartesian, 557
    spherical components from cylindrical, 557
Displacement vector, 3−4, 32
    distribution plotting, 565b−566b
Distortional deformation, 127
Distortional strain energy, 67
Divergence, 20
Dot product, 16
Doublet mechanics (DM), 477−478, 508−512
    classical elasticity and, 512f
    geometry, 509f
    solution of Flamant problem, 510b, 511f
Duhamel−Neumann constitutive relation, 380
Duhamel−Neumann thermoelastic constitutive law, 92
Dummy subscripts, 5
Dyadic notation, 10

E

Edge dislocation, 478f, 480b
    models, 479f
    schematics of edge and screw dislocations, 479f
Eigenstrain technique, 477−478
Eigenvalue, 12−13
Elastic constants, bounds on, 129−130
    hydrostatic compression, 129−130
    simple shear, 129
    uniaxial tension, 129
Elastic constitutive law. *See* Hooke's law
Elastic continuum theory, 477, 495

Elastic cylinders
    extension formulation, 242−243
    flexure formulation, 270−274
    flexure problems without twist, 274−277
    general formulation, 242
    prismatic bar subjected to end loadings, 241f
    torsion
        of circular shafts of variable diameter, 267−270
        of cylinders with hollow sections, 264−267
        formulation, 243−253
        solutions derived from boundary equation, 253−258
        solutions using Fourier methods, 259−263
Elastic limit, 84
Elastic moduli, 85−86, 433, 439
    physical meaning of, 88−92
    typical values of, 91t
Elastic solid, 83−84, 123
Elasticity. *See also* Anisotropic elasticity; Nonhomogeneous elasticity
    boundary-value problem, 128
    field equations, 145
    problems, 523
    tensor, 85−86, 90
Elasticity theory, 16. *See also* Anisotropic elasticity; Nonhomogeneous elasticity
    with distributed cracks, 491−494
        elastic solid containing a distribution of cracks, 491f
        isotropic crack distribution using self-consistent model, 494b
        isotropic dilute crack distribution, 492b
        planar transverse isotropic dilute crack distribution, 492b
    with voids, 503−507
        elastic continuum with distributed voids, 503f
        elastic material with voids, 508f
        stress concentration around circular hole, 505b−507b
Elementary theory, 577, 589
Elements, 113, 523−524
Elliptical hole, stressed infinite plane with, 318b−319b
Elliptical section, 254b−255b
Elongational microstress, 508−509
Energy equation, 379−380
Energy source term, 379−380
Engineering shear strain, 36
Equilateral triangular section, 256b, 257f, 257b
Equilibrated stress vector, 504
Equilibrium analysis, 582
Equilibrium equations, 71−73, 75, 146, 152, 386, 389, 421−423, 434
    body and surface forces acting on arbitrary portion of continuum, 71f
    Cartesian coordinates, 549
    cylindrical coordinates, 549

spherical coordinates, 549
in terms of displacements, 551
Essential boundary conditions, 134
Euler–Bernoulli beam theory, 136f, 136b–138b,
    167b–168b
Euler–Bernoulli relation, 169b–173b
Extension formulation, 242–243
Extensional deformation of rods and beams, 578, 578f

F

Fabric tensor, 331–332
FDM. *See* Finite difference method (FDM)
FEM. *See* Finite element method (FEM)
FGMs. *See* Functionally graded materials (FGMs)
Fiber strain, 587–588
Field equations, 97–98
    equilibrium equations, 550
        in displacements, 553
    Hooke's law, 551–552
    strain–displacement relations, 549–550
Finer mesh, 524
Finite difference method (FDM), 113, 523
Finite domains, 419–421
Finite element method (FEM), 113, 140, 523–525
    analysis of elastic plate under uniform tension, 532f
    code applications, 535–539
        circular and elliptical holes in plate, 535b–537b
        circular disk under diametrical compression, 537b
        torsion problem examples, 538b–539b
    problem application, 532–535
        elastic plate under uniform tension, 532b–535b
    solution
        of disk under diametrical compression, 538f
        of plate under uniform tension, 536f–537f
        to three torsion problems, 539f
Finite length problem, 147–148
Finite multiply connected domains, 304
Finite simply connected domains, 303
    typical domains of interest, 303f
Fixed boundaries, 98
Fixed ends, 147
Flamant problem, 195–200, 196f, 312b–313b
    displacement field for, 200f
    DM solution of, 510b, 511f
    gradient elasticity solution to, 516b, 517f
    normal and shear stress distributions, 197f
    radial stress contours for, 198f
Flexure formulation, 270–274. *See also*
    Torsion—formulation
Flexure problems without twist, 274–277
    circular section, 274b–275b
    rectangular section, 276b–277b

Force doublet, 486f, 486b
    with moment, 487f, 487b
Force vector, 530
Force–moment system in infinite plane, 311b–312b
Formulation and solution strategies
    boundary conditions and fundamental problem
        classifications, 98–103
    displacement formulation, 104–106
    field equations, 97–98
    general solution strategies, 109–114
    Saint-Venant's principle, 107–109
    singular elasticity solutions, 114–116
    stress formulation, 103–104
    superposition principle, 106–107
FORTRAN, 559
Fourier analysis, 307
Fourier cosine series, 178–182
Fourier integral form, 459
Fourier integral theory. *See* Fourier series
Fourier law of heat conduction, 379
Fourier methods, 112, 312b–315b
    Cartesian coordinate solutions using, 174–182
        applications involving Fourier series, 177–182
        rectangular domain with arbitrary boundary loading,
            179b–182b
    torsion solutions using, 259–263
        rectangular section, 259b–263b
Fourier series, 112
    theory, 259b–263b
Fourier sine series, 178–182
Fourier trigonometric series, 178
Fourth-order elasticity, 85
Fracture mechanics, 206, 320–323
    central crack in infinite plane, 321f
    crack geometry, 322f
Fracture mechanics applications, 360–364
    central crack in infinite anisotropic plane, 364f
    crack in infinite anisotropic plane, 363f
Fredholm integral equation, 461–462
Frictionless ends, 147
Functionally graded materials (FGMs), 86, 433,
    449
Fundamental invariants, 13

G

Galerkin strain potential. *See* Love's strain potential
Galerkin vector representation, 409–414
    Boussinesq's problem, 413f, 413b
    Cerruti's problem, 414f, 414b
    Kelvin's problem, 411b–412b, 413f
Gamma function, 452
Gauss theorem, 20

General solution strategies, 109—114
  analytical solution procedures, 111—112
  approximate solution procedures, 112—113
  direct integration, 109b
    prismatic bar under self-weight, 110f
  direct method, 109
  inverse method, 110
  numerical solution procedures, 113—114
  semi-inverse method, 111
Generalized Hooke's law for linear isotropic elastic solids, 409
Generalized plane stress, 151—153, 347—348
Geomechanics, 86
Geometric construction, 34—39
  deformations of rectangular element, 34f
  determination of displacements from strains, 38b—39b
  line length and angle changes from strain tensor, 39b
  strain and rotation, 38b
  2D
    deformation, 40f
    geometric strain deformation, 35f
    rigid-body rotation, 37f
Gradation model, 440, 466
Graded materials, 433
Graded shear modulus, 461—462
Gradient, 18
Gradient elasticity, 478, 513—518, 514b—517b, 570b
Gradient elasticity theory, 513
Granular interpretation, 508—509
Green's theorem, 20, 247, 273

H

Half-space
  under concentrated surface force system, 195—200
  examples, 194
  under surface concentrated moment, 200
  under uniform normal loading, 201—205
  under uniform normal stress, 194—195
Harmonic functions, 293—294
Heat conduction, 379—380
Helmholtz displacement vector representation, 407—408
Helmholtz theorem, 407
Help package, 559
High surface hardness/stiffness, 449
Higher grade continua theories, 513
Higher gradient elasticity theories, 513—518
  gradient elasticity solution
    to Flamant problem, 516b, 517f
    to screw dislocation in z-direction, 514b—515b
Higher-order approximations, 524
Higher-order boundary polynomials, 258b
Higher-order effects, 61—62

Hollow elliptical section, 264b—265b
Hollow thin-walled sections, 265b, 266f
Homogeneous assumption, 86
Homogeneous equation, 446
Homogeneous solution, 441
Hooke's law, 85—88, 98, 100—101, 123—124, 126, 146, 165b—166b, 245, 271, 383—384, 394, 433, 529
  Cartesian coordinates, 551
  cylindrical coordinates, 552
  spherical coordinates, 552
Hydrostatic compression, 89—92, 129—130
Hydrostatic state of stress, 308b—309b
Hydrostatic tension, 130
Hyperelasticity, 127

I

Index notation, 4—6, 6b
Indirect tension test. *See* Indirect tension test
Infinite domains, 305
Infinite plane with central crack, 320b—323b
Infinite regions, 419—421
Influence functions, 543
Inhomogeneity, 434, 441—442, 449
Inhomogeneous elasticity problem, 449
Inhomogeneous isotropic case, 433
Inhomogeneous model, 433
Inhomogeneous stress field, 436b—437b
Inner multiplication, 5
Integrability, 42—43
Integral formulation of elasticity, 131—132
Integral transform method, 112
Interface conditions, 101—102
Internal boundary, 304
Internal microstructural defects, 478
Interpolation, 113
  functions, 523—524, 526—527, 540
Intrinsic equilibrated body force, 504
Inverse method, 110
  pure beam bending, 110f, 110b
Irrotational part. *See* Lamellar part
Isochromatic fringe patterns, 203, 204f
Isoclinics, 68
Isothermal case, 395—401
Isothermal problem, 385
Isotropic/isotropy, 86
  crack distribution, 494b
  dilute crack distribution, 492b
    planar transverse, 492b
  limit, 358—360
  materials, 86, 127, 338—339
  plane problem, 434
  tensors, 10

K

Kelvin state with unit loads in coordinate directions,
484b–485b
generalized Kelvin state, 486f
unit concentrated loadings, 484f
Kelvin's problem, 411b–412b, 413f, 483b
Kinematic displacement boundary condition,
132
Kinematic microdeformation, 477
Kolosov parameter, 299
Kolosov–Muskhelishvili potentials, 300–301
Kronecker delta, 7

L

Lagrange interpolation functions, 527
Lamé's equations, 105
Lamé's potential, 409
strain potential, 408–409
Lamellar part, 407
Laplace transform theory, 505b–507b
Laplace's equation, 153, 419
Laurent series, 295
Learn-by-example approach, 559–561
Legendre differential equation, 419
Legendre functions, 419
Legendre polynomials, 419
Line of dilatation, 490f, 490b–491b
Linear approximation, 524–526
Linear elastic fracture mechanics, 320
Linear elastic solids, 83–84
hydrostatic compression, 89–92
relations among elastic constants, 91t
typical values of elastic moduli, 91t
linear elastic materials, 85–88
material characterization, 83–85
physical meaning of elastic moduli, 88–92
special characterization states of stress,
88f
pure shear, 89
simple tension, 89
thermoelastic constitutive relations, 92
Linear elasticity theory, 477
Linear equations, 106
Linear isotropic elastostatic problems, 524
Linear small deformation theory, 92
Linear triangular element, 532–535
Local extremum value, 134
Local minimum value, 134
Local stress concentration, 443–444
Localization theorem, 21
Love's strain potential, 410

M

m-files, 559
Maclaurin series, 295
Maple software, 173
code, 114
Mapping function, 315–316
Material characterization, 83–85
Material gradation, 469
Material gradation functions, 459
Material inhomogeneity, 444
Material symmetry, 86, 333–339
axis of symmetry, 336–337
axis of symmetry for transversely isotropic material,
337f
complete symmetry, 338–339
group, 333
hydrostatic compression of monoclinic cube, 338b–339b
perpendicular planes of symmetry, 335–336
plane of symmetry, 334–335
plane of symmetry for monoclinic material, 334f
Mathematica software, 114, 173, 318b–319b
Mathematical preliminaries
calculus of Cartesian tensors, 17–21
Cartesian tensors, 10–12
coordinate transformations, 8–9
index notation, 4–6
Kronecker delta and alternating symbol, 7
orthogonal curvilinear coordinates, 21–26
principal values and directions for symmetric second-order
tensors, 12–15
scalar, vector, matrix, and tensor, 3–4
MATLAB® Primer, 559
calculation of invariants, eigenvalues, and eigenvectors of
matrix, 562b–563b
displacement vector distribution plotting, 565b–566b
examples, 559–575
MATLAB Command window, 560f
MATLAB Editor/Debugger window, 560f
PDE Toolbox, 569b–570b
plotting
of maximum shear stresses below flat punch problem,
566b–567b
three-dimensional warping displacement surface,
568b–569b
of warping displacement contours, 567b–568b
polar contour plotting of hoop stress around circular hole,
564b–565b
roots determination for orthotropic materials, 564b
transformation of second-order tensors,
561b–562b
XY plotting of stresses, 563b
MATLAB® software, 63, 163, 220, 223, 535, 559

Matrix, 3—4
  algebra, 16—17
  variables, 3—4
Maxwell stress functions, 407, 424
Mechanics of materials, 167b—168b, 577
  bending deformation of beams under moments and shear
    forces, 580—587
  curved beams, 587—589
  extension, torsion, and bending/shear deformation of beam-
    type structures, 577f
  extensional deformation of rods and beams, 578
  thin-walled cylindrical pressure vessels, 589—591
  torsion of circular rods, 579—580
Membrane analogy, 251—253
Mesh geometry, 532—535
Michell solution, 307
  in polar coordinates, 182—183
Micromechanical theories, 518
Micromechanics applications
  dislocation modeling, 478—482
  doublet mechanics, 508—512
  elasticity theory
    with distributed cracks, 491—494
    with voids, 503—507
  higher gradient elasticity theories, 513—518
  micropolar/couple-stress elasticity, 494—503
  singular stress states, 482—491
Micromorphic theories, 482
Micropolar elasticity, 499b—502b
  stress concentration behavior for micropolar theory, 503f
Micropolar theory, 58, 482, 495
Micropolar/couple-stress elasticity, 494—503
  heterogeneous materials with microstructure, 495f
  2D couple-stress theory, 496—503
Microrotation vector, 495
Minimum potential and complementary energy, principles of,
  134—138
Mode I deformation. *See* Opening mode deformation
Modified westergaard stress function formulation, 324
Modulus, 291
  of elasticity, 87—88
  of rigidity. *See* Shear modulus
Mohr's circles of stress, 65—66, 65f
Monoclinic material, 334—335
Morera stress functions, 407, 425
Multiply connected cross-sections, 46, 249—251

N

Naghdi-Hsu solution, 407, 416—417
Natural boundary conditions, 134
Navier's equations, 105—106, 146, 184, 298—299, 414, 434,
  440

Cartesian coordinates, 552
  cylindrical coordinates, 552
  spherical coordinates, 551
Navier's relations, 382
Neumann principle, 331—332
Newton's third law, 59
Nodal values, 526
Nodes, 523—524
Nonhomogeneous elasticity, 433—437. *See also* Anisotropic
  elasticity; Elasticity theory; Nonhomogeneous
  elasticity
  antiplane strain problems, 458—462, 460f
  continuously graded material in single direction, 435f
  modulus of elasticity gradation, 438f
  plane problem of hollow cylindrical domain, 437—444
  point force on free surface of half-space, 449—458
    inhomogeneous half-space with graded modulus, 450f
    nondimensional normal stress distribution, 457f
    point load on three-dimensional inhomogeneous half-
      space, 456f
    radial displacement distribution, 453f
    radial stress contour comparisons for Flamant problem, 455f
    radial stress distribution, 453f
    tangential displacement distribution, 454f
  rotating disk problem, 444—449, 445f
  solutions, 469
  stress intensity factor as function of inhomogeneity, 462f
  torsion problem, 462—469
  uniaxial tension
    of graded sheet, 436b—437b
    of inhomogeneous sheet, 436f
Nonhomogeneous stress, 450
Nonlinear elastic material, 135f
Nonzero equilibrium equation, 444—445
Normal stresses, 59—60
Notch and crack problems, 205—207
Numerical methods, 223
Numerical solution methods, 523
  boundary element formulation, 540—545
  FEM, 524—525, 525f
    code applications, 535—539
    problem application, 532—535
  virtual work formulation for plane elasticity, 527—532
Numerical solution procedures, 113—114
  boundary element method, 113—114
  finite difference method, 113
  finite element method, 113
Numerical stress analysis, 523

O

Octahedral plane, 67
Octahedral shear stress, 128

Octahedral stress, 66–68
Opening mode deformation, 320b–323b
Orthogonal curvilinear
    coordinates, 21–26
    system, 90
Orthogonal relations, 9
Orthotropic material, 335–336
Outer multiplication, 4

P

Papkovich–Neuber representation, 414–418
    Boussinesq's problem revisiting, 416b–417b
    comparison of two-and 3D stress concentrations, 423f
    displacement potential solutions, 418f
    spherical cavity in infinite medium under tension, 422f
    stress concentration factor behavior, 422f
Papkovich–Neuber solution scheme, 482
Parabolic displacement distribution, 139b–140b
Particular failure theories of solids, 128
PDE Toolbox, 535, 569b–570b
    boundary specification window, 574f
    draw mode screen for rectangle with circular hole, 573f
    graphical user interface for, 571f
    plot selection window, 575f
    selection of problem type in, 572f
Perfectly bonded interface, 101–102
Permutation symbol, 7
Perpendicular planes of symmetry, 335–336
Physically nonlinear elasticity, 142
Piola–Kirchhoff stress tensor, 61–62
Planar approximation. *See* Linear approximation
Planar transverse isotropic dilute crack distribution, 492b
    cracked elastic solid with common crack orientation, 493f
    elastic moduli for transversely cracked solid, 493f
Plane and half-plane problems, 310–315
    concentrated force system
        on half-space, 312f
        in infinite medium, 311f
    stress-free hole under general far-field loading, 314f
    typical indentation problem, 314f
Plane deformation problems, 347–360
    concentrated force system
        in infinite plane, 352b–353b
        on surface of half-plane, 354b–355b
    infinite plate with an elliptical hole, 355b–357b
        elliptical hole in an infinite anisotropic plane, 356f
    uniform pressure loading case, 357–360
    uniform tension
        of anisotropic plane, 352f
        of anisotropic sheet, 351b–352b
Plane elasticity, 527
    virtual work formulation for, 527–532

Plane elasticity problems, 320
    complex formulation of, 298–301
Plane equilibrium equations, 154
Plane of symmetry, 334
Plane problem of hollow cylindrical domain, 437–444
    equivalent stress concentration factor for small stress-free hole, 444f
    gradation in Young's modulus, 440f
    hollow cylindrical domain under uniform pressure, 439f
    nondimensional radial stress distribution, 441f
    nondimensional tangential stress distribution, 442f
Plane problem of hollow cylindrical domain under uniform pressure, 437–444
Plane strain, 145–148, 298, 347–348, 381–383
    long cylindrical body representing plane strain conditions, 146f
    typical domain for plane elasticity problem, 147f
Plane stress, 145, 148–151, 298, 347–348, 381, 383–384
    conditions, 149–150
    elastic moduli transformation, 151t
        relations, 384t
    elastic plate representing plane stress conditions, 148f
    generalized plane stress, 151–153
Poisson equation, 245
Poisson's ratio, 87–88, 90, 130, 150–151, 416b–417b, 435
Polar contour plotting of hoop stress around circular hole, 564b–565b
Polar coordinates, 145, 350, 354f, 354b–355b
    example solutions, 184–217
        biaxial and shear loading cases, 191–192
        curved cantilever under end loading, 208–217
        half-space examples, 194
        half-space under concentrated surface force system, 195–200
        half-space under surface concentrated moment, 200
        half-space under uniform normal loading, 201–205
        half-space under uniform normal stress, 194–195
        notch and crack problems, 205–207
        pressurized hole in infinite medium, 186–187
        pure bending example, 207–208
        quarter-plane example, 193–194
        rotating disk problem, 215b–217b
        stress-free hole in infinite medium under equal biaxial loading, 187–190
        thick-walled cylinder under uniform boundary pressure, 184b–186b
    formulation, 156–158, 388–389
    solutions in, 182–184
        axisymmetric solution, 183–184
        general Michell solution, 182–183
Polar moment of inertia, 579
Polar-orthotropi, 365

Pole, 295
Polynomials, Cartesian coordinate solutions using, 163—173
Positive definite quadratic form, 126
Potential function, 154
Potential theory, 249, 482
Power series method, 112
Power-law variation, 86, 435
Prandtl's function, 423
    stress function, 245, 425, 463
Pressurized hole in infinite medium, 186—187
Principal axes, 13—14
Principal minors, 340—341
Principal strains, 41—42
Principal value, 12—13
    problem, 14b—15b
Programming language, 559
Proportional limit, 84
Pure shear, 89
Pythagorean theorem, 64

R

Radial distribution, 196
Radial inhomogeneity, 441—442
Radial stress, 447
    contours, 452
    distribution, 448, 451
Radially orthotropic material, 365—366
Radially symmetric problems, 389—394
    annular plate geometry, 391f
    circular plate problems, 391b—393b
    stress distribution in an annular plate, 393f
    temperature distribution in an annular plate, 392f
Rayleigh—Ritz method, 138—140, 139b—140b
Real-world problems, 523
Reciprocal theorem, 130—132, 540—541
Rectangular section, 259b, 260f
Rectilinear anisotropy, 338—339
Reference temperature, 379—380
Regular elastic state, 482
Reissner's principle, 136—138
Restrictions on elastic moduli, 339—341
    positive definite restrictions on elastic moduli, 340t
Resultant boundary conditions, 301—302
Rigid-body motions, 165b—166b, 303
Ritz approximation, 139b—140b
Ritz method, 113
Ritz/Galerkin scheme, 524—525
Rotating disk problem, 215f, 215b—217b, 444—449, 445f
    gradation in Young's modulus, 446f
Rotation tensor, 33

S

Saint-Venant
    approximation, 165b—166b
    compatibility equations, 44
    relations, 147
    semi-inverse method, 342
Saint-Venant's principle, 107—109, 108f, 164—173, 241—242, 371—372
    statically equivalent loadings, 107f
Scalar, 3—4, 16
    field, 19b
    quantities, 3—4
    triple product, 16
Screw dislocation, 481f, 481b
Second potential function, 308b—309b
Second stress relation, 303
Second-order tensors transformation, 561b—562b
Self-consistent model, isotropic crack distribution using, 494b
Self-equilibrated forms, 423
Self-equilibrated uniform, 459
Semi-infinite domain problems, 192b
Semi-inverse method, 111, 242
Shear
    loading, 191—192
    microstress, 508—509
    modulus, 87, 455
    strain, 36
    stress—shear strain curve, 89
Shearing
    modes, 320b—323b
    stresses, 59—60
Simple plane contact problems, 217—223
    elastic half-space
        subjected to flat rigid indenter, 220f
        subjected to rigid indenter, 218f
    maximum shear stress distribution
        under circular rigid indenter, 222f
        under flat rigid indenter, 221f
Simple shear, 129
Simple tension, 89
Simply connected domains, 46
Single complex potential, 323
Single vector function, 409
Single-valued continuous displacements, 83
Singular elasticity solutions, 114—116
Singular points, 293, 320—323
Singular stress states, 482—491
    center of compression/dilatation, 488f, 488b
    center of rotation, 489b
    concentrated force in infinite medium, 483f, 483b
    force doublet, 486f, 486b
    force doublet with moment, 487f, 487b

half line of dilatation, 490f, 490b–491b
Kelvin state with unit loads in coordinate directions,
    484b–485b
Singularities. *See* Singular points
Skew-symmetric crack problems, 324
Slip interface, 101–102
Small deformation theory, 61–62
Solenoidal, 407
Somigliana's identity, 131–132, 540–541
Spherical components
    from Cartesian, 557–558
    from cylindrical, 556–557
Spherical coordinates, 48–49, 73–76, 418
    equilibrium equations, 549
        in displacements, 551
    formulations, 418–421
    Hooke's law, 550–551
    strain–displacement relations in, 548
Spherical harmonics, 419
Spherical strains, 42
Spherical stress tensor, 66–68
Spherical-orthotropic material, 368–369
Static equilibrium, 132
Static finite element analysis, 524
Stationary energy, 134
Steady-state conditions, 380
Stiffness matrix, 530
Stokes theorem, 20
Strain, 84
    tensor, 33
Strain compatibility, 42–47
    continuity of displacements, 45f
    domain connectivity, 47f
    physical interpretation of strain compatibility, 43f
Strain energy, 123–128
    bounds on elastic constants, 129–130
        hydrostatic compression, 129–130
        simple shear, 129
        uniaxial tension, 129
    deformation under uniform uniaxial stress, 124f
    density, 124
    principle
        of minimum potential and complementary energy,
          134–138
        of virtual work, 132–134
    Rayleigh–Ritz method, 138–140
    related integral theorems, 130–132
        Betti/Rayleigh reciprocal theorem, 130–131
        Clapeyron's theorem, 130
        integral formulation of elasticity, 131–132
    shear deformation, 125f
    strain energy for uniaxial deformation, 125f

uniqueness of elasticity boundary-value problem, 128
Strain transformation, 34–39
    two-dimensional rotational transformation, 41f
Strain–displacement, 48
    equations, 149
Strain–displacement relations, 36, 42–43, 73, 97, 145,
    388–390, 454–455
    Cartesian coordinates, 549
    cylindrical coordinates, 550
    spherical coordinates, 550
Strength of materials. *See* Mechanics of materials
Stress
    amplification, 366
    analysis problems, 379
    combinations, 300, 320b–323b
    concentration factor, 535b–537b
    contours, 452
    disappears, 61–62
    formulation, 103–104, 381
    intensity factor, 320b–323b, 461
    matrix, 3–4
    principal, 63–66
        general and principal stress states, 63f
        traction vector decomposition, 64f
    tensor, 58–62
    trajectories, 68
Stress and equilibrium
    body and surface forces, 57–58
    curvilinear cylindrical and spherical coordinates, 73–76
    equilibrium equations, 71–73
    principal stresses, 63–66
    spherical, deviatoric, octahedral, and von mises stresses,
        66–68
    stress distributions and contour lines, 68–69
    stress transformation, 62–63
    traction vector and stress tensor, 58–62
Stress components, 59–60
    cylindrical coordinates, 74f
    spherical coordinates, 75f
Stress distributions, 68–69
    in disk under diametrical compression, 69, 70f
Stress functions, 104, 421–425
    formulation, 385–388
    Maxwell stress function, 424
    Morera stress function, 425
    technique, 154
    thermal stresses in elastic strip, 387b–388b
Stress transformation, 62–63
    cylindrical components from Cartesian, 556
    spherical components
        from Cartesian, 558
        from cylindrical, 557

Stress-decreasing effects, 443–444
Stress-free hole, 443
    circular hole, 314b–315b
    in infinite medium, 187–190
        under uniform far-field tension loading, 188b–190b
Stress–strain curve, 135
Stress–strain relations, 127
Stress–stress function
    formulation, 244–248
    relations, 164
Structural materials, 83–84
Structural metals, 84
Superposition principle, 106–107
    2D superposition, 107f
Surface forces, 57–58, 58f
    density function, 57–58
Symbolic manipulation, 114
Symmetric second-order tensors, principal values and
        directions for, 12–15
Symmetric symbol, 5
Symmetric tensor, 33
Symmetry axis, 336–337
Symmetry of order, 336

T

Taylor series expansion, 295
Tensor, 3–4
    algebra, 16–17
    formalism, 4
Tensorial form, 86
Tensorial-based equations, 65–66
Thermal conductivity tensor, 379
Thermal effects, 379
Thermal stresses in elastic strip, 387b–388b
    thermoelastic rectangular strip, 388f
Thermodynamic theory, 379–380
Thermoelastic constitutive relations, 92
Thermoelastic plane problem, 394
Thermoelasticity
    complex variable methods for plane problems, 394–401
    displacement potential solution, 384–385
    general uncoupled formulation, 381
    heat conduction and energy equation, 379–380
    polar coordinate formulation, 388–389
    radially symmetric problems, 389–394
    stress function formulation, 385–388
    two-dimensional formulation, 381–384
Thin-walled cylindrical pressure vessels, 589–591, 590f
Three-dimension (3D)
    case, 36, 62
    constitutive law, 85
    elasticity problems, 407

equations, 496
    Navier equations, 408
    problems, 145
    warping displacement surface, plotting, 568b–569b
Time-dependent response, 504–505
Toolbox, 535–539
Torsion
    of circular rods, 579–580, 579f
    of circular shafts of variable diameter, 267–270
    of cylinders with hollow sections, 264–267
        hollow elliptical section, 264b–265b
        hollow thin-walled sections, 265b–267b
    formulation, 243–253
        displacement formulation, 248–249
        membrane analogy, 251–253
        multiply connected cross-sections, 249–251
        stress–stress function formulation, 244–248
    of solid possessing plane of material symmetry, 341–347
        displacement formulation, 344–345
        of elliptical orthotropic bar, 346b–347b
        general solution to governing equation, 345–347
        stress formulation, 342–344
    solutions
        derived from boundary equation, 253–258
        using Fourier methods, 259–263
Torsion problem, 462–469, 538b–539b
    shear modulus behavior, 465f
    shear stress distribution for torsion problem, 467f–468f
    torsion of nonhomogeneous circular cylinder, 463f
Torsional loading, 89
Torsional microstress m_α, 508–509
Torsional rigidity, 248, 464
Total potential energy, 133
Tractions, 306
    form, 153–154
    problem, 102
    vector, 58–62
Transformation matrix, 510–512, 556
Transformation of field variables
    cylindrical components from Cartesian, 556
    spherical components
        from Cartesian, 557–558
        from cylindrical, 556–557
    stress components in cylindrical and spherical coordinates,
        555f
Transformation relations, 41, 62
Transverse sinusoidal loading, beam subject to, 175b–177b
Transversely isotropic material, 336–337
Triclinic material, 333
True stress, 84–85
Two-dimensional biharmonic equation, 155
Two-dimensional case, 62, 496–497, 510–512

Two-dimensional couple-stress theory, 496—503
    micropolar elasticity, 499b—502b
Two-dimensional domain of interest, 385
Two-dimensional elasticity problems, 524
Two-dimensional formulation, 381—384
    airy stress function, 154—156
    antiplane strain, 153—154
    generalized plane stress, 151—153
    plane strain, 145—148, 381—383
    plane stress, 148—151, 383—384
    polar coordinate formulation, 156—158
Two-dimensional linear triangular elements, 525—527
Two-dimensional plane strain model, 434
Two-dimensional problem solution
    Cartesian coordinate solutions
        using Fourier methods, 174—182
        using polynomials, 163—173
    example polar coordinate solutions, 184—217
    general solutions in polar coordinates, 182—184
    simple plane contact problems, 217—223
Two-dimensional regions, 46
Two-dimensionality, 145

U

Uncoupled conduction equation, 380
Uncoupled formulation, 381
Uniaxial gradation, 436b—437b
Uniaxial stress—strain curve, 84—85
Uniaxial tension, 129
    of beam, 165b—166b
Uniform
    compression, 308b—309b
    hydrostatic compression, 129
    simple shear, 129
    stress, 449
    tension
        of anisotropic plane, 352f
        of anisotropic sheet, 351b—352b
    transverse loading, 169b—173b

V

Vector, 3—4, 16
    algebra, 16—17
    field, 19b
    quantities, 3—4
Virtual displacement, 132
Virtual stresses, 134
Virtual work
    formulation, 527
        for plane elasticity, 527—532
    principle of, 132—134
Voigt matrix notation, 332—333
Volume fraction, 477
Volumetric change, 127
Volumetric strain energy, 127
Von mises stresses, 66—68

W

Warping displacement contours, plotting of, 567b—568b
Weak form, 524—525
Wedge domain problems, 192b
Weighted residual method, 140, 524—525
Westergaard method for crack analysis, 323—324
Westergaard stress function, 320, 324
Work and energy, 123

Y

Yield point, 84
Young's modulus. *See* Modulus of elasticity

Z

Zero body forces, 298—299
Zero-traction boundary condition, 447
Zero-value theorem, 21, 72